Frontiers of Nonequilibrium Statistical Physics

NATO ASI Series

Advanced Science Institutes Series

A series presenting the results of activities sponsored by the NATO Science Committee, which aims at the dissemination of advanced scientific and technological knowledge, with a view to strengthening links between scientific communities.

The series is published by an international board of publishers in conjunction with the NATO Scientific Affairs Division

A	**Life Sciences**	Plenum Publishing Corporation
B	**Physics**	New York and London
C	**Mathematical and Physical Sciences**	D. Reidel Publishing Company Fordrecht Boston, and Lancaster
D	**Behavioral and Social Sciences**	Martinus Nijhoff Publishers
E	**Engineering and Materials Sciences**	The Hague, Boston, and Lancaster
F	**Computer and Systems Sciences**	Springer-Verlag
G	**Ecological Sciences**	Berlin, Heidelberg, New York, and Tokyo

Recent Volumes in this Series

Volume 128—Techniques and Concepts of High-Energy Physics III
edited by Thomas Ferbel

Volume 129—Transport in Nonstoichiometric Compounds
edited by George Simkovich and Vladimir S. Stubican

Volume 130—Heavy Ion Collisions: *Cargèse* 1984
edited by Paul Bonche, Maurice Lévy, Philippe Quentin, and Dominique Vautherin

Volume 131—Physics of Plasma-Wall Interactions in Controlled Fusion
edited by D. E. Post and R. Behrisch

Volume 132—Physics of New Laser Sources
edited by Neal B. Abraham, F. T. Arecchi, Aram Mooradian, and Alberto Sona

Volume 134—Scaling Phenomena in Disordered Systems
edited by Roger Pynn and Arne Skjeltorp

Volume 134—Fundamental Processes in Atomic Collision Physics
edited by H. Kleinpoppen, J. S. Briggs, and H. O. Lutz

Volume 135—Frontiers of Nonequilibrium Statistical Physics
edited by Gerald T. Moore and Marlan O. Scully

Series B: Physics

Frontiers of Nonequilibrium Statistical Physics

Edited by

Gerald T. Moore
University of New Mexico
Albuquerque, New Mexico

and

Marlan O. Scully
Max-Planck Institute for Quantum Optics
Garching, Federal Republic of Germany
and University of New Mexico
Albuquerque, New Mexico

Plenum Press
New York and London
Published in cooperation with NATO Scientific Affairs Division

Proceedings of a NATO Advanced Study Institute on
Frontiers of Nonequilibrium Statistical Physics,
held June 3–16, 1984,
in Santá Fe, New Mexico

Library of Congress Cataloging in Publication Data

NATO Advanced Study Institute on Frontiers of Nonequilibrium Statistical Physics (1984: Santa Fe, N.M.)

Frontiers of nonequilibrium statistical physics.
(NATO ASI series. Series B, Physics; vol. 135)
"Proceedings of a NATO Advanced Study Institute on Frontiers of Nonequilibrium Statistical Physics, held June 3–16, 1984, in Santa Fe, New Mexico"—T.p. verso.
"Published in cooperation with NATO Scientific Affairs Division."
Includes index.
1. Statistical physics—Congresses. 2. Quantum mechanics—Congresses. 3. Nonequilibrium thermodynamics—Congresses. 4. Quantum optics—Congresses. I. Moore, Gerald T. II. Scully, Marlan O. (Marlan Orvil), 1939- . III. North Atlantic Treaty Organization. Scientific Affairs Division. IV. Title. V. Series: NATO ASI series. Series B, Physics; v. 135.
QC174.7.N38 1984 530.1′3 85-31171

ISBN-13: 978-1-4612-9284-5 e-ISBN-13: 978-1-4613-2181-1
DOI: 10.1007/978-1-4613-2181-1

Softcover reprint of the hardcover 1st edition 1986
A Division of Plenum Publishing Corporation
233 Spring Street, New York, N.Y. 10013

PREFACE

The four-week period from May 20 to June 16, 1984 was an intensive period of advanced study on the foundations and frontiers of nonequilibrium statistical physics (NSP). During the first two weeks of this period, an advanced-study course on the "Foundations of NSP" was conducted in Albuquerque under the sponsorship of the University of New Mexico Center for High-Technology Materials. This was followed by a two-week NATO Advanced Study Institute on the "Frontiers of NSP" in Santa Fe under the same directorship. Many Students attended both meetings.

This book comprises proceedings based on those lectures and covering a broad spectrum of topics in NSP ranging from basic problems in quantum measurement theory to analogies between lasers and Darwinian evolution.

The various types of quantum distribution functions and their uses are treated by several authors. Other tools of NSP, such as Langevin equations, Fokker-Planck equations, and master equations, are developed and applied to areas such as laser physics, plasma physics, Brownian motion, and hydrodynamic instabilities. The properties and experimental detection of squeezed states and antibunching are described, as well as experimental tests of the violation of Bell's inequality. Information theory, mean-field theory, reservoir theory, entropy maximization, and even a novel nonlinear generalization of quantum mechanics are used to discuss nonequilibrium phenomena and the approach toward thermodynamic equilibrium.

The areas of bifurcation theory, fractals, strange attractors, and routes to chaos in nonlinear dynamical systems have been very active recently, and these concepts are dealt with at length, including applications to optical multistability and phase-transition critical phenomena. A geometrical classification of bifurcations based on catastrophe theory is also presented.

The opening lecture at the ASI was delivered by Professor Willis E. Lamb Jr. on the approach to equilibrium. The ideas presented in this talk were based on his 1973 paper "Approach to Thermodynamic Equilibrium (and other Stationary States)", which we are pleased to reprint here as the opening chapter of this book.

This book also contains two other reprinted articles. One is a paper by A. O. Barut and P. Meystre on a classical model of the EPR experiment. This generated considerable discussion at the ASI which led Professor A. Aspect to write a comment on this paper for these proceedings. The other is a paper by M. O. Scully on the relation between quantum mechanics and hidden-variable theory.

ACKNOWLEDGEMENTS

We are grateful for the generous financial support provided by the Scientific Affairs Division of NATO, and in particular to Dr. Craig Sinclair. We also are grateful to the University of New Mexico Center for High Technology Materials, which provided partial support for the ASI, as well as funding for the two-week meeting in Albuquerque preceeding the ASI. We are indebted to the lecturers whose scholarly and pedagogical efforts made these meetings successful. We are grateful to Ms. Jeanne Williams and Ms. Gayla Angel, the conference secretaries, for their organizational help prior to and during the meetings. Ms. Williams also provided valuable help in the final assembly of the book.

DEDICATION

We dedicate this book to the memory of two pioneers in nonequilibrium statistical physics, Marc Kac of the University of Southern California and Stanislaw M. Ulam of Los Alamos National Laboratory. Though death prevented these good friends from contributing directly to this book, the legacy of their work can be found here in the writings of all those who learned so much from them.

Gerald T. Moore and Marlan O. Scully
Albuquerque, July 1985

CONTENTS

FOUNDATIONS

The first seven chapters which open this book are concerned with concepts and analytical machinery which are used to understand and make predictions about the statistical behavior of physical systems. These basic ideas include: measurement and meaning, space-time, probability, information, the principle of maximum entropy, quantum processes and distribution functions, and catastrophe (in the mathematical sense). The opening chapter by Professor Lamb, which is reprinted from proceedings of an earlier (1973) conference, examines the approach to thermodynamic equilibrium (and other stationary states). This topic remains an active area of investigation, and is a theme developed further throughout the book.

APPROACH TO THERMODYNAMIC EQUILIBRIUM

(AND OTHER STATIONARY STATES)

Willis E. Lamb, Jr.

Department of Physics and
Optical Sciences Center
University of Arizona
Tucson, Arizona 85721

This Symposium on the Physicist's Conception of Nature celebrates the 70th birthday of Paul Dirac. Before beginning the subject of my talk, I would like to make some remarks about ways in which Dirac has influenced me personally. The first edition of his book, *The Principles of Quantum Mechanics*, was published in 1930. In that same year I entered the University of California at Berkeley where I majored in Chemistry. I spent the summer of 1932 at home in Los Angeles, and came across Dirac's book on the shelves of the public library. I remember that it made a great impression on me, although I could not really understand very much of it. In part, due to its influence, I changed to Physics for graduate study, and during the years 1934 to 1938 worked with J. R. Oppenheimer on theoretical physics. Every spring, accompanied by most of his students, Oppenheimer went to Caltech in Pasadena, and around 1936 I had there the opportunity to hear a lecture by Dirac on magnetic poles. This was unfortunately not the occasion on which a member of the audience, addressing the speaker during the discussion period, was told of the difference between a question and a statement.

The Symposium's Sponsoring Committee had invited me to give a lecture on the history of electrodynamic level shift experiments and theory. My experiments with Retherford[1] on the fine structure of hydrogen were carried out at Columbia University in 1946/47. I first met Dirac personally at the Pocono Conference of 1948, where fine structure was one of the main topics of discussion.

During the Lindau Conference of 1959, Dirac invited me to accompany him on a long walk around the island of Mainau. He told me that he had first learned of the measurements on the hydrogen fine structure from the front page of *The New York Times* in 1947. I apologized for not sending him the news more directly, but explained that I had never thought of this work as reflecting any criticism of the Dirac equation.

This article is reprinted with permission from The Physicist's Conception of Nature, J. Mehra (ed.), pp. 527-547, copyright © 1973 by D. Reidel Publishing Company, Dordrecht, Holland.

After a pause, he asked whether I had enjoyed making the discovery of the fine structure anomaly. I replied that I had, but that I would much rather have discovered the Dirac equation. Dirac said kindly that things like that are much harder now.

Indeed they are. On the calculational side, in the late forties Norman Kroll[2] and I were able to make a relativistic extension of Bethe's non-relativistic quantum electrodynamic level shift calculation of 1947. Although Kroll went on to more complicated calculations of this kind, this represented the limit of my own participation on the theoretical side of this subject. On the experimental side, my colleagues Retherford, Dayhoff, and Triebwasser at Columbia reached an accuracy of about 0.1 MHz in the hydrogen level shift of about 1059 MHz. There has been experimental work by others which slightly improved on the accuracy of our work of the early 1950's but it is still very difficult to greatly increase the accuracy. After coming to Yale in 1962 I participated in attempts to make more accurate measurements on hydrogen as well as on various states of singly ionized helium. This work turned out to be rather disappointing with respect to quantum electrodynamic level shifts, but it did lead to very interesting results on highly excited states of the helium atom.

For my contribution to this Symposium, however, I have looked for another subject related to Dirac's contributions to physics. It is: 'Approach to Thermodynamic Equilibrium (and other Stationary States)'. The announced title had an obvious connection with Dirac's contributions to physics. As a matter of fact, the subject I shall discuss also has many connections with these contributions. Dirac[3] wrote major papers on statistical mechanics even before he began his famous series on quantum mechanics in 1925. The work under discussion makes essential use of quantum mechanics and of the time-dependent perturbation theory developed by Dirac. Furthermore, the density matrix[4] of Landau, von Neumann and Dirac plays an essential role in it.

This work grew out of research on the theory of the maser and the laser. I began the former in 1955 after the development of the ammonia beam maser by Townes, and the latter in 1961 with the gas laser of Javan. There are very close connections with the quantum theory of the laser formulated by Scully and myself in 1966/67.

Dirac[3] studied electrical engineering at Bristol before turning to physics. I asked him recently in Miami whether he had published any papers on electrical engineering. His reply was negative. I didn't tell him then, and don't know whether he realizes it now that his quantum theory of radiation is of basic importance to an important branch of electrical engineering called 'Quantum Electronics'.

Since there are professional experts here among the lecturers on equilibrium and non-equilibrium statistical mechanics (Uhlenbeck, Kac, Cohen and Prigogine), I hope I may be forgiven for poaching on their field.

Newtonian mechanics developed into the theory of analytical dynamics. This subject deals primarily with the problem of an isolated system, described by a conservative Hamiltonian function. The concept of an isolated system is an abstraction, for which one must pay later on by considering more complicated situations. It is relatively easy to allow for the presence of external forces acting on an otherwise isolated system. Non-conservative forces or dissipative forces represent greater complications, but are still manageable. Quantum mechanics is most easily applied to conservative dynamical systems, but even then it is essential to consider the disturbance of the sys-

tem caused by the process of measurement. Several sessions of this symposium are to be devoted to this still poorly understood subject.

On a higher level of complexity, we come to systems which are in thermodynamic equilibrium, and can be characterized by a temperature T. The system is in intimate contact with a thermal reservoir. An interplay of forces between system and reservoir occurs, with corresponding exchange of energy. In the theory of equilibrium statistical mechanics, one learns how to describe phase transitions. In some problems one finds that a system can develop spatial order, and in others there are long range quantum mechanical effects such as super-fluidity.

At the next level of complication, we find systems in non-thermodynamic stationary states. One example of this is heat conductivity where two different regions of the system of interest are in contact with thermal reservoirs, one at temperature T_1 and the other at temperature T_2. In some approximation, the thermal state of the system of interest may be described by a temperature which is a function of position. Another example of a non-thermodynamic stationary state is found in the theory of the laser oscillator. Such a device may display a well-defined temporal order. The field of biology presents us with still more complicated situations. It is doubted by some people that these are describable by present-day physical laws. Even if they are, we must clearly allow the biological system of interest to be in interaction with a more complicated environment than the thermal reservoir discussed above. It is necessary to have an interchange of matter, as well as energy. I doubt that the method of the grand canonical ensemble will suffice to describe the intake of food and processes of waste elimination.

For the discussion of equilibrium problems, the transition from thermodynamics to statistical mechanics makes it possible to treat many problems in a much more quantitative fashion. The central task is the evaluation of the partition function, or the sum-of-states

$$Z = \mathrm{spur}\,\{\exp(-H/kT)\}\,. \tag{1}$$

A great deal of current interest centres on the theory of phase transitions. This usually involves an attempt to perform an exact evaluation of the sum-of-states. Much work is being done on various forms of the Ising model. Such problems are over idealized, but they can approximate physical reality for suitably chosen crystals.

We are concerned today with the problem of the approach to thermodynamic equilibrium. This subject[5] has been given intensive consideration for over 100 years. I mention the names of some of the great workers in this field: Maxwell, Boltzmann, Rayleigh, Gibbs, P. and T. Ehrenfest, Fokker and Planck, Pauli, Onsager, Uhlenbeck, Chandrasekhar, Bloch, van Hove, Bogolubov, and Bergmann and Lebowitz. Among the topics discussed were the H theorem, Loschmidt's and Zermelo's paradoxes, ergodic and quasi-ergodic hypotheses, Poincaré cycles and randomization of phases. A key question is: what makes the approach to thermodynamics equilibrium occur? Suppose that the system of interest is an isolated conservative dynamical system. Can this come to thermodynamic equilibrium? Clearly not, for a simple system, but perhaps so in some approximation for a very complicated system, and probably so for a small part of such a complicated system. Some authorities feel that a 'speck of dust' is

needed, while others believe that the essential feature is the interaction of the system with a thermal reservoir. I strongly favour the last point of view.

One question which has received much attention is the following 'How does a causal and time reversible dynamical theory give irreversible behaviour?' I remind you that one of the earliest specific models for treating such questions is the Rayleigh problem where a heavy molecule moves into a gas of light molecules which are assumed to be maintained in thermodynamic equilibrium.

My object in this talk is to discuss the simplest possible non-trivial model for the approach to thermodynamic equilibrium. It may be regarded as playing the role of the Ising model in equilibrium statistical mechanics. The system of interest will be taken to be a single atom. It may be well to state explicitly at the outset that one is necessarily going to be considering an ensemble of such systems. This atom can be a simple harmonic oscillator, a non-linear oscillator or a two level atom. For the present discussion the system will be taken to be a quantum mechanical simple harmonic oscillator.

It is usual to think that a thermodynamic reservoir has to be a very large and complicated system which is somehow maintained in thermodynamic equilibrium. When the reservoir interacts with the system, it is changed. Some approximation or idealization has to be made in order to insure that the reservoir is not disturbed by its action on the system. The interaction of system and reservoir is not necessarily a small one, and in my view most discussions tend to sweep some difficulties under the carpet. A simpler model of a reservoir is suggested by calculations on the theory of the ammonia beam maser which were carried out at Stanford by Helmer[6] and myself in the mid-1950's. The reservoir consists of a stream of atoms characterized by a temperature T. These reservoir atoms may be simple harmonic oscillators, non-linear oscillators or two level atoms. One may use quantum mechanics for any of these models, or for the case of simple harmonic oscillators of non-linear oscillators one may use classical mechanics. For the present discussion we will use quantum mechanical two level atoms.

As stated above, the system of interest will be a quantum mechanical simple harmonic oscillator. Its cartesian coordinate will be denoted by E. This notations is carried over from the discussion of the maser in which the system is a one mode cavity electromagnetic resonator. Such a problem is dynamically equivalent to the problem of a mechanical simple harmonic oscillator. The energy eigenvalues of the system are given by the equation

$$\hbar W_n = (n + \tfrac{1}{2})\hbar\omega, \qquad (2)$$

(see Fig. 1), the eigenfunctions are denoted by $h_n(E)$ where the quantum number n can have the values 0, 1, 2, ..., ∞.

At $t=0$, we imagine that the system is described by a wave function

$$\Psi(E) = \sum_{n=0}^{\infty} c_n h_n(E). \qquad (3)$$

The probability for the state n to be found is given by

$$P_n = |c_n|^2. \qquad (4)$$

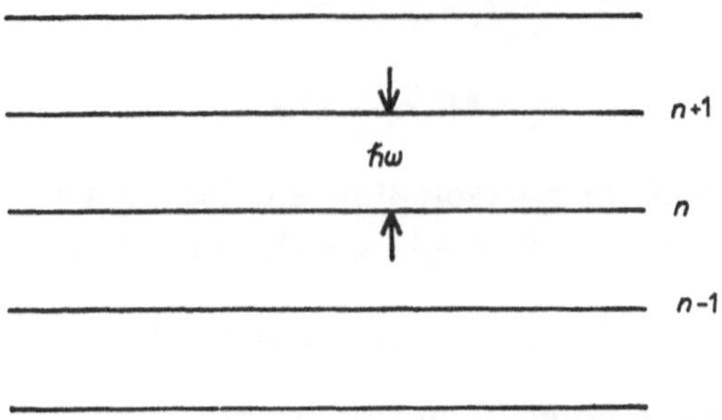

Fig. 1. Energy levels of the simple harmonic oscillator system atom as given by Equation (2).

It is important to recall that the description by a wave function is not really of a single system but of an ensemble of similarly prepared systems. For a more general description of the system we must use the density matrix[4] ϱ. For a pure case, where there is a wave function $\Psi(E)$, $\varrho[\Psi]$ has elements

$$\varrho_{nm} = c_m^* c_n \tag{5}$$

in the n representation. The diagonal elements ϱ_{nn} are probabilities P_n for finding the nth state.

More generally, we may have a mixture of states Ψ_i with weights w_i. The density matrix for a mixture is given by

$$\varrho = \sum_i w_i \varrho[\Psi_i]. \tag{6}$$

If we are given the density matrix ϱ, we can find the weights w_i and the constituent wave functions Ψ_i by transforming the density matrix ϱ to diagonal form. The eigenvalues are the w_i and the eigenfunctions are the Ψ_i. For a physical case, we need to have all of the weights $w_i \geqslant 0$ and $\sum_i w_i = 1$.

The reservoir consist of a stream of two level atoms, with energy levels shown in Fig. 2. The transition frequency ω is defined by the equation

$$\omega = W_a - W_b. \tag{7}$$

At most, only one of these two level reservoir atoms interacts with the system at any one time. Let the duration of the coupling be denoted by T. For the present, we assume that the reservoir atom is initially in its lower state b. Its wave function is $u_b(x)$ where x is a cartesian coordinate characterizing the reservoir atom. Let the system have the starting wave function $\Psi(E)$. The perturbation energy is taken in simple dipole–

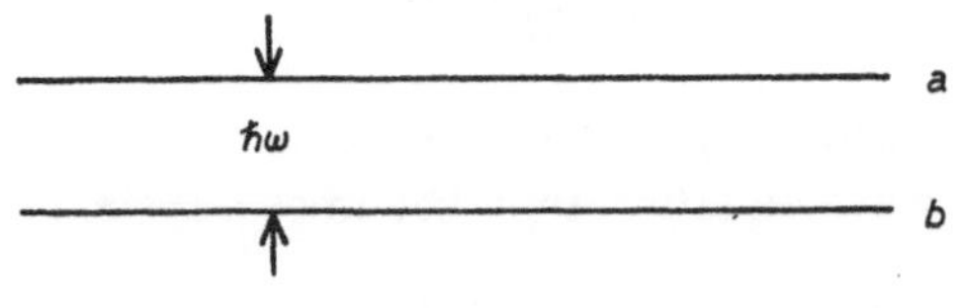

Fig. 2. Energy levels of a two-level reservoir atom with separation $\hbar\omega$ given by Equation (7).

dipole form

$$\hbar V = - eEx. \tag{8}$$

The combined system of the reservoir atom and the atom of interest is described by two cartesian coordinates E, x. At $t=0$ its wave functions is given by

$$\Psi(E, x, t = 0) = \Psi(E) u_b(x). \tag{9}$$

The state of the composite system is still a pure case.

During the time of interaction $0<t<T$ the wave function for the combined system has the form

$$\begin{aligned}\Psi(E, x, t) = &\sum_n a_n(t) h_n(E) u_a(x) e^{-i(W_n+W_a)t} \\ &+ \sum_n b_n(t) h_n(E) u_b(x) e^{-i(W_n+W_b)t}.\end{aligned} \tag{10}$$

At time T, we remove the reservoir atom from interaction with the system of interest. The joint probability distribution in E and x, given by

$$P(E, x, T) = |\Psi(E, x, T)|^2, \tag{11}$$

can describe correlations between E and x. One can learn something about the state of the system after the interaction has ceased by investigating the state of the reservoir atom, and vice versa. The 'difficulties' with the interpretation of quantum mechanics pointed out by Einstein, Podolsky and Rosen are met at this point but they do not bother one who believes that a probabilistic description of microscopic phenomena is inescapable.

We do not know whether the reservoir atom leaves in the state 'a' or in the state 'b' unless we choose to look at it. If we ignore the reservoir atom, the probability for the simple harmonic oscillator system to be in the nth energy eigenstate is

$$\begin{aligned}P_n(T) &= \int |a_n(T) u_a(x) e^{-i\omega T} + b_n(T) u_b(x)|^2 \, dx = \\ &= |a_n(T)|^2 + |b_n(T)|^2.\end{aligned} \tag{12}$$

The system of interest is no longer describable by a wave function. The pure case has been converted into a mixture, because we don't know, or care, whether the reservoir atom departed in the state 'a' or in the state 'b'.

In order to find the time development of the wave function during the interaction we have to integrate the Schrödinger equation

$$-\frac{\hbar}{i}\frac{\partial \Psi}{\partial t} = H\Psi. \tag{13}$$

It is a convenient and valid simplification to make the resonance approximation indicated in Fig. 3. The wave equation then becomes a system of equations

$$\begin{aligned} i\dot{a}_{n-1} &= V_n b_n \\ i\dot{b}_n &= V_n a_{n-1} \end{aligned} \tag{14}$$

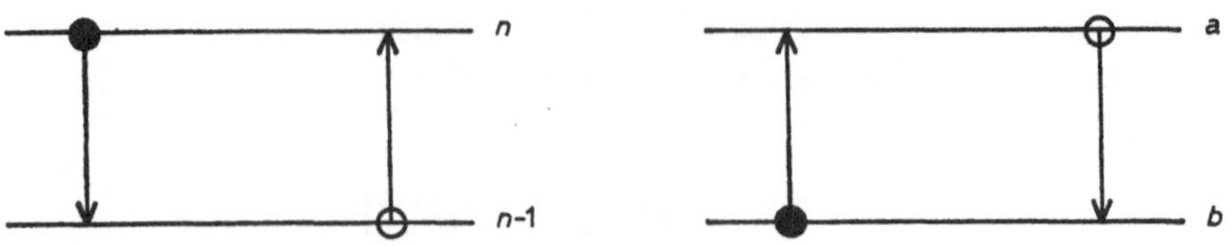

Fig. 3. Schematic representation of resonance approximation leading to Equation (14). When the reservoir atom makes a transition from state b to state a (or a to b), the system atom may make only transitions of the type n to $n-1$ (or $n-1$ to n).

where n takes on all values $1, 2, \ldots, \infty$, and the matrix elements of the perturbation are given by

$$\hbar V_n = -\, e\langle n-1|\, E\, |n\rangle \langle a|\, x\, |b\rangle . \tag{15}$$

This system of equations decouples into pairs of equations and the solution is given by

$$\begin{aligned} a_{n-1}(T) &= -\, i b_n(0) \sin V_n T \\ b_n(T) &= b_n(0) \cos V_n T \end{aligned} \tag{16}$$

for all n, where $|\, b_n(0)\,|^2 = P_n(0)$. It follows that

$$a_n(T) = -\, i b_{n+1}(0) \sin V_{n+1} T . \tag{17}$$

The probability that the system is in the nth state at the end of the interaction is given by

$$P_n(T) = P_n(0) \cos^2 V_n T + P_{n+1}(0) \sin^2 V_{n+1} T , \tag{18}$$

so that the change in probability becomes

$$\delta P_n = P_n(T) - P_n(0) = P_{n+1}(0) \sin^2 V_{n+1} T - P_n(0) \sin^2 V_n T . \tag{19}$$

The result of the interaction of one reservoir atom with the system has been a small but finite change in the probability distribution for the states of the harmonic oscillators system. If we inject atoms at an average rate r, we can write a differential equation for the probability distribution P_n given by

$$\frac{\mathrm{d}P_n}{\mathrm{d}t} = r\left[P_{n+1}(t) \sin^2 V_{n+1} T - P_n(t) \sin^2 V_n T\right]. \tag{20}$$

The transition from a small change in P_n to a differential equation for its average behaviour is sometimes described as 'coarse graining' but this term is also used in statistical mechanics to describe a very different concept where a number of stationary states of a complex system are grouped together in clusters, or in cells of phase space.

Our present discussion considers that each interaction lasts a definite time T. It is very easy now to average over a distribution of possible interaction times. We will take an exponential weight function

$$W(T) = \gamma e^{-\gamma T} . \tag{21}$$

This makes it possible to consider a different model of the reservoir, appropriate for

gas lasers, in which the active atoms are excited by some random process and allowed to decay in an exponential fashion by radiative decay. In this case the function $\sin^2 VT$ is replaced by its average

$$\langle \sin^2 VT \rangle = \tfrac{1}{2}V^2/[V^2 + (\gamma/2)^2]. \tag{22}$$

Let

$$V_n^2 = nV^2 \tag{23}$$

(simple harmonic oscillator)

$$\begin{aligned} A &= \tfrac{1}{2}rV^2 \\ B &= \tfrac{1}{2}rV^2(4V^2/\gamma^2). \end{aligned} \tag{24}$$

The differential-difference equation for P_n becomes

$$\frac{\mathrm{d}P_n}{\mathrm{d}t} = \frac{(n+1)A}{1 + \frac{B}{A}(n+1)} P_{n+1} - \frac{nA}{1 + \frac{B}{A}n} P_n. \tag{25}$$

Up to this point we have injected b atoms. Let us also inject a atoms at some rate r. We use primes to denote terms corresponding to b atoms (damping atoms) and omit primes for a atoms (pumping atoms). The differential equation for P_n then becomes

$$\begin{aligned} \frac{\mathrm{d}P_n}{\mathrm{d}t} = &- \frac{(n+1)A}{1 + \frac{B}{A}(n+1)} P_n + \frac{nA}{1 + \frac{B}{A}n} P_{n-1} + \\ &+ \frac{(n+1)A'}{1 + \frac{B'}{A'}(n+1)} P_{n+1} - \frac{nA'}{1 + \frac{B'}{A'}n} P_n. \end{aligned} \tag{26}$$

These are a special case of equations derived by Scully[7] and Lamb in 1966 in their quantum theory of the laser. However, as we will show, the equations can also be used to describe a realistic model for the approach to thermodynamic equilibrium for our system. We now proceed to consider a number of special cases of these equations.

Case 1. The 'a' and 'b' atoms are of the same type but their rates of injection are different so that r is not equal to r'. Then $B'/A' = B/A$ and $A/A' = r/r'$. We set $C = A'$ and $A/C = r/r'$. Equations (26) have a steady-state solution which is pictorially represented in Fig. 4. We have detailed balance if the rates represented by arrow 1 and 3 are equal and also those represented by arrow 2 and 4. These conditions are satisfied if

$$P_n = \frac{A}{C} P_{n-1} \quad \text{for} \quad n = 1, 2, 3, \ldots, \infty. \tag{27}$$

Our solution can be written in the form

$$P_n = (A/C)^n P_0 \tag{28}$$

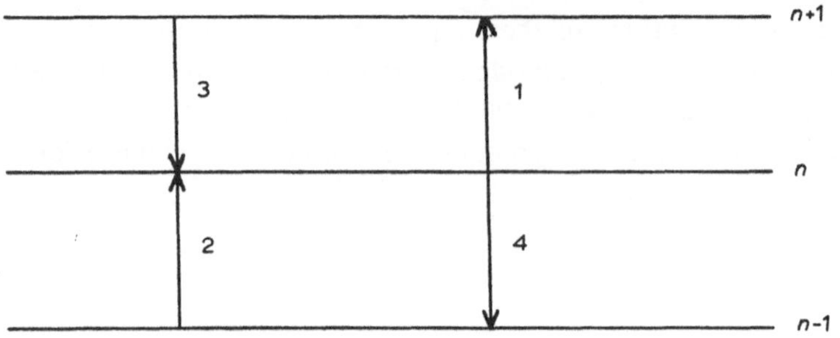

Fig. 4. Pictorial representation of the four terms on the right hand side of Equation (26), each represented by a numbered arrow.

where

$$P_0 = [1 - (A/C)] \tag{29}$$

if $A < C$. Write

$$P_n = P_0 e^{-n\hbar\omega/kT} \tag{30}$$

where the 'temperature' T is given by

$$e^{-\hbar\omega/kT} = A/C = r/r'\,. \tag{31}$$

We see that our reservoir of two level atoms at temperature T is able to bring the simple harmonic oscillator system to the single mode Planck distribution. This result is obtained independently of the strength of the coupling.

Case 2. $B = B' = 0$. The calculation is carried out only to first order in the coupling strength V^2. This is sometimes called the 'linear' theory. (The equations for $P_n(t)$ are linear in any case.) The equations for this case were discussed by Shimoda, Takahashi and Townes,[8] and Feller[9] devotes a chapter to similar equations.

These equations have the same steady-state solution as we had in Case 1. In both cases, there are also solutions which decay exponentially in time. They allow us to describe the approach to thermodynamic equilibrium. These solutions have the form

$$P_n(t) = \varrho_n e^{-\lambda t}\,. \tag{32}$$

The possible values of the decay constant λ are called decay eigenvalues and denoted by λ_r. The corresponding eigenfunctions are $\varrho_n^{(r)}$. The steady-state solution previously found corresponds to $\lambda_0 = 0$. It can be shown that the eigenvalues are of the form

$$\lambda_r = r(C - A)\,. \tag{33}$$

(Note that the eigenvalue spectrum becomes highly degenerate when $A = C$, i.e., when T approaches infinity.) The general solution of the system of equations is given by

$$P_n(t) = \sum_{r=0}^{\infty} \varrho_n^r\, e^{-\lambda_n t}\,. \tag{34}$$

Some steady-state solutions of these equations are illustrated by figures prepared using a small computer with a cathode ray oscilloscope display. Fig. 5 shows P_n as a function of n for the ratio $A/C=0.96$. The horizontal scale has discrete spacing at intervals of 1 from 0 to 250. The points representing the P_n values should be seen as

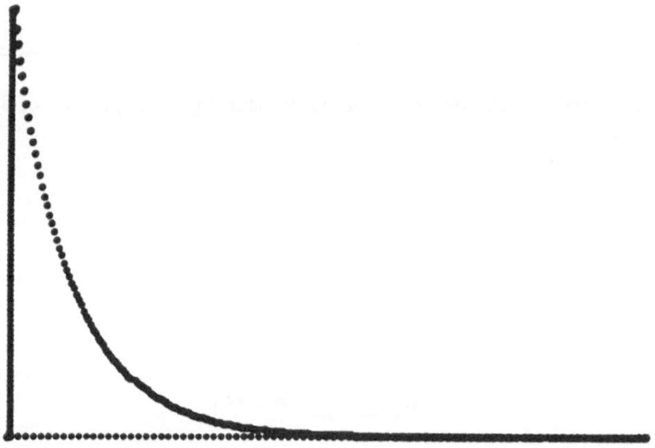

Fig. 5. Steady state (Boltzmann) distribution for the relative probability P_n for finding the n-th harmonic oscillator stationary state as a function of quantum number n, for a pumping to damping ratio $A/C=0.96$ and saturation parameter $B=0$. The horizontal scale runs from $n=0$ to $n=250$. The distribution P_n is unnormalized.

Fig. 6. Plot of relative values of $P_n=\varrho_{nn}$ for $0\leqslant n\leqslant 250$ with $A/C=1.0$ (corresponding to an 'infinite temperature') and $B=0$. The distribution is unnormalizable.

discrete points, but because of the crudeness of the display this feature can only be seen on the steeper parts of the curves. The curve in this case is a simple exponential function corresponding to the Planck distribution law. In all of the figures, one will see a point slighly to the left of the lower extremity of the vertical axis. This should be ignored, as it represents a minor deficiency in the plotting routine. Fig. 6 shows the corresponding plot for $A/C=1.0$, i.e., $T=\infty$. In this case the distribution of probabil-

ities for finding various number of photons is uniform, and hence we have an unnormalizable distribution. As in all the other cases, the point corresponding to $n=0$ is placed at a vertical height of 1.0 for convenience. Fig. 7 shows the corresponding graph for the physically unrealistic value $A/C=1.04$, which will correspond to a 'negative temperature'. Such a concept is perfectly reasonable for a two level atom, but is absurd for a simple harmonic oscillator with an infinite number of equally spaced energy levels. The figure indicates the higher one goes in n the greater the probability P_n, and the distribution is thoroughly unnormalizable.

Let us now see computer generated plots of the first two eigenmodes of decay for $A/C=0.96$. The corresponding eigenvalues are $\lambda_1=0.04$ and $\lambda_2=0.08$. Fig. 8 corresponds to λ_1. The eigenfunction is positive for small n values and negative for larger n values. There is one node. It may well be to comment briefly on the physical significance of this curve. Suppose that the distribution of photons in the cavity resonator is given by the exponential curve of Fig. 5 which represents the stationary state. Now

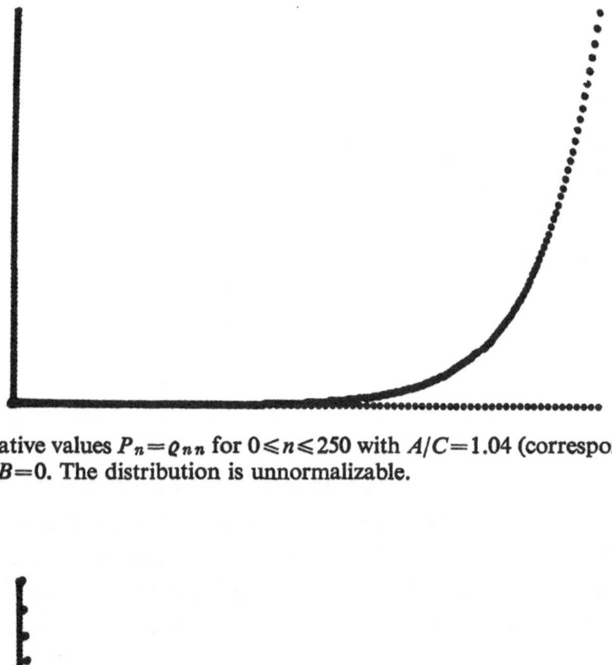

Fig. 7. Plot of relative values $P_n=\varrho_{nn}$ for $0\leqslant n\leqslant 250$ with $A/C=1.04$ (corresponding to a 'negative temperature') and $B=0$. The distribution is unnormalizable.

Fig. 8. First decay eigenmode of Equation (26) for $A/C=0.96$ and $B=0$, with decay constant $\lambda_1=0.04$. The physical significance of decay eigenmodes is discussed in the text.

contemplate a distortion of this curve produced by adding to P_n the distribution shown in Fig. 8 with a small positive coefficient. The resulting distribution will be one which has more photons for small n values and fewer photons for large n values than for the equilibrium steady-state of Fig. 5. In the course of time, the resulting distribution will relax to thermodynamic equilibrium in such a way that the departure from equilibrium values of P_n decay exponentially in time with a rate constant λ_1. If, instead of adding to P_n the distribution of Fig. 8, it were subtracted with a small coefficient, the approach to equilibrium would take place in such a way that P_n for small n increase and P_n for large n decrease. Fig. 9 shows the decay eigenmode for $\lambda_2=0.08$.

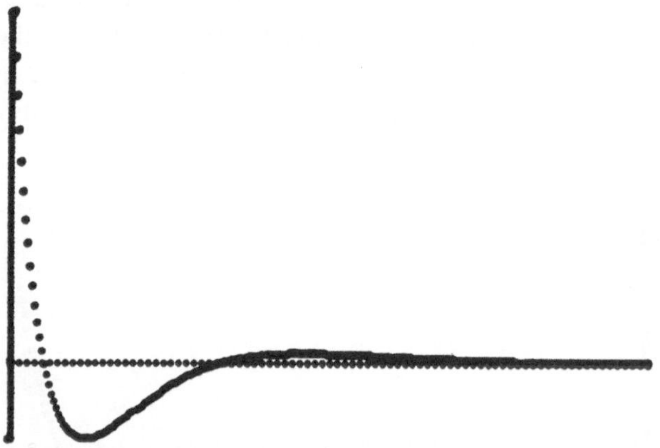

Fig. 9. Second decay eigenmode of Equation (26) for $A/C=0.96$ and $B=0$, with decay constant $\lambda_2=0.08$. The physical significance of decay eigenmodes is discussed in the text.

This curve has two nodes and represents a more complicated way in which the distribution of photons can be deformed. There are an infinite number of higher eigenmodes of decay; that with eigenvalues λ_r has r nodes. These are needed in the expansion (34) of $P_n(t)$ to express more rapidly decaying departures from thermal equilibrium.

We now turn briefly to a discussion of the behaviour of the non-diagonal elements of the density matrix. A partial idea of their significance can be had from the observation that the ensemble average value of the coordinate E is given by the equation

$$\langle E\rangle = \mathrm{spur}\,(\varrho E) \tag{35}$$

which is different from 0 if and only if we have non-zero elements $\varrho_{nn\pm1}$ of the density matrix. In this case we speak of an off-diagonality index of one unit. More generally, we may have elements k units off-diagonal of the form $\varrho_{nn\pm k}$, which, for a given fixed k value, will be denoted by σ_n. In Case 2, σ_n obey a differential equation

$$\begin{aligned}\frac{d\sigma_n}{dt} &= -A\,(n+1+\tfrac{1}{2}k)\,\sigma_n + A\sqrt{n\,(n+k)}\,\sigma_{n-1} + \\ &+ C\sqrt{(n+1)(n+1+k)}\,\sigma_{n+1} - C\,(n+\tfrac{1}{2}k)\,\sigma_n .\end{aligned} \tag{36}$$

We now have more general decay eigenmodes with decay eigenvalues $\lambda(r, k)$ depending on two indices r and k. The second of these measures the off-diagonality and the first labels the possible eigenvalues in order of increasing size beginning with $r=0$. It can be shown that the decay eigenvalues of (36) are given by

$$\lambda(r, k) = (C - A)(r + \tfrac{1}{2}k) \tag{37}$$

for general values of r and k. If $k=0$, $r=0$, $\lambda(0, 0)=0$ corresponding to the steady-state distribution of photon probabilities from the diagonal elements of the density matrix in the n representation. If $k=1$, $r=0$ we get the lowest eigenvalue for unit off-diagonality $\lambda(0,1)=\frac{1}{2}(C-A)$. In Fig. 10 we show a plot of σ_n as a function of n for $k=1$ and $r=0$ when $A/C=0.96$. The corresponding plot for $r=1$ is shown in Fig. 11 These distributions do not represent stationary-states but decaying elements of the density matrix of off-diagonality $k=1$. As in the earlier plots the horizontal scale runs

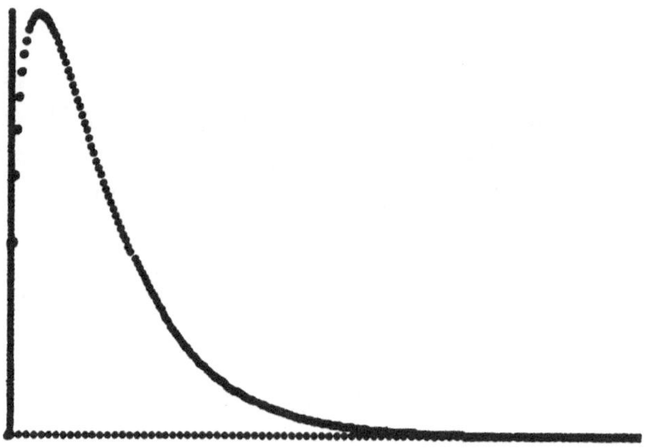

Fig. 10. A plot of the relative values of $\sigma_n = \varrho_{nn+1}$ as a function of $n(0 \leqslant n \leqslant 250)$ for the lowest decay eigenmode of Equation (36) when $A/C=0.96$ and $B=0$. The eigenvalue is $\lambda(0, 1)=0.02$.

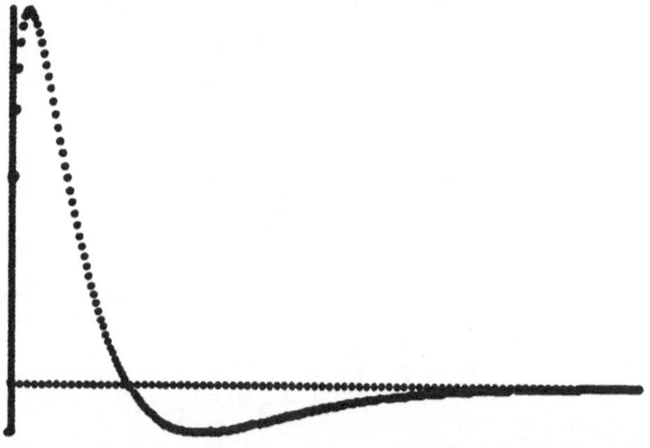

Fig. 11. A plot of $\sigma_n = \varrho_{nn+1}$ as a function of $n(0 \leqslant n \leqslant 250)$ for the second lowest eigenmode of Equation (36) for the parameters of Fig. 10. The eigenvalue $\lambda(1, 1)=0.06$.

through the integers from 0 to 250. The vertical scale is so arranged that the point for $n=0$ is taken to be at unit height.

A number of comments are in order. It is seen that even for a system with one degree of freedom there are an infinite number of decay eigenmodes. The average coordinate $\langle E \rangle$ decays exponentially to zero at a rate $\frac{1}{2}(C-A)$ in Case 2. If one thinks of analogy between the problem of the simple harmonic oscillator and the problem of electrical oscillations in a circuit containing inductance, capacitance and resistance, the decay of the amplitude at half the rate of the decay of the non-thermal energy stored in the cavity will seem reasonable. (It can be shown that terms corresponding to higher decay eigenvalues cancel out completely in Case 2.) The linear theory explodes if A is greater than or equal to C. The equation

$$e^{-\hbar\omega/kT} = A/C > 1 \tag{38}$$

means that T is less than 0. We have already mentioned that a negative temperature is a valid concept for two level atoms but absurd for simple harmonic oscillators. The trouble occurs even in the non-linear theory if we take $B/A = B'/A'$.

Case 3. Here we take $B'=0$, and again use the notation $A'=C$. This means that we are injecting non-linear pumping atoms in their upper state and linear damping atoms in their lower state. The constant B describes the 'saturation' property of the pumping atoms. In this case our differential equations takes the form

$$\begin{aligned}\frac{dP_n}{dt} = -\frac{A(n+1)}{1+\frac{B}{A}(n+1)}P_n + \frac{An}{1+\frac{B}{A}n}P_{n-1} + \\ + C(n+1)P_{n+1} - CnP_n.\end{aligned} \tag{39}$$

This case describes a simple quantum mechanical theory of a laser. The damping atoms simulate the ohmic resistance of the cavity resonator, and the pumping atoms the effect of the population inversion produced by a gas discharge or an injected beam of state selected atoms. These equations for the diagonal elements of the density matrix have a steady-state solution, and the same sort of detailed balance pairing of terms in the equations we met before. The steady-state is characterized by the equations

$$P_n = \frac{A}{C}\,\frac{1}{1+\frac{B}{A}n}\,P_{n-1} \tag{40}$$

so that

$$P_n = P_0 \prod_{j=1}^{n} \frac{A/C}{1+(B/A)j}. \tag{41}$$

This defines a 'truncated Poisson distribution'. If $A<C$, the distribution P_n falls off with increasing n, much as it did in the linear case. When $A>C$ the peak value will be

for an integer close to n_p given by

$$A/C = 1 + (B/A)\, n_p$$

or

$$n_p = \left(\frac{A}{B}\right)\frac{(A - C)}{C}\,. \tag{42}$$

The quantity P_n as a function of n gives the probability distribution for finding n photons in a laser cavity. We now show a number of plots of P_n for $C = 1.0$, and a succession of A values. Fig. 12 has $A = 0.96$, $B = 6 \times 10^{-4}$. It is seen that the distribution for P_n is just a little sharper than it was in the linear case. If the number of pumping atoms is now increased, keeping the number of damping atoms constant, both the quantities A and B will increase in the same proportion. Fig. 13 is shown for $A = 1.0$

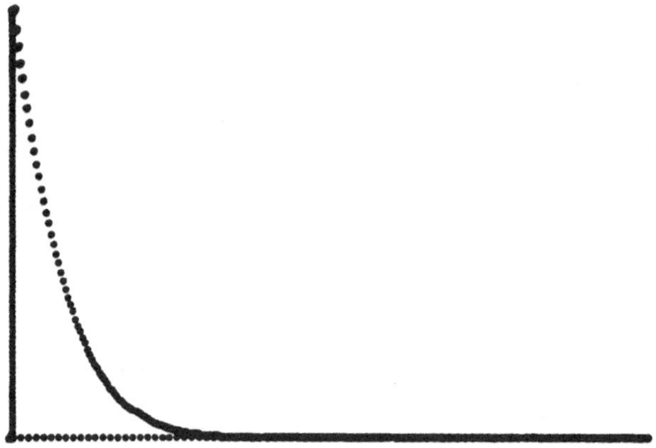

Fig. 12. Relative steady state probability distribution P_n for finding n photons in a laser below threshold, as given by Equation (40) for parameters $A/C = 0.96$ and $B/A = 6.25 \times 10^{-4}$. The horizontal scale runs from $n = 0$ to $n = 250$.

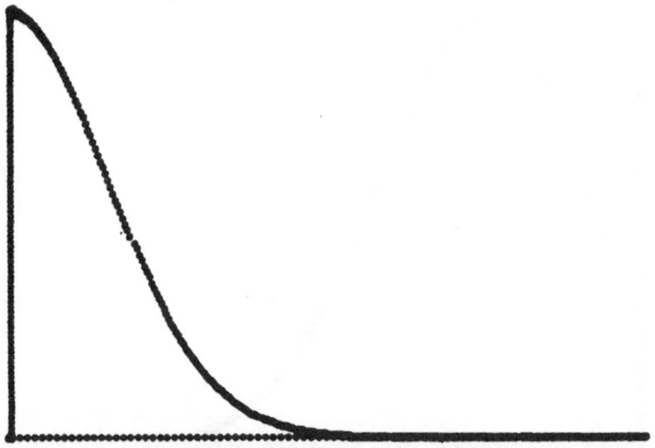

Fig. 13. Relative steady state probability distribution P_n for finding n photons in a laser at threshold as given by Equation (40) for parameters $A/C = 1.0$ and $B/A = 6.25 \times 10^{-4}$ with the horizontal scale used in Fig. 12.

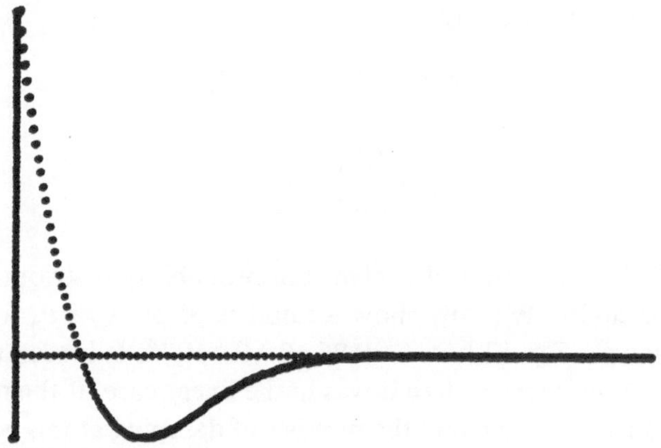

Fig. 14. First decay eigenmode of Equation (26) for parameters $A/C = 1.0$ (laser at threshold) and $B/A = 6.25 \times 10^{-4}$. The physical significance of decay eigenmodes is discussed in the text. The horizontal scale runs from $n = 0$ to $n = 250$.

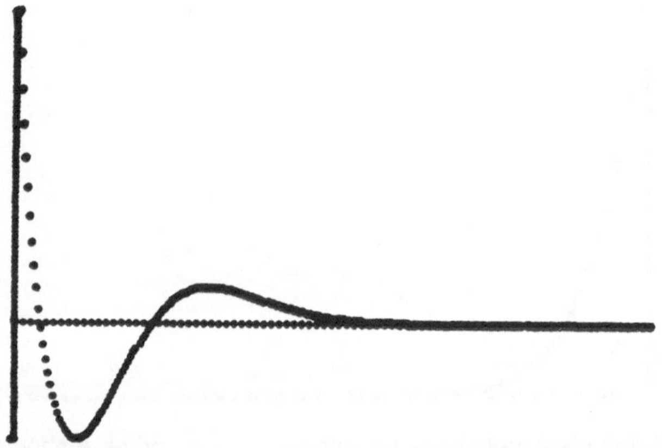

Fig. 15. Second decay eigenmode of Equation (26) for parameters $A/C = 1.0$ (laser at threshold) and $B/A = 6.25 \times 10^{-4}$. The physical significance of decay eigenmodes is discussed in the text. The horizontal scale runs from $n = 0$ to $n = 250$.

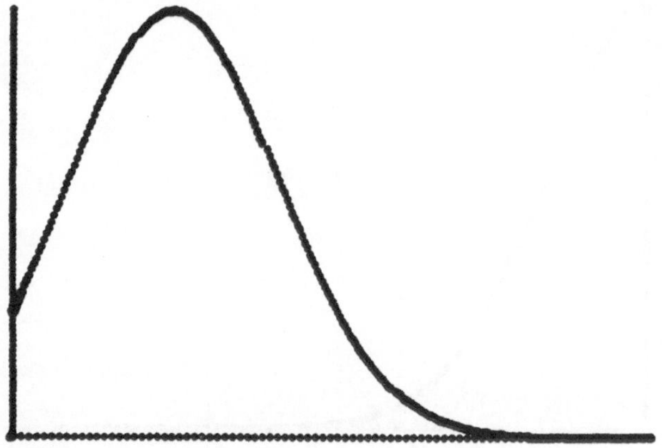

Fig. 16. Relative steady state probability distribution P_n for finding n photons in a resonator of a laser 4% above threshold, as given by Equation (40) for $A/C = 1.04$ and the same value of the saturation parameter B/A used in Figs. 12 to 15.

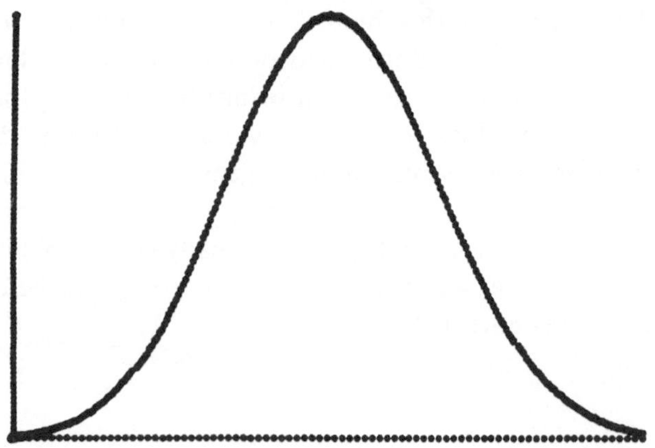

Fig. 17. Relative steady state probability distribution P_n for finding n photons in the resonator of a laser 8 % above threshold, as given by Equation (40) for $A/C = 1.08$ and the same value of the saturation parameter B/A used in Figs. 12 to 16.

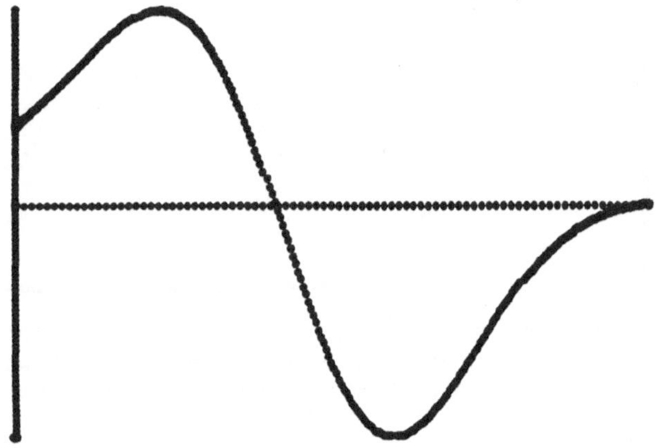

Fig. 18. First decay eigenmode of Equation (26) for $A/C = 1.08$ and $B/A = 6.25 \times 10^{-4}$ corresponding to the steady state photon distribution of Fig. 17. The physical significance of decay eigenmodes is discussed in the text.

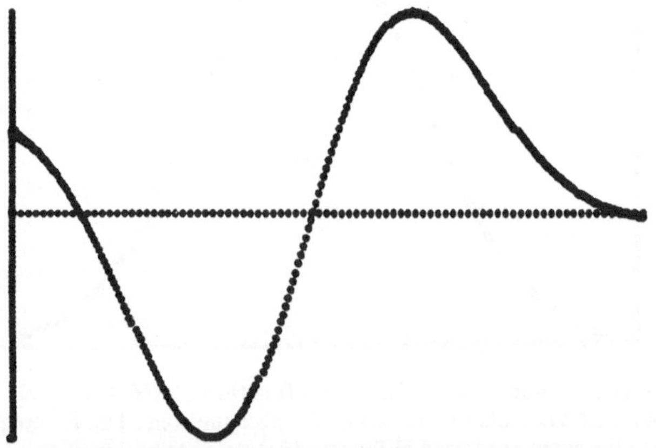

Fig. 19. Second decay eigenmode of Equation (26) for $A/C = 1.08$ and $B/A = 6.25 \times 10^{-4}$ corresponding to the steady state photon distribution of Fig. 17. The physical significance of decay eigenmodes is discussed in the text.

which corresponds to the laser threshold. This distribution has an approximately Gaussian form instead of the uniform distribution met in the linear theory. The saturation properties of the non-linear atom brings the photon probability distribution down with increasing n values. We also show in Figs. 14 and 15 the first two decay eigenmodes for this case. The corresponding eigenvalues are $\lambda_1=0.04983$ and $\lambda_2=$ $=0.12716$.

Fig. 16 corresponds to $A=1.04$ and gives the steady-state probability distribution P_n. Notice the beginning of a well-defined peak at $n=n_p$. As before, the curve is normalized by setting the value of $P_0=1$.

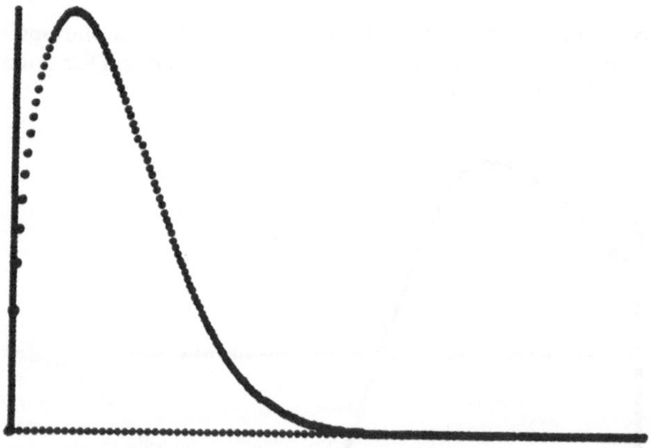

Fig. 20. A plot of relative values of $\sigma_n=\varrho_{nn+1}$ as a function of $n(0\leqslant n\leqslant 250)$ for the lowest decay eigenmode of off-diagonal elements of the Scully–Lamb equations. The parameters, $A/C=1.0$ (laser threshold), and $B/A=6.25\times10^{-4}$, are the same one used in the steady state distribution shown in Fig. 13. The decay eigenvalue $\lambda(0,1)=0.014753$.

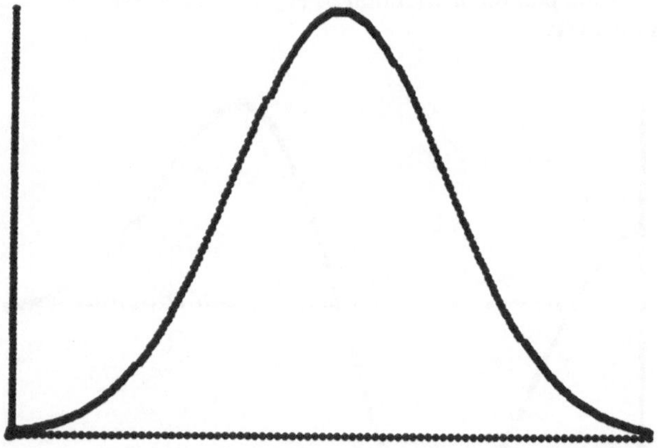

Fig. 21. A plot of relative values of $\sigma_n=\varrho_{nn+1}$ as a function of $n(0\leqslant n\leqslant 250)$ for the lowest decay eigenmode of off-diagonal elements of the Scully–Lamb equations. The parameters $A/C=1.08$ and $B/A=6.25\times10^{-4}$ are the same ones used in the steady state distribution shown in Fig. 17. The decay eigenvalue $\lambda(0,1)=0.002483$ is much smaller than that of Fig. 20 indicating that the average electric field in a laser well above threshold decays only slowly because of phase diffusion arising from spontaneous emission.

Fig. 17 shows the steady-state distribution for a value of $A=1.08$. The peak is now very well developed. The curve is similar to, but broader than, a similar Poisson distribution. Figs. 18 and 19 show the corresponding decay eigenmodes for $\lambda_1=0.05644$ and $\lambda_2=0.09565$. As before, if the steady-state distribution is perturbed by adding a small amount of one of these decay eigenmodes, the resulting non-stationary distribution relaxes to the stationary laser distribution by the exponential decay of the admixed decay eigenmode.

We also show examples of decay eigenmodes for off-diagonal elements of the density matrix for $k=1$. Fig. 20 corresponds to $A=1.0$ and $\lambda(0, 1)=0.014753$. As before, this curve gives statistical information about the electromagnetic field in a cavity resonator which is excited just at threshold. The decay constant is relatively small. Fig. 21 gives the corresponding curve for $A=1.08$. The corresponding eigen-

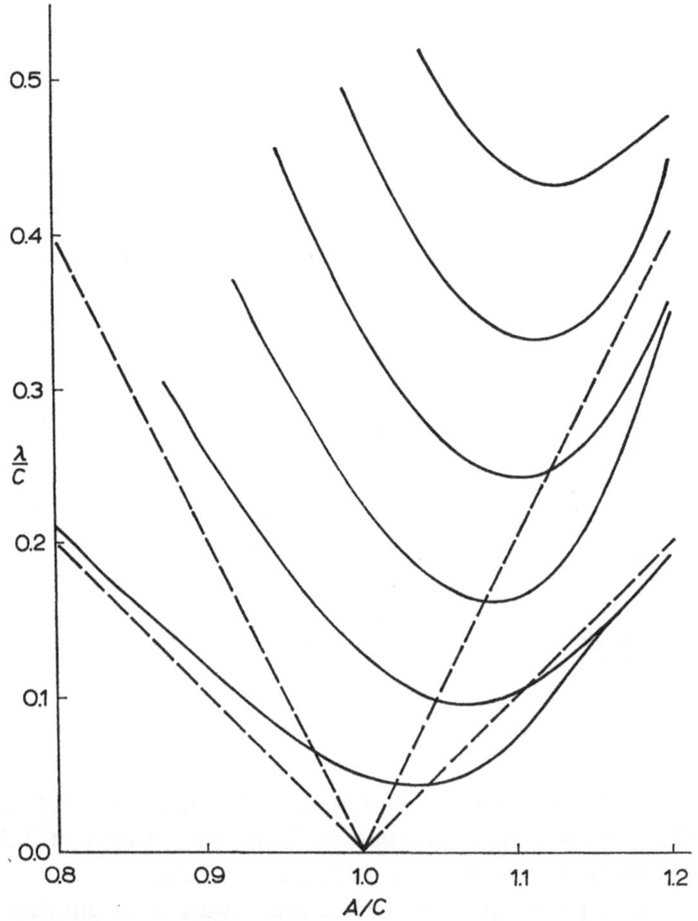

Fig. 22. Summary of numerical calculations for the eigenvalue spectrum in Case 3 for the diagonal elements of the density matrix. The steady state solution is indicated by the axis of abscissae. The higher eigenvalues are shown by solid curves. The dashed curves correspond to the linear theory. $B/A=6.25\times10^{-4}$.

value, $\lambda(0, 1) = 0.002483$, becomes much smaller. Note that the shape of the curve is almost indistinguishable from that for the diagonal elements of the density matrix for the same A value.

Fig. 22 summarizes a number of results of Y. K. Wang[10] and myself in Case 3 for the eigenvalue spectrum for the diagonal elements of the density matrix. The steady-state is indicated by the axis of abscissae. The higher eigenmodes are shown by solid curves. The dashed curves correspond to the linear theory. Above the threshold, $A/C = 1.0$, the absolute value of the difference $A - C$ has been used in the equation for $\lambda(r, 0)$, following a suggestion of P. Mandel.[11] Fig. 23 shows the smallest eigenvalue $\lambda(0, 1)$ for unit off-diagonality.

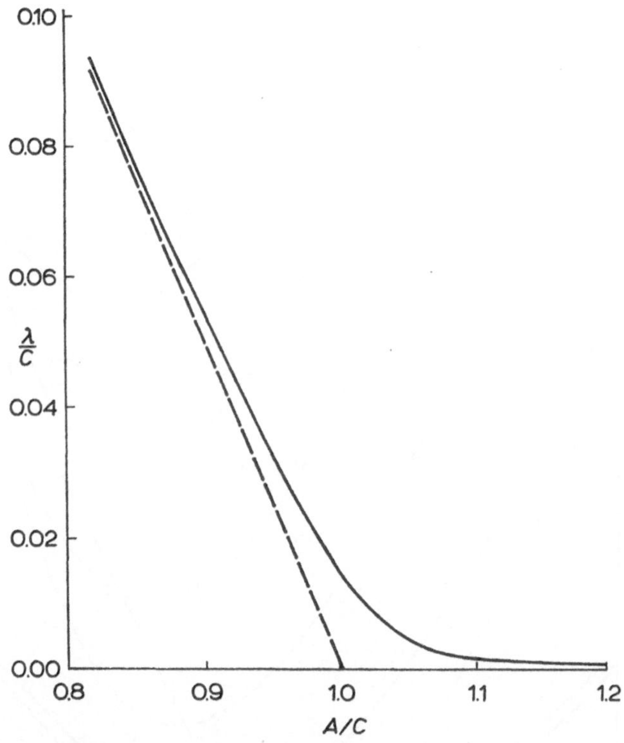

Fig. 23. Numerical calculations for the smallest eigenvalue $\lambda(0, 1)$ for unit off-diagonality of the density matrix. $B/A = 6.25 \times 10^{-4}$.

I do not have the time to discuss all of the physical ramifications of these equations. Among the topics neglected are the amplitude decay, phase diffusion, Johnson noise, and the effects of spontaneous emission from the 'a' atoms.

Only fluctuations of purely quantum mechanical nature enter our theory. They are associated with the making and breaking of contact between the system and reservoir atoms. If the system were described by a wave function before the interaction, it would be described by a mixed case density matrix after the interaction. I should

mention that it would be possible to introduce equivalent random noise forces to simulate the quantum mechanical fluctuations, in a way similar to that used by Uhlenbeck and Goudsmit[12] in their treatment of the Brownian motion of a galvanometer mirror in 1929.

I have given some simple and explicit calculations, and I hope they are compatible with the more general discussion of non-equilibrium phenomena.

REFERENCES

1. W. E. Lamb, Jr., and R. C. Retherford, *Phys. Rev.* **72**, 241 (1947). For a fuller account, see W. E. Lamb, Jr., *Rep. Progr. Phys.* **14**, 19 (1951).
2. N. M. Kroll and W. E. Lamb, Jr., *Phys. Rev.* **75**, 388 (1949).
3. See J. Mehra, 'The golden age of theoretical physics: P. A. M. Dirac's scientific work from 1924 to 1933' in *Aspects of Quantum Theory* (edited by A. Salam and E. P. Wigner), Cambridge University Press, 1972.
4. See D. ter Haar, *Rep. Progr. Phys.* **24**, 304 (1961).
5. See T. Y. Wu 'On the nature of theories of irreversible processes', *Int. J. Theor. Phys.* **2**, 325 (1969).
6. J. C. Helmer, 'Maser Oscillators', Ph.D. thesis, Stanford University (1957). (Obtainable from University Microfilms, Ann Arbor, Michigan.) Also see W. E. Lamb, Jr., 'Quantum mechanical amplifiers' in *Lectures in Theoretical Physics*, vol. II, Boulder, Colorado, 1959 (edited by W. E. Britten and B. W. Downs), Interscience Publishers, New York (1960).
7. M. O. Scully and W. E. Lamb, Jr., *Phys. Rev. Letters* **16**, 853 (1966) and *Phys. Rev.* **159**, 208 (1967).
8. Shimoda, Takehashi and Townes, *J. Phys. Soc. Japan* **12**, 686 (1957).
9. W. Feller, *An Introduction to Probability Theory and its Applications*, Vol. I, 2nd ed., John Wiley and Sons, Inc. New York (1957), chapter XVII.
10. Y. K. Wang, Ph.D. thesis, Yale University (1971). (Obtainable from University Microfilms, Ann Arbor, Michigan.) Also, Y. K. Wang and W. E. Lamb, Jr., to be published in *Phys. Rev.*
11. P. Mandel, private communication.
12. G. E. Uhlenbeck and S. A. Goudsmit, *Phys. Rev.* **34**, 145 (1929).

"PHYSICS AS MEANING CIRCUIT": THREE PROBLEMS

John Archibald Wheeler

Center for Theoretical Physics
The University of Texas at Austin
Austin, TX 78712

I. WHY THE QUANTUM?

Existence, yes; but why existence ruled, everywhere we look, by the quantum principle? What is the necessity of the quantum? Out of what deeper foundation can we recognize that the quantum supplies absolutely the only conceivable way to construct what we call "reality"?

At least one working hypothesis suggests itself by way of answer: "Existence is a meaning circuit"[1] (Fig.1). On this view the elementary quantum phenomenon, in the sense of Bohr, is the "atomistic" element in the establishment of meaning. "Brought to a close", as he put it, "by an irreversible act of amplification"--such as the click of a counter or the blackening of a grain of photographic emulsion--"Bohr's phenomenon" provides the elementary "yes" or "no" that one person can communicate to another in plain language[2]. And in the operation of the eye or the measurement of a field strength or in the decibel level of a sound we have many a chance "yes, no" adding up to the definiteness of "how much".

What have "how much" determinations to do with the construction of meaning? The results of philosophy are too important to be left to the philosophers. No topic has engaged workers in that field more intensively over recent decades than the subject of meaning. No conclusion out of their investigations is more relevant than the sentence of Føllesdal, "Meaning is the joint product of all the evidence that is available to those who communicate"[3].

The step is small from meaning to the working hypothesis of existence as a meaning circuit ("MCH"). The part of this circuit that runs "above ground," in plain view, is well known. Physics provides light, pressure and sound, the means of communica<u>tion</u>. Physics also gives rise to chemistry and biology and, through them, to communica<u>tors</u>. The communicators and the communications between them generate meaning. That is the first part of the meaning circuit. It makes meaning the child of physics.

Physics itself, however, according to this view, is also the child of meaning. The return portion of the meaning circuit, the connection that leads back from meaning to physics, runs "underground," out of

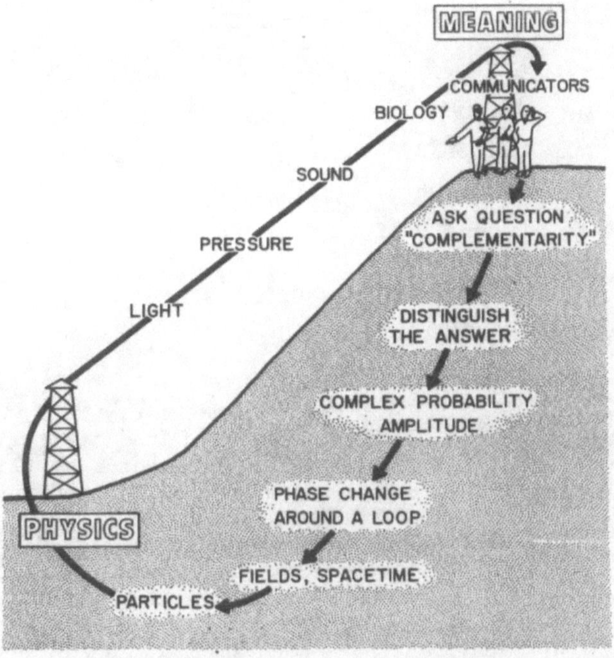

Fig. 1. The "meaning circuit." Physics gives light and sound and pressure, tools of communication. It gives biology and chemistry and, through them, communicators. Communication between communicators gives meaning. Meaning calls for the asking of questions, but the asking of one question stands in a complementary relation in the asking of another. The reception of an answer demands distinguishability. Mathematical analysis of distinguishability demands probability amplitudes. Complementarity demands that these probability amplitudes be complex. A complex probability amplitude has a phase. The change of phase around a closed loop can be regarded as the definition and measure and even the sole form of existence of the "flux of field" through that loop. Fields so defined -- electrodynamic, geometrodynamic and chromodynamic -- give rise to particles and physics, thus closing the circuit.

sight, like the return circuit of the early electric railway lines, that also ran underground, through channels fathomless to man.

How does the return circuit operate? The establishment of meaning demands the asking of questions and the distinguishability of the answers to those questions. Freedom to ask questions, one question or the other but not two complementary questions at the same time, is encapsulated in the Heisenberg principle of indeterminism. Distinguishability of the answer to this question requires for its proper quantification--as shown by R. A. Fischer in 1922 for gene pools[4] and by William Wootters later in broader contexts[5]--not probabilities themselves, but the square roots of probabilities; that is to say, probability amplitudes. These probability amplitudes are real numbers when we are asking questions always of the same kind, such as frequency of gray eyes, brown eyes, and blue eyes--and as we do, for example, when we are seeking to distinguish one tribe from another. In contrast, making a physical observation requires the choice between one or other complementary feature of the system under study, such as its coordinate or its momentum. This complementarity, as expressed in the principle of indeterminism, demands, according to Stueckelberg[6], that the probability amplitudes should be complex numbers.

Such a complex number Feynman associates[7] with each conceivable history that leads from the starting configuration, A, of a system to its final configuration, B. He goes on to express the total probability amplitude to transit from A to B as the sum of all the individual probability amplitudes for all such histories with democratically equal weights. A special case of this addition of probability amplitudes is seen in the double-slit electron-interference experiment in the version of Aharonov and Bohm[8]. The interference fringes experience a phase shift proportional to the flux of magnetic field through the loop formed by tracing out in imagination the path through the one slit in the forward sense and the path through the other slit in the reverse sense.

This A-B shift in the interference pattern is customarily viewed as the consequence of the presence of the magnetic field. The meaning-circuit hypothesis reverses this point of view. It regards the magnetic field - and, by extension, the electromagnetic, geometrodynamic[9], and chromodynamic fields, and the whole world of particles built upon these fields--as having no function, no significance, no existence, except insofar as they affect wave phase; or, more generally, affect the probability of registration of elementary quantum phenomena.

All particles and all fields, according to this hypothesis, have a pure information-theoretic significance and origin, going back for their foundation to "meaning". All of physics derives, on this view, from "meaning", via the "yes, no" of multitudinous elementary quantum phenonema, past, present, and future. This link completes the underground pathway envisaged as leading from meaning to physics. This is the concept of the meaning circuit in brief.

The MC hypothesis, right or wrong, has the merit that it poses for study more than one interesting and analyzable question. We take up here three problems: Section II, how to translate A-B phase shift to Zeeman level displacement when the "loop" is as small as a wavelength; Section III, how to translate energy-level alteration into the registration of an elementary quantum phenomenon; Section IV, in what direction to look to express "time" as of statistical and information-theoretic origin.

II. WHAT HAPPENS TO THE CONCEPT OF A-B PHASE SHIFT FOR A "SMALL" LOOP?

How can we deal with the concept of "Aharonov-Bohm phase shift" under circumstances where the dimensions of the region enveloped by the two sets of de Broglie waves are comparable to the wavelength itself? What then is to be meant by the "area" pierced by the magnetic field?

The answer is simple in principle. The key idea translates itself from "phase shift" for an electron in an interferometer to phase shift for an electron going around in an atomic orbit; and from the "might-have-been" phase shift for this de Broglie wave to the "Zeeman energy-level displacement" that compensates and kills this phase shift.

This translation from A-B effect to Zeeman effect can be seen nowhere more clearly than in the hydrogen atom, for a circular orbit of large quantum number, n, with its plane of motion perpendicular to the magnetic field.

The total phase change around the circuit is $2\pi n$. The Aharonov-Bohm phase shift for such a loop is $2\pi\delta n=\delta\phi=(e/\hbar c)\Phi$, where Φ is the flux of magnetic field, B, through the circuit, $\Phi=B \times \text{Area} = B\pi(n^2\hbar^2/me^2)^2$.

Obviously no such phase shift does or can take place. The number of waves around the circuit has to remain an integer. The energy alters by just enough to compensate the A-B phase shift. The energy itself being given by the standard Bohr formula, $E = -(me^4/2\hbar^2 n^2)$, the necessary alteration in energy is

$$\delta E = (me^4/\hbar^2 n^3)\delta n.$$

When we insert here for δn the A-B phase shift divided by 2π, we end up with the standard large-n result for the Zeeman energy level shift,

$$\delta E = (eB/2mc)n\hbar.$$

In other words, the displacement of the energy level by the magnetic field provides a direct measure of the flux of magnetic field through the loop, and this even when the loop has dimensions comparable with the wavelength of the electron.

The same result can be seen more generally from first-order perturbation theory. The Hamiltonian is obtained by replacing the momentum in the usual expression by $p + (e/c)A$, with the magnetic potential given by $A_r = A_z = 0$ and $A_\phi = (B_z/2)r$. Squaring this revised momentum, dividing by twice the mass of the electron and dropping the term of the second order in the perturbing field, we have for the perturbation in the Hamiltonian the familiar expression,

$$\delta H = (e/mc)A_\phi p_\phi = (e/2mc)m_z\hbar B_z,$$

reproducing again the standard result for the Zeeman energy-level displacement. The occurence of the area-proportional factor in the magnetic potential shows how the magnetic flux does its work "behind the scenes" in this derivation of the level shift.

In the foregoing analysis, for simplicity, we have neglected any effect of the intrinsic or spin angular momentum of the electron, as is reasonable when the magnetic field is strong enough to decouple spin and orbital angular momenta.

III. REDUCING THE "HOW MUCH" OF THE FIELD MEASUREMENT TO THE "MANY A CHANCE YES-NO" OF ELEMENTAL QUANTUM PHENOMENA

It might appear that we have now satisfactorily disposed of the problem of defining the flux over a region comparable to a wavelength, disposed of it by measuring the Zeeman shift of an energy level instead of the A-B shift in the location of an interference fringe. Those fringes, however, were registered by individual blackened grains of photographic emulsion; that is, by elementary quantum phenomena; in other words, by "atoms of information" of the kind that build meaning. Where is any information-theoretic feature to be seen in the energy-level shift?

No elementary quantum phenomenon *is* a phenomenon, is a registered bit of information, is a potential contributor to meaning, until it has been brought to a close by an irreversible act of amplification. The electron in its orbit is like the electron in the interferometer on its way to the photographic plate, but not yet registered by the plate. It has to be got out and into a situation where it can register, or where some equivalent entity--such as a photon that has interacted with it--can register.

A half-dozen experimental procedures suggest themselves. It is enough to pursue one to its end, to the click of a counter or the registration of a photodetector, to see how field becomes level-shift, and how level-shift becomes information; and to appreciate by the inverse line of reasoning how the appropriate "yes, no" bits of information translate into level-shift, and how this level-shift translates into, not merely measurer of field-in-a-region, but the very definer of that field.

To accomplish the desired registration, it is enough in principle to illuminate the atom with monochromatic radiation of a precisely chosen frequency, too small to eject the electron from the atom when the system is in a state of the normal Bohr energy, but just enough for photoionization when the magnetic field is "on". The electron, once freed, is drawn by an electric field into a device like a photomultiplier. A pulse registers on a counter. The elemental quantum phenomenon is brought to a close by this irreversible act of amplification. An item of information is generated which one person can communicate to another "in plain language"[2]. Out of sufficiently many such items of "yes, no" information obtained for enough slightly different values of the frequency of the photoionizing radiation we determine more and more precisely the value of the Zeeman level shift and measure--and define--the magnetic field.

IV. TIME

Of all obstacles to thinking of the whole of physics as of information-theoretic origin, none bulks larger than "time". There is hardly a word spoken, a paper written or a book produced that does not take the concept of time as heaven-sent, provided free from outside physics and beyond atomistic dissection. The status of time in the last part of this century can only be compared to the status of elasticity in the last part of the preceding century. To know the elasticity of bronze was to be able to figure the speed of sound in a bronze bar, the bending of a bronze beam and the tones of a bronze bell. But where did one get the value of that elasticity? No one knew how to get it from a deeper,

atomistic, theory. Only measurement would do. Today elasticity can be calculated from first principles. With that gain, however, goes a loss. "Elasticity" is deprived of all meaning in the space between the electron and the atomic nucleus. In brief, it falls in status from primordial and precise to secondary, approximate and derived. Not except by a like dethronement from primordial and precise to secondary, approximate and derived can "time" ever be reduced, like all the rest of physics, to an information-theoretic foundation.

No difficulty in the way of this reduction is so great as the human-created difficulty of the word itself, an invention immensely useful but all too self-defeating in the problems it creates for a deeper understanding. Long distant though that deeper understanding may be, and revolutionary the deeper concepts that take the place of time, it is nevertheless a happy circumstance that at least three clues stand out for making progress into that unknown territory.

First, the very simplest considerations of quantum geometrodynamics tell us that there is no such thing as spacetime. "Spacetime"--the deterministic evolution of space geometry with time--violates the Heisenberg principle of indeterminism no less violently than does "world line"--the deterministic evolution of particle position with time. "World line" means position and velocity specified at every instant--an impossibility. "Spacetime" means geometrodynamic field coordinate (3-geometry) and geometrodynamic field momentum (or extrinsic curvature) simultaneously specified on every spacelike slice through that 4-geometry-- an equal impossibility. The proper description of the motion of the particle uses the concept of wave packet or, more generally, wave function, $\psi(x,t)$. The proper description of the dynamics of geometry uses the probability amplitude, $\Psi({}^3G)$, for this, that, and the other 3-geometry, ${}^3G$, or an equivalent description in terms of partly or wholly complementary field quantities[10]

In the appropriate semiclassical limit,[11] this quantum description reduces to classical geometrodynamics and gives back classical spacetime. However, when the distances under consideration are comparable to the Planck length, fluctuations dominate[12]. Then the very ideas of "before" and "after" lose all applicability. "Time" makes no sense at the Planck scale of distances.

A second insight into time comes from a look at the elemental quantum phenomenon, say the delayed choice between a "which-route" experiment and a "both-routes interference" experiment performed on a photon that had its origin, long before there was any life on earth, in a gravitationally-lensed quasar[13]. Vast though its reach is in space and time, the elemental phenomenon has neither meaning nor existence until it is brought to a close in the here and now by a registration. It hadn't happened. This conclusion appears contrary to normal notions of causality. Equally contrary to those everyday presuppositions about causality appear at first sight the results of the Einstein-Podolsky-Rosen experiment[14]. Nevertheless, the role of present choice in the "making" of what we often call the "past" is battle-tested and an inescapable consequence of quantum theory. Event-like as these elementary quantum phenomena are, they do make an undeniable linkage between present choice and any atomistically-information-theoretic picture of the past. That linkage runs contrary to the normal order of time as presupposed in everyday discussions of causality.

Won't those backward-running linkages destroy every possibility of the familiar everyday causal account of existence? No, we can well believe, if we judge by the lesson of half-advanced, half-retarded electrodynamics[15]. A collection of charged particles coupled according

to that algorithm present to view an unbelievable maze of light-like connections running backward and forward in time. Nevertheless, when the number of particles is sufficiently great to guarantee absorption of all radiation, the maze translates itself into the familiar full-strength pure retarded electromagnetic interactions, plus the familiar force of radiative reaction. No contradiction with everyday notions of causality is to be seen. No less has to be expected of an information-theoretic account of physics.

V. ACKNOWLEDGMENT

The author thanks Yakir Aharonov, David Albert, Jeeva Anandan, Aage Bohr, Jørgen Kalckar, Arkady Kheyfets, Warner Miller, William Wootters, Donald Page, Wolfgang Schleich, E. C. G. Stueckelberg and W. H. Zurek for helpful discussions, and the Center for Theoretical Physics at The University of Texas at Austin and NSF Grant PHY 8205717 for its assistance in preparing this report for publication.

REFERENCES

1. J. A. Wheeler, Bits, Quanta, Meaning, in Caianiello Celebration volume, A. Giovannini, M. Marinaro, F. Mancini and A. Rimini, eds., 1984.

2. N. Bohr, selected papers reprinted in J. A. Wheeler and W. H. Zurek, eds., Princeton University Press, Princeton, New Jersey (1983).

3. D. Føllesdal, Meaning and experience, in "Mind and Language", S. Gutenplan, ed., Clarendon Press, Oxford (1975), p. 254.

4. R. A. Fisher, Proc. Roy. Soc. Edinburgh 42: 321 (1922); cf. also R. A. Fisher, "Statistical Methods and Statistical Inference", Hafner, New York (1956), pp. 8-17.

5. W. K. Wooters, The acquisition of information from quantum measurements, Ph.D. dissertation, University of Texas at Austin (May 1980), available from University Microfilms, Inc., Ann Arbor, Michigan 48106; also Statistical distribution and Hilbert space, Phys. Rev.: 357-362 (1981).

6. E. C. G. Stueckelberg, Quantum theory in real Hilbert space, Helv. Phys. Acta 33: 727-752 (1960); M. Guenin and E. C. G. Stueckelberg, Quantum theory in real Hilbert space II (addenda and errata), Helv. Phys. Acta 34 621-628 (1934); and E. C. G. Stueckelberg, M. Guenin, C. Piron and H. Ruegg, Quantum theory in real Hilbert space III: fields of the first kind (linear field operators), Helv. Phys. Acta 34: 675-698 (1961).

7. R. P. Feynman, The principle of least action in quantum mechanics, Ph.D. dissertation, Princeton University, Princeton, New Jersey (May 1942), available from University Microfilms, Inc., Ann Arbor, Michigan 48106; also A. R. Hibbs and R. P. Feynman, "Quantum Mechanics and Path Integrals", McGraw-Hill, New York (1965).

8. For references on the A-B effect, see S.N.M. Ruijsenaars: Ann. of Phys. 146 (1983).

9. Jeeva S. Anandan, Relativistic Electromagnetic-Gravitational Effects in Quantum Interference, "Proceedings of the International Symposium Foundations of Quantum Mechanics In The Light ofNew Technology", S. Kamefuchi et al, eds., Physical Society of Japan, Tokyo, pp. 65-73, (1984).

10. J. A. Wheeler, Superspace and the nature of quantum geometrodynamics in: "Battelle Rencontres: 1967 Lectures in Mathematics and Physics," C. M. DeWitt and J. A. Wheeler, eds., Benjamin, New York (1968), pp. 242-307; C. W. Misner, K. S. Thorne and J. A. Wheeler, "Gravitation," Freeman, San Francisco (1973), chap. 43, Superspace: arena for the dynamics of geometry, pp. 1180-1195.

11. U. Gerlach, Derivation of the ten Einstein field equations from the semiclassical approximation to quantum geometrodynamics, Phys. Rev. 177: 1929-1941 (1969).

12. J. A. Wheeler, On the nature of quantum geometrodynamics, Ann. of Phys. 2: 604-614, (1957).

13. J. A. Wheeler, Delayed-choice experiments and the Bohr-Einstein dialog, in: "The American Philosophical Society and the Royal Society: Papers Read at a Meeting June 5, 1980," J. E. Rhoads, ed., American Philosophical Society, Philadelphia (1981), pp. 9-40.

14. A. Einstein, B. Podolsky, N. Rosen, N. Bohr, J. Bell, A. Aspect and others, articles on the EPR experiment, delayed-choice experiments and related issues in: "Quantum Theory and Measurement," J. A. Wheeler and W. H. Zurek, eds., Princeton University Press, Princeton, New Jersey (1983).

15. J. A. Wheeler and R. P. Feynman, Interaction with the absorber as the mechanism of radiation, Rev. Mod. Phys. 17, pp. 157-181. (1945). J. A. Wheeler and R. P. Feynman, Classical electrodynamics in terms of direct inter-particle action, Rev. Mod. Phys. 21, pp. 425-434. (1949).

PREDICTIVE STATISTICAL MECHANICS

E. T. Jaynes*

St. John's College and Cavendish Laboratory

Cambridge CB2 1TP, United Kingdom

INTRODUCTION

This workshop is concerned with two topics, foundations of quantum theory and of irreversible statistical mechanics, which might appear quite different. Yet the current problems in both fields are basically the same, two different aspects of a deep conceptual hangup that permeates not only physics, but all fields that use probability theory.

A different way of thinking about these problems is expounded, which has had useful results recently in statistical mechanics and more general problems of inference, and which we hope may prove useful in quantum theory. An adequate account of all the technical details alluded to in the writer's five talks would require a volume in itself, but much of this is now in print or in the publication pipeline. Therefore we try to explain here the original motivation in quantum theory, the formalism that evolved from it, and some recent applications, with references to further details.

QUANTUM THEORY

We think it unlikely that the role of probability in quantum theory will be understood until it is generally understood in classical theory and in applications outside of physics. Indeed, our fifty-year-old bemusement over the notion of state reduction in the quantum-mechanical theory of measurement need not surprise us when we note that today, in all applications of probability theory, basically the same controversy rages over whether our probabilities represent real situations, or only incomplete human knowledge.

If the wave function of an electron is an "objective" thing, representing a real physical situation, then it would be mystical--indeed, it would require a belief in psychokinesis--to suppose that the wave function can change, in violation of the equations of motion, merely because information has been perceived by a human mind.

*Visiting Fellow, 1983-84. Permanent address: Department of Physics, Washington University, St. Louis, Missouri 63130, U.S.A.

If the wave function is only "subjective", representing a state of knowledge about the electron, then this difficulty disappears; of course, by definition, it will change with every change in our state of knowledge, whether derived from the equations of motion or from any other source of information. But then a new difficulty appears; the relative phases of the wave function at different points have not been determined by our information; yet they determine how the electron moves.

There is no way quantum theory could have escaped this dilemma, short of avoiding the use of probability altogether. Not only in Physics, but also in Statistics, Engineering, Chemistry, Biology, Psychology, and Economics, the nature of the calculations you make, the information you allow yourself to use, and the results you get, depend on what stand you choose to take on this surprisingly divisive issue: are probabilities "real"?

But in quantum theory the dilemma is more acute because it does not seen to be merely a choice between two alternatives. The "subjective" and "objective" aspects are scrambled together in the wave function of an electron, in such a way that we are faced with a paradox like the classical paradoxes of logic; whatever stand you take about the meaning of the wave function, it will lead to unacceptable consequences.

To achieve a rational interpretation we need to disentangle these aspects of quantum theory so the "subjective" things can change with our state of knowledge while the "objective" ones remain determined by the equations of motion. But to date nobody has seen how to do this; it is more subtle than merely separating into amplitudes and phases.

WHAT IS "REALITY"?

As many have pointed out, starting with Einstein and Schrödinger fifty years ago and continuing into several talks at this Workshop, the Copenhagen interpretation of quantum theory not only denies the existence of causal mechanisms for physical phenomena; it denies the existence of an "objectively real" world.

But surely, the existence of that world is the primary experimental fact of all, without which there would be no point to physics or any other science; and for which we all receive new evidence every waking minute of our lives. This direct evidence of our senses is vastly more cogent than are any of the deviously indirect experiments that are cited as evidence for the Copenhagen interpretation.

Perhaps our concern should be not with hidden variables, but hidden assumptions; not only about the theory, but about what we are measuring in those experiments. Consider a cascade decay experiment. As soon as we say something like "In this experiment we observe two photons emitted from the same atom", we have already assumed the correctness of a great deal of the theory that the experiment was supposed to test. This initial stacking of the cards then affects how we analyze the data.

Bell's theorem was a great, and startling, advance, because it showed us that von Neumann's analysis contained hidden assumptions that had escaped the notice of physicists for over 30 years. But it is not so startling when we note that relativity theory owes its existence to Einstein's perception of a hidden assumption that nobody had noticed in the 300 years since Kepler. In 1949 he wrote, concerning phenomena in two reference frames: "Today everyone knows, of course, that all attempts to resolve this paradox were doomed to failure as long as the axiom of the

absolute character of time, viz., of simultaneity, was anchored unrecognized in the subconscious. Clearly to recognize this axiom and its arbitrary character really implies already the solution of the problem."

We hope it will not take another 300 years to locate the hidden assumptions in Bell's analysis. This is unlikely to be accomplished within the confines of present orthodox thinking; we can hardly expect that a viewpoint which denies the very existence of causal mechanisms will suggest the proper experiment to reveal them.

But to escape from the irrationality of present quantum theory, it is idle merely to complain about its philosophy. The onus is on the dissenter to move outside its area of thought and offer a constructive, experimentally testable, alternative.

The writer's neoclassical theory of electrodynamics is one such effort; but it has not yet been applied to experiments of the type reported here by Aspect, in a sufficiently careful way to draw any conclusions. In any event, many more such efforts may be needed before we can ferret out the crucial hidden assumption, now anchored unrecognized in our subconscious, which is preventing us from tying the mathematics of quantum theory to a rational view of the world.

Do we need more hidden variables? Perhaps eventually, but maybe our immediate problem is the opposite; first we need to get rid of some. To define the state of a classical particle we must specify three coordinates and three velocities. Quantum theory, while denying that even these degrees of freedom are meaningful and claiming to "remove unobservables", replaces them with an infinite number of degrees of freedom defining a continuous wave field. To specify a classical wave field, we need one complex amplitude for each mode; for a quantized field we need an infinity of complex amplitudes for each mode.

So perhaps quantum theory, far from removing unobservables, has introduced infinitely more mathematical degrees of freedom than are actually needed to represent physical phenomena. If so, it would not be surprising if a few infinities leak out into our calculations.

Neoclassical theory was a preliminary attempt to remove the hidden variables that are not actually used (at least, not up to second order) in calculation of the Lamb shift, the anomalous moment, and vacuum polarization. After removing an infinity of irrelevant degrees of freedom, it might be clearer how to add a few new relevant ones. It is not yet completely formulated, and can still be modified in many ways.

Distant correlations, so prominent in all these discussions from EPR to Aspect, are not in themselves difficult conceptually, since a state of knowledge may well be of the form; "if A, then B; but if C, then D". Experiments which merely confirm that these correlations exist hardly surprise us (they would be sensational if they showed that the correlations do not exist). As EPR emphasized, it is not the experimental facts, but the claim of the Copenhagen theory that the wave function containing such correlations is at the same time a complete description of reality, on which one chokes.

Part of Bohr's defense of his position was on the grounds that EPR did not show that his theory contradicts any experimental fact; but that was not the point. Indeed, the difficulty with the Copenhagen theory is most acute just when the experimental facts agree exactly with its predictions; for then we are in the following situation. Since the experimenter can decide, after the interaction with a system S has ceased, which of two

non-commuting quantities (x,p) in S he will be able to predict with certainty, we are again in the realm of psychokinesis if we deny reality to either of them before his decision. Then, since the wave function does not determine either x or p we do not see, any more than did EPR, how it can be called a complete description of reality.

The word "reality" must have had a different meaning to Niels Bohr than it has to others. On this topic, we note a quotation from Heisenberg (1958) which will surprise many: "The conception of objective reality--has thus evaporated--into the transparent clarity of a mathematics that represents no longer the behavior of elementary particles, but rather our knowledge of this behavior."

OUR MIND-BOGGLING PROBLEM

The difficulty in finding a rational interpretation of quantum theory can be illustrated a little more specifically as follows. If one wishes to do so, the amplitudes $|a(t)|$ of stationary states in a wave function

$$\psi(x,t) = \sum a_n(t)u_n(x)$$

can be interpreted, with at least some success, as "subjective" probability amplitudes expressing human information; but they are combined with phases which seem to have no such interpretation. Yet the relative phases strongly affect our predictions in ways that are necessary for agreement with experiment (for example, they determine the polarization of resonance radiation). We seem obliged to consider the relative phases as "objectively real" things.

But then those same phases cause conceptual difficulties; in Schrödinger's paradox, if we observe whether the cat is alive or dead, quantum theory gives reasonable probability statements. But if we choose to measure something that does not commute with "liveness", our predictions will depend on the relative phase with which the live cat and dead cat are superposed in the wave function.

There must be some limit on how far the superposition principle can be applied; at the atomic level we must have it, but at the macroscopic level we would like to get rid of it. Even at the molecular level, it is puzzling; if we take it literally, what prevents us from having a new kind of air in which each molecule is neither nitrogen nor oxygen, but a linear combination of them?

As noted, a mere separation into amplitudes and phases is not enough; further thought convinces us that the amplitudes must also be in part "objective", reviving the difficulty with state reduction. What seems mind-boggling at present is: how can we find a separation into "subjective" and "objective" parts that is invariant under changes in the representation--or at least under a sufficiently wide group of transformations to make physical sense? We do not have a clue about how to do this.

Yet we are optimistic, think that it will surely be solved eventually, and the answer will be simple and obvious (although it may require us to renounce our linear Hilbert space). As Seneca wrote long ago, "Posterity will be astonished that truths so clear had escaped us."

But for the present, to paraphrase Gibbs: difficulties of this kind have deterred the writer from trying to explain the mysteries of quantum theory, and forced him to be content with a more modest goal.

WHAT WAS BOHR TRYING TO DO?

Now let's look at that mind-boggling problem from a different side. A single mathematical quantity ψ cannot, in our view, represent incomplete human knowledge and be at the same time a complete description of reality. But it might be possible to accomplish Bohr's objective in a different way. What he really wanted to do, we conjecture, is only to develop a theory which takes into account the fact that the necessary disturbance of a system by the act of measurement limits the information we can acquire, and therefore the predictions we can make. This was the point always stressed in his semipopular expositions. Also, in his reply to EPR he noted that, while there is no physical influence on S, there is still an influence on the kinds of predictions we can make about S.

With all of this, we can agree entirely. The fact of disturbance of one measurement by another was equally true in classical physics (for example, one cannot use a voltmeter and ammeter to measure the current and voltage of a resistor simultaneously, because of this "complementarity": however you connect them, either the voltmeter reads the potential drop across the ammeter or the ammeter reads the current through the voltmeter).

But in classical physics such limitations on our knowledge could be recognized and taken into account in our predictions without losing our hold on reality; for the separation into what was "objective" and what was "subjective" was never in doubt. The coordinates and velocities remained "objective", while the "subjective" human information resided entirely in the probability distributions over them. The probabilities could vary in any way as our state of knowledge changed for whatever reason; while the coordinates and velocities continued to obey the equations of motion.

Furthermore, the limitations on our ability to make measurements at the microscopic level did not prevent us from discovering the microscopic equations of motion, or from checking them accurately enough to discover their failure in quantum effects. There are lessons in this for the present.

Could we make a theory of microscopic phenomena more like this which, while keeping a firm hold on what is "objective", also recognizes and represents explicitly the role of limited human information in the predictions it can make? Such a theory need not, we think, contradict the successful parts of Bohr's theory; rather it would remove the contradictions that still mar it, thus fully realizing Bohr's goal (indeed, Bohr himself may have had such thoughts; it is known that toward the end of his life he showed an interest in information theory).

Not knowing how to make this separation in quantum theory, we sought first a simpler model in the hope that it might be useful in its own right, and also give some clues for the big problem. The first hope is now fully realized; the second is still in too delicate a condition to be put on public view.

We want to reformulate statistical mechanics in a way that reflects Bohr's thinking by introducing explicitly the role of human information in determining the kind of predictions we can make. This is also closely related to Einstein's thinking; for he regarded present quantum theory as incomplete, related to a complete microscopic theory in much the same way that classical statistical mechanics is related to classical mechanics.

Thus, to enter into pure conjecture (whistling in the dark, some will say); if we could eventually see the predictions of present quantum theory as resulting from the "statistical mechanics" of a deeper theory, this might

realize both Bohr's goal and Einstein's. Of course, in view of Bell's theorem we do not expect this to be possible exactly, in terms of the same variables used in present quantum theory.

NATURE OF PROBABILITY THEORY

In trying to use probability theory for this purpose, we note that one can find very different views as to what probability theory is. It has been termed:

"the theory of additive measure" (Kolmogorov)
"the theory of rational belief" (Jeffreys, Good, Savage)
"the theory of frequencies in random experiments" (Fisher)
"the calculus of inductive reasoning" (de Morgan)
"common sense reduced to calculation" (Laplace)
"the art of conjecture" (Jacob Bernoulli)
"the exact science of mass phenomena and repetitive events" (von Mises)

For over a Century, controversy has raged between those who want us to adopt one of these views and reject others. Clearly, however, each of the above definitions merely reflects the particular problems that the author was concerned with; to insist that we reject all views that are not helpful in his problems is to insist that we work only on his problems.

Here we take the view that all of the above are valid and useful in different contexts, and we are free to choose whichever seems appropriate in ours. But some views are more general than others. Consider, for example, that of von Mises. It is true that probability theory is used successfully in dealing with mass phenomena and repetitive events; but if we insist that it may be used only for that purpose we shall be prohibited from using it for our purpose of representing human information.

On the other hand, the probability theory of Jeffreys (1961) in which probability expresses basically a state of knowledge, applies to a much wider range of problems, automatically including those of von Mises and Fisher; for a state of knowledge may refer to any context. If our problem of interest happens to involve random experiments or mass phenomena, then Jeffreys' theory applies in a natural way, yielding the same or domonstrably better results than those of Fisher and von Mises. For Jeffreys' approach deals easily with technical problems (nuisance parameters, nonexistence of sufficient or ancillary statistics, rectangular likelihood functions, cogent prior information) on which narrower views fail (Jaynes, 1976).

Therefore, while we should not consider it "wrong" to adopt a narrow view if it happens to be adequate for our problem, we have nothing to lose, and may avoid unnecessary technical difficulties, if we always adopt the broadest view of Jeffreys. In our present problem it is necessary to do this, for our explicit purpose is to represent human information of any kind, whether or not it happens to refer to random experiments or mass phenomena.

This choice will avoid not only technical difficulties, but even more disturbing conceptual ones, if our long-range goal is to clarify quantum theory. In our discussions here we have heard such questions as:

"Does the measurement create the state?"
"Does the act of human perception do it?"
"What is the role of consciousness and free will?"
"Are objects real?"

But astonishingly, we have heard also:

"What is the <u>true</u> nonequilibrium ensemble?"

At this point, it is clear that theoretical physics has gone berserk. In quantum theory we have got ourselves into a situation where the objects have become unreal, but the probabilities have become real!

But this too is not peculiar to quantum theory. On deep thought, it will be seen that whenever we allow probabilities to become "physically real" things, logical consistency will force us, eventually, to regard the objects as "unreal". If we are to reach Bohr's goal while at the same time keeping our objects real we must recognize, with Laplace, Maxwell, and Jeffreys, that whenever we use probability it must be as a description of incomplete human knowledge, as it was in classical statistical mechanics.

Therefore, we must go back much further in first principles than merely re-hashing EPR and the QM theory of measurement. We need a basically different way of thinking than scientists are now taught. The conventional attitude asks the question: "How does Nature behave, as determined by the laws of mechanics?" Or, as we heard it put more dramatically here: "I don't like thermo--let's solve the n-body problem and see if all this stuff is really true."

Here we want to ask a different question: "How shall we best think about Nature and most efficiently predict her behavior, given only our incomplete knowledge?" Of course, although it would be impossibly difficult, it would not be "wrong" to solve the n-body problem; the point is that it is unnecessary and would contribute almost nothing to understanding thermo.

To understand and like thermo we need to see it, not as an example of the n-body equations of motion, but as an example of the logic of scientific inference, which by-passes all the detail by going directly from our macroscopic information to the best macroscopic predictions that can be made from that information; a model of what we would like to do in quantum theory.

That this must be possible, at least for thermo, is seen as follows. The fact that thermodynamic and other macroscopic experiments are reproducible shows that most details of those initial microscopic conditions are irrelevant. If control of a small number of macroscopic quantities is sufficient, in the laboratory, to yield a reproducible result, then <u>information</u> about those quantities must suffice for theoretical <u>prediction</u> of that result; there must be an algorithm that goes directly from one to the other.

Predictive Statistical Mechanics is not a physical theory, but a method of reasoning that accomplishes this by finding, not the particular things that the equations of motion say in any particular case, but the general things that they say, in "almost all" cases consistent with our information; for those are the reproducible things.

We are not, however, throwing away the conventional methods or results. Quite the contrary, the statistical mechanics of Gibbs and conventional quantum statistics are contained in it as special cases for certain particular kinds of information, just as the probability theory of von Mises is contained in that of Jeffreys for particular kinds of information.

This concludes our lengthy sermon; the technical problem now before us is: how shall we use probability theory to help us do plausible reasoning in situations where, because of incomplete information, we cannot use deductive reasoning?

A fairly complete treatment of these questions, with extensive physical applications, is given by Jeffreys (1961); we sketch a few of the details. The propositions about which we reason are denoted by letters A, B, C, etc. For example, we might choose

A = "The pressure of the gas is in the range (P, P + dP)"

B = "Its kinetic energy is E"

I = "It has N molecules in a volume V"

As in the usual Boolean algebra, we may construct new propositions by conjunction, disjunction, and negation:

AB = "Both A and B are true"

A + B = "At least one of the propositions A, B is true"

$\overline{A}$ = "A is false"

The symbol $p(A|B)$ stands for "the probability that A is true, given that B is true", and is a real number in $0 \leq p \leq 1$. Thus $p(A+B|CD)$ is the probability that at least one of the propositions A, B, is true, given that both C, D are true, and so on. Since in this system probabilities are a means of representing incomplete information, all probabilities are of necessity conditional on some information; there is no such thing as an "absolute" probability.

The rules for plausible reasoning are simply the familiar product and sum rules:

$$p(AB|I) = p(A|I)p(B|AI) = p(B|I)p(A|BI) \quad , \tag{1}$$

$$p(A|B) + p(\overline{A}|B) = 1 \quad . \tag{2}$$

All other relations can be deduced from these. In particular, if $p(B|I) \neq 0$, we have from (1)

$$p(A|BI) = p(A|I)p(B|AI)/p(B|I) \tag{3}$$

where, since "I" occurs as a condition in all terms, we shall call it the "prior information". Equation (3), usually called Bayes' theorem, represents the process of learning from experience. We start with the prior probability of A, $p(A|I)$ when we know only the prior information, and (3) shows how this is converted into the posterior probability $p(A|BC)$ as a result of acquiring new information B.

Bayes' theorem is undoubtedly the most fundamental principle of scientific inference; by its repeated use we can incorporate long chains of evidence into our reasoning, the posterior probability for one application becoming the prior probability for the next. We readily verify its consistency for this; the final result is independent of the order in which different pieces of information were taken into account. But there has been long controversy about it also, so it is important to dwell a moment on the status of these rules and others derived from them.

Historically, they were given by Laplace in the 18'th Century, on intuitive grounds, and applied by him in many problems of data analysis in astronomy, geodesy, meteorology, population statistics, etc. He had great success with them, using Bayes' theorem to help him decide which astronomical problems were worth working on. That is, are the discrepancies

between calculation and telescopic observation so small that they might well be accounted for by measurement errors; or are they so large that they indicate, with high probability, the existence of some unknown systematic cause not included in the calculations? If so, Laplace would undertake to find that cause. This process (what would be called today a "significance test" by statisticians) led him to some of the most important discoveries in celestial mechanics.

In spite of this success, a reaction set in after Laplace's death as others questioned the validity and uniqueness of the above rules. However, they would be true trivially if we interpret p as an "objective" frequency in a random experiment instead of a mere "subjective" measure of plausibility. So for over a Century probability theory went off into the Fisher/von Mises views that only the frequency interpretation was respectable. In fact, this view succeeded rather well for many years because there were actually many scientific problems where it was adequate; the information available and the questions of interest could be expressed solely in terms of frequencies, and there was no other cogent information.

But as scientific problems became more sophisticated, it became increasingly difficult to adapt frequency interpretations to them. Eventually one had to resort to inventing imaginary universes of repetitions of experiments that could in fact be performed only once, in order to maintain the illusion of a frequency interpretation. More serious, anomalous results began to appear, which could be traced to the failure to take into account cogent information that common sense could see was relevant to the inference, but which the frequency theory could not use because it did not consist of frequency data.

In 1946, R. T. Cox cut through the confusion by a marvelous argument. He had the good sense to ask a constructive question; whether or not Laplace's methods were sound, would it be possible today to make a consistent "calculus of plausible reasoning" along those lines? Cox found that the conditions for consistency of such a theory could be stated in the form of functional equations, whose general solutions could be found. The result was: any method of plausible reasoning in which we represent degrees of plausibility by real numbers is necessarily either equivalent to Laplace's, or inconsistent (in the sense that two methods of calculation, each permitted by the rules, would yield different results). Since 1946, there has been no excuse for anyone to reject the use of equations (1)-(3) as valid scientific inferences. For the details of beautiful applications, see Jeffreys (1961).

THE MASS OF SATURN

Many of these points are illustrated by the famous example of Laplace's analysis of the accuracy with which the mass of Saturn was known at the end of the 18'th Century. Let A stand for the proposition: "the mass of Saturn is in M, M + dM", and denote by D a set of observational data to be taken into account, while I stands for whatever prior information Laplace had about M before the data D were known. Then we may define prior and posterior probability density functions:

$$p(A|I) = f(M)dM \quad ; \qquad p(A|DI) = F(M)dM \tag{4}$$

Before the data, one did not know much about M, so f(M) was a very broadly spread out function. But Laplace knew at least that M was not zero, else Saturn would not hold its rings and moons or perturb Jupiter; and there would be no data to analyze. Also, he knew that M could not be an appreciable fraction of the solar mass, else Saturn would totally disrupt the

solar system. So f(M) must go to zero at extreme values; but this left an intermediate range of several orders of magnitude, within which f(M) could be taken to be essentially constant.

Of course, f(M) represents only Laplace's state of prior knowledge about M; it can be regarded as a frequency only in a grotesque collection of imaginary universes in which the mass of Saturn takes on all conceivable values. So frequency views of probability would not allow us to do this calculation at all.

In its dependence of M the term $L(M) = p(D|MI)$ in Bayes' theorem is called the likelihood function, proportional to the probability that the data D would be observed if M was the true mass of Saturn. All the evidence from the data resides in this factor, which represents our knowledge of the likely errors in the data due to the imperfection of telescopes and clocks.

This term is worth dwelling on a bit, because it shows both the usefulness and the shortcomings of the (probability = frequency) view. On that view, $p(D|MI)$ means the limiting frequency with which the data set D would be obtained in infinitely many repetitions of the measurement, if the mass of Saturn were held constant at the particular value M. But obviously, if one could actually measure that frequency directly, we would be far beyond any need to estimate M; so how do we, in practice, choose the function L(M)?

For this it is necessary to do a little sub-problem of plausible inference, within the original problem. We suppose that the errors in angles depend only on properties of the telescope, and would be the same whatever the true mass of Saturn and whatever object we are observing. So we shall find the telescope errors by making repeated measurements on some more fixed object like Sirius. The frequency distribution that we find for Sirius is inferred to hold also for Saturn.

This is an inference that common sense leads us to accept at once (although logically it is just the kind of inference that the frequency theory holds to be invalid; frequentists are masters of the art of concealing this). So, in practice we do indeed usually choose the likelihood factor on the basis of frequencies of errors; and the frequentist's procedure does indeed serve our purpose.

But this is so only because the feasible source of information about errors usually happens to come from repeated measurements. What is fundamentally required in (3) is the probability of various errors in the specific case of Saturn now before us. This may or may not be the same as the frequency of various errors in other cases that we are not reasoning about.

Indeed, strictly speaking, the specific case at hand is always in some ways unique and not comparable to others. There may be other cogent information pertaining to the errors in the specific case before us, in addition to frequencies. Then the frequentist's procedure is incomplete, allowing us to take into account only part of the information relevant to our problem. So attempts to uphold frequentist views in all cases can lead to anomalous results, which have become increasingly troublesome in recent applications. The only way to deal with the full problem is to use the Laplace-Jeffreys method, which can in principle take any kind of information into account, because it interprets the concept of probability more broadly.

Likewise, if there is cogent prior information in addition to the data, the Laplace-Jeffreys methods have the means of taking it into account by use of a "informative" prior density f(M) that is not flat, but specifies what values of M are already indicated or contra-indicated by the prior information. Again, frequentist methods which do not admit the existence of a prior probability are helpless to take such information into account.

In recent applications these shortcomings of frequentist methods of inference have become more and more serious. With the advent of the "Generalized Inverse" problem frequentist methods have become totally unusable; yet the Laplace-Jeffreys methods continue to deal with them without any difficulty of principle but with a new technical problem whose solution leads us into predictive statistical mechanics.

GENERALIZED INVERSE PROBLEM

If the likelihood function $L(M) = p(D|MI)$ does not peak sharply at any definite value of M but has a very broad maximum, the frequentist maximum likelihood principle becomes unstable, a small change in the data leading to a large change in the estimate, which common sense rejects as unreasonable. This makes the determination of the prior probabilities a more exacting task; the prior information is no longer overwhelmed by the data, and so it must be considered more carefully than Laplace needed to.

In the limit when the likelihood function develops a flat top, frequentist methods break down entirely; only an informative prior probability can locate a definite estimate somewhere within that flat region. This means that we are faced with a new technical problem: how do we translate verbally stated prior information into a quantitative prior probability assignment?

A typical kind of problem where this occurs is that where we are trying to invert a singular matrix; our data are

$$d_k = \sum_{i=1}^{n} A_{ki}\, x_i \;, \qquad 1 \le k \le m < n \tag{5}$$

where the $\{x_i\}$ represent the "state of Nature" that we are trying to estimate, and A is a known matrix with rank less than n, so there is no inversion of the form $x = A^{-1}d$. Then the likelihood function $L(x) = p(d|xI)$ is rectangular; the data merely partition the set of all x into subsets of possible and impossible values, with nothing to choose within the possible subset. L(x) is only the indicator function of the set of possible states of Nature. So we shall call any problem with a flat-topped likelihood a "generalized inverse problem".

Such problems have proved to be very common in recent applications; after a talk in 1983 the writer was approached by a statistician in Government who said "I suddenly realized that every problem my agency is trying to solve is a generalized inverse problem."

On further reflection it is seen that, from the standpoint of principle, this technical problem is present in every application of probability theory. For, as Cox's derivation showed clearly, the rules (1), (2) express only the consistency of our reasoning, telling us how probabilities of different propositions are related to each other. That is, given some probabilities, they tell us how to calculate others consistent with them. They do not tell us what initial probabilities should be assigned so the calculation can get started.

But for decades probability theory has concerned itself only with building upward from (1), (2), deducing their consequences in the large body of mathematics that fills our libraries. The problem of assigning the initial probabilities represents fully half of probability theory as it is needed for applications; yet its very existence remains unrecognized in most of those books. So we have a great deal of catching up to do; it will require decades to bring this neglected bottom half of probability theory up to the level of power and generality of the top half. But we have made enough progress to date so that the range of useful applications of probability theory is already extended far beyond the confines of the frequentist viewpoint.

ASSIGNING PRIOR PROBABILITIES

Today a number of principles are known by which prior information can be translated, or encoded, into prior probabilities. The simplest and most obvious is symmetry; from the earliest pre-mathematical gropings toward a set of rules for plausible reasoning it was clear to gamblers that if coins, dice, playing cards, roulette wheels, etc. were made with perfect symmetry there was no known cause tending to produce one outcome more than any other; so we had no reason to consider one more likely than another. The only honest way to express this state of knowledge is to assign equal probabilities to all of them.

We hasten to add that this is not to assert that all outcomes must occur equally often, as frequentists invariably accuse us of doing; for of course there may be unknown symmetry-breaking causes (very skillful tossing, shuffling, spinning, etc.). It does, however, mean according to the rules (1), (2) that if we are obliged to predict the frequencies from the incomplete information we have, our "best" estimates, by almost any criterion of "best", will be uniform.

If we have no information about the specific kind of symmetry-breaking influence at work, neither we nor the frequentist can make use of it to alter our predictions. Even if we know that some symmetry breaking is present but do not know which outcome it favors, this information cannot change our estimates, but only increases their probable error.

This again shows the two basically different attitudes noted above; we are not asking how Nature must behave, but only what are the best predictions we can make on our incomplete information. But this is always the real problem before the scientist, as Niels Bohr saw.

The idea of symmetry became more abstract, and more general, in the work of Jacob Bernoulli (1713). He envisaged an underlying population of N inherent possibilities from which we draw (the famous "urn" of elementary probability theory, with its N labelled but otherwise identical balls). If M of the balls are labelled "A" and we draw from the urn blindfolded, the probability that we shall draw an "A" is $p(A) = M/N$, the basic definition of probability used by Laplace. Note that Bernoulli's definition is just the frequency with which we would find "A" if we sampled the entire population without replacement.

But it was recognized already by Bernoulli that in many real problems we do not see how to analyze the situation into ultimate "equally likely" cases. As he put it, "What mortal will ever determine the number of diseases?" A principle was still needed for dealing with cases where we have prior information that makes the known possibilities not equally likely. It is surprising that Laplace did not find this principle, since he saw the problem so clearly.

Now the scene shifts to Boltzman (1877). To determine how gas molecules distribute themselves in a conservative force field such as gravitation, he divided the accessible 6-dimensional phase space of a single molecule into equal cells, with N molecules in the i'th cell. Noting that the number of ways this distribution can be realized is the multinomial coefficient

$$W = N!/N_1!N_2! \ldots N_n! \tag{6}$$

he concluded that the "most probable" distribution is the one that maximizes W subject to the known constraints of his prior knowledge; in this case the total number of particles and total energy:

$$N = \sum N_i = \text{const.} \quad , \qquad E = \sum N_i E_i = \text{const.} \tag{7}$$

where E is the energy of a molecule in the i'th cell. If the number are large, Stirling's approximation gives asymptotically

$$N^{-1} \log W \longrightarrow -\sum (N_i/N) \log (N_i/N) \tag{8}$$

today usually called the "Shannon entropy", although it was in use by von Neumann before Shannon entered the field and by Boltzmann and Gibbs before Shannon was born. This argument, repeated in every statistical mechanics textbook, led to the famous Boltzmann distribution law: our best estimate of N_i is

$$\hat{N}_i = NZ^{-1} \exp(-\beta E_i) \tag{9}$$

where Z is a normalizing factor, and the Lagrange multiplier is found to have the meaning $\beta = (kT)^{-1}$. This is the distribution that can be realized in more ways than can any other that agrees with the information (7). Gibbs generalized this result to the continuous phase space of N interacting molecules, leading to his canonical and grand canonical ensembles, later extended easily to quantum statistics.

At this point, then, (1902) the Principle of Maximum Entropy (PME) was fully in hand and operational, and it has been the de facto foundation tool of statistical mechanics ever since. So why did it take another fifty years to recognize it, ten more to generalize it to nonequilibrium cases, and yet another five to apply it to a problem of inference outside thermodynamics?

The barrier that has held up progress throughout this Century is just that philosophical position that tries to make probabilities "physically real" things. Not until the work of Shannon (1948), which showed that the quantity Boltzmann and Gibbs had maximized is at the same time a unique measure of "amount of uncertainty" in a probability distribution, could the general rationale underlying their procedure be seen. As should be obvious by now, Boltzmann and Gibbs had been, unwittingly, solving the prior probability problem of Bernoulli and Laplace, the principle of symmetry generalized to the principle of maximum entropy.

THE MAXIMUM ENTROPY PRINCIPLE

But for all this time writers on statistical mechanics had interpreted the work of Boltzmann and Gibbs in an entirely different frequentist context, which sought to justify the canonical ensemble as a physical fact, a provable consequence of the equations of motion--that n-body problem--via ergodic theorems.

In the case of a body in equilibrium with a heat bath or isolated, such theorems might conceivably be true in many cases, although one can easily invent counter-examples in which the Hamiltonian is too "simple" for ergodic behavior to be possible in an isolated body. But even when ergodicity can be proved, it cannot ensure that ensemble averages are equal to experimental values, for reasons that were pointed out already by Boltzmann, and discussed in some detail in Jaynes (1967).

When we try to extend the theory to irreversible processes, ergodic theorems are worse than useless, for if the theory is successful the ensemble averages must still be equal to experimental values; but the very phenomena to be predicted express the fact that these are not equal to time averages. The probabilities we use to predict irreversible processes are necessarily only descriptions of human information; i.e., the probability is spread smoothly over the range of possible microstates, compatible with our information.

If a process is reproducible, it cannot be because the probabilities are "real", but only because most of those possible microstates would all lead to the same macroscopic behavior. The maximum-entropy distribution is the safest, most "conservative" distribution to use for prediction, because it spreads the probability out over all the states consistent with our information; and not some arbitrary subset of them. Thus it takes the fullest possible "majority vote", and prevents us from making arbitrary assumptions not justified by our information.

In this sense the maximum-entropy principle expresses nothing more than intellectual honesty; it frankly acknowledges the full extent of our ignorance. Indeed, it measures that ignorance, since the "physical entropy" S of Clausius, which is the "Shannon information entropy" (8) after maximization, is essentially the logarithm of the number of microstates consistent with our information (as Boltzmann, Einstein, and Planck had all noted before 1907).

But in fact the microstates leading to the reproducible behavior are such an overwhelming majority of the possible set that in practice "almost any" probability distribution that concentrates its probability on some subset of them, will lead to substantially the same macroscopic predictions. And any probability distribution--whatever its high-probability domain of microstates--that gives the same covariance functions for the macroscopic quantities of interest, will lead to exactly the same predictions. The maximum-entropy ensemble is sufficient, but very far from necessary, to make correct predictions. Yet some still ask: "What is the true non-equilibrium ensemble?"

But the question can be construed as meaningful in a different sense. If PME predictions are successful, that does not prove its ensemble is "correct" since many other algorithms might also be successful. But if PME fails, the ensemble must have been, in some sense, "wrong". Supposing the data used as constraints were not in error, there seem to be two possibilities: (A) There may be further constraints, essential to determine the phenomena, which we failed to take into account; or (B) Our enumeration of the physically possible microstates (the place where our knowledge of the laws of physics comes in) was wrong; Nature actually uses more states, or fewer, than we supposed.

On possibility (A), if we have included as constraints all the macroscopic conditions that are found, in the laboratory, to be sufficient to determine a reproducible outcome, then there seems to remain only the possibility that the equations of motion have new constants of the motion that we didn't know about. The nature of the error gives us a clue as to

what these new constants might be. This happened in the case of ortho- and para-hydrogen, where an approximate constant of the motion prevented the system from reaching the equilibrium predicted by the canonical ensemble in the time of experimental measurements; and the in existence of allotropic forms like red and white phosphorus, which are stable over many years.

On possibility (B), the failure of PME based on a particular state enumeration would give us a clue to new laws of physics. This, too, has happened more than once. First, the failure ofGibbs' classical statistical mechanics to predict specific heats correctly was the first clue pointing to the discrete energy levels of quantum theory; Nature did not use all the energy values permitted by classical mechanics. There remained a failure to predict vapor pressures and equilibrium constants correctly; this clue was found to indicate that Nature does not use all the mathematical solutions of the Schrödinger equation, only those that are symmetric or antisymmetric under permuations of identical particles.

We point this out to emphasize the completely different way of thinking about the relation of statistical theory to the real world, that is being expounded here. Repeatedly, conventional thinking has led to attacks on the PME viewpoint on the grounds that we make no appeal to ergodic theorems; ergo, there might be unknown constants of the motion which prevent a system from getting to the macroscopic state that PME predicts. Therefore, our prediction might be wrong; and this is seen as a terrible calamity that invalidates our approach.

This point of logic is discussed in more detail in Jaynes (1985), where we note that this conventional thinking is like that of a chess player who thinks ahead only one move. If we think ahead two moves we see that, while the usual success of PME predictions makes them useful in an "engineering" sense, an occasional failure is far from being a calamity.

If we see the formalism of statistical mechanics as a physical theory, then we worry about whether it is "right" or "wrong". If we recognize it instead as a method of reasoning which makes the best predictions possible from the information we put into it, its range of application is seen as vastly greater, not limited to thermodynamics or to physics; and its "failures" are seen to be even more valuable than its "successes", because they point the way to basic advances in science. My colleague Steve Gull has termed this change in attitude, "The Leap" in our understanding of the role of statistical inference in s cience. Instead of fearing "failure", we look eagerly for it.

This change in attitude is so great that it seemed misleading to use the same term "statistical mechanics" for both viewpoints. To avoid confusion we coined the name "Predictive Statistical Mechanics" to distinguish our line of thought from others.

For some time, this issue could not progress beyond being a mere philosophical difference, because to find a difference in pragmatic results one had to go beyond equilibrium thermodynamics; but the nonequilibrium PME calculations proved to be just as difficult, differing only in details, as the equilibrium ones. Indeed, most current work in statistical mechanics is still struggling with equilibrium calculations, which are hard enough.

It required the advent of the computer before the merit of PME as a predictive tool outside equilibrium theory could be demonstrated; and the impressive recent successes have been in applications outside of thermo-

dynamics. These are quite different from the thermodynamic applications, where it would be impractical to do pencil-and-paper calculations with more than a few simultaneous constraints.

With computers it is possible to locate maximum entropy points with almost any number of simultaneous constraints; spectrum analyses with dozens of constraints have been produced routinely for some fifteen years, and image reconstructions with over a million constraints (i.e., generalized canonical distributions with over a million Lagrange multipliers) have been produced routinely for three years now.

Finally, thanks to these developments, the real powers of the PME method could be seen in a way that transcended all philosophical arguments about the "meaning of probability". It is the computer printouts, not the philosophy or theorems, that is making the converts.

Lacking the space to present all the technical details now in print, in the following we survey the main new applications and results, with references. The field is presently in a phase of rapid growth, spreading into new fields with the number of workers appearing to double every year; and it is no longer possible for one person to keep up with all that is being done. Therefore the following is only a partial list, of applications known to the writer.

SPECTRUM ANALYSIS

The first breakthrough occurred in 1967, when John Parker Burg produced power spectrum estimates from incomplete geophysical data, by maximizing the entropy of the underlying time series $\{y_1, y_2, \ldots y_n\}$ defined at discrete times $t = 1,2,\ldots$, subject to the constraints of the data. More specifically, one has measured values of the autocovariance

$$R = n^{-1} \sum_t y_t y_{t+k} \tag{10}$$

for $m+1$ lags, $0 \le k \le m < n$. The true power spectrum is

$$P(f) = \sum_{k=-n}^{n} R_k \cos 2\pi kf \tag{11}$$

but there is nothing "random" about this. The problem is that $m < n$; our data are incomplete. On frequentist views, one would not see how probability theory is applicable to such a problem; but from our viewpoint we see that this is a classic example of the generalized inverse problem discussed above; the class C of possible spectra is given by (11) in which the R_k are taken equal to the data when $|k| \le m$, and are arbitrary but for the nonnegativity requirement $P(f) \ge 0$ when $|k| > m$. Picking out a particular "best" estimate $P(f)$ from class C thus amounts to finding the "best" extrapolation of $\{R_k\}$ beyond the data.

Previous to this, power spectra had been estimated by the Blackman-Tukey (1958) method, which led to an estimate, for this problem, of:

$$[\hat{P}(f)]_{BT} = \sum_{k=-m}^{m} R_k W_k \cos(2\pi kf) \tag{12}$$

where W_k is a "taper" or "window" function, typically chosen as

$$W_k = \frac{1}{2}(1 + \cos \pi k/m) \tag{13}$$

which avoids spurious "side-lobes" in $\hat{P}(f)$ by tapering the data smoothly to zero at $k = m$, which would otherwise be an abrupt discontinuity. But it is at once evident that (12) does not lie in the class C of possible solutions; for (12) disagrees with the data at every data point where $W_k \neq 1$. Furthermore, (12) is not in general nonnegative. Therefore we know, not as a plausible inference, but as a demonstrated fact, that a time series with the spectrum (12) could not have produced our data! In addition, the estimate (12) failed to resolve sharp spectrum lines satisfactorily; the window function, in removing the side lobes, threw away half the resolution that we would have had without it.

Burg pointed out these shortcomings of (12) and proceeded to find a spectrum estimate by a totally different argument. Find the probability distribution $p(y_1, \ldots, y_n) = p(y)$ that has maximum entropy

$$S = -\int p(y) \log p(y) d^n y \tag{14}$$

subject to the constraints that the expectations of the R_k agree with the data (10). The resulting generalized canonical distribution

$$p(y) \propto \exp[-\Sigma \lambda_k R_k] \tag{15}$$

is, in view of of (10), a multivariate gaussian for which one can show that the entropy is also, to within an additive constant,

$$S = \int df \log P(f) \tag{16}$$

which is coming to be called the "Burg entropy" although they are substantially the same thing, only expressed in different variables. One could say equally well that he is maximizing (16). By a curious algebraic twist, the power spectrum estimate obtained as an expectation over (15) turns to have the form

$$\hat{P}(f) = (\Sigma \lambda_k \cos 2\pi k f)^{-1} \tag{17}$$

Burg used a computation algorithm given in the 1940's by N. Levinson to find the Lagrange multipliers λ_k in (15), (17) that agree with the data.

The resulting computer printouts were a revelation. The maximum-entropy spectra were clean and sharp; spurious artifacts like sidelobes were eliminated, but at the same time the resolution was greatly increased rather than sacrificed. No linear processing of the data could have produced such results.

The field then grew rapidly; to cite a few of the key references, the thorough mathematical analysis in the review article of Smylie, Clarke, and Ulrych (1973) and in the thesis of John Parker Burg (1975) are still required reading. By 1978 the literature had grown to the point where the IEEE issued a special volume of reprints on PME spectrum analysis (Childers, 1978) and Haykin (1979) edited another volume. Currie (1981) demonstrated its application to detection of small geophysical/meteorological effects. Haykin (1982) is a Special Issue of the IEEE Proceedings devoted to spectrum analysis, with numerous articles on the theory and practice of PME methods. Many more are in the Proceedings of the Laramie and Calgary Workshops on Bayesian/Maximum Entropy methods (Smith and Grandy, 1985; Justice, 1985).

IMAGE RECONSTRUCTION

The writer does not know exactly when the first attempt to use PME in image reconstruction occurred. Early discussions by Frieden et al.

(1972) and Ables (1974) gave some theory and computer simulations. Perhaps the major landmark is the article of Gull and Daniell (1978), which has real results supported by a very clear, simple rationale. Image reconstruction was seen as another generalized inverse problem based on (5) as follows. Since anything we can actually compute is digitized, we break the true scene into n picture elements, or "pixels" and imagine that scene generated by distributing N small "quanta" of intensity, the i'th pixel receiving N_i of them. Let $x_i = N_i/N$ be the fraction in the i'th pixel, and denote the resulting scene by $X = \{x_1, x_2, \ldots x_n\}$. It has an entropy S(X) given by (8).

But we do not know these numbers; we have available only a blurred scene consisting of $m < n$ pixels, with intensities d_k given by (5) in which the matrix A is the digitized point-spread function that describes the imperfections of our telescope (we are still not far, in either topic or basic rationale, from Laplace's problem of the mass of Saturn). Given the data $D = \{d_1, \ldots d_m\}$ of the blurred image, the reconstructed estimate of the true scene X that can be realized in the greatest number of ways is the one that maximizes the entropy S(X) subject to the constraints (5). This is a pure generalized inverse problem; the likelihood L(X) is constant on the set of possible scenes consistent with (5), zero elsewhere.

Gull and Daniell gave the resulting computer printouts for some real problems in radio and x-ray astronomy. They were just as impressive as were those of Burg's spectrum analysis. Again, the spurious artifacts of previous linear data analysis methods disappeared, while the resolution and dynamic range improved. This beautifully concise article is also required reading; it can be read and comprehended fully in an hour.

Frieden (1980) gave an equally impressive reconstruction of a galaxy so distant that the original optical photograph appears to the eye as a featureless blob; yet the maximum entropy reconstruction reveals five clearly resolved spiral arms. Having seen this reconstruction, one can go back and look at the original blurred image and see that there was, indeed, evidence for those arms in the data. The variational principle that generates it ensures that the maximum entropy reconstruction cannot show any detail for which there is no evidence in the data that were used as constraints.

In image reconstruction, maximum entropy is doing something much like what a skilled x-ray diagnostician learns to do. Knowing just what to look for, he can perceive details in a blurred picture that are quite invisible to the untrained eye. But at present most maximum entropy algorithms are doing this without being told in advance what to look for. If we can put the kind of prior knowledge that the x-ray diagnostician is using, into the underlying "hypothesis space" on which our entropy is defined, maximum entropy reconstructions will become still better. Two examples, for astronomical scenes about which we already know some features in advance, are given by Horne (1982) and Skilling (1983); we do not yet know how to do this in general.

Again the method spread rapidly to other applications. Gull and Skilling (1984) give many more examples, including reconstruction of a blurred image of an auto license plate for the London Police (whom we understand now have their own in-house maximum entropy facilities, with program supplied by Gull and Skilling).

NOISY DATA

A new feature forced itself upon us in these practical implementations of PME. The "pure PME formalism" illustrated in Eqs. (7)-(9) above has

assumed that data used as constraints are noiseless. With noisy data, pure PME or any other method must inevitably start showing spurious detail which is only an artifact of the noise, but which is fundamentally impossible to remove entirely (since the information needed to distinguish signal from noise is not there). But one can remove most of this by making allowance for noise in the same way that Laplace did; by using the full Bayes' theorem (3) for inference, in which we take the prior probability proportional to the multiplicity factor W in (6). The prior probability of a scene X becomes proportional to exp[NS(X)]. Then the scene with the highest posterior probability is the one that maximizes not S(X) but

$$NS(X) + \log L(X) \tag{18}$$

where L(X) is its likelihood, no longer rectangular, in the light of the probable errors in the data. If the blurred image data (5) have independent gaussian errors of RMS value h, then

$$L(X) = \exp\left\{-\sum_k (d_k - m_k)^2/2h^2\right\} \tag{19}$$

where $\{m_k\}$ are the "mock data", RHS of (5), that we would have obtained if X were the true scene but the noise were absent.

Maximizing (18) replaces the "hard" constraints of pure PME with "soft" constraints determined by the shape of the likelihood function. The reconstruction moves up a little higher on the entropy hill, giving us a slightly smoother picture. In effect, fine details in the data below the noise level, that we would have to accept as real if we knew there was no noise, are reinterpreted as more likely due to noise and ignored. A higher noise level makes softer constraints and a reconstruction showing less detail. Strongly represented features are those for which the details in the data rise above the noise and give strong evidence for a real thing; just Laplace's original significance test, now applied to every pixel.

The question of the proper choice of N has given rise to a great deal of discussion, into which we cannot go here. Some of this--and a great deal more--will appear in Justice (1985).

OTHER APPLICATIONS

After the success of the one-dimensional PME spectrum analysis and two-dimensional image reconstructions, one was encouraged to try reconstructing three-dimensional objects from incomplete data. Medical tomography (Minerbo, 1979) seems to be the first example. But the classical three-dimensional generalized inverse problem, crystallographic structure analysis from x-ray scattering data, had been calling out for such a method for decades.

Here the data are notoriously incomplete, consisting of magnitudes, but not phases, of only a few of the Fourier coefficients of the electron density function d(x). The analogy to Burg's original problem is clear; one must extrapolate the data to higher wavenumbers. But it is more complicated because one must also estimate phases; this is discussed in some detail by Bricogne (1982,1984) and Wilkins et al. (1983,1984). The Proceedings of the recent Orsay workshop (Bricogne, 1985) will have many more articles on it, including one by Skilling on the computation problem.

A PME structure analysis of biological macromolecules is given by Bryan et al. (1983). It appears that this is to become one of the major areas of application in view of the number of biological structures in need of analysis, and the resulting large efforts going into it at several places.

It seems that every success of PME in one area points to a new application in a related area. Sibisi et al. (1983,1984) apply it to estimation of line frequency and decay rate from free decay NMR signals, Livesey et al. (1984) to analysis of EXAFS data, Gburski et al. (1984) to inferring line shapes from a few calculated moments. Mead and Papanicolaou (1984) show that it has advantages over previous methods for estimating mathematical functions from incomplete information, such as the first few terms of a series expansion. Applications in econometrics are beginning to appear.

CONCLUSION

We have tried to present a general survey, rather than the technical details, of a line of thought that has been evolving slowly and painfully, impeded by philosophical differences, for some thirty years. But, as will be apparent from the above, we are now far beyond the stage of philosophical debate. Wherever it has been applied in a competent way to computationally feasible problems, PME has not only succeeded, but led to substantive improvements over previous methods of inference from incomplete information. To acknowledge and represent explicitly the incompleteness of our information is just what these problems had needed.

We are thus encouraged to renew our efforts in the applications originally envisaged in irreversible statistical mechanics and, eventually, to try to see Bohr's and Einstein's views of quantum theory finally unified as an example of the same kind of reasoning. We have as yet no new results on explicit experimental numbers but there are many results of a more general nature (Jaynes, 1978,1980).

It is easy to show that Predictive Statistical Mechanics leads automatically to such known correct results as the Onsager reciprocities, the Kubo formulas for transport coefficients, the Wiener prediction algorithm, and in a short-memory approximation, to Fokker-Planck equations. But these always appear in more generally applicable form; the reciprocities require no assumptions about short memory or regression of spontaneous fluctuations, the Kubo formulas are no longer limited to the quasi-stationary, long-wavelength regime, Fokker-Planck equations hold not only in momentum space, but in any thermodynamic state space, whose coordinates are the macroscopic quantities of interest.

Einstein's view of fluctuations as providing the "driving force" that makes an irreversible process go, and Onsager's view of the entropy gradient in thermodynamic state space as providing the "steering" telling it in which direction to go, appear automatically, but no longer restricted to situations close to equilibrium. A generalized "Bubble Dynamics" shows how a bubble of probability moves up the entropy hill far from equilibrium, at a velocity proportional to

(mean square fluctuation) x (local entropy gradient)

while constantly readjusting its size and shape to the local curvature of the entropy function. At a local loss of entropy convexity, the stabilizing forces are lost and the bubble stretches out leading to a bifurcation or other instability corresponding to a phase change; a time dependent version

of what Gibbs showed in 1873 for static conditions. The various "catastrophes" of Rene Thom appear as consequences of different kinds of local loss of entropy convexity.

Indeed, such results have been appearing in such quantity that we have fallen far behind in getting them written up for publication. But with the new applications outside thermodynamics off to a good start, we can now return to the original goal, with strenuous efforts to correct this.

REFERENCES

J. G. Abels (1974) "Maximum Entropy Spectral Analysis", Astron. Astrophys. Suppl. 15; 383.

G. Bricogne (1982), in Computational Crystallography, D. Dayre, Editor, pp. 258-264; Oxford University Press, New York.

G. Bricogne (1984), "Maximum Entropy and the Foundations of Direct Methods", Acta Cryst. (in press).

G. Bricogne, ed. (1985), Proceedings of the EMBO Workshop on Maximum-Entropy Methods, Orsay, April 24-28, 1984. Applications in crystal and biological macromolecular structure determination from x-ray/neutron scattering data.

R. K. Bryan, M. Bansal, W. Folkhard, C. Nave, and D. A. Marvin, (1983), "Maximum-Entropy calculation of the electron density at 4A resolution of Pf1 filamentous bacteriophage", Proc. Nat. Acad. Sci. U.S.A. 80, 4728.

S. F. Burch, S. F. Gull & J. Skilling (1983), "Image Restoration by a Powerful Maximum Entropy Method", Comp. Vision, Graphics & Image Processing, 23, 118-123.

John Parker Burg (1967), "Maximum Entropy Spectral Analysis", Proc. 37th Annual Meeting, International Society of Exploration Geophysicists, Oklahoma City. Reprinted in Childers (1978).

John Parker Burg (1975), Maximum Entropy Spectral Analysis, Ph.D. Thesis, Stanford University.

D. G. Childers, Editor (1978); Modern Spectrum Analysis, IEEE Press and J. Wiley & Sons, N.Y. A reprint collection.

R. T. Cox (1946), "Probability, Frequency, and Reasonable Expectation", Am. J. Phys. 14, 1-13. This is greatly expanded in R. T. Cox, The Algebra of Probable Inference, Johns Hopkins University Press, Baltimore (1961).

R. G. Currie (1981), "Solar Cycle Signal in Earth Rotation: Nonstationary Behavior", Science, 211, 386.

R. G. Currie (1981), "Evidence for 18.6 year Mn Signal in Temperature and Drought Conditions in North America Since AD 1800", J. Geophys. Res. 86, #C11, pp. 11055-11064.

R. G. Currie (1981), "Solar Cycle Signal in Air Temperature in North America: Amplitude, Gradient, Phase and Distribution", J. Atmos. Sci., 38, 809-818.

G. J. Daniell & S. F. Gull (1980), "Maximum Entropy Algorithm Applied to Image Enhancement", IEE Proc. 127E, 170-172.

A. C. Fabian, R. Willingale, J. P. Pye, S. S. Murray and G. Fabbiano (1980), "The X-Ray Structure and Mass of the Cassiopeia A Supernova Remnant", Mon. Not. Roy. Astr. Soc., 193, 175-188.

B. R. Frieden (1972), "Restoring with Maximum Likelihood and Maximum Entropy", J. Opt. Soc. Am. 62, 511-518.

B. R. Frieden (1980), "Statistical Models for the Image Restoration Problem", Computer Graphics and Image Processing, 12, 40-59.

Z. Gburski, C. G. Gray, and D. E. Sullivan, (1984), Chem. Phys. Lett. 106, 55.

J. Willard Gibbs (1902) Elementary Principles in Statistical Mechanics, reprinted in The Collected Works of J. Willard Gibbs, Vol. 2, by Yale University Press, New Haven, Conn., 1948 and by Dover Publications, Inc., New York, 1960.

S. F. Gull & G. J. Daniel (1978), "Image Reconstruction from Incomplete and Noisy Data", Nature 272, 686-690.

S. F. Gull & G. J. Daniell (1979), "The Maximum Entropy Method", in Image Formation from Coherent Functions in Astronomy, C. van Schooneveld, Ed., D. Reidel Pub. Co.

S. F. Gull & J. Skilling (1983), "The Maximum Entropy Method", IAU/URSI Symposium on Indirect Imaging, Sydney Australia.

S. Gull and J. Skilling (1984), "The Entropy of an Image", in J. A. Roberts, ed., Indirect Imaging, Cambridge U. Press.

S. Haykin, Editor (1979), Nonlinear Methods of Spectral Analysis, Topics in Applied Physics, Vol. 34; Springer-Verlag, New York.

S. Haykin, Editor (1982). The September 1982 IEEE Proceedings, Vol. 70, is a special issue devoted to spectrum analysis.

W. Heisenberg (1958), Daedalus 87, 100.

K. D. Horne (1982), "Eclipse Mapping of Accretion Disks in Cataclysmic Binaries", Ph.D. Thesis, CalTech. PME with "prior prejudice" favoring circular symmetry.

E. T. Jaynes (1967), "Foundations of Probability Theory and Statistical Mechanics", in Delaware Seminar in the Foundations of Physics, M. Bunge, Ed., Springer-Verlag, Berlin.

E. T. Jaynes (1976), "Confidence Intervals vs. Bayesian Intervals", in Foundations of Probability Theory, Statistical Inference, and Statistical Theories of Science, W. L. Harper and C. A. Hooker, Eds. Reidel Publishing Co., Dordrecht-Holland. Reprinted in Jaynes (1983).

E. T. Jaynes (1978), "Where do we Stand on Maximum Entropy?", in The Maximum Entropy Formalism, R. D. Levine and M. Tribus, Eds., M.I.T. Press, Cambridge Mass. Reprinted in Jaynes, (1983).

E. T. Jaynes (1980), "The Minimum Entropy Production Principle", in Annual Review of Physical Chemistry, S. Rabinovitch, Ed., Annual Reviews, Inc., Palo Alto Calif. Reprinted in Jaynes, (1983).

E. T. Jaynes (1983), Papers on Probability, Statistics, and Statistical Physics, R. D. Rosenkrantz, Editor, D. Reidel Publishing Co., Dordrecht-Holland. Contains reprints of the preceding four articles and nine others.

E. T. Jaynes, (1984), "Prior Information and Ambiguity in Inverse Problems", SIAM-AMS Proceedings, 14, 151-166.

E. T. Jaynes, (1985), "The Evolution of Carnot's Principle", in Bricogne (1985).

H. Jeffreys (1961), Theory of Probability, Oxford University Press.

R. W. Johnson & J. E. Shore (1983), "Minimum-Cross-Entropy Spectral Analysis of Multiple Signals", IEEE Trans. Acoust. Speech & Signal Processing ASSP-31, 574-582.

J. H. Justice, Ed., (1985), Proceedings of the Workshop on Bayesian/Maximum Entropy Methods, University of Calgary, August 1984. In Press.

M. C. Kemp (1980), "Maximum Entropy reconstructions in emission tomography", Medical Radionuclide Imaging, 1, 313-323.

P. S. Laplace (1814), Essai Philosophique sur les Probabilites, Courcier Imprimeur, Paris; reprints of this work and of Laplace's much larger Theorie Analytique des Probabilites are available from Editions Culture et Civilisation, 115, Ave. Gabriel Lebron, 1160 Brussels, Belgium.

A. K. Livesey (1984), "Structural Investigations of Metallic Glasses", Ph.D. Thesis, Cambridge University.

A. K. Livesey & J. Skilling (1984), "Maximum Entropy Theory", Submitted to Acta Cryst.

L. R. Mead & N. Papanicolaou (1984), "Maximum Entropy in the Problem of Moments", J. Math. Phys. 25, 2404-2417.
G. Minerbo (1979), "MENT: A maximum entropy algorithm for reconstructing a source from projection data", Comp. Grap. & Image Processing, 10, 48-68.
L. H. Schick and R. Inguva (1981), "Information Theoretic Processing of Seismic Data", Geophys. Res. Lett. 8, 1199.
P. F. Scott (1981), "A 31 GHz map of W3(OH) with a resolution of 0.3 arsec", Mon. Not. R. Astr. Soc. 194, 25P-29P.
J. E. Shore & R. W. Johnson (1980), "Axiomatic Derivation of Maximum Entropy and the Principle of Minimum Cross-Entropy", IEEE Trans. Inform. Theory IT-26, 26-37.
J. E. Shore (1981), "Minimum Cross-Entropy Spectral Analysis", IEEE Trans. Acoust. Speech & Signal Processing ASSP-29, 230-237.
S. Sabisi (1983), "Two-Dimensional Reconstructions from One-Dimensional Data by Maximum Entropy", Nature 301, 134-136.
S. Sibisi, J. Skilling, R. Brereton, E. D. Laue, J. Staunton (1984), "Maximum Entropy Method in 13C-NMR Spectroscopy" (in preparation for Chemical Communications).
J. Skilling, A. W. Strong, K. Bennett (1979), "Maximum Entropy Image Processing in Gamma-Ray Astronomy", Mon. Not. Roy. Astron. Soc., 187, 145-152.
J. Skilling (1983), "Maximum Entropy Image Reconstruction from Phaseless Fourier Data", presented at Optical Soc. Am. meeting on Signal Recovery with Incomplete Information and Partial Constraints Incline Village, Nevada, Jan. 1983.
J. Skilling and S. Gull (1983), "Algorithms and Applications", in C. R. Smith & W. T. Grandy (1985).
J. Skilling & S. Gull (1984), "The Entropy of an Image", Am. Math. Soc. SIAM Proceedings, Vol. 14.
J. Skilling & R. K. Bryan, (1984), "Image Reconstruction by Maximum-Entropy: General Algorithms", Mon. Not. Roy. Astr. Soc. In Press.
C. R. Smith & W. T. Grandy, Editors (1985); Proceedings of the 1981, 1982 1983 Workshops on Maximum Entropy and Bayesian Methods in Inverse Problems; University of Wyoming, Laramie WY 92071 (2 Volumes; in press).
D. E. Smylie, G. K. C. Clarke, and T. J. Ulrych, (1973), "Analysis of Irregularities in the Earth's Rotation", in Computational Physics, B. A. Bolt, B. Alder, & S. Feinbach, Eds., Academic Press, New York.
S. Steenstrup & S. W. Wilkins (1984), "On Information and Complementarity in Crystal Structure Determination", Acta Cryst. (in press).
R. J. Tuffs (1984), Secular Changes in the Supernova Remnant Cassiopeia A", Ph.D. Thesis, Cambridge University, U.K.
S. W. Wilkins, J. N. Varghese & M. S. Lehmann (1983), "Statistical Geometry. I: A Self-Consistent Approach to the Crystallographic Inversion Problem Based on Information Theory", Acta Cryst. A39, 49-60.
S. W. Wilkins, J. N. Varghese, & S. Steenstrup (1984), "X-Ray Structure Determination using Information Theory--the Statistical Geometric Approach", in Proceedings, IAU/URSI Symposium on Indirect Imaging, Sydney, Australia, Aug. 1983 (in press).
R. Willingale, "Use of the Maximum Entropy Method in X-Ray Astronomy", Mon. Not. R. Astr. Soc. 194, 359-364.

BIFURCATION GEOMETRY IN PHYSICS

Werner Güttinger

Institute for Information Sciences
University of Tübingen

Tübingen, Fed. Rep. of Germany

SUMMARY

The application of bifurcation theory to nonlinear physical problems is reviewed. It is shown that the topological singularities and bifurcation processes deriving from the concept of structural stability determine the most significant phenomena observed in both structure formation and structure recognition. From this emerges a unifying geometrical framework for the description of nonlinear physical systems which, when passing through instabilities, exhibit analogous critical behavior both at the microscopic and macroscopic levels. After a survey on the basic concepts of singularity and bifurcation theory some new developments are outlined. These include nonlinear conservation laws in various physical fields, the relation between analytical and topological singularities in the inverse scattering problem and in phonon focusing, interacting Hopf and steady-state bifurcations in nonlinear evolution equations and applications to optical bistability and neuronal activity.

I. INTRODUCTION

Bifurcation geometry aims at an understanding of the fascinating analogies which have been discovered in the critical behavior of systems of various genesis and at a unifying description of structure formation and recognition processes. Nature surprises us with the fact that the tremendous amount of physical data and results can be condensed into a few simple laws that summarize our knowledge. These laws are essentially qualitative. In mathematical terms "qualitative" means not "poorly quantitative" but topologically invariant, i.e., independent of a coordinate description. Common to critical physical phenomena, besides their qualitative similarity, is their universality expressed by the fact that the details of the interaction of a system undergoing spontaneous transitions are often almost irrelevant. This calls for a topological description of the phenomena under consideration. When Feynman discussed turbulence he observed that "the next great era of awakening of human intellect may well produce a method for understanding the qualitative content of complex equations." The concept of structural stability and the ensuing framework of singularity and bifurcation theory have undoubtedly opened the door to it.

The basic role physics plays in the sciences may be traced to the fact that most systems and structures in nature enjoy an inherent physical stability property: They preserve their quality under slight perturbations, i.e., they are structurally stable. Otherwise we could hardly think about or describe them, and today's experiment would not reproduce yesterday's results. We do not know how it got that way. But accepting structural stability as a fundamental principle that complements the known physical laws, universal critical physical phenomena have a common topological origin. They are describable and classifiable by stable unfoldings of singularities, i.e., in terms of topological normal forms, that organize the bifurcation processes exhibited by dynamical systems. We use the term bifurcation to refer to changes in the qualitative structure of solutions to differential equations. A phenomenon is said to be structurally stable if it persists under all allowed perturbations in the system. Most of the nonlinear equations of physics are not amenable to a quantitative analysis and few, if any, are completely known, so that it is often not clear for which particular effects one ought to look. Since, however, they derive from geometrical invariance principles and conservation laws they must possess structurally stable solutions. To determine these provides us with conceptual guidance to single out the most significant phenomena in complex systems and to answer the basic questions of structure formation and structure recognition which we address in this paper.

It is incontestable that, in spite of its complexity, our universe is not chaos but a ceaseless creation, evolution and destruction of forms in space which last for some period of time. One of the central problems of science is to explain this change of form and, if possible, to predict it. Since the formation of spatio-temporal structures and modes of behavior is a geometrical phenomenon, uncovering its common roots is a topological problem. The existence of unifying topological principles may be inferred from the wealth of analogies found in the critical behavior of systems which, when passing through instabilities, suddenly exhibit new and apparently universal spatio-temporal patterns. Recognizing analogies is an important source of knowledge, indeed the only one besides intuition. In the course of evolution it constantly happens that, independently of each other, two different systems take similar, parallel paths in adapting themselves to the same external circumstances. It is this parallel adaptation that gives rise to the phenomenon of analogy and, since a recognition process is involved, it is conceivable that structure formation and structure recognition must have a common topological origin. Most of this paper pretends to demonstrate just this.

The phenomenological picture looks indeed seductively general. There is a striking similarity among the instabilities that lead to convection patterns in fluids, cellular solidification fronts in crystal growth, geophysical textures, vortex formation in superconductors, phase transitions in condensed matter and particle physics, biological and chemical patterns or rhythms, and so forth. Their common characteristic is that one or more significant behavior variables or order parameters undergo spontaneous, large and discontinuous changes or cascades of these if slow, competing but continuously driving control parameters or forces cross a bifurcation set and enter conflicting regimes. As a consequence, an initially quiescent system becomes unstable at a critical value of some control variable and then restabilizes into a more complex space- or time-dependent configuration. Primary bifurcations induce limit cycles, spatial patterns and spatio-temporal patterns in the form of standing waves when the bifurcation branches remain disjoint. If other controls cause these branches to interact, multiple degenerate bifurcation points produce higher instabilities. Then the system undergoes additional transitions into more complex states, giving rise to travelling waves, hysteresis, resonance and entrainment effects. These ulti-

mately lead to states which are intrinsically chaotic. In the vicinity of degenerate bifurcation points a system becomes extremely sensitive to small ambient factors like imperfections, external fields or fluctuations that lead to symmetry breaking. This in turn enhances the system's ability to perceive its external environment and, adapting to it by capturing its asymmetry, to form preferred patterns or modes of behavior. Most prominent among the theoretical programs venturing into the area of general principles are Prigogine's concept of dissipative structures [1], Haken's synergetics [2] and Thom's catastrophe theory [3]. Among these Thom's program on structural stability and morphogenesis has both the potential to provide a geometrical explanation for the variety of analogies encountered in the critical behavior of systems of different genesis and, as we shall see, also the power to predict new phenomena.

After a survey of the basic principles of bifurcation theory in Section II we discuss in Section III some recent applications to wave motion, including the inverse scattering problem of structure recognition, phonon focusing, dispersive wave phenomena and nonlinear conservation laws. In Section IV, basic concepts underlying structure formation are discussed. The bifurcation geometry of interacting time-periodic and steady-state solutions of nonlinear evolution equations is analysed and applied to problems in optical bistability and neural activity.

II. STRUCTURAL STABILITY, SINGULARITY AND BIFURCATION THEORY

In this section the basic principles of singularity and bifurcation theory are outlined, referring to [4]-[7] for further details. A physical system is said to be structurally stable if it preserves its quality under slight distortions. For example, a nonlinear dynamical system $\dot{x}(t)=F(x,\lambda)$, $x \in R^n$, $\lambda \in R$, is structurally stable if the phase portrait of the perturbed system $\dot{x}=F+\delta$ is "qualitatively," i.e., topologically equivalent to that of the unperturbed system. We are interested in those unstable systems whose perturbation gives rise to new modes of behavior due to bifurcations. Bifurcation occurs in a parametrized dynamical system when a variation of a parameter causes a qualitative change in the behavior of solutions, e.g., when an equilibrium splits into two. Catastrophe occurs when the stability of an equilibrium breaks down, causing the system to jump into another state.

The pitchfork bifurcation provides a familiar example: The stationary states $\dot{x}=0$ of $\dot{x}=G=x^3-\lambda x$ can be represented by the bifurcation diagram $x(\lambda)$ of Figure 1. The trivial solution $x=0$ branches at the bifurcation point $\{x=0,\lambda=0\}$ which is called a singularity of G. The two branches of the pitchfork may, e.g., be interpreted as the primary right or left displacements of a vertical elastic column buckling under a load λ, or as the two magnetization directions in a ferromagnet below critical temperature $T=\lambda=0$. The bifur-

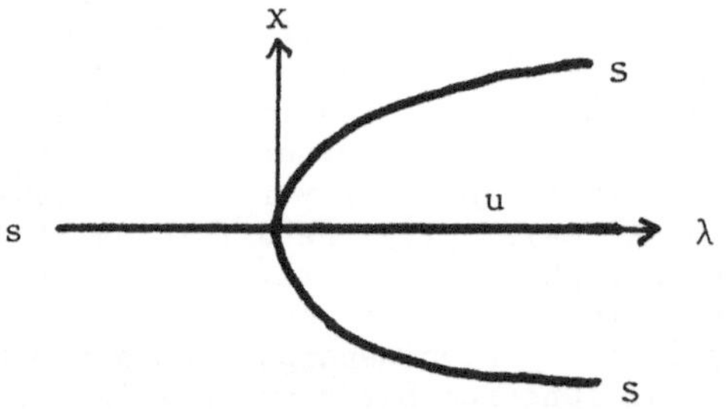

Fig. 1. Pitchfork bifurcation

cation diagram of the so-called Hopf bifurcation of a time-periodic solution from a stationary one follows from Figure 1 by rotating the parabola around the λ-axis above a (x,y)-plane with y orthogonal to (x,λ). Introducing polar coordinates r,θ in the (x,y)-plane the canonical equations are $\dot{\theta}=1$, $\dot{r}=-(r^3-\lambda r)$. When $\lambda<0$, the origin is an attractor in the (x,y)-plane with all trajectories spiralling into it; when $\lambda>0$, the radial component is outward for $r<\sqrt{\lambda}$ and inward for $r>\sqrt{\lambda}$; the origin becomes a repellor and there is a stable limit cycle thrown off with $r=\sqrt{\lambda}$. The point $\lambda=0$ is a "Hopf-bifurcation point," the amplitude grows on a paraboloid and the origin is an "organizing center" for the singularity.

A general bifurcation problem consists of the solution $x=x(\lambda)$ of a system of equations $G(x,\lambda)=0$ with $G(0,0)=0$, $d_xG(0,0)=0$ where G is a germ of a C^∞ mapping of $R^n\times R$ into R^m. The point $(x=0,\lambda=0)$ is called a bifurcation point or a singularity of G. This does not necessarily mean that solutions $x(\lambda)$ of $G=0$ branch at the bifurcation point but merely implies that the solution diagram changes its qualitative form when G is slightly perturbed. For example, a hysteresis point, i.e., a point where $x(\lambda)$ has a vertical tangent, is a bifurcation point since an appropriate perturbation changes the solution type of $G=0$ from a monostable to a bistable one. At a bifurcation point a system becomes structurally unstable. The main result of bifurcation theory is that, in the course of possible bifurcations, a system does not become unstable in an arbitrary way but that the bifurcations occur in certain definite and classifiable ways. In other words a structurally unstable event can be embedded into a structurally stable family of maps or systems. This family is parametrized by a number of unfolding or control parameters. The number of parameters that is necessary to embed a given bifurcation into a stable family is called the codimension of the bifurcation or singularity. Thus, a structurally unstable event or bifurcation occurs within a stable process.

Thom's elementary catastrophe theory [3] is at the origin of what is now called topologically invariant bifurcation theory. It can be interpreted as describing the stable bifurcations of the stationary states of a gradient system $\dot{x}_i=\partial f(x,a)/\partial x_i$ with $x=(x_1,\ldots,x_n)\in R^n$ when the parameters $a=(u,v,w,\ldots)\in R^k$ in the "potential" function f vary. Catastrophe theory deals with the classification of degenerate singular points of smooth real-valued, parametrized families of functions f of n real variables. The singular points of a smooth function $f:R^n\to R$ are those points x where the differential vanishes: grad $f(x)=0$. The function f has a nondegenerate (or Morse) singularity at x if the Hessian $H=\det(\partial^2 f(x)/\partial x_i \partial x_k)$ does not vanish. Then, in suitable local coordinates the tail of the Taylor expansion of f can be made to vanish so that the function f can in some neighborhood of such a nondegenerate singular point be represented in the Morse normal form $f=\text{const}+\Sigma(\pm x_i^2)$. Morse singularities are stable in the sense that a small perturbation of f also has a Morse singularity. For example, a paraboloid preserves its shape under small deformations. On the other hand, every degenerate singular point x of f, for which grad $f=0$ and $H=0$, bifurcates into some nondegenerate points after an arbitrarily small deformation. The function $f(x)=x^3/3\,(x\in R)$ and its perturbation $f_t(x)=x^3/3+tx$ furnish an example: f is not stable at $x=0$ because a small perturbation of it, generated by adding tx, radically changes its shape when t is varied: for $t<0$, $f_t(x)$ has two extrema while for $t>0$ it has none. Thus, the family $f_t(x)$ is stable as a (t-)parametrized family near $x=0$. We call $f_t(x)$ the unfolding of f, t an unfolding parameter, and $t=0$ a catastrophe or bifurcation point of the family because it separates the stable regions $t\lessgtr 0$. This shows that degenerate singular points appear naturally if a function depends upon parameters, i.e., if one considers not an individual function but a parametrized family of functions. Then it is possible for a non-Morse function to appear as a member of a stable family. Formally, one may continue playing by considering the function $f=x^4/4$, $x\in R$, and its unfolding or perturbation $f_{uv}(x)=x^4/4+ux^2/2+vx$ near its

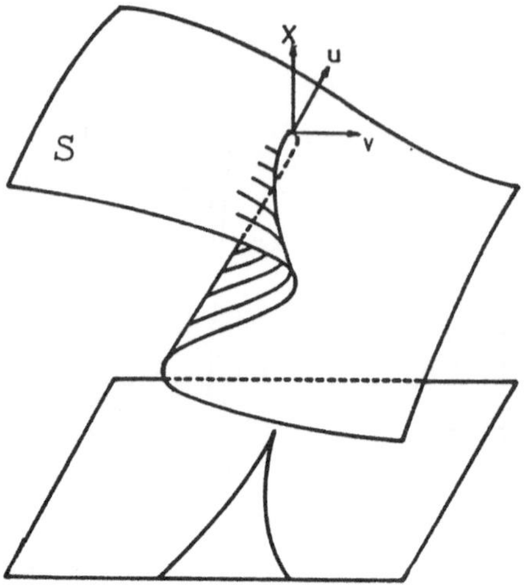

Fig. 2. Cusp catastrophe manifold

singularity x=0. The singular points are given by the overhanging cliff S of Figure 2, given by $f'_{uv}(x)=x^3+ux+v=0$, in (x,u,v)-space. Projecting the tangents of the two edges of S, given by $f''_{uv}=3x^2+u=0$, vertically onto the (u,v)-plane by eliminating x from the two equations, one obtains the familiar cusp equation $B:4u^3+27v^2=0$, viz., the cusped bifurcation set of Figure 3 on which two stationary points (a minimum and a maximum) of f_{uv} coalesce. The shapes of f_{uv} remain basically the same inside or outside the cusp, respectively (two minima of f_{uv} if u and v vary inside B, one minimum of f_{uv} if u and v stay outside of B): f_{uv} remains stable under small variations of (u,v) not crossing B. B is a singularity of the map of S onto the plane made up by (u,v). Intuitively, deforming S smoothly does not change the qualities of the "cusp catastrophe" which is a very fundamental object. Take a two-dimensional manifold $M \subset R^3$, i.e., a piece of elastic fiber. Lift it into R^3, deform it slightly and map it vertically down onto the plane R^2. One discovers three sorts of points on the fiber M: regular points for which the sheet lies smooth and flat over R^2, and singular points which lie on a foldline of the sheet or form a pleat which marks the origin of an overhanging cliff S, and, finally, messy points which resemble none of the former. Pulling the sheet slightly near any of these points shows that the "quality" of the regular and singular points remains the same and they cannot be made to disappear under small deformations of the fiber while the messy points turn under small deformations into points of the former type. This is an experimental proof of the Whitney theorem: The only structurally stable singularities of the projection $M \to R^2$ are folds and cusps. The physical meaning of this geometrical fact is obvious: If a system with one behavior variable or order parameter x is slowly driven from one phase to another by a control variable v, and if an orthogonal drive u sets in to split the quality of the phases (represented by the two sheets of the cliff) and if a phase may persist for a while with the transition to the other delayed, then the "cusp catastrophe"

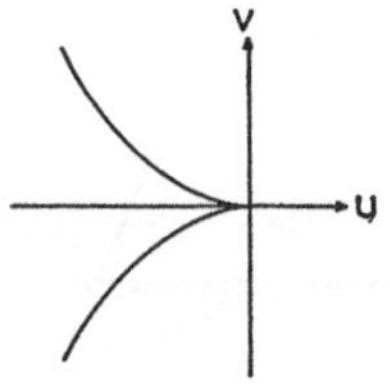

Fig. 3. Cusp bifurcation set

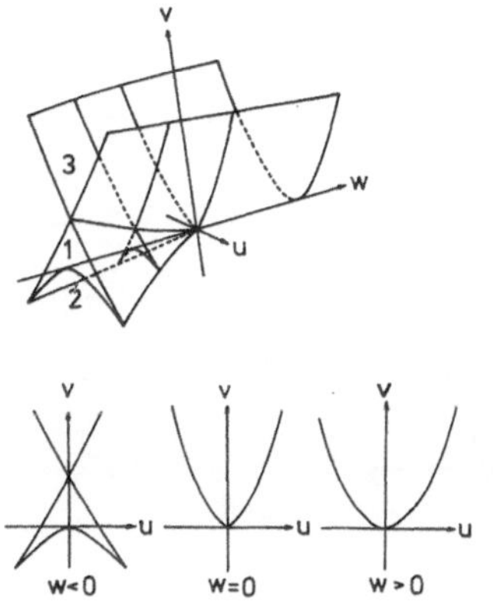

Fig. 4. Swallowtail bifurcation set

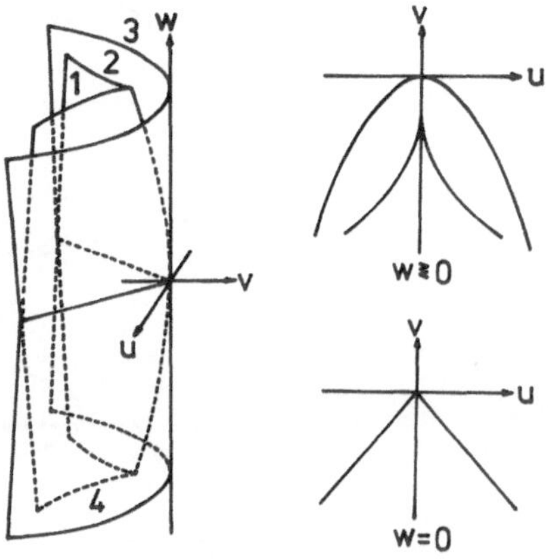

Fig. 5. Hyperbolic umbilic bifurcation set

of Figure 2 is intuitively the simplest model. As we have seen, this cusp catastrophe is the only structurally stable one and, in three dimensions, the most complicated singularity that can occur locally.

Catastrophe theory generalizes this result to n dimensions by classifying the singularities of maps f into equivalence classes. Two functions f and g in the same class are called (right-) equivalent if $f(x)=g(\phi(x))+\text{const}$, where ϕ is a diffeomorphism. Functions from different classes differ by the number of control, perturbation, or unfolding parameters a and by the form of the unfolding terms which have to be added to a function in order to embed the degenerate singular point into a stable family. The number of terms is the codimension of the singularity of f. Then Thom's theorem states: Let $f: R^n \times R^r \to R$ be a parametrized family of smooth functions $f_a: R^n \to R$, where $a \in R^r$. Let $r \leq 3$, $a=(u,v,w)$, $(x_1,x_2,\ldots)=(x,y\ldots)$. Then, near a singularity at $(x=0,a=0)$, almost all such f are (up to the addition of a Morse function) equivalent to a polynomial $P(x,a)$ out of the following list; $[A_2]$ Fold: $P=x^3+ux$; $[A_3]$ cusp: $P=x^4+ux^2+vx$; $[A_4]$ Swallowtail: $P=x^5+ux^3+vx^2+wx$; $[D_4^+]$ Hyperbolic umbilic: $P=x^3+xy^2+uy^2+vx+wy$; $[D_4^-]$ Elliptic umbilic: $P=xy^2-x^3+ux^2+uy^2+vx+wy$. The list can be extended to $r>3$ [8]. The polynomials $P(x,0)$ are called normal forms of the singularities and the associated families $P(x,a)$ are their unfoldings. We denote by $M=\{(x,a)\,|\,\text{grad } f=df(x,a)=0\}$ the catastrophe manifold of f. $\Sigma=\{(x,a)\,|\,df=0, H=\text{Det } d^2f=0\}$ is the singularity set. The projection $\chi: M \to R^r$ of M on the bifurcation set $B=\{a\,|\,(x,a)\in\Sigma \text{ for } x\in R^n\}$ gives the familiar Thom diagrams of Figs. 3-6. When a varies transversely through B, the topology of the extrema of f changes in a neighborhood of the origin.

Bifurcation is an idealized, nongeneric phenomenon because diagrams such as the pitchfork in Figure 1 are structurally unstable, and small perturbations will break the diagram. Since imperfections are present in any real physical system we should require that a description of bifurcations include the effects of variations (impurities, imperfections etc.) in the problem

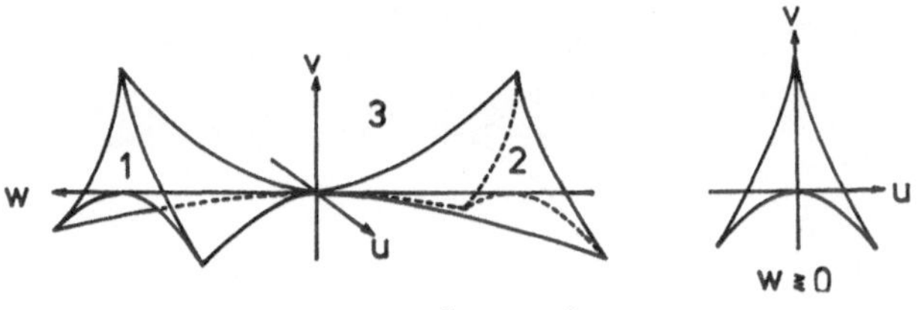

Fig. 6. Elliptic umbilic bifurcation set

other than the distinguished bifurcation parameter λ If, for example, the idealized bifurcation problem is determined by the pitchfork equation $G(x,\lambda)=x^3-\lambda x=0$, one wishes to know the most general form of the solution structure when G is slightly perturbed. Moreover, Thom's theory must be generalized to non-gradient systems. Golubitsky and Schaeffer [9] have analyzed these questions. This has led to "imperfect bifurcation theory."

Consider a bifurcation problem

$$G(x,\lambda)=0 \tag{2.1}$$

with $d_xG(0,0)=0$, where $G:R^n\times R\to R^m$, $x\in R^n$, and $\lambda\in R$ is a distinguished, externally controllable bifurcation parameter -- distinct from others describing perturbations -- which represents the control variable in a physical experiment. We may, for example, assume that (2.1) is the equation for the amplitudes x of the solutions of a system of nonlinear evolution equations $\partial u/\partial t=F(u,\lambda)$ obtained by a Lyapunov-Schmidt reduction (Sec. IV). In [9] a technique has been developed which permits the determination of all possible qualitatively different bifurcation diagrams in a neighborhood of the singularity ($x=0$, $\lambda=0$) which are stable against perturbations of G. This is achieved by changing coordinates in such a way that G reduces to polynomial normal forms. Since the reduction of a system of evolution equations to an algebraic system (2.1) is not unique, the possibility arises that two bifurcation problems G and G' assume different forms although they possess qualitatively the same solutions. Therefore, G and G' are called contact equivalent if there exists a smoothly parametrized family of invertible matrices $T(x,\lambda)$ on R^m and diffeomorphisms $x\to X(x,\lambda)$, $\lambda\to\Lambda(\lambda)$ such that

$$G'(x,\lambda) = T(x,\lambda)\cdot G(X(x,\lambda),\Lambda(\lambda)) \tag{2.2}$$

The distinguished parameter λ is not allowed to mix with the order parameters or state variables x and the topology of the bifurcation diagram $B(G)=\{(x,\lambda)|G=0\}$ remains invariant under (2.2) relative to the λ-coordinate. A representative example is this: Let $G(x,\lambda)$ with $x\in R$ be a bifurcation problem such that at $x=\lambda=0$ we have $G=G_x=G_{xx}=G_\lambda=0$ and $G_{xxx}G_{x\lambda}<0$. Then G is contact equivalent to the pitchfork $x^3-\lambda x$.

To characterize all perturbations of G which respect the special role of λ amounts to finding a universal unfolding of G and then enumerating all qualitatively distinct bifurcation diagrams. Their number is called the codimension of G. Denoting an unfolding of G by $G_\alpha=G+H_\alpha$ with a perturbation term H_α with $H_0=0$, we call the unfolding parameters α imperfection parameters. G_α is a stable unfolding of G if, for any other unfolding G_β, there exists a smooth map $\alpha=\psi(\beta)$ such that $G_{\psi(\beta)}$ is contact equivalent to G_β. G_α is universal if it is stable and possesses the minimum number of unfolding parameters needed for stability. With these definitions we have the following theorem [9]: A bifurcation problem of finite codimension is contact equivalent to one described by polynomial normal forms whose universal unfoldings are also polynomials. The list of normal forms presently known is not yet as complete as is the Thom-Arnold list. As an example we show in Figure 7 the bifurcation diagrams for the perturbed pitchfork with universal unfolding $G_\alpha=x^3-\lambda x+\alpha_2\lambda+\alpha_1=0$ with unfolding parameters α_1,α_2. These diagrams can be obtained directly by cutting the cusp catastrophe cliff of Figure 2 by vertical planes, i.e., by following straight lines in the (u,v)-plane through the cusp of Figure 3. For, setting $\lambda=-u$ and $\alpha_1+\alpha_2\lambda=v$, $G_\alpha=0$ is just the equation of the overhanging cliff S of Figure 2. Fixing (α_1,α_2) defines a straight line in the (u,v)-plane as λ varies, and the bifurcation diagrams of the perturbed pitchfork are the intersection curves of S with vertical planes having these lines as basis. Crossing a fold branch by a line through the cusp corresponds to a limit point L in the bifurcation diagram $x(\lambda)$ of $G_\alpha=0$, while a contact of the line with the cusp point produces a hysteresis point H in $x(\lambda)$. This is shown in Figure 8. Another example is provided by the unfolding of the

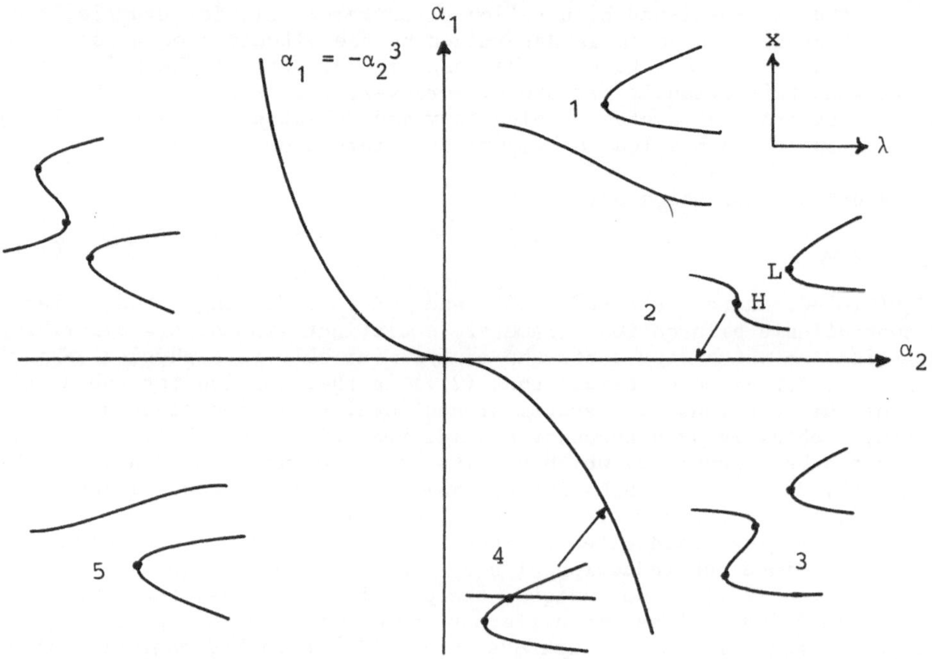

Fig. 7. Perturbed pitchfork for various α_1, α_2

asymmetric or winged cusp $G_\alpha = x^3 + \lambda^2 + \alpha_2 x + \alpha_3 x\lambda + \alpha_1 = 0$. Typical bifurcation diagrams $x(\lambda)$ are shown in Figure 9 together with the generating paths through the cusp in Figure 10. The observed formation of mushroom-like bifurcation diagrams and of an isola is of considerable physical importance (Sec. IV). The imperfect bifurcation theory outlined above has been generalized to problems with symmetry [10]. Let G be a bifurcation problem with m=n and Γ a compact Lie group acting orthogonally on R^n. Then G is a bifurcation problem with symmetry group Γ if for all $\gamma \in \Gamma$ we have $G(\gamma x, \lambda) = \gamma G(x, \lambda)$. One has then to analyze the problem in which a perturbation breaks symmetry. Stable ways to

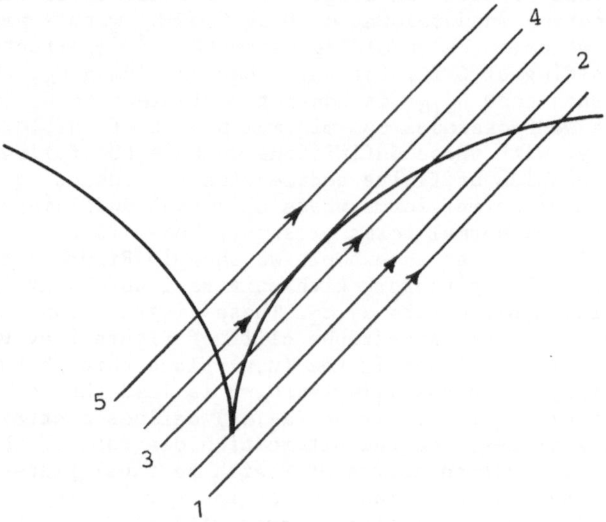

Fig. 8. Straight paths through the cusp producing the perturbed pitchforks

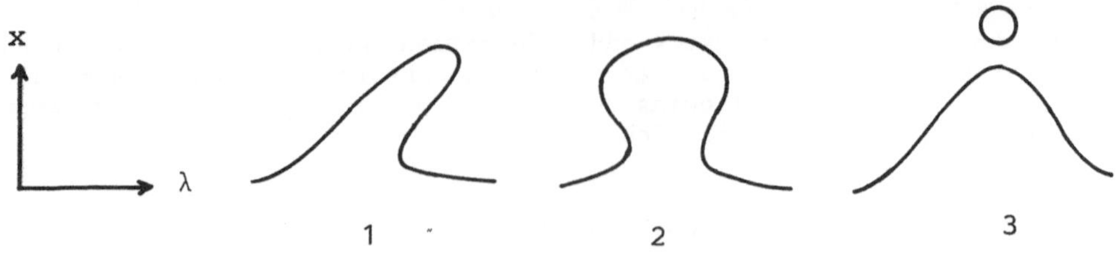

Fig. 9. Bifurcation diagrams of the perturbed winged cusp

break symmetry in the pitchfork seem to be obvious but a general theory of symmetry breaking is still missing [10], [11].

In a given physical system different types of bifurcations may occur which interact when λ reaches a critical value (Sec. IV). In addition, in many instances only partial information about the system's behavior is available. It is therefore pertinent to ask if a highly singular configuration, an "organizing center" [12], can be found, in which all singularities are pushed together in such a way that its unfoldings reproduce the experimentally established bifurcations. Such a center also bridges the gap between these bifurcations in the sense that different sections through the space of unfolding parameters reveal new stable bifurcations which have so far not been observed. An organizing center is defined as follows. Let G and G' be bifurcation problems and let G_α be a stable unfolding of G. To any set $\mu=(x_o,\lambda_o,\alpha)$ such that $G_\alpha(x_o,\lambda_o)=0$ we associate a bifurcation problem defined by

$$G_\mu(x,\lambda) = G_\alpha(x_o+x,\lambda_o+\lambda).$$

Then G is called an organizing center of G' (i.e., G organizes G') if there is a μ such that G_μ is contact equivalent to G'. G is an organizing center of G' if the bifurcation diagram of G' appears as a subdiagram of the universal unfolding of G for some value of the unfolding parameter α. This implies that the bifurcation diagram of G_α is qualitatively the same as that of G'. Therefore, the singularities of G' are compressed into one highly singular configuration of G whose perturbation reveals all possible imperfection effects. Such organizing centers have been found in thermal chainbranching models [13], in the Hodgkin-Huxley model of neurons [14] and in optical bistability [15] (cf. Sec. IV). In these cases the critical value λ_o of the organizing center usually lies outside the physical region. So one may expect, for example, that in cosmology time values before the big bang are relevant for describing an expanding universe.

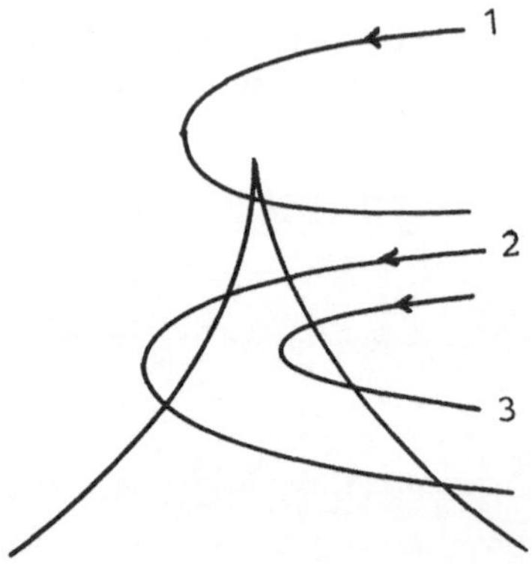

Fig. 10. Parabolic paths through the cusp producing Fig. 9

The concepts of singularity and bifurcation theory discussed above have given rise to a wealth of applications in the physical sciences, [4]-[7], [35]-[37], [42]. In the following sections we outline some recent developments in this field of bifurcation physics.

III. BIFURCATION GEOMETRY IN WAVE MOTION

In this section we show that the concept of structural stability and the bifurcation processes deriving from it govern the inverse scattering problem, phonon focusing in condensed matter physics, dispersive wave phenomena and the geometry of a variety of physical phenomena related to nonlinear conservation laws.

1. Topological Approach to Inverse Scattering

In all fields of physics wave systems are used as a tool for the nondestructive investigation of unknown structures. Such a structure or object impresses geometrical singularities upon a smooth incident wavefield. Thus the question arises what information about that structure can be inferred from these singularities. This view of the inverse scattering problem in remote sensing geometrizes classical S-matrix theory in the sense that the observed singularities in scattering amplitudes and echo recordings are linked with the topological singularities of the scatterer. Since, as we shall see, the former can be classified, so can the latter, and a new structural approach to the inversion problem emerges. In order that the reconstruction of surfaces or structures from backscattered or transmitted waves be physically feasible, i.e., repeatable, the scattering process underlying remote sensing has to be structurally stable, i.e., qualitatively insensitive to slight perturbations of the sensing wave system. Imposing this principle on the inversion process permits us to classify the geometric singularities that an unknown surface or structure generically impresses on a sensing wavefield into a few universal topological normal forms described by catastrophe polynomials. These geometrical singularities produce the dominant analytic singularities and typical configurations that are observed in recorded signals, travel-time curves, contour maps, and in the associated diffraction patterns. As the source position varies, the patterns change their morphologies according to typical universal bifurcation sets. These and power laws for the intensity permit a reconstruction of profiles in a genuine zero-offset survey. Moreover, the topological singularities provide an explanation for the similarity and universality of the high-intensity patterns encountered in surface sensing [19], [20], [21], seismology [22], ocean acoustics [23], electromagnetic sensing [24], [25], ultrasound tomography, phonon spectroscopy (Sec. II.2, [26]), and so forth.

When a point source at $\underline{x}_0$ in space emits a spherical wave pulse that propagates through a layered medium made up of partially reflecting surfaces,

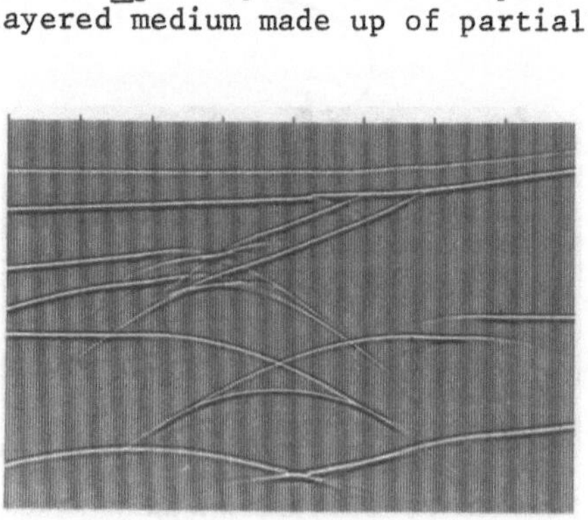

Fig. 11. Caustics in an echo recording

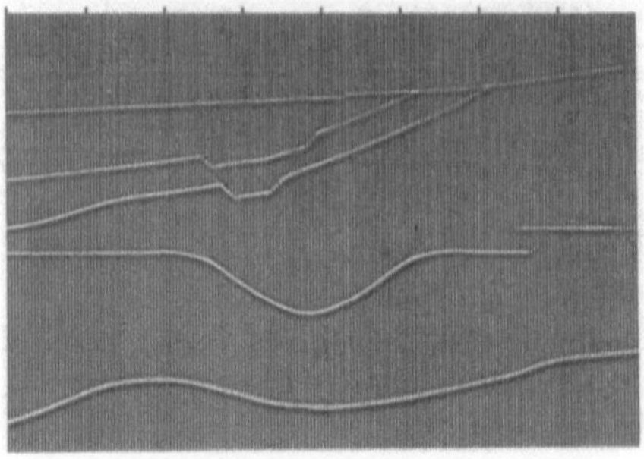

Fig. 12. Layered medium which produces the caustics in Fig. 11

the scattered wave $\psi(\underline{x}_o,t)$ that is received back at the source, the echo, exhibits as a function of time t a number of strong peaks which vary with the source position. In this echogram (a seismogram in a geological survey) caustic structures are discernible, e.g., those shown in Figure 11 where the intensity $|\psi|^2$ of the echo is plotted as a function of the vertical time or depth, and of the horizontal source position. These caustic events and the diffraction patterns around them are the singularities which the medium shown in Figure 12 impresses on the incident wavefield. We confine ourselves here to scattering by a single reflecting surface S. Since source and receiver are at the same place, only the geometrical optics specular reflection points $\underline{x}=\underline{x}_s$ of the surface whose distance vectors $\underline{R}=\underline{x}_o-\underline{x}$ to the source are normal to the surface contribute to the echo. Then, as an observer at $\underline{x}_o$ moves along a line on an observation surface Σ above the lowest curvature center of a concave part of the surface, a reversal in the direction in which the reflecting point $\underline{x}$ moves occurs when $\underline{x}_o$ on Σ crosses the surface's evolute E. E is the envelope of the surface normals where neighboring rays, normal to S, touch and focusing occurs (Fig. 13a). When the source-receiver crosses E, the number of rays going through $\underline{x}_o$ suddenly changes by two. So does the number of specularly reflecting points, and the number of arrival peaks in the echo also changes by two. Consequently, while there is but one echo when $\underline{x}_o$ is at A, there are three when $\underline{x}_o$ is at B and the multivaluedness is recorded as a cusp in the $(R,\underline{x}_o)$-diagram. In three-dimensional space the observation surface Σ intersects the evolute sheet generically in smooth curves L, called fold (A_2) lines, and in isolated cusp (A_3) points $\hat{C}$

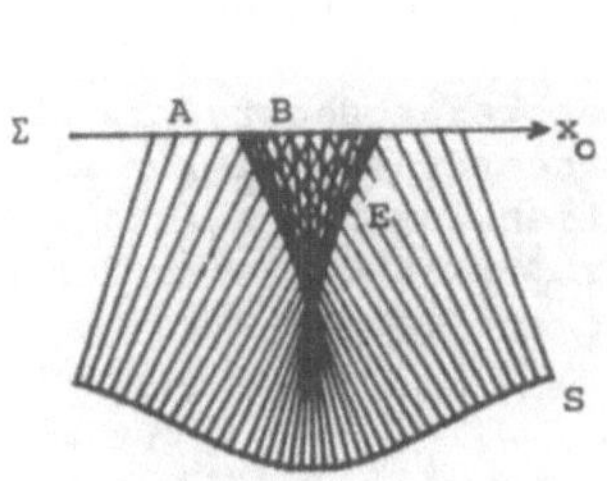

(a)

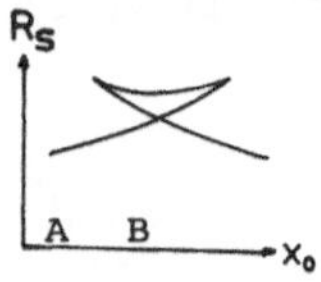

(b)

Fig. 13. (a) Surface S and evolute E, (b) Travel-time recording

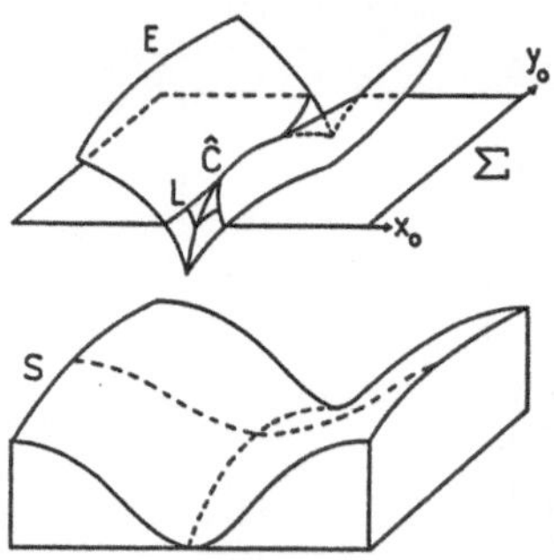

Fig. 14. Saddle surface S, with evolute E, cusps and rib

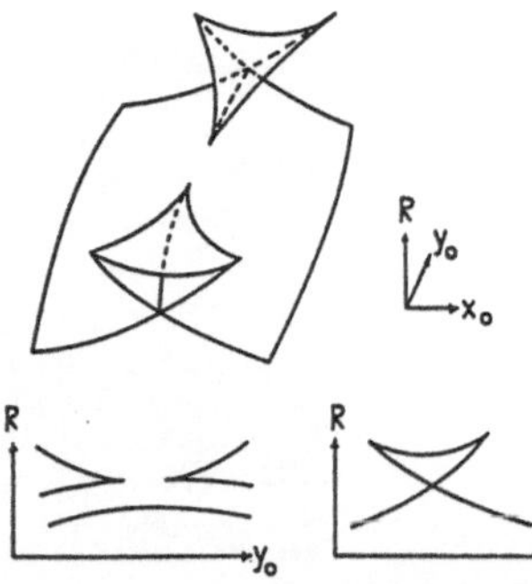

Fig. 15. Travel-time surface for Figure 11

(Fig. 14). As the height of Σ varies, the fold lines evolve into evolute sheets while the cusp points evolve into lines called ribs. If the rib is curved upwards (or downwards) and Σ is nearly tangent to a point on the rib, then the observation surface intersects the evolute in two beak-to-beak cusp singularities (or in a lip). This is shown in Fig. 15 for the winged evolute sheets E of a saddle-surface S. Let $R=R(\underline{x}_o,\underline{x})=|\underline{x}_o-\underline{x}|$ be the distance from $\underline{x}_o$ to a point $\underline{x}$ on S and $T=2R/c$ the two-way travel-time (with speed c). The signals that are received back at the source after reflections by the specular points $\underline{x}=\underline{x}_s(\underline{x}_o)$ on S have travel-times $T_s=T(\underline{x}_o,\underline{x}_s(\underline{x}_o))$ or distances $R_s=R(\underline{x}_o,\underline{x}_s(\underline{x}_o))$. These form a multi-sheeted 3-dimensional hypersurface in 4-dimensional $(T_s,\underline{x}_o)$-space. For example, in case of a plane observation surface Σ, with $\underline{x}_o=(x_o,y_o)$, the saddle-surface S of Fig. 14 gives rise to the travel-time surface on top of Fig. 15 with two-dimensional sections shown below. A rib is smooth except for isolated points where either of the following singularities can occur. The rib itself may have a cusp point, called a swallowtail point (A_4) because near it the evolute has the shape shown in Fig. 4. Or a rib may touch a fold or two other ribs in which case two evolute sheets have a contact in a common curvature center and are either hyperbolic or elliptic umbilics (D_4^+ or D_4^-) shown in Figs. 5 and 6, respectively, and the surface has an umbilic point. More complex caustic morphologies will arise if a surface possesses edges [27], [28] or if source and receiver are different [29].

Since the observed singularities in echo recordings derive from the focal surfaces which constitute surface evolutes, one expects that the latter and the singularities a surface impresses on an incident wavefield or ray family can be classified by catastrophe theory. To see this let the point $\underline{x}$ on S be parametrized by surface coordinates $\underline{x}=\underline{x}(x,y)$. The distance vector $\underline{R}_s=\underline{x}_o-\underline{x}_s$ from a specular reflection point $\underline{x}=\underline{x}_s(\underline{x}_o)$ to the source is normal to S, $\underline{R}_s=R_s(\underline{x}_o)\underline{n}(\underline{x}_s)$ where $R_s=|\underline{x}_o-\underline{x}_s|$. The distance R has an extremum R_s at $\underline{x}_s$: $\underline{t}\cdot\nabla_{\underline{x}}R=0$ for vectors $\underline{t}$ tangent to S, or

$$\nabla R = 0 \tag{3.1}$$

with $R=R(\underline{x}_o,\underline{x}(x,y))$ and $\nabla=(\partial/\partial x,\partial/\partial y)$. Hence, a solution $\underline{x}=\underline{x}_s(\underline{x}_o)$ of (3.1) is a singularity of R in the sense of Sec. II whose type depends on the Hessian determinant

$$H(R) = \begin{vmatrix} R_{xx} & R_{xy} \\ R_{xy} & R_{yy} \end{vmatrix} \tag{3.2}$$

evaluated at $\underline{x}_s$. If $H\neq 0$, $\underline{x}_s$ is nondegenerate and R can be represented by a

stable Morse quadratic form. As the source-receiver $\underline{x}_o$ varies above the surface, each point of S eventually becomes specular. In virtue of the specularity condition (3.1), the surface normal at $\underline{x}_s$ is given by $\underline{n}(\underline{x}_s(\underline{x}_o))=\nabla_{\underline{x}_o}R_s(\underline{x}_o)$. Then we obtain from $\underline{x}_o-\underline{x}_s=R_s\underline{n}$ the equation for the surface in terms of the observable R_s, viz.,

$$\underline{x} = \underline{x}_o - R_s(\underline{x}_o)\nabla_{\underline{x}_o}R_s(\underline{x}_o) \tag{3.3}$$

Since $|\underline{n}|=1$, only two components of $\nabla_{\underline{x}_o}R_s$ are needed and it suffices therefore to vary $\underline{x}_o$ on a surface Σ, say, on a plane. Inferring R_s from the arrivals or peaks the echo ψ possesses at times $t=T_s=2R_s/c$, the surface profile can be reconstructed directly from (3.3). This remains still true if H=0, i.e., if $\underline{x}_s=\underline{x}_{sc}$ is a degenerate specular point for which R_s becomes multivalued. In that case $\underline{x}_o=\underline{x}_c$ lies in a curvature center $\underline{x}_c$ on the surface's evolute E. The equation for the evolute follows by eliminating x and y from (3.1) and H=0, leaving one equation for $\underline{x}_o=\underline{x}_c$ alone.

At this point we require that the scattering process be structurally stable, i.e., that the observed images preserve their quality under slight perturbations of the system (caused, e.g., by small variations of the source). Then Thom's theorems (Sec. II) assert that in an appropriate curvilinear surface coordinate system (x,y) and in a curvilinear source coordinate system $\underline{x}_o-\underline{x}_c=(u,v,w)$ the structurally stable distance function R can take on but five "catastrophe" polynomial normal forms near a degenerate specular point $\underline{x}_{sc}$ inside a smooth part of S, viz.,

$$R = R_o + P(x,y;\underline{x}_o) + Dy^2/2 \tag{3.4}$$

where $R_o=|\underline{x}_c-\underline{x}_{sc}|$ and $D=\kappa_1-\kappa_2$ where κ_i are the surface's principal curvatures in $\underline{x}_{sc}$. If $D\neq 0$, P is one of the cuspoid polynomials A_2, A_3, A_4 and if D=0, P is one of the umbilic polynomials $D_4^{\pm}$. The bifurcation set of R is just the evolute E which, therefore, is classified into the five types mentioned above. They were shown in Figs. 3-5. Thus, there are but five structurally stable travel-time singularities observable in an echogram, all others being concatenations of these. If source and receiver are at different positions, there are 14 genuine singularities [19], but the above five describe completely the observed high intensity focal surfaces. For A_3, Equ. (3.1) yields with (3.4) the equation $4x^3+2ux+v=0$, i.e., the "overhanging cliff" of Fig. 2 where $v=-x_o$, X=x and u=h are identified with the source position, surface point and height h of the observation line. Projecting the cliff's edges onto the (x_o,h)-plane gives a cusp which is precisely the evolute E of Fig. 13a.

The travel-times recorded in the echo as a function of the source position are obtained by eliminating x and y from the equations (3.1) and (3.4). One example is shown in Fig. 15. A bifurcation sequence of travel-times coming from an umbilical surface point of elliptic type, when $\underline{x}_o$=u varies on a line above the curvature center, is shown in Fig. 16, where $\rho_s=R_s-R_o$. The spherical wavefronts of constant radius r=R, centered on $\underline{x}_o$, cut the surface

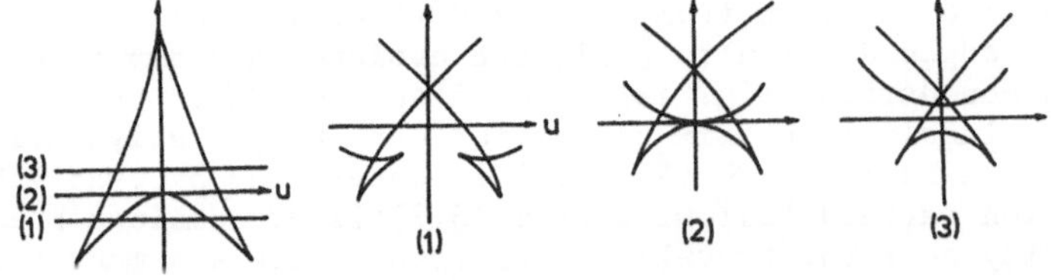

Fig. 16. Travel-time bifurcation sequence from an elliptic umbilic surface point

Fig. 17. Contour map for a hyperbolic surface point

in a series of contours $C(r,\underline{x}_o)$. An observer moving with $\underline{x}_o$ sees the Fresnel-zone contour topography changing according to universal bifurcation sets. A typical topographic change is shown in Fig. 17 for an umbilical surface point of hyperbolic type. The contours follow by setting $R-R_o=r=\text{const}$ in (3.4).

The wave-type diffraction patterns around the caustics follow from Kirchhoff's diffraction formula [19] for the backscattered wave ψ in the shortwave limit,

$$\psi(t,\underline{x}_o) = (1/8\pi^2 c)\int_o^\infty dr G(r,\underline{x}_o)\dot{F}(t-2r/c) \qquad (3.5)$$

with the incident spherical wave $F(t-R/c)/4\pi R$. Here, the surface structure function G -- the classical analog of the scattering matrix -- is given by the integral

$$G(r,\underline{x}_o) = \int_S dS R^{-3}(\underline{n}\cdot\underline{R})\delta(r-R) = r^{-3}\int_{C'} ds\sqrt{|g|}\,(\underline{n}\cdot\underline{R})/|\nabla R| \qquad (3.6)$$

where the contour C' is the projection of C onto the (x,y)-plane. From (3.6) the main contributions to G and ψ are seen to come from those values of r for which the integrand is infinite, i.e., from the geometrical optics specular reflection points given by $\nabla R=0$, Equ. (3.5), and (3.6) establishes the relation between the topological singularities of S and the analytical singularities of ψ. The Fourier transform of G is

$$\tilde{G}(k,\underline{x}_o) = \int_S dS(\partial R^{-1}/\partial n)\exp(ikR) \qquad (3.7)$$

where $k=\omega/c$. Hence, $\tilde{\psi}(\omega,\underline{x}_o) = -i\omega\tilde{G}(2\omega/c,\underline{x}_o)\tilde{F}(\omega)/8\pi^2 c$. The generic diffraction patterns associated with the travel-time singularities follow by substituting (3.4) into (3.7). Asymptotic evaluation of (3.8) gives high-intensity diffraction patterns of Airy, Pearcey and higher-order type [19, 29]. For an input signal F(t) with basic source pulse frequency ω_o the echogram near a degenerate singularity $\underline{x}_s=\underline{x}_{sc}$ has the form

$$\psi(t,\underline{x}_c)\propto\omega_o^\alpha\{AF(t-T_{sc})+B\bar{F}(t-T_{sc})\} \qquad (3.8)$$

where $\alpha=\{1/6,\ 1/4,\ 3/10,\ 1/3\}$ are the singularity indices for $\{A_2,A_3,A_4,D_4^\pm\}$ and $\bar{F}$ is the Hilbert transform of F. Varying ω_o and measuring the rate of change of the received power permits one to identify and to classify the structural details of the reflecting surface near travel-time reversals. The ω_o-dependence in (3.8) follows from (3.7) and (3.4) by scaling. If the surface possesses edges and faults [27], the evolutes are amputated by the surface's shadow boundaries (Fig. 18). The effects produced by such discontinuities can be classified by six constraint catastrophe polynomials [30], [31]. For example, a fault in S (Fig. 18) is determined by $P=x^3+vx^2+ux$ $(x>0)$ and the diffraction pattern following from (3.7) is of combined Fresnel-Airy type: The intensity near the travel-time reversal decreases much faster than the one produced by a slope. Figs. 19 (a) and (b) show the travel-time curves when $\underline{x}_o$ varies on a line above and below the point B in Fig. 18, respectively. This effect is also present in Fig. 12.

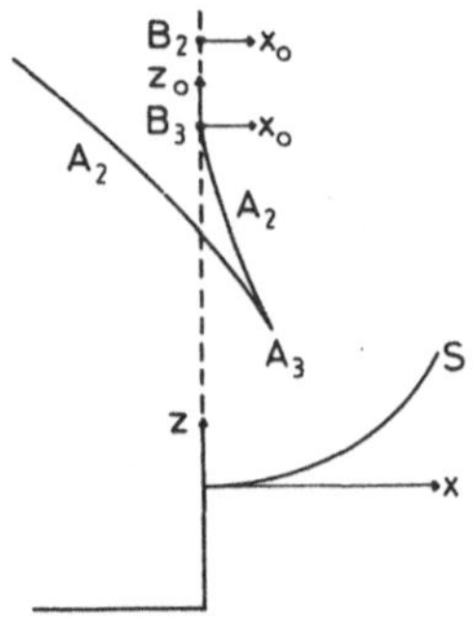

Fig. 18. Surface S with a fault

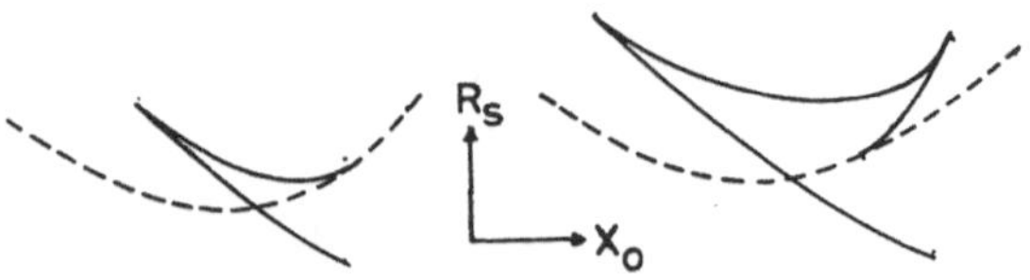

Fig. 19. Travel-times produced by the fault of Fig. 18

2. Nonlinear Phonon Focusing

The topological principles described above also govern the phenomenon of focusing of ballistic phonons in anisotropic media [32]-[34]. Suppose a burst of heat energy is created at a point of a crystal surface with a laser or electron beam. The heat produces phonons radiating in all directions. For a single crystal cooled to low temperatures (2K) the phonons travel in it ballistically, i.e., without scattering, at the speed of sound over distances of the order of centimeters. Since the elastic properties in a crystal are anisotropic, the constant-frequency (or energy) surface S_ω in wave-vector ($\underline{k}$-)space is in general not spherical and displays pronounced anisotropy depending on the material. Consequently, the group velocity $\underline{v}$, which defines the direction of the energy flux of the elastic (or phonon) wave, is also very anisotropic. The group velocity is normal to S_ω and thus $\underline{k}$ and $\underline{v}$ are not colinear. Therefore, considering an incoherent phonon source of frequency ω with isotropic angular distribution of $\underline{k}$-vectors, the phonon flux is channeled into intense beams along preferred crystal directions. This channeling of high-frequency phonons is called phonon focusing because time-integrated signals are detected as two-dimensional high intensity images exhibiting caustic effects. Singularity theory classifies the singular events which generically occur if the frequency ω varies in a nonlinear dispersion relation. The changes in that surface induce the topological changes in the caustics. The reason for such a classification being possible is that the image produced by ballistic phonons must be insensitive to small perturbations in the system, i.e., structurally stable.

Suppose a monochromatic point source of frequency ω generates phonons with wave-vectors $\underline{k}$ that propagate ballistically in a crystal whose anisotropy is described by a dispersion relation $\omega=\Omega(\underline{k})$. Then only those $\underline{k}$ contribute to the phonon field $u(\underline{r},\omega)$ at a point $\underline{r}$ in space which make up the constant-frequency surface $S=S_\omega:\omega=\Omega(\underline{k})=\text{const}$, i.e.,

$$u(\underline{r},\omega) \propto \int d\underline{k}\,\delta(\omega-\Omega(\underline{k}))e^{i\underline{k}\underline{r}} = \int dS\frac{e^{ir\phi}}{|\nabla\Omega(\underline{k})|} \tag{3.9}$$

for a given polarization mode. Here, $\underline{r}=r\hat{\underline{r}}$ with unit vector $\hat{\underline{r}}$, $\phi=\hat{\underline{r}}\cdot\underline{k}$ and the second integral is taken over S. Suppose first that the phonon's group velocity $\underline{v}=\nabla\Omega(\underline{k})$, with $\hat{\underline{v}}=\underline{v}/|\underline{v}|=\underline{n}$ the unit normal to S, has no zeros. Then the phonon flux is in the direction $\hat{\underline{r}}=\hat{\underline{v}}(\underline{k})$. The corresponding wave-vectors $\underline{k}$ on S are (for large r) those for which ϕ is stationary, $\underline{t}\cdot\nabla_{\underline{k}}\phi=0$ for vectors $\underline{t}$ tangent to S. Phonon focusing directions, i.e., angular caustics, come from the inflection points of S along a principal curvature line where the Gaussian curvature vanishes. These are the stationary points where the Hessian determinant of ϕ vanishes. Since the caustics are structurally stable, ϕ is equivalent to a Thom catastrophe polynomial $\phi=P$. This implies that the

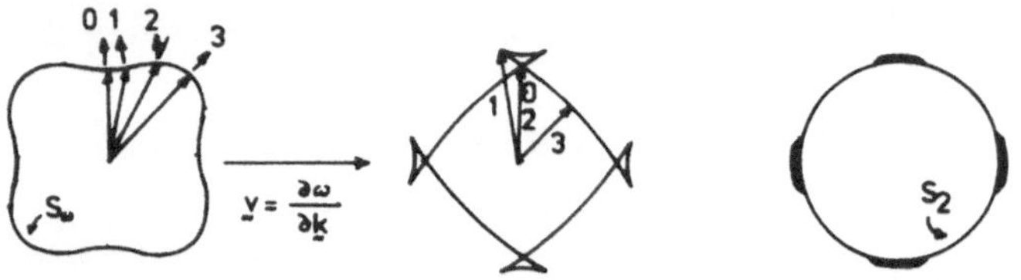

Fig. 20. Phonon focusing: constant frequency surface S_ω, group velocity plot, and multicovered unit sphere

caustics can be classified by the topological singularities of the Gaussian map G of the frequency surface S on the multiply covered sphere S^2 of the group velocity directions $\hat{\underline{v}}(\underline{k})$ on which the caustic images are observed (Fig. 20).

The fact that the structurally stable caustics are the images on S^2 of all points on S with zero Gaussian curvature has the following consequences. By Whitney's theorem (Sec. II), the only generic singularities on S^2 are folds and cusps. They are produced, respectively, by the line L on S with zero Gaussian curvature and its points of contact with the principal curvature line that has zero curvature on L. When the fold lines, which separate dark from bright regions on S^2, are crossed, the number of phonon $\underline{k}$-vectors changes by two while three $\underline{k}$-vectors coalesce at a cusp point. If the dispersion relation is linear, $\omega=c(\hat{\underline{k}})k$, the surfaces S are similar for different ω, so that folds and cusps are the only possible caustics. If, however, the dispersion relation $\omega=\Omega(\underline{k})$ is nonlinear, a variation of ω may change the topological type of S and that of the caustics when ω crosses a critical value ω_c. These sudden transitions are called critical events. The following events can occur: Swallowtail, i.e., the caustics of Fig. 21; Hyperbolic and elliptic umbilics when S posseses an umbilic point which is a point $\underline{k}$ with equal principal curvatures where the surface is locally spherical, i.e., the caustics of Fig. 22. Finally, there are but two possible degeneracies in a line L for $\omega=\omega_c$, viz., one that produces a caustic beak-to-beak event (Fig. 23a) and one giving rise to a caustic lip event (Fig. 23b). When crystal symmetries are involved, new types of caustics are generated. They have been described in [34]. A C_6-symmetric singular event is shown in Fig. 24. The Airy, Pearcey etc. diffraction patterns around the caustics follow by substituting P into (3.9) and evaluating the integral asymptotically [19]. By scaling one finds that on the caustic $|u|\propto r^{-\beta}$ for large r with the singularity indices $\beta=(5/6,\ 3/4,\ 7/10,\ 2/3)$ for the fold, cusp, swallowtail and umbilic events and $\beta=3/4$ for lips and beak-to-beak events. If the dispersion relation $\omega=\Omega(\underline{k})$ has stationary (Morse) points, where the group

Fig. 21. Swallowtail caustics

Fig. 22. (a) Hyperbolic and (b) elliptic umbilic caustics

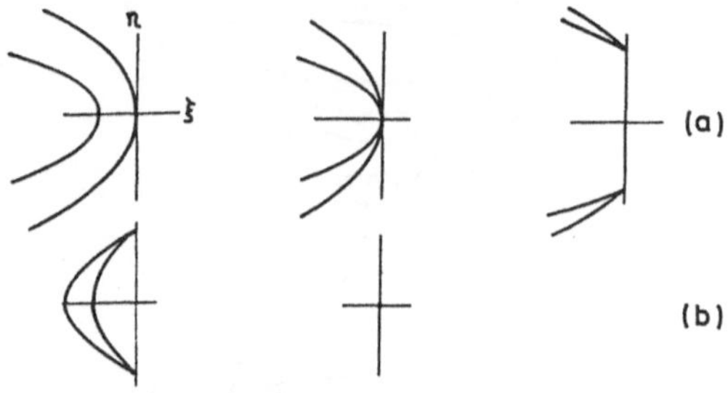

Fig. 23. (a) Beak-to-beak event, (b) lip event

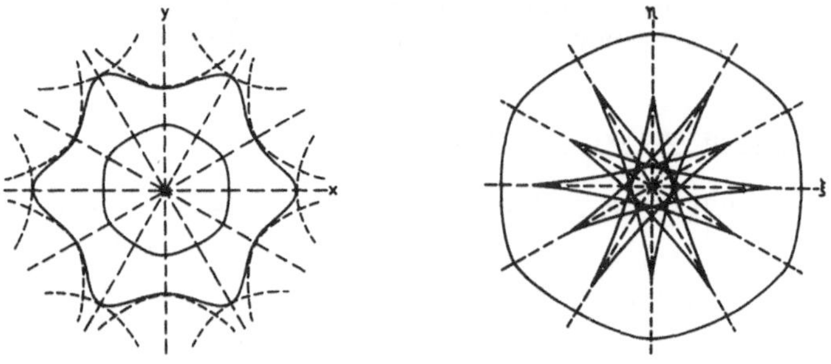

Fig. 24. Caustics in a crystal with C6-symmetry

velocity $\underline{v}$ vanishes, a new phenomenon springs up. Suppose that in Fig. 25 $(\omega_1,\underline{k}_1)$ is a minimum and $(\omega_2,\underline{k}_2)$ a saddle. Then the constant-frequency surface develops a second disconnected sheet when ω goes through ω_1, which gives rise to an overall flux enhancement. If ω goes through ω_2, the two surfaces coalesce in a way shown in Fig. 26. The caustic originating from the vertex of the cones is a great circle on S^2, but since $\underline{v}$ is zero at the point $\underline{k}_2$, the intensity is zero and, therefore, this circle must appear as a dark ring ("anticaustic"). On either side of this dark ring one expects two diffuse bright small circles whose positions correspond to the semiangles of the distorted cone near its vertex. We expect that with the recent development of tunable phonon sources the caustic singularities occuring at the specific critical frequencies ω_c will become accessible to detailed experimental investigation. The same methods can be applied to magnon focusing and so forth.

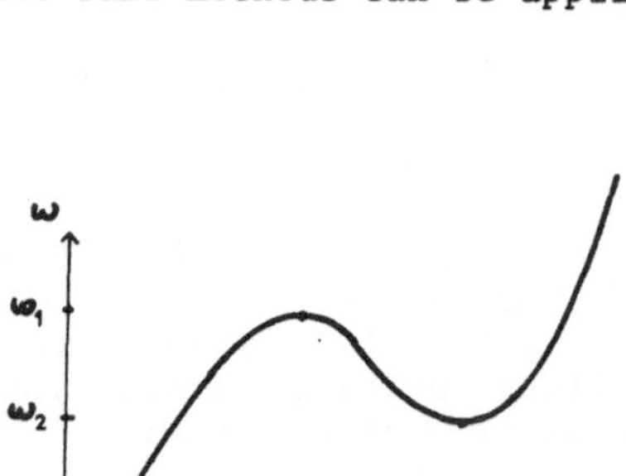

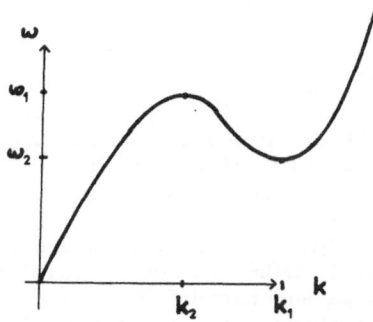

Fig. 25. Nonlinear phonon dispersion relation

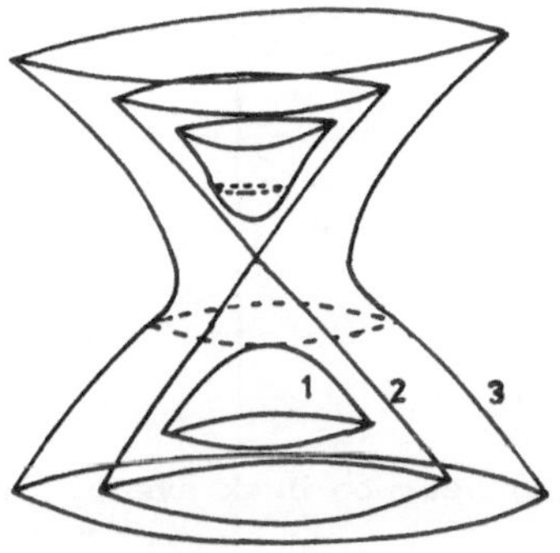

Fig. 26. Caustic cone produced by a nonlinear dispersion relation

3. Topological versus Analytical Singularities

Let $f(\underline{x},\lambda)=0$ define a parametrized hypersurface S in R^n. Then the well-known formula

$$I(\lambda) := \int_{R^n} \delta(f(\underline{x},\lambda))\psi dx = \int_{S:f=0} \frac{dS\ \psi}{|\nabla f(\underline{x},\lambda)|} \tag{3.10}$$

with a smooth $\psi(\underline{x})$ establishes a relation between the topological singularities of S (with $\nabla f=0$ where the integrand is largest) and the analytical singularities which $I(\lambda)$ will possess at those values of λ for which both f and ∇f vanish. Imperfect bifurcation theory discussed in Sec. II can now readily be applied to a variety of physical problems. (i) Spectral densities: Consider a vibrating system with a dispersion relation $f(\underline{k},\omega)=0$. Then, Equ. (3.10) represents e.g. the distribution function for the density of modes. In the Morse case this gives rise to van Hove singularities, while in the degenerate case one can classify the spectra by means of the normal forms of imperfect bifurcation theory. The same argument may be applied to orbitals in an electron plasma, to magnon interactions and so forth [17]. (ii) Dispersive wave phenomena: In a dispersive medium with a time-harmonic source with frequency ω and a dispersion relation $f(\omega,\underline{k})=0$, I is the wave function at point $\underline{x}$, $\psi=\exp(i\underline{k}\underline{x})$ and the integral goes over the surface S_ω in $\underline{k}$-space defined by f=0. Imperfect bifurcation theory yields then readily a classification for the bifurcation modes of waves in a dispersive medium [17], [18].

4. Nonlinear Conservation Laws

Consider the quasilinear wave equation

$$\psi_t + q_x(\psi) = \psi_t + c(\psi)\psi_x = 0 \tag{3.11}$$

for the amplitude $\psi(t,x)$, t being the time and $x \in R$, with initial condition $\psi(x,o)=f(x)$. The flux q depends nonlinearly on ψ and $q_x=c$ is the speed. The solution of (2.3) is given implicitly by $F(\psi):=\psi-f(x-c(\psi)t)=0$. Suppose that $c'(\psi)>0$. Then larger amplitudes propagate faster along the x-axis than lower ones and, if $f'<0$ on some section of the x-axis, one sees from $F'=0$ that the wave ψ starts to break at a critical time t_c. Consequently the shape of the wave is given by the catastrophe manifold S of Figure 2 with the identification $(x,u,v)\to(\psi,-t,x)$. The cusp in the (u,v)-plane is the envelope of the characteristics of (3.11). If $x \in R^3$, the only singularities the wave function can exhibit are folds, cusps and butterfly catastrophes. This model can serve as a basis for Zeldovich's theory of the large scale distribution of matter in the universe. When one asks for a single-valued but discontinuous solution

ψ of (3.11), the familiar Rankine-Hugoniot (RH) jump condition $s\cdot[\psi]=[q]$ must be invoked where $s=\dot{x}$ is the speed of the discontinuity and $[g]$ the jump of g across the discontinuity line (Maxwell convention). Equ. (3.11) also describes two phases in equilibrium thermodynamics when the volume V is kept constant. In this case $\psi=V$, q is the entropy, t the temperature, x the pressure and the RH condition becomes the Clausius-Clapeyron formula. In mechanics, the Hamiltonian H satisfies (3.11) if we identify $(t,x,\psi,c)=(p,q,H,-\partial_H J)$ where J is the action J(q,H). Then the RH condition reads $(H_2-H_1)/(J_2-J_1)=dp/dq\equiv\nu$. If we assume that $J_2-J_1=h$, the RH condition becomes just the Ritz combination principle. It is likely that also the geometric nature of phase transitions in gauge theories and problems of cosmology may be captured by singularity and bifurcation theory [16].

IV. BIFURCATION GEOMETRY AND STRUCTURE FORMATION

In this section we discuss topological problems posed by the formation of patterns and modes of behavior in physical systems whose order parameters or state variables undergo spontaneous changes if slowly varying and competing driving forces and imperfections enter conflicting regimes by crossing a bifurcation set. The general physical situation has been discussed in the Introduction. The objective here is to classify the transitions between interacting time-periodic and steady-state solutions of nonlinear evolution equations by using the normal forms of imperfect bifurcation theory (Sec. II). Two types of bifurcation of solutions of general evolution equations are fundamental: The bifurcation of a steady state with amplitude x and the Hopf bifurcation of a time-periodic solution with amplitude y from a stationary one. Interactions between them lead to secondary bifurcations of periodic solutions and to tertiary bifurcations of double-periodic motions lying on tori and eventually to chaotic motions. Such interactions occur if a control parameter λ in the evolution equations crosses some critical values. By a Lyapunov-Schmidt reduction we show below that x and y satisfy two algebraic normal form equations $a(x,\lambda,y^2)=0$ and $yb(x,\lambda,y^2)=0$. The solution of the first is a two-dimensional multisheeted surface $y=\phi(x,\lambda)$ in (x,y,λ)-space, that of the second equation is another surface $y=\psi(x,\lambda)$. The lines along which both surfaces intersect are the bifurcation diagrams of the evolution equations from which the behavior of the system can be inferred as λ varies. The intersection of the two surfaces may either be transversal so that any perturbation of a and b, i.e., a slight deformation or shifting of the surfaces causes no new types of intersections. In this case structural stability of the bifurcation diagram is ensured at the outset. However, if the two surfaces intersect with tangential contact, or just touch, then a slight deformation or shifting of them produces new intersections and so gives rise to new bifurcation diagrams which then, however, are stable against any further perturbations. These perturbations can be thought to be induced by variations of system-immanent imperfection parameters (impurity of material parameters etc.) in the original evolution equations. Since the forms of the perturbed polynomials a and b can be classified into a finite set by imperfect bifurcation theory, the problem of interacting spatial and temporal patterns has thus been reduced to a concatenation of the possible basic bifurcation diagrams. Then a variety of new phenomena, such as gaps in Hopf branches, periodic motions not stably connected to steady states, and the formation of islands are discovered which one can expect to find in general systems of evolution equations [9], [12], [35], [36]. Many of the new phenomena, predicted here on topological grounds alone, still await experimental confirmation.

1. Interacting Spatio-Temporal Patterns

Suppose that the dynamics of a physical system are governed by the system of evolution equations

$$\frac{\partial v}{\partial t} = F(v,\lambda) \tag{4.1}$$

where v is an element of an appropriate Banach space, $\lambda \in \mathbb{R}^1$ is a real bifurcation parameter, and F is a nonlinear operator defined on a neighborhood of the origin satisfying F(0,0)=0. We assume that the linearized operator

$$A = D_v F(0,0) \tag{4.2}$$

has a simple zero eigenvalue and, in addition, a simple pair of imaginary eigenvalues $\pm i\omega_o (\omega_o > 0)$. The remaining spectrum of A is assumed to be to the left of the imaginary axis. Equation (4.1) has the stationary solution v=0 for $\lambda=0$. When the externally controllable bifurcation parameter λ is varied away from zero, then, because of the nonlinearity of F, two basic types of solutions bifurcate from the trivial one, viz., steady-state solutions associated with the zero eigenvalue, and time-periodic or Hopf solutions associated with the eigenvalues $\pm i\omega_o$ of A. The nonlinearity of F causes these two solutions to interact and, since they tend to the trivial solution v=0 for $\lambda \to 0$, F has a degenerate bifurcation at (0,0). The degeneracy can be removed by subjecting F to small perturbations, representable, e.g., by additional imperfection parameters σ in F itself, $F \to F_\sigma$. This is achieved by stably unfolding the algebraic bifurcation equations to which (4.1) will be reduced. Then, as the unfolding parameters (which are functions of σ) are varied, zero and imaginary eigenvalues occur for different values of λ and, with the degeneracy so removed, new bifurcation phenomena are springing up which are structurally stable.

Since the linearization of (4.1) at (0,0) has $(2\pi/\omega_o)$-periodic solutions, we seek for periodic solutions of (4.1) near (0,0) with period $2\pi/\omega$ where ω is close to ω_o. Setting $s=\omega t$, $u(s)=v(s/\omega)$ so that u has period 2π in s, we rewrite (4.1) as

$$N(u,\lambda,\tau) \equiv \tau\frac{du}{ds} + Lu - R(u,\lambda) = 0 \tag{4.3}$$

where $\tau=\omega-\omega_o$, $L=\omega_o\frac{d}{ds}-A$ and $R(u,\lambda)=F(u,\lambda)-Au$. In the space of (2π)-periodic vector-valued functions u=u(s) the linear operator L has a three-dimensional nullspace spanned by the eigenfunctions $\phi=(\phi_1,\phi_2,\phi_3)$. We reduce the bifurcation problem (4.3) to an algebraic one by the Lyapunov-Schmidt method. To this end we set

$$u = u(s,z) = \sum_{i=1}^{3} z_i\phi_i + w \equiv z\phi + w(s) \tag{4.4}$$

with amplitudes $z=(z_1,z_2,z_3)$ and $\langle w,\phi_i\rangle=0$. Hence, $z\phi=Pu$ and $w=Qu$ with the projections $Pu=\sum_{i=1}^{3} \langle u,\phi_i^*\rangle\phi_i$, $Q=I-P$. Then, solving (4.3) is equivalent to solving the two equations PN=0 and QN=0. By standard implicit function arguments QN=0 has a smooth unique solution $w(z,\lambda,\tau)$, $w=O(z^2)$ which, when substituted into PN=0, yields the bifurcation equations

$$g_i \equiv \langle -\tau\frac{d}{ds}(z\phi+w) + R(z\phi+w,\lambda),\phi_i^*\rangle = 0 \tag{4.5}$$

Equs. (4.5) express the fact that, for (4.3) to be solvable, R must be orthogonal to the eigenfunctions of L (Fredholm alternative). Since (4.1) is invariant under time-translation, it suffices to choose $z_2=0$. Therefore, solving $g_2=0$ for $\tau=\tau(x,\lambda,y^2)$ and substituting into the remaining two equations, (4.5) reduces with $z_1=y$, $z_3=x$ to the degenerate algebraic system of bifurcation equations

$$G(x,\lambda,y) := \begin{pmatrix} a(x,\lambda,y^2) \\ yb(x,\lambda,y^2) \end{pmatrix} = 0 \qquad (4.6)$$

with $a(0,0,0)=b(0,0,0)=0$, $a_x(0,0,0)=0$. Equ. (4.6) describes the Z(2)-covariant interaction between the Hopf and steady-state solutions of (4.1).

The multivalued solutions of (4.6) are the bifurcation diagrams in (x,λ,y)-space. We classify them together with their stable perturbations by means of imperfect bifurcation theory (Sec. II). Equs. (4.6) possess two coupled types of solutions, viz., pure steady-state solutions with amplitude x determined by $\{a(x,\lambda,0)=0, y=0\}$, and periodic solutions with $y \neq 0$ obtained by the simultaneous solution of the equations $a(x,\lambda,y^2)=0$, $b(x,\lambda,y^2)=0$. The periodic solutions branch from the steady-state at a secondary Hopf bifurcation point and may further undergo tertiary bifurcations to tori [35], [37]. By the method of Sec. II we first change coordinates so that the qualitative topology of the bifurcation diagram G=0 is preserved (contact equivalence), the special role of the externally controllable bifurcation parameter λ is respected and G takes the simple polynomial forms given in [12], [35] from which the solutions of G=0 may easily be determined. To classify all the possible, stable and inequivalent, i.e., qualitatively different, bifurcation diagrams that may arise when a given $G(x,\lambda,y)$ is subjected to small perturbations which correspond to imperfections in F, the universal unfoldings of G are determined. The unfolding parameters of a universal unfolding of G are functions of the imperfection parameters σ in $F_\sigma(v,\lambda)$. Their number, the contact codimension of G, is a measure for the degree of complexity of the singularity. Hence, unfolding G displays the effects of all imperfections. The result is a finite list of generic perturbed bifurcation diagrams describing interacting Hopf (H) and steady-state problems (S) [35]. Of major interest for applications are special points in the bifurcation diagrams, viz., limit points and secondary bifurcation points (SB) which are here all Hopf bifurcation points, and tertiary bifurcations (T) from the Hopf branch to a torus. The stability properties are indicated by assigning to each branch of a diagram its stability symbol ((--)=stable, etc.), i.e., the signs of the real parts of the eigenvalues of the Jacobian DG. Fig. 27 shows the simplest secondary bifurcation of a Hopf branch (H) from a steady state in the (x,λ)-plane associated with the normal form $a=x^2+\varepsilon_2 y^2+\lambda=0$, $b=y(x-\gamma)=0$. Figure 28 shows two Hopf branches bifurcating from a steady state with bistability. In Figure 29 a tertiary bifurcation point T appears where a transition to a double-periodic solution occurs. There are many more diagrams exhibiting a variety of new phenomena such as gaps in Hopf branches, hysteresis between Hopf and steady-state branches, periodic solutions coming "out of nothing," i.e., not connected to steady states, and so forth. As an application we discuss the problem of optical bistability [39].

2. Optical Bistability and Self-Pulsing

Optically bistable systems are of interest not only because of their potential applications but also as laboratories for the study of self-pulsing,

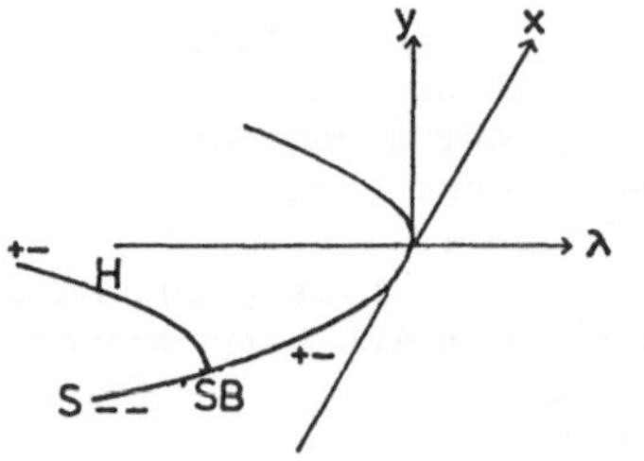

Fig. 27. Secondary Hopf bifurcation (SB) emerging from a steady state

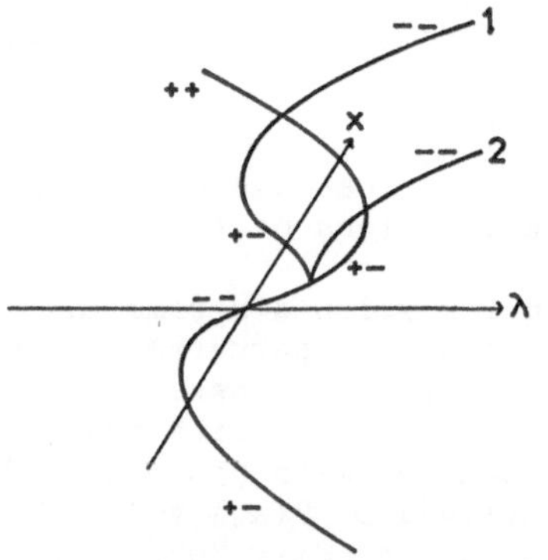

Fig. 28. Simultaneous Hopf bifurcations in y-direction originating at a hysteresis branch in the (x,λ)-plane

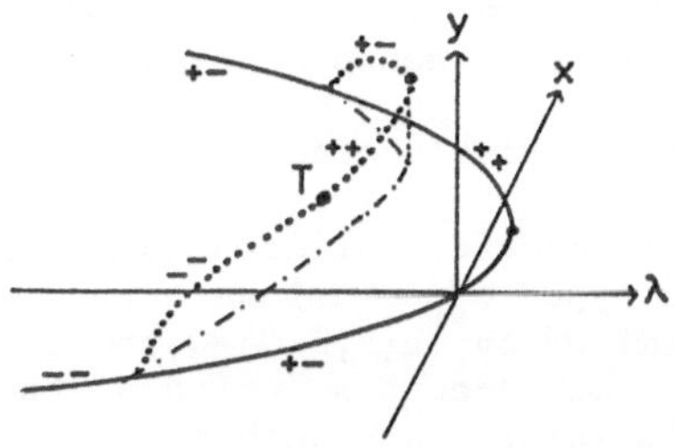

Fig. 29. Hopf-steady-state interaction leading to a torus point T

interacting spatio-temporal patterns and chaotic behavior [38]. Since the Maxwell-Bloch equations which govern optical bistability cannot be solved analytically except in limiting cases, a geometrical approach along the lines outlined above appears necessary. By a reinterpretation of known numerical facts, one can find a degenerate self-pulsing situation on the high-transmission branch, which in virtue of structural stability leads to new bifurcation diagrams [39]. Among them are first and second order transitions to the self-pulsing mode and the formation of an isola (Sec. II). Evidence is given for the existence of a codimension four organizing center (cf. II) which organizes both the bistable switching (which itself is organized by a hysteresis point) and the self-pulsing whose degeneracy is an H(7) degenerate [9] Hopf bifurcation (cf., Sec. IV.2). Unfolding this organizing center shows, apart from all known interaction phenomena in optical bistability, several new ones whose experimental or numerical detection is to be expected.

In optical bistability experiments a coherent laser beam is injected into an optical resonator filled with a homogeneously broadened two-level atomic system between two partially transmitting mirrors within a feedback loop. The refraction index of the medium depends on the incident intensity I which plays the role of the externally controllable parameter λ, $I=\lambda$. The transmitted intensity x becomes a nonlinear function of I described by the S-shaped hysteresis loops of Fig. 30. Jumps from the high to the low transmission branch and vice versa describe the phenomenon of optical bistability. It has been shown that for pure absorption a section of the upper branch of the hysteresis curve can become unstable (the (+ -)-part in Fig. 30). Then the system exhibits a limit cycle, i.e., an undamped time-periodic sequence of pulses branches from a cw-transmission. This is the Hopf bifurcation producing self-pulsing in optically bistable systems. We represent the amplitude of this time-periodic solution by y. Furthermore, there exists a region for the cavity mistuning and atomic detuning and values for the cooperation parameter such that the hysteresis curve can possess a point with a vertical tangent (a hysteresis point). Consequently there are two singularities in optical stability, viz., a hysteresis point in the stationary transmission branch and a degenerate Hopf bifurcation point on the upper branch. The simplest normal form $G(x,\lambda,y)$ in which both degeneracies coalesce is

$$G=\begin{pmatrix} x^3 + y^2 - \lambda + \beta x + \alpha \\ y(x^2+\lambda^2) + y(\delta+\gamma x) \end{pmatrix} \tag{4.7}$$

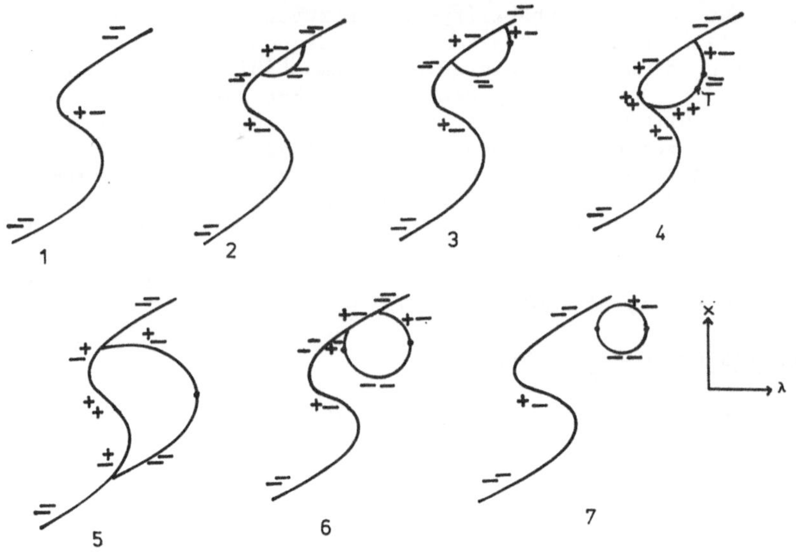

Fig. 30. Self-pulsing and island formation processes in optical bistability (from [15])

with unfolding parameters $\alpha,\beta,\gamma,\delta$ which depend on the system parameters [39]. G is an organizing center for optical bistability bifurcation in the sense of Section II. The bifurcation diagrams of G=0 have been discussed in [15]. Some of these are shown in Figure 30 in which the loops emerging from the hysteresis curve are projections onto the (x,λ)-plane of branches in (x,λ,y)-space. We see that among the new phenomena which one should expect to find in optical bistability experiments are the following: Bifurcation to self-pulsing from the high transmission branch (Fig. 30.2); two hysteresis loops between self-pulsing and the cw-regime (Fig. 30.6), i.e., periodic motions not stably connected to a stationary solution; torus bifurcation points marked by T or dots on the loops of Fig. 30.3-5 which may lead to chaotic behavior. The most surprising phenomenon to be encountered is the appearance of a stable island (cf. also Sec. II) representing a self-confined light-pulsing that is trapped between fixed incident intensities (Fig. 30.7). Most of these phenomena still require experimental confirmation. Their theoretical existence has been proven here on pure geometric grounds under the

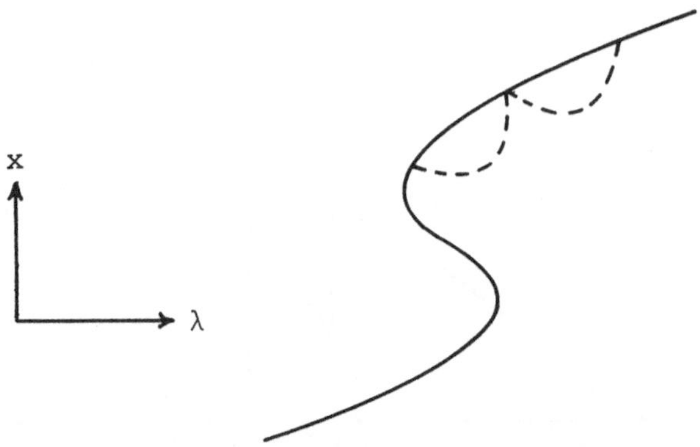

Fig. 31. Two simultaneous Hopf branches (dotted lines) emerging from a steady state (solid line)

hypothesis that optical bistability is a structurally stable phenomenon. Computer simulations indicate a Feigenbaum sequence to chaos [41] through period-doubling in optically bistable systems [38]. In addition a Ruelle-Takens scenario [40] may take place when the 2-torus T turns into a 3-torus by another Hopf bifurcation, thus leading to a strange attractor. We see now that another non-Feigenbaum route to chaos might spring up: A 2-torus is not structurally stable under perturbations breaking the SO(2) symmetry [42] but leads to transverse homoclinic orbits in whose neighborhood chaotic behavior takes place. Such unstable tori also arise when two simultaneous Hopf bifurcations emerge from a point on the steady-state curve and interact as shown in Figure 31. It would be interesting to discover these non-Feigenbaum routes to chaos in optically bistable systems.

3. Bifurcation Phenomena in Neurons

Suppose that A in (4.2) has only a pair of imaginary eigenvalues for $\lambda=0$. Then $a_x(o)\neq 0$ and $a=0$ has a unique solution whose substitution into b gives the single algebraic bifurcation equation

$$G(y,\lambda) = yb(\lambda,y^2) = 0, \tag{4.8}$$

with $b_\lambda(0,0)=0$, for the amplitude y of a periodic solution of (4.1) branching from a steady state. Equ. (4.8) describes the problem of degenerate Hopf bifurcations [9]. One of its most interesting bifurcation diagrams describes the formation of mushrooms and isolae (Sec. II). In fact, the H(7) bifurcation sequence used above in optical bistability is described by a special though highly important normal form into which G, (4.8), can be transformed.

Because of the topological universality of these bifurcation diagrams it is not surprising, though very remarkable, that they describe also the spike production mechanism in nerve cells. Indeed, it can be shown that basic aspects of the Hodgkin-Huxley equations can be reduced to the algebraic equation (4.8) where y is the amplitude of the time-periodic membrane potential and λ the input current [14]. Imperfection parameters are the temperature and material concentrations. The organizing center for the bifurcation sequence is found to be at a critical temperature T_c which is beyond the experimental range in which the neuron operates. A typical bifurcation diagram is shown in Figure 32. When, for $T<T_c$, λ increases, a temporal oscillation with amplitude $y>0$ starts from the steady state $y=0$ and then exhibits hysteresis at λ_c. As λ increases further, a gap occurs in the Hopf bifurcation diagram followed by an indefinitely continuing periodic branch "coming from nowhere." This implies that perturbations or fluctuations in the material concentrations can,

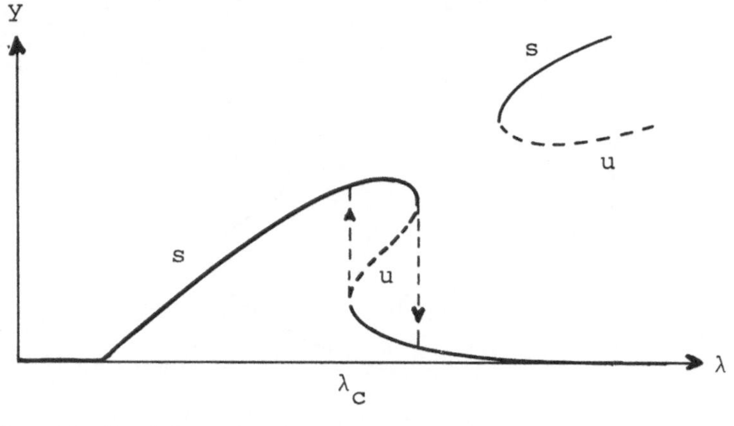

Fig. 32. Hopf hysteresis and gap between Hopf branches in neuronal activity

for sufficiently large λ, cause the system to jump from the rest state $y=0$ to an oscillatory state with increasing amplitude. This phenomenon still awaits experimental confirmation.

V. OUTLOOK

We have shown -- in terms of representative examples -- that, at the macroscopic and microscopic levels, the topological singularities and bifurcation processes deriving from the principle of structural stability determine the dominating phenomena and features observed in both structure formation and structure recognition. Because of their universality and classifiability these geometrical bifurcation processes also provide the basis for our understanding of the qualitative analogies that have been discovered in the critical behavior of nonlinear systems of various genesis. In particular, the theory of interacting Hopf-steady-state bifurcations outlined above describes the creation, evolution, interaction and destruction of spatio-temporal structures in physical, chemical and biological systems together with a variety of new phenomena. Their appearance in solidification and melting processes, in hydrodynamics, laser physics and so forth is presently in the center of interest. The fact that many of these phenomena, which were predicted on geometrical grounds alone, still await experimental confirmation, demonstrates both the power of this geometric-physical reasoning and the limits of our present experimental possibilities.

In pursuing this goal of a geometrization of physics further, a number of questions arise in which we have hardly scratched the surface. Among these is the problem of bifurcations with symmetry breaking. A typical example is the Benard problem in which symmetry breaking is caused by non-Boussinesq terms in the Navier-Stokes equations. A major problem is, of course, provided by bifurcation processes in particle physics and cosmology, and by the problem of structural stability in chaotic dynamics. These problems raise difficult but fascinating questions that promise a great challenge for future research.

ACKNOWLEDGMENTS

It is a pleasure to acknowledge many helpful discussions with D. Armbruster, G. Dangelmayr, C. Geiger, P. Haug, D. Lang, M. Neveling and F. J. Wright, and the support by the Stiftung Volkswagenwerk.

REFERENCES

[1] G. Nicolis and I. Prigogine, "Self-organization in nonequilibrium systems," Wiley (1977).

[2] H. Haken, "Synergetics" (2nd Edition), Springer (1982) and "Advanced Synergetics," Springer (1983).

[3] R. Thom, "Structural Stability and Morphogenesis," Benjamin (1975).

[4] T. Poston and I. Stewart, "Catastrophe Theory and Applications," Pitman, London (1978). I. Stewart, Physica 2D:245 (1981).

[5] E. C. Zeeman, "Catastrophe Theory," Addison-Wesley (1977). R. Gilmore, "Catastrophe Theory," Wiley (1981).

[6] M. V. Berry, "Les Houches Lectures," (1980).

[7] W. Güttinger and H. Eikemeier (Eds.), "Structural Stability in Physics," Springer (1979).

[8] V. I. Arnold, "Mathematical Methods of Classical Mechanics," Springer (1978).

[9] M. Golubitsky and D. Schaeffer, Comm. Pure Appl. Math. 32:21 (1979). M. Golubitsky and W. F. Langford, J. Diff. Equs. 41:375 (1981).

D. G. Schaeffer and M. Golubitsky, Arch. Rat. Mech. Anal. 75:315 (1981).
[10] M. Golubitsky and D. Schaeffer, Comm. Math. Phys. 67:205 (1979).
[11] D. H. Sattinger, Bull. Am. Math. Soc. 3:779 (1980).
[12] G. Dangelmayr and D. Armbruster, Proc. London Math. Soc. 46 (3):517 (1983).
[13] M. Golubitsky, B. L. Keyfitz and D. Schaeffer, Comm. Pure Appl. Math. 34:433 (1981).
[14] I. S. Labouriau, "Applications of Singularity Theory to Neurobiology," Univ. of Warwick preprint (1983).
[15] D. Armbruster, Z. Phys. B. 53:157 (1983).
[16] C. Geiger and W. Güttinger, in preparation (1984).
[17] D. Armbruster and W. Güttinger, in preparation (1984).
[18] G. Dangelmayr, J. Phys. A. L337-342 (1982).
[19] G. Dangelmayr and W. Güttinger, Geophys. J. Roy. Astro. Soc. 71:79 (1982).
[20] W. Güttinger and F. J. Wright, Topological Approach to inverse scattering in remote sensing, NATO-ARW proceedings, Reidel (1984).
[21] W. Berry, J. Phys. A, 5:272 (1972). F. J. Hilterman, Geophysics 40:745 (1975).
[22] N. Bleistein and J. Cohen, US Nav. Res. Rep. MS-R-7806 (1978).
[23] J. B. Keller and I. Papadakis, "Wave Propagation and underwater acoustics," Springer (1977).
[24] H. Baltes, "Inverse Scattering in Optics," Springer (1980).
[25] W. M. Boerner in [24].
[26] H. Eisenmenger, K. Lassmann and S. Döttinger (Eds.), "Phonon Scattering in Condensed Matter," Springer (1984).
[27] J. Achenbach and A. Norris, J. Acoust. Soc. Am. 70:165 (1981).
[28] F. J. Wright, preprint (1984), see also [20].
[29] M. V. Berry and C. Upstill in: "Progress in Optics," E. Wolf (Ed.), (1980).
[30] T. Poston and I. Stewart in [4].
[31] F. J. Wright in [28], G. Dangelmayr and W. Güttinger in [19].
[32] J. P. Wolfe, Phys. Today, 33:44 (1980).
[33] P. Taborek and D. Goodstein, Solid State Comm., 33:1191 (1980). Cf. also Ref. [26].
[34] D. Armbruster and G. Dangelmayr, Z. Phys. B, 52:87 (1983). D. Armbruster, G. Dangelmayr and W. Güttinger in [26].
[35] D. Armbruster, G. Dangelmayr and W. Güttinger, Imperfection Sensitivity of interacting Hopf and Steady-State Bifurcations and their Classification," Physica D (1984).
[36] D. Armbruster in Ref. [15].
[37] J. Guckenheimer and P. Holmes, "Nonlinear Oscillations, Dynamical Systems and Bifurcations of Vector Fields," Springer (1983).
[38] R. Bonifacio (Ed.), "Dissipative Systems in Quantum Optics," Springer (1982).
[39] See Ref. [15].
[40] F. Takens, Singularities of vector fields, Publications of the IHES, 43:47 (1974).
[41] J. P. Eckmann, Rev. Mod. Phys., 53:643 (1981). E. C. Zeeman, in: "New Directions in Applied Mathematics," Case Western Reserve University (1980).
[42] J. Guckenheimer, SIAM J. Math. Anal., 15:1 (1984).

QUANTUM DISTRIBUTION FUNCTIONS IN NON-EQUILIBRIUM STATISTICAL MECHANICS

R. F. O'Connell

Department of Physics and Astronomy
Louisiana State University
Baton Rouge, LA 70803

I. INTRODUCTION

Quantum distribution functions provide a means of expressing quantum mechanical averages in a form which is very similar to that for classical averages. Also, the Bloch equation for the density matrix for a canonical ensemble is replaced by a classical equation and, turning to dynamics, the von Neumann equation describing the time development of the density matrix is replaced by a classical equation which is similar in form to the Liouville equation but contains exactly the same information as the quantum von Neumann equation.

These distribution functions have been applied in essentially all areas of non-equilibrium statistical mechanics, from the more traditional branches to more recent branches such as quantum optics and synergetics. We have recently given an overview of such applications[1,2] and, in addition, participated in the presentation of a detailed review of the fundamentals underlying the theory of distribution functions.[3] Thus, to avoid repetition, we will concentrate here on a subject not treated in detail in our review viz. the inter-relationships between some of the more commonly used quantum distribution functions.

The original quantum distribution function was that introduced by Wigner[4] and which we will designate as $P_w(q,p)$, where q and p refer to position and momentum coordinates, respectively. Many authors have considered other distribution functions. In q-p phase space language the more commonly discussed functions include the Kirkwood function, the Husimi function, as well as the standard and anti-standard functions.

With applications to quantum optics in mind, Glauber[5] and Sudarshan[6] introduced a function, commonly designated as $P(\alpha)$, where α is a complex variable. This function has enjoyed wide usage. In complex variable language, other functions often considered include the $Q(\alpha)$ distribution function, the normal and anti-normal functions and the Cahill-Glauber[7] class of generalized functions. In addition, the New Zealand group has made extensive use of a function of two complex variables,[8] but this will not concern us here.

We have recently concerned ourselves with the general question of what criteria one should use in the selection of a quantum distribution function when one is faced with carrying out a specific calculation.[2,9] In a similar vein, one might have intermediate results pertaining to some calculation and might want to consider switching from the use of one distribution function to another. For this and other reasons it is of interest to obtain the relationships between the more commonly used functions. This will form the focus of our present considerations.

As we have emphasized in our previous investigations of this question[2,9] the optimum approach to the problem is via the use of characteristic functions, which are nothing more than Fourier transforms of distribution distributions. The next step is to write $P_w(q,p)$ in terms of the complex variable α. Then we obtain relationships between the various distribution functions, which are readily converted into relationships between the distribution functions, either in q-p or α language.

For simplicity, we treat a one-dimensional system in a pure state since the case of a mixture in multi-dimensions presents no essential complications. Since the touchstone of our considerations is the Wigner distribution, we will write down its properties in Section II, repeating of necessity some of the material in Ref. 9 since it is necessary for the purpose of establishing our notation. In Section III, we summarize the properties of the generalized distribution functions of the Cahill-Glauber[7] class and their relation to the Wigner distribution. In addition, we discuss the Husimi (smoothed) distribution function and point out its exact equivalence with the anti-normal distribution. In Section IV, we discuss the Glauber-Sudarshan $P(\alpha)$ function as well as the $Q(\alpha)$ function and also the relation between them and P_w. Then we show that the $Q(\alpha)$, Husimi, and anti-normal distributions are identical to each other and, similarly, the Glauber-Sudarshan and normal distributions are identical. Finally, in Section V, we introduce a new class of generalized distribution functions, with emphasis on two particular cases of same viz. the standard of anti-standard distributions, and we also demonstrate the equivalence of the latter to the Kirkwood distribution. In Section VI we discuss results.

II. THE WIGNER DISTRIBUTION FUNCTION

This is the original distribution function and is given by[4]

$$P_W(q,p) = (\pi\hbar)^{-1} \int \psi(q+y)^* \psi(q-y) e^{2ipy/\hbar} dy. \tag{1}$$

If we now introduce the characteristic function

$$C_W(\sigma,\tau) = \langle\psi| \exp[\tfrac{i}{\hbar}(\sigma\hat{q} + \tau\hat{p})]|\psi\rangle, \tag{2}$$

where the hats denote operators, then it follows[3] that P_w is the Fourier transform of C_w i.e.

$$P_W(q,p) = (2\pi\hbar)^{-2} \iint d\sigma d\tau \exp[-\tfrac{i}{\hbar}(\sigma q + \tau p)]\, C_W(\sigma,\tau). \tag{3}$$

Next we introduce the language of creation and annihilation operators by defining, as usual.

$$\hat{a} = \tfrac{1}{2}\left[\frac{\hat{q}}{q_o} + i\,\frac{\hat{p}}{p_o}\right] \tag{4a}$$

$$\hat{a}^{+} = \tfrac{1}{2}\left[\frac{\hat{q}}{q_o} - i\,\frac{\hat{p}}{p_o}\right], \tag{4b}$$

where

$$q_o = (\hbar/2m\omega)^{1/2} \tag{5a}$$

and

$$p_o = (m\hbar\omega/2)^{1/2} = m\omega q_o = (\hbar/2q_o). \tag{5b}$$

In addition, we define

$$\alpha = \tfrac{1}{2}\left[\frac{q}{q_o} + i\,\frac{p}{p_o}\right] \tag{6a}$$

$$\alpha^{*} = \tfrac{1}{2}\left[\frac{q}{q_o} - i\,\frac{p}{p_o}\right], \tag{6b}$$

and

$$\eta = (\sigma q_o - i\,\tau\,p_o)/\hbar \tag{7a}$$

$$\eta^{*} = (\sigma q_o + i\,\tau\,p_o)/\hbar, \tag{7b}$$

from which it follows that

$$\hbar^{-1}(\sigma\hat{q} + \tau\hat{p}) = \eta\hat{a} + \eta^{*}\hat{a}^{+} \tag{8a}$$

and

$$\hbar^{-1}(\sigma q + \tau p) = \eta\alpha + \eta^{*}\alpha^{*}. \tag{8b}$$

Hence

$$C_W(\sigma,\tau) = \langle\psi|\, \exp[\, i\,(\eta\hat{a} + \eta^{*}\hat{a}^{+})\,|\psi\rangle$$

$$\equiv C_W(\eta,\eta^{*}). \tag{9}$$

Substituting this result in Eq. (3), and using Eq. (8b) and the fact (see Eq. (7)) that $d\sigma\, d\tau = \hbar d\eta d\eta^{*}$, it follows that

$$P_W(q,p) = \hbar(2\pi\hbar)^{-2} \iint d\eta d\eta^* \exp[-i(\eta\alpha + \eta^*\alpha^*)] C_W(\eta,\eta^*)$$

$$\equiv \hbar^{-1} P_W(\alpha,\alpha^*) \equiv (2\hbar)^{-1} P_W(\alpha). \tag{10}$$

The normalization factors of $\hbar^{-1}$ and $(2\hbar)^{-1}$ in this equation are introduced to ensure that

$$\iint P_W(q,p)dq\, dp = 1 \tag{11a}$$

$$\iint P_W(\alpha,\alpha^*)d\alpha\, d\alpha^* = 1 \tag{11b}$$

$$\int P_W(\alpha)d^2\alpha = 1 \quad , \tag{11c}$$

where we have used the fact that (see Eqs. 5 and 6)

$$dq\, dp = \hbar\, d\alpha\, d\alpha^*$$
$$= 2\hbar\, d^2\alpha \quad , \tag{12}$$

where

$$\int d^2\alpha \equiv \iint d(\mathrm{Re}\alpha)\, d(\mathrm{Im}\alpha). \tag{13}$$

Similar relations to Eqs. (10) and (11) exist for the other distribution functions to be considered below.

III. THE CAHILL-GLAUBER CLASS OF GENERALIZED DISTRIBUTION FUNCTIONS

A generalized distribution function $P_g(q,p,s)$, where s is a parameter, is defined by replacing $C_W(\eta,\eta^*)$ in Eq. (9) by $C_g(\eta,\eta^*)$, defined as follows[7]

$$C_g(\eta,\eta^*) \equiv \langle \psi \mid \exp\left[\frac{s|\eta|^2}{2} + i(\eta\hat{a} + \eta^*\hat{a}^+)\right] |\psi\rangle. \tag{14}$$

It follows that for s = 0 we get the Wigner (or as oft-times called, the symmetric) characteristic function C_W given in Eq. (9). Also, making use of the Baker-Hausdorff theorem:

$$\exp(\hat{A})\exp(\hat{B}) = \exp(\tfrac{1}{2}[\hat{A},\hat{B}])\exp(\hat{A}+\hat{B}) \tag{15}$$

provided $[\hat{A},[\hat{A},\hat{B}]] = [\hat{B},[\hat{A},\hat{B}]] = 0$, and the fact that $[\hat{a},\hat{a}^+] = 1$, it follows that

$$C_a(\eta,\eta^*) = \langle\psi|\exp[i\eta\hat{a}]\exp[i\eta^*\hat{a}^+]|\psi\rangle \tag{16}$$

and

$$C_n(\eta,\eta^*) = \langle\psi|\exp[i\eta^*\hat{a}^+]\exp[i\eta\hat{a}]|\psi\rangle \ , \tag{17}$$

where C_a and C_n are the so-called anti-normal and normal characteristic functions, corresponding to taking s = - 1 and s = 1, respectively.

Using these relations, we have recently shown[9] that the generalized distribution function may be written in terms of P_w, as follows:

$$P_g(q,p) = \exp\left[-\frac{s}{2}q_0^2\frac{\partial^2}{\partial q^2} - \frac{s}{2}p_0^2\frac{\partial^2}{\partial p^2}\right] P_w(q,p). \tag{18}$$

Thus, in particular, for s=1 and -1 (corresponding to normal and anti-normal respectively) we obtain

$$P_{n,a}(q,p) = \exp\left[\mp\tfrac{1}{2}\left(q_0^2\frac{\partial^2}{\partial q^2} + p_0^2\frac{\partial^2}{\partial p^2}\right)\right], \tag{19}$$

a result which we have presented previously[2] without proof.

In addition, we have shown that

$$P_{-1}(q,p) \equiv P_a(q,p) = P_s(q,p), \tag{20}$$

where P_a is the anti-normal distribution and P_s is Husimi's distribution function.[10] The latter is a smoothed Wigner function, the smoothing being the Wigner function of the ground-state of a harmonic oscillator[11,12] It is everywhere non-negative and is defined as follows:

$$P_s(q,p) = (\pi\hbar)^{-1}\iint_{-\infty}^{\infty} P_w(q',p')\exp[-(q'-q)^2/\lambda]\exp[-\lambda(p'-p)^2/\hbar^2]\,dq'dp'$$

$$\text{where } \lambda = (\hbar/m\omega). \tag{21}$$

It is also of interest to re-write the phase-space operator appearing in Eq. (18) in terms of α. From Eq. (6) it follows that

$$\frac{\partial}{\partial q} = \frac{1}{2q_0}\left(\frac{\partial}{\partial\alpha} + \frac{\partial}{\partial\alpha^*}\right) \tag{22a}$$

$$\frac{\partial}{\partial p} = \frac{i}{2p_0}\left(\frac{\partial}{\partial\alpha} - \frac{\partial}{\partial\alpha^*}\right), \tag{22b}$$

so that

$$0(q,p,s) \equiv \exp[-\frac{s}{2}(q_0^2 \frac{\partial^2}{\partial q^2} + p_0^2 \frac{\partial^2}{\partial p^2})]$$

$$= \exp[-\frac{s}{2} \frac{\partial^2}{\partial\alpha\partial\alpha^*}]$$

$$\equiv 0(\alpha,\alpha^*,s). \tag{23}$$

Hence, using Eqs. (10) and (18), we obtain

$$P_g(\alpha,\alpha^*) = \exp[-\frac{s}{2} \frac{\partial^2}{\partial\alpha\partial\alpha^*}] P_W(\alpha,\alpha^*) \tag{24a}$$

which for the specific cases of s=1 and s=-1 gives

$$P_{n,a}(\alpha,\alpha^*) = \exp[\mp 1/2 \frac{\partial^2}{\partial\alpha\partial\alpha^*}] P_W(\alpha,\alpha^*), \tag{24b}$$

a result first obtained by Agarwal and Wolf [13] (noting that their superscripts A and N correspond to our subscripts "n" and "a", respectively).

Finally, we use our defining Eq. (21) for $P_s(q,p)$ in conjunction with Eqs. (6) and (10), to obtain

$$P_s(\alpha) = 2\hbar\, P_s(q,p)$$

$$= (2/\pi)\int d^2\beta\, P_W(\beta)\exp[-2|\beta-\alpha|^2], \tag{25}$$

with

$$\int P_s(\alpha)d^2\alpha = 1. \tag{26}$$

IV. THE P(α) AND Q(α) FUNCTIONS

As a preliminary to the introduction of these functions, we write down some well-known properties[14-16] of the coherent states $|\alpha\rangle$. They are defined in terms of the number states $|n\rangle$ as follows:

$$|\alpha\rangle = \sum_{n=0}^{\infty} \frac{\hat{a}^n}{(n!)^{1/2}} \exp[-1/2\ |\alpha|^2]|n\rangle. \tag{27}$$

It may be verified that

$$|\langle\alpha|\alpha\rangle|^2 = 1, \tag{28}$$

$$|\langle\alpha|\beta\rangle|^2 = \exp[-|\beta-\alpha|^2], \tag{29}$$

$$\pi^{-1}\int d^2\alpha\ |\alpha\rangle\langle\alpha| = I, \tag{30}$$

and

$$\pi^{-1}\int d^2\beta\ \exp[-|\beta-\alpha|^2] = \pi^{-1}\int d^2\beta\ |\langle\alpha|\beta\rangle|^2 = 1. \tag{31}$$

The Glauber-Sudarshan $P(\alpha)$ distribution function is defined by expanding the density matrix $\hat{\rho}$ as follows:[5-6]

$$\hat{\rho} = \int d^2\alpha\ P(\alpha)|\alpha\rangle\langle\alpha| \tag{32}$$

It then follows that, for an arbitrary operator $\hat{A}$,

$$\begin{aligned}\langle\hat{A}\rangle &= \mathrm{Tr}(\hat{\rho}\hat{A}) \\ &= \int d^2\alpha\ P(\alpha)\langle\alpha|\hat{A}|\alpha\rangle \\ &\equiv \int d^2\alpha\ P(\alpha)\ A(\alpha).\end{aligned} \tag{33}$$

Turning now to $Q(\alpha)$, it is a non-negative function, defined by Glauber[14] and Kano[15] according to the relation

$$Q(\alpha) = \pi^{-1}\langle\alpha|\hat{\rho}|\alpha\rangle. \tag{34}$$

Substituting Eq. (32) into Eq. (34), and using Eq. (29), we obtain a well-known relation between the Glauber-Sudarshan and $Q(\alpha)$ functions:[14]

$$Q(\alpha) = \pi^{-1}\int d^2\beta\ P(\beta)\exp[-|\alpha-\beta|^2]. \tag{35}$$

From Eqs. (35), (31), and (33) it may be verified that

$$\int P(\alpha)d^2\alpha = 1 \tag{36a}$$

$$\int Q(\alpha)d^2\alpha = 1. \tag{36b}$$

Next we wish to prove that $P(\alpha)$ is the same as $P_n(\alpha)$. First of all, we define

$$\hat{C}_n \equiv \exp[i\eta^* \hat{a}^+]\exp[i\eta\hat{a}], \tag{37}$$

so that, from Eqs. (17) and (33), we may write

$$C_n(\eta,\eta^*) = \langle\hat{C}_n\rangle$$

$$= \int d^2\alpha \, P(\alpha) \, C_n(\alpha) \quad , \tag{38}$$

i.e. C_n is the Fourier transform of $(1/2)P(\alpha)$. But, by definition, C_n is the Fourier transform of $P_n(\alpha,\alpha^*)$, which from Eq. (10) is the same as $(1/2)P_n(\alpha)$. Thus

$$P(\alpha) = P_n(\alpha), \tag{39}$$

i.e. the Glauber-Sudarshan and normal distributions are identical.

There also exists a well-known relation between P_w and $P(\alpha)$, which is most easily obtained by starting with the following relation between their corresponding characteristic functions (see Eq. (14))

$$C_w(\eta,\eta^*) = \exp[-\frac{|\eta|^2}{2}] \, C_n(\eta,\eta^*). \tag{40}$$

First, we take Fourier transforms on both sides of the equation and then we use the Fourier Convolution Theorem i.e.

$$F[f_1(x)] \, F[f_2(x)] = F[\int_{-\infty}^{\infty} f_1(x-y)f_2(y)dy]. \tag{41}$$

Then using the fact that $P_w(\alpha)$ and $P(\alpha)$ are the Fourier transforms of $C_w(\eta,\eta^*)$ and $C_n(\eta,\eta^*)$, respectively, and the fact that the Fourier transform of a Gaussian is also a Gaussian, we obtain

$$P_w(\alpha) = (2/\pi)\int d^2\beta \, P(\beta)\exp[-2|\alpha-\beta|^2], \tag{42}$$

a result first obtained by Glauber.[14)]

In a similar manner, analogous to Eq. (40), we may write

$$C_a(\eta,\eta^*) = \exp[-\frac{|\eta|^2}{2}]\, C_W(\eta,\eta^*), \tag{43}$$

from which it follows that

$$P_a(\alpha) = (2/\pi) \int d^2\beta\, P_W(\beta)\exp[-2|\alpha-\beta|^2]. \tag{44}$$

Thus, by comparison with Eq. (25), we see immediately that

$$P_a(\alpha) = P_s(\alpha). \tag{45}$$

An alternative proof of this relation may be found in Ref. 9. However, here we were also interested in obtaining the relations given by Eqs. (42) and (44), which we now combine together by substituting Eq. (42) into Eq. (44) to obtain

$$\begin{aligned} P_a(\alpha) &= (2/\pi)^2 \int\int d^2\beta\, d^2\gamma\, P(\gamma)\, \exp[-2|\beta-\alpha|^2 - 2|\beta-\gamma|^2] \\ &= (2/\pi)^2 \int\int d^2\beta\, d^2\gamma\, P(\gamma)\exp[-4|\beta|^2 + 2(\gamma^* + \alpha^*)\beta + 2(\gamma+\alpha)\beta^*] \\ &= (2/\pi)^2 \int d^2\gamma\, P(\gamma)\exp[-2|\gamma|^2 - 2|\alpha|^2] \\ &\times \{\int d^2\beta\, \exp[-4|\beta|^2 + 2(\gamma^* + \alpha^*)\beta + 2(\gamma+\alpha)]\} \ . \end{aligned} \tag{46}$$

The β integration may be carried out to give the result $(\pi/4)\exp(|\gamma+\alpha|^2)$ so finally we obtain

$$P_a(\alpha) = \pi^{-1} \int d^2\gamma\, P(\gamma)\, \exp[-|\gamma-\alpha|^2]. \tag{47}$$

But this is identical with the expression given for $Q(\alpha)$ in Eq. (35). Combining this result with Eq. (45) we deduce that

$$Q(\alpha) = P_a(\alpha) = P_s(\alpha). \tag{48}$$

In summary, the $Q(\alpha)$ distribution as defined in Eq. (34), the anti-normal distribution, and the Husimi (smoothed) distribution are identical. In addition, the Glauber-Sudarshan distribution, as defined in Eq. (32), is identical to the normal distribution.

V. A NEW CLASS OF GENERALIZED DISTRIBUTION FUNCTIONS

We define a new class of generalized distribution functions $P_G(q,p,b)$, where b is a parameter, by replacing $C_w(\sigma,\tau)$ in Eq. (2) by $C_G(\sigma,\tau)$, defined as follows

$$C_G(\sigma,\tau) \equiv \langle\psi|\exp[-\frac{ib\sigma\tau}{2\hbar} + \frac{i}{\hbar}(\sigma\hat{q}+\tau\hat{p})]|\psi\rangle. \tag{49}$$

Again using the Baker-Hausdorff theorem, and the fact that $[\hat{q},\hat{p}] = i\hbar$, we obtain

$$C_{ST}(\sigma,\tau) = \langle\psi|\exp[\frac{i}{\hbar}\sigma\hat{q}]\exp[\frac{i}{\hbar}\tau\hat{p}]|\psi\rangle, \tag{50}$$

$$C_{AS}(\sigma.\tau) = \langle\psi|\exp[\frac{i}{\hbar}\tau\hat{P}]\exp[\frac{i}{\hbar}\sigma\hat{q}]|\psi\rangle, \tag{51}$$

where C_{ST} and C_{AS} are the Fourier transforms of the so-called standard and anti-standard distributions, as discussed by Agarwal and Wolf,[13)] corresponding to taking b=1 and b=-1, respectively. Hence

$$P_G(q,p) = \exp[\frac{i\hbar}{2} b \frac{\partial^2}{\partial q\partial p}] P_w(q,p). \tag{52}$$

In addition, making use of Eq. (22), it follows that

$$P_G(\alpha,\alpha^*) = \exp[-\frac{b}{4}(\frac{\partial^2}{\partial\alpha^2} - \frac{\partial^2}{\partial\alpha^{*2}})] P_w(\alpha,\alpha^*). \tag{53}$$

From either Eqs. (52) or (53) it is apparent that

$$P_{ST} = P^*_{AS} . \tag{54}$$

We turn now to a demonstration of the equivalence of P_{AS} and P_K, where P_K is the Kirkwood distribution function[16)] given by

$$P_k(q,p) = (\pi\hbar)^{-1}\int\langle q|\hat{\rho}|q+2y\rangle e^{2ipy/\hbar}dy, \tag{55}$$

which for the case of a pure state reduces to

$$P_k(q,p) = (\pi\hbar)^{-1}\int\psi^*(q+2y)\ \psi(q)\ e^{2ipy/\hbar}dy . \tag{56}$$

If we now change variables from y to q' according to the relation $2y = q'-q$ then Eq. (56) may be re-written as

$$P_k(q,p) = (2\pi\hbar)^{-1}\int\psi^*(q')\psi(q)e^{ip(q'-q)/\hbar}dq' , \tag{57}$$

which is the form originally presented by Kirkwood.[16] Defining the wave-function in momentum space

$$\phi(p) = (2\pi\hbar)^{-1}\int\psi(q)e^{-iqp/\hbar}dq , \tag{58}$$

it is clear that Eq. (57) may also be re-expressed as

$$P_k(q,p) = \phi^*(p)\ \psi(q)e^{-ipq/\hbar}. \tag{59}$$

Still another form for $P_K(q,p)$ is obtained if in Eq. (55) we write $\hat{\rho} = |\psi\rangle\langle\psi|$. Then using the fact that

$$|q+2y\rangle = e^{-2iy\hat{p}/\hbar}|q\rangle, \tag{60}$$

and

$$|q\rangle\langle q| = (2\pi\hbar)^{-1} \int_{-\infty}^{\infty} d\sigma \exp[\tfrac{i}{\hbar}\sigma(\hat{q} - q)], \tag{61}$$

and replacing y by $(-\tau/2)$, we obtain

$$P_k(q,p) = (2\pi\hbar)^{-2} \int\int d\sigma d\tau \exp[-\tfrac{i}{\hbar}(\sigma q+\tau p)]\langle\psi|\exp[\tfrac{i}{\hbar}\tau\hat{p}]\exp[\tfrac{i}{\hbar}\sigma\hat{q}]|\psi\rangle \tag{62}$$

Comparing Eq. (51), for the Fourier transform of P_{AS}, with this result, we conclude that

$$P_k(q,p) = P_{AS}(q,p),$$

i.e. the Kirkwood and anti-standard functions are equivalent.

This concludes our discussion of the inter-relationship between the most commonly used quantum distribution functions used in non-equilibrium statistical mechanics, with the exception of distribution functions on four-dimensional phase space.[8] It is clear that they are all encompassed by the generalized functions P_g and P_G defined in Eqs. (18) and (52), respectively. Regarding the properties and use of these generalized functions, we have already obtained[9] an analytic result for the time development of P_g in the case of a non-zero potential V(q). The analogous result for P_G is presently under study as well as the corresponding classical equivalents of the Bloch equation for both P_g and P_G.[17] In addition, we remark that the definition of the Wigner distribution function has been extended recently to the case of spin one-half particles.[18] A similar extension for the generalized distribution functions considered here is readily obtainable.

VI. DISCUSSION OF OUR RESULTS

We have presented a detailed treatment of the relationships between the more commonly used distribution functions in quantum optics. In addition, we have introduced a new class of generalized distribution functions.

The relationships between the functions are of two types--integral and differential. In general, the emphasis in the literature has been on integral relations and these have been used particularly to derive Fokker-Planck equations for, say, the Wigner distribution from Fokker-Planck equations for the Glauber-Sudarshan or other distributions.[19] In our view, the latter goal may be achieved more easily with the use of the differential relationships (it's easier to differentiate than integrate!) and we plan to illustrate this point in a future publication.

ACKNOWLEDGMENTS

This research was partially supported by the Department of Energy under Contract No. DE-AS05-79ER10459 and Grant No. DE-FG05-84ER45135. The author would like to thank Mr. Lipo Wang for a critical reading of the manuscript.

REFERENCES

1. R. F. O'Connell, Found. Phys. 13, 83 (1983).
2. R. F. O'Connell in Laser Physics, ed. by J. D. Harvey and D. F. Walls (Springer-Verlag, 1983), p. 238. In the third line from the bottom of p. 244 and in the first line of Section 4 on p. 245, the subscript "a" should be replaced by "n".
3. M. Hillery, R. F. O'Connell, M. O. Scully and E. P. Wigner, Phys. Reports 106 (3), 121 (1984).
4. E. P. Wigner, Phys. Rev. 40, 749 (1932).
5. R. J. Glauber, Phys. Rev. 131, 2766 (1963); ibid. Phys. Rev. Lett. 10, 84 (1963).
6. E. C. G. Sudarshan, Phys. Rev. Lett. 10, 277 (1963).
7. K. E. Cahill and R. J. Glauber, Phys. Rev. 177, 1857, 1882 (1968).
8. P. D. Drummond and C. W. Gardiner, J. Phys. A 13, 2353 (1980); P. D. Drummond, C. W. Gardiner, and D. F. Walls, Phys. Rev. A24, 914 (1981).
9. R. F. O'Connell, L. Wang, and H. A. Williams, Phys. Rev. A, in press.
10. K. Husimi, Proc. Phys. Math. Soc. Japan 22, 264 (1940).
11. R. F. O'Connell and E. P. Wigner, Phys. Lett. 85A, 121 (1981).
12. N. D. Cartwright, Physica 83A, 210 (1976).

13. G. S. Agarwal and E. Wolf, Phys. Rev. D2, 2161 (1970); G. S. Agarwal, Phys. Rev. A4, 739 (1971).
14. R. J. Glauber, in Quantum Optics and Electronics, ed. by C. deWitt, A. Blanden, and C. Cohen-Tannoudji (Gordon and Breach, New York, 1965), p. 65.
15. Y. Kano, Proc. Phys. Soc. (Japan) 19, 1555 (1964); ibid. J. Math. Phys. 6, 1913 (1965).
16. J. G. Kirkwood, Phys. Rev. 44, 31 (1933), Eq. (22).
17. R. F. O'Connell and L. Wang, in preparation.
18. R. F. O'Connell and E. P. Wigner, Phys. Rev. A, in press.
19. See, for example, R. Graham in Quantum Statistics in Optics and Solid-State Physics, Vol. 66 of "Springer Tracts in Modern Physics", edited by G. Höhler (Springer-Verlag, New York, 1973), p. 80.

POSITIVE AND NEGATIVE JOINT QUANTUM DISTRIBUTIONS

Leon Cohen

Hunter College of The City University
695 Park Ave.
New York, N.Y. 10021

1. INTRODUCTION

There have been many attempts to write a proper joint quantum probability distribution that is positive and gives the correct individual quantum distributions of position and momentum (marginals). The general impression that has prevailed is that there is something in quantum mechanics which prevents the writing of such a distribution. The two most common reasons given why proper distributions cannot exist is the uncertainty principle and the fact that position and momentum do not commute. None the less we shall show that positive distributions do exist, are easy to obtain, and an infinite number of them can be generated readily.

The basic requirements that a proper joint distribution, $F(q,p)$, must satisfy are the marginals and positivity:

$$\int F(q,p)\, dp = |\psi(q)|^2 \quad , \tag{1.1}$$

$$\int F(q,p)\, dq = |\varphi(p)|^2 \quad , \tag{1.2}$$

$$F(q,p) \geq 0 \; , \tag{1.3}$$

where $\psi(q)$ and $\varphi(p)$ are the position and momentum wave functions respectively.⁺

In the next section we shall show how to obtain distributions satisfying the above properties. Subsequent to that we discuss the class of distributions which are bilinear in the wave function and which do not satisfy the positivity condition. The bilinear distributions are the ones

⁺ Notation: All integrals go from $-\infty$ to ∞ unless stated otherwise. We will restrict ourselves to the one dimensional case, extension to the multidimensional situation is straightforward. Operators of position and momentum will be denoted by capital Q and P. The time variable will be suppressed unless explicitly needed.

which have been most extensively studied. We shall give a general treatment for them and show how quantum mechanics can be formulated in terms of them. Although they are not positive definite, they have been very useful, and the study of their properties will guide us in developing the positive ones. After that we return to the positive distributions and examine some of their properties in detail.

2. POSITIVE DISTRIBUTIONS I

Consider the distributions[1-5] given by

$$F(q,p) = |\psi(q)|^2 \; |\varphi(p)|^2 \; \left[1 + c \; \rho(u,v)\right] \tag{2.1}$$

where c is a numerical constant and $\rho(u,v)$ is a function of two variables, u and v, which depend on position and momentum in the following way:

$$u(q) = \int_{-\infty}^{q} |\psi(q')|^2 \, dq' \tag{2.2}$$

$$v(p) = \int_{-\infty}^{p} |\varphi(p')|^2 \, dp' \tag{2.3}$$

The range of u and v are zero to one as q and p vary from minus infinity to plus infinity.

We now show how to obtain $\rho(u,v)$ so that the correct quantum mechanical marginal conditions are satisfied. Integrating with respect to momentum we have

$$\int F(q,p) \, dp = |\psi(q)|^2 + c|\psi(q)|^2 \int \rho(u,v) |\varphi(p)|^2 \, dp \; . \tag{2.4}$$

Therefore, $\rho(u,v)$ must be chosen such that

$$\int \rho(u,v) |\varphi(p)|^2 \, dp = 0 \; . \tag{2.5}$$

From Eq. (2.3) we have,

$$dv = |\varphi(p)|^2 \, dp \tag{2.6}$$

and the condition on $\rho(u,v)$ becomes

$$\int_0^1 \rho(u,v) \, dv = 0 \; . \tag{2.7}$$

Similarly, integration with respect to position yields

$$\int_0^1 \rho(u,v) \, du = 0 \; . \tag{2.8}$$

Hence, $\rho(u,v)$ can be any function of two variables which integrates to zero when integrated with respect to either variable. Such ρ's are easily obtained, an example being

$$\rho(u,v) = (2u-1) \; (2v-1) \; . \tag{2.9}$$

A particularly convenient way to generate them is as follows. Choose any function h(u,v) and normalize it to one,

$$\int_0^1 \int_0^1 h(u,v)\ du dv = 1 \ . \tag{2.10}$$

Define

$$h_1(u) = \int_0^1 h(u,v)\ dv \ , \tag{2.11}$$

$$h_2(v) = \int_0^1 h(u,v)\ du \ . \tag{2.12}$$

Then, taking

$$\rho(u,v) = h(u,v) - h_1(u) - h_2(v) + 1 \tag{2.13}$$

assures that Eqs. (2.7) and (2.8) are satisfied. To see this, integrate with respect to v,

$$\int_0^1 \rho(u,v) dv = \int_0^1 h(u,v)\ dv - h_1(u) - \int_0^1 h_2(v) dv + 1 = h_1(u) - h_1(u) - 1 + 1 = 0 \ , \tag{2.14}$$

and similarly for integration with respect u.

We now show how c can be chosen to make F positive. Of course the choice c equal to zero is the correlationless case which is positive but in general we want to take c not equal to zero. For F to be positive we must have

$$1 + c\rho(u,v) \ \geq \ 0 \ . \tag{2.15}$$

Now, suppose that the minimum and maximum values of ρ are $-\ell_1$ and ℓ_2 respectively. Then c can chosen to be any number such that

$$-1/\ell_2 \ \leq \ c \ \leq \ 1/\ell_1 \tag{2.16}$$

so that (2.15) will be satisfied.

We now summarize the method for obtaining the positive distributions. Choose any function h(u,v) and calculate $\rho(u,v)$, via Eq. (2.13). Use any constant c from the range given by Eq. (2.16). These two steps assure that the proper quantum distribution of position and momentum are satisfied and makes the joint distribution positive.

Finch and Groblicki[6] have shown that this method generates all possible positive distributions. We note that $\rho(u,v)$ can be a functional of the wave function by which we mean dependence beyond the ones reflected through u and v. Indeed, the choice of ρ will reflect what correlations we want to impose on F(q,p) and in general will depend on the properties of the system and circumstances under consideration.

3. BILINEAR DISTRIBUTIONS

Before continuing the development of the positive distributions we develop a general theory of distributions that depend on the wave function in a bilinear way.[7-12] The most widely used bilinear distribution is the Wigner[13-16] distribution. A number of others have been proposed and studied.[17-20] A unified treatment can be achieved by considering the following class of functions,[7-12]

$$F(q,p) = \frac{1}{4\pi^2} \iiint e^{-i\theta q - i\tau p + i\theta u} f(\theta,\tau)\ \psi^*(u - \tfrac{1}{2}\tau\hbar)\ \psi(u + \tfrac{1}{2}\tau\hbar)\ du\,d\tau\,d\theta \quad (3.1)$$

where the kernel, $f(\theta,\tau)$, is any function of two variables such that

$$f(0,\tau) = f(\theta,0) = 1\ . \quad (3.2)$$

In principle f can be a functional of the wave function and indeed the f which leads to the positive distribution given in the previous section does depend on as will be demonstrated later. For simplicity we will assume in this section that f is not a functional of the wave function which makes the distributions bilinear in the state, although some of the properties derived will apply even if f is state dependent.

The main condition that $F(q,p)$ should satisfy is that it yield correct individual quantum mechanical distributions of position and momentum, i.e. the marginals. This is assured by Eq. (3.2) as can be seen by integration with respect to p,

$$\int F(q,p)\ dp = \frac{1}{2\pi} \int e^{-i\theta q + i\theta u} f(\theta,0)\ |\psi(u)|^2\ du = |\psi(q)|^2 \quad (3.3)$$

and similarly for intergration with respect to position.

The general form can be cast in terms of the momentum wave function,

$$F(q,p) = \frac{1}{4\pi^2} \iiint e^{-i\theta q - i\tau p + i\tau u} f(\theta,\tau)\ \varphi^*(u + \tfrac{1}{2}\theta\hbar)\ \varphi(u - \tfrac{1}{2}\theta\hbar)\ du\,d\tau\,d\theta\ . \quad (3.4)$$

Also, by a change of variables it can be written in the following form

$$F(q,p) = \iint R(q,p,x,y)\ \psi^*(x)\ \psi(y)\ dx\,dy \quad (3.5)$$

where

$$R = \frac{e^{-i(y-x)p/\hbar}}{4\pi^2 h} \int e^{i\theta(2q-x-y)/2} f(\theta,(y-x)/\hbar)\ d\theta\ . \quad (3.6)$$

The general form given by Eq. (3.1) includes all possible $F(q,p)$'s since for any given $F(q,p)$ one can always find an $f(\theta,\tau)$ to generate it,

$$f(\theta,\tau) = \frac{\iint F(q,p)\ e^{+i\theta q + i\tau p}\ dq\,dp}{\int e^{+i\theta u}\ \psi^*(u - \tfrac{1}{2}\tau\hbar)\ \psi(u + \tfrac{1}{2}\tau\hbar)\ du}\ . \quad (3.7)$$

There have been many specific cases that have been investigated extensively. We now summarize them and show which kernels generate them. The most widely used is the Wigner[13] distribution

$$F_W(q,p) = \frac{1}{2\pi} \int e^{-i\tau p}\ \psi^*(q - \tfrac{1}{2}\tau\hbar)\ \psi(q + \tfrac{1}{2}\tau\hbar)\ d\tau \quad (3.8)$$

which is obtained from Eq. (3.1) by taking

$$f(\theta,\tau) = 1\ . \quad (3.9)$$

A number of authors[17-20] have studied

$$F_A(q,p) = \frac{1}{2\pi} \text{Real}\, \psi(q) \int e^{-i\tau p} \psi^*(q - \tau\hbar)\, d\tau$$

$$= \frac{1}{\sqrt{2\pi h}} \text{Real}\, \psi(q)\, \varphi^*(p)\, e^{-iqp/\hbar} \quad , \tag{3.10}$$

which is obtained from (3.1) by taking

$$f(\theta,\tau) = \cos(\theta\tau h/2) \quad . \tag{3.11}$$

The real part of (3.10) does not have to be taken. The distribution given by

$$F_B(q,p) = \frac{1}{\sqrt{2\pi\hbar}} \psi(q)\, \varphi^*(p)\, e^{-iqp/\hbar} \quad , \tag{3.12}$$

also satisfies the marginals and has the kernel

$$f(\theta,\tau) = e^{i\theta\tau\hbar/2} \quad . \tag{3.13}$$

This distribution, although complex, is particularly simple and has a number of interesting properties.[20] Its complex conjugate is also a distribution whose kernel is the complex conjugate of (3.13)

Another distribution[7] is

$$F_C(q,p) = \frac{1}{4\pi^2} \iiint e^{-i\theta q - i\tau p + i\theta u} \frac{\sin(\theta\tau\hbar/2)}{(\theta\tau\hbar/2)} \psi^*(u - \tfrac{1}{2}\tau\hbar)\, \psi(u + \tfrac{1}{2}\tau\hbar)\, du\, d\tau\, d\theta \tag{3.14}$$

which is obtained by using the kernel

$$f(\theta,\tau) = \frac{\sin(\theta\tau\hbar/2)}{(\theta\tau\hbar/2)} \quad . \tag{3.15}$$

New distributions can be obtained which are combinations of old ones since the product or ratio of two kernels is a kernel. Also, the sum of kernels weighted with coefficients a_i's

$$f(\theta,\tau) = \sum_i a_i f_i(\theta,\tau) \tag{3.16}$$

is also a kernel if

$$\sum_i a_i = 1 \quad . \tag{3.17}$$

In this case the distribution is given by the sum of the distributions weighted with the a_i's.

$$F(q,p) = \sum_i a_i F_i(q,p) \quad . \tag{3.18}$$

<u>Characteristic Function</u>

As in ordinary probability theory one can define the characteristic function or moment generating function for a distribution by

$$M(\theta,\tau) = \iint F(q,p)\, e^{+i\theta q + i\tau p}\, dq\, dp = \langle e^{+i\theta q + i\tau p} \rangle \quad . \tag{3.19}$$

A knowledge of $M(\theta,\tau)$ gives the distribution by inversion,

$$F(q,p) = \frac{1}{4\pi^2} \iint M(\theta,\tau)\, e^{-i\theta q - i\tau p}\, d\theta d\tau \ . \tag{3.20}$$

By expanding the exponent in Eq. (3.19), $M(\theta,\tau)$ can be expressed in terms of the moments

$$M(\theta,\tau) = \sum_{n,m=0}^{\infty} \frac{(i\theta)^n\,(i\tau)^m}{n!\ m!} \langle\, q^n p^m \,\rangle \ . \tag{3.21}$$

Also, the moments can be obtained from the characteristic function by

$$\langle\, q^n p^m \,\rangle = \frac{1}{i^{n+m}} \frac{\partial^{n+m}}{\partial\theta^n\, \partial\tau^m} M(\theta,\tau) \Big|_{\theta,\tau=0} \ . \tag{3.22}$$

Comparing Eq. (3.19) with Eq. (3.1) we see that for the quantum distributions

$$M(\theta,\tau) = f(\theta,\tau) \int e^{+i\theta u}\, \psi^*(u - \tfrac{1}{2}\tau\hbar)\, \psi(u + \tfrac{1}{2}\tau\hbar)\, du \ . \tag{3.23}$$

This can be rewritten to be the expected value of an operator. Making the change of variables $q = u + \tau h/2$ and using the fact that

$$e^{i\theta Q + i\tau P} = e^{i\theta\tau\hbar/2}\, e^{i\theta Q}\, e^{i\tau P} \tag{3.24}$$

and

$$e^{i\tau P}\, \psi(q) = \psi(q+\tau\hbar) \ , \tag{3.25}$$

we have

$$\begin{aligned} M(\theta,\tau) &= f(\theta,\tau) \int \psi^*(q)\, e^{i\theta(Q+\tau\hbar/2)}\, e^{i\tau P}\, \psi(q)\, dq \\ &= f(\theta,\tau)\ \langle\, e^{i\theta Q\, +\, i\tau P} \,\rangle \ . \end{aligned} \tag{3.26}$$

It should be noted that the characteristic functions defined by Eq. (3.23) are not proper since they are not derivable from a positive joint distribution. The characteristic function we will derive for the positive distribution will be proper.

Recovery of Wave Function from Distribution

For an arbitrary $F(q,p)$ a wave function may not exist (see section on representability) but if it does it can be obtained from the distribution uniquely. From Eq. (3.1) we have

$$\psi^*(u - \tfrac{1}{2}\tau\hbar)\, \psi(u - \tfrac{1}{2}\tau\hbar) = \frac{1}{2\pi} \iiint \frac{F(q,p)\, e^{i\theta q + i\tau p - i\theta u}}{f(\theta,\tau)}\, dq dp d\theta \tag{3.27}$$

or

$$\psi^*(x)\, \psi(y) = \frac{1}{2\pi} \iiint \frac{F(q,p)\, e^{i\theta(q-(x+y)/2) + i(y-x)p/\hbar}}{f(\theta,(y-x)/2)}\, dq dp d\theta \ . \tag{3.28}$$

Since x and y are arbitrary we can choose $x = 0$ to obtain

$$\psi(y) = \frac{1}{2\pi R} \iiint \frac{F(q,p)\, e^{i\theta(q-y/2)+iyp/\hbar}}{f(\theta,y/2)}\, dq dp d\theta \,, \tag{3.29}$$

where R is a normalizing factor,

$$R^2 = \int F(0,p)\, dp \quad . \tag{3.30}$$

For various f's considerable simplification occurs. For the Wigner distribution,

$$\psi^*(x)\, \psi(y) = \int F((x+y)/2,p)\, e^{i(y-x)p/\hbar}\, dp \,, \tag{3.31}$$

$$\psi(y) = \frac{1}{2\pi R} \int F(y/2,p)\, e^{iyp/\hbar}\, dp \,. \tag{3.32}$$

Relationship between Different Distributions

Suppose we have two distributions corresponding to two different kernels $f_1(\theta,\tau)$ and $f_2(\theta,\tau)$ respectively. Their moment generating functions are related in a simple way. From Eq. (3.23)

$$\frac{M_1(\theta,\tau)}{f_1(\theta,\tau)} = \frac{M_2(\theta,\tau)}{f_2(\theta,\tau)} \quad . \tag{3.33}$$

The distributions can be related by using Eq. (3.20)

$$F_1(q,p) = \frac{1}{4\pi^2} \iiiint \frac{f_1(\theta,\tau)}{f_2(\theta,\tau)}\, e^{i\theta(q'-q)+i\tau(p'-p)}\, F_2(q',p')\, d\theta d\tau dq' dp' \tag{3.34}$$

Expressions similar to the above arise often and they can usually be put in an operator form. We prove two general theorems relating to these types of expressions. We show that

$$\frac{1}{4\pi^2} \iiiint g(\theta,\tau)\, e^{ia\theta(q-q')+ib\tau(p-p')}\, F(q',p')\, d\theta d\tau dq' dp'$$

$$= \frac{1}{|ab|}\, g\left(-\frac{i}{a}\frac{\partial}{\partial q},\ -\frac{i}{b}\frac{\partial}{\partial p}\right) F(q,p) \,, \tag{3.35}$$

where g is any function and a and b are constants. Expanding g in a Taylor series

$$g = \sum_{rs} a_{rs}\, \theta^r\, \tau^s \tag{3.36}$$

the left hand side of Eq. (3.35) equals

$$= \frac{1}{4\pi^2} \iiiint \sum a_{rs}\, \theta^r\, \tau^s\, e^{ia\theta(q-q')+ib\tau(p-p')}\, F(q',p')\, d\theta d\tau dq' dp'$$

$$= \frac{1}{4\pi^2} \iiiint \sum a_{rs} \left(-\frac{i}{a}\frac{\partial}{\partial q}\right)^r \left(-\frac{i}{b}\frac{\partial}{\partial p}\right)^s e^{ia\theta(q-q')+ib\tau(p-p')}\, F(q',p')\, d\theta d\tau dq' dp'$$

$$= \frac{1}{|ab|} \iiiint \sum a_{rs} \left(-\frac{i}{a}\frac{\partial}{\partial q}\right)^r \left(-\frac{i}{b}\frac{\partial}{\partial p}\right)^s \delta(q-q')\ \delta(p-p')\ F(q',p')\ d\theta d\tau dq' dp' \tag{3.37}$$

which equals the right hand side of (3.35). Similarly, we also show that

$$\frac{1}{4\pi^2} \iiiint g(q',p')\ e^{i\theta(q-q')+i\tau(p-p')}\ h(q + \tfrac{1}{2}\tau\hbar, p - \tfrac{1}{2}\theta\hbar)\ d\theta d\tau dq' dp'$$

$$= \quad g\left(q + \tfrac{1}{2}i\hbar\frac{\partial}{\partial p}, p - \tfrac{1}{2}i\hbar\frac{\partial}{\partial q}\right) h(q,p) \ . \tag{3.38}$$

The left hand side can be written as

$$\frac{1}{4\pi^2} \iiiint g(q',p')e^{+i\theta(q-q')+i\tau(p-p')}\ \exp(-\tfrac{1}{2}\hbar\theta\frac{\partial}{\partial p})\ \exp(+\tfrac{1}{2}\hbar\tau\frac{\partial}{\partial q})\ h(q,p)\ d\theta d\tau dq' dp'$$

$$= \iint g(q',p')\ \delta(q - q' + \tfrac{1}{2}i\hbar\frac{\partial}{\partial p})\ \delta(p - p' - \tfrac{1}{2}i\hbar\frac{\partial}{\partial q})\ dq'dp'\ h(q,p) \tag{3.39}$$

which equals the right hand side of Eq. (3.38).

Returning to expression (3.34) which connects two different distributions and using Eq. (3.35) we have

$$F_1(q,p) = \frac{f_1(i\partial/\partial q, i\partial/\partial p)}{f_2(i\partial/\partial q, i\partial/\partial p)}\ F_2(q,p) \ . \tag{3.40}$$

If we take F_2 to be the Wigner distribution then

$$F(q,p) = f(i\partial/\partial q, i\partial/\partial p)\ F_W(q,p) \ . \tag{3.41}$$

We should also note that by taking

$$f(\theta,\tau) = 1 + c\ \theta^n \tau^m \tag{3.42}$$

where c is any constant we obtain

$$F(q,p) = F_W(q,p) + c\ \frac{\partial^{n+m}}{\partial q^n\ \partial p^m}\ F_W(q,p) \tag{3.43}$$

Hence, one can add any derivative to the Wigner function and still have a joint distribution.

Constraints on the Kernel

One may wish to impose certain other seemingly reasonable properties on F(q,p) beyond the requirement that it yield the correct marginals. Since the bilinear distributions can never be interpreted as proper distributions perhaps one should question whether the imposition of other conditions, as if the distributions were proper, is wise or necessary. Perhaps the best approach is to use the distribution that is easiest for the particular problem at hand. Indeed, for most of the practical calculations done with these distributions one eventually obtains an average by integrating a function of position or momentum with the distribution. In such a case it makes no difference which distribution is used as long as

the marginals are satisfied and indeed even a complex one may be used because the final answer will be real. Of course if one calculates expected values of mixed quantities, then, which distribution is used does make a difference but in that case distributions which can go negative will almost always give an absurd result anyway. For example, the expected value of a square of a mixed quantity can be negative.[3,12] None the less, it is interesting to see what properties of F can be imposed by constraining the kernel.[14,21]

Marginals. We have already shown that the marginals are satisfied if

$$f(0,\tau) = f(\theta,0) = 1 \ . \qquad (3.44)$$

The above condition also implies that

$$\frac{\partial^n f(\theta,\tau)}{\partial\theta^n}\Big|_{\theta=0} = \frac{\partial^n f(\theta,\tau)}{\partial\tau^n}\Big|_{\tau=0} = 0 \qquad (3.45)$$

Positivity. Wigner[14] has proven that bilinear forms cannot be positive definite and hence for F to be positive f must be a functional of the wave function. As we will demonstrate in section 5 the kernel for the positive distributions is a functional of the wave function.

Reality. Taking the complex conjugate of Eq. (3.1) and making a change of variables we obtain

$$F^*(q,p) = \frac{1}{4\pi^2}\iiint e^{+i\theta q+i\tau p-i\theta u}\, f^*(\theta,\tau)\, \psi(u - \tfrac{1}{2}\tau\hbar)\, \psi^*(u + \tfrac{1}{2}\tau\hbar)\, du\,d\tau\,d\theta$$

$$= \frac{1}{4\pi^2}\iiint e^{-i\theta q-i\tau p+i\theta u}\, f^*(-\theta,-\tau)\, \psi^*(u - \tfrac{1}{2}\tau\hbar)\, \psi(u + \tfrac{1}{2}\tau\hbar)\, du\,d\tau\,d\theta \qquad (3.46)$$

Hence, for reality we must have

$$f(\theta,\tau) = f^*(-\theta,-\tau) \ . \qquad (3.47)$$

Normalization. This is automatically satisfied if the marginals are satisfied. If the marginals are not satisfied then we must have

$$f(0,0) = 1 \ . \qquad (3.48)$$

Galilei invariance. If $\psi_b(q) = \psi_a(q+d)$ then $F_b(q,p)$ should equal $F_a(q+d,p)$. From Eq. (3.1)

$$F_b(q,p) = \frac{1}{4\pi^2}\iiint e^{-i\theta q-i\tau p+i\theta u}\, f(\theta,\tau)\, \psi_a^*(u+d - \tfrac{1}{2}\tau\hbar)\,\psi_a(u+d + \tfrac{1}{2}\tau\hbar)\, du\,d\tau\,d\theta$$

$$= \frac{1}{4\pi^2}\iiint e^{-i\theta q-i\tau p+i\theta(u-d)}\, f(\theta,\tau)\, \psi_a^*(u - \tfrac{1}{2}\tau\hbar)\psi_a(u + \tfrac{1}{2}\tau\hbar)\, du\,d\tau\,d\theta$$

$$= F_a(q+d,p) \ . \qquad (3.49)$$

Therefore any $f(\theta,\tau)$ will do. Also, if $\psi_b(q) = e^{-iqp'/\hbar}\,\psi_a(q)$ then $F_b(q,p)$ should equal $F_a(q,p+p')$. Direct substitution in Eq. (3.1) leads again to the conclusion that any $f(\theta,\tau)$ will work.

Space reflection. If $\psi_b(q) = \psi_a(-q)$ then $F_b(q,p)$ should equal $F_a(-q,-p)$. From Eq. (3.1)

$$F_b(q,p) = \frac{1}{4\pi^2} \iiint e^{-i\theta q - i\tau p + i\theta u} f(\theta,\tau) \psi_a^*(-u + \tfrac{1}{2}\tau\hbar) \psi_a(-u - \tfrac{1}{2}\tau\hbar) \, du d\tau d\theta$$

$$= \frac{1}{4\pi^2} \iiint e^{-i\theta q - i\tau p + i\theta u} f(-\theta,-\tau) \psi_a^*(u + \tfrac{1}{2}\tau\hbar) \psi_a(u - \tfrac{1}{2}\tau\hbar) \, du d\tau d\theta \qquad (3.50)$$

and hence we must have,

$$f(\theta,\tau) = f(-\theta,-\tau) \ . \qquad (3.51)$$

Time reflection. If $\psi_b(q) = \psi_a^*(q)$ then $F_b(q,p)$ should equal $F_a(q,-p)$

$$F_b(q,p) = \frac{1}{4\pi^2} \iiint e^{-i\theta q - i\tau p + i\theta u} f(\theta,\tau) \psi_a(u - \tfrac{1}{2}\tau\hbar) \psi_a^*(u + \tfrac{1}{2}\tau\hbar) \, du d\tau d\theta$$

$$= \frac{1}{4\pi^2} \iiint e^{-i\theta q + i\tau p + i\theta u} f(\theta,-\tau) \psi_a^*(u - \tfrac{1}{2}\tau\hbar) \psi_a(u + \tfrac{1}{2}\tau\hbar) \, du d\tau d\theta \qquad (3.52)$$

and hence

$$f(\theta,\tau) = f(\theta,-\tau) \ . \qquad (3.53)$$

4. FORMULATION OF QUANTUM MECHANICS IN TERMS OF JOINT DISTRIBUTIONS

We shall now discuss a number of topics which shows that quantum mechanics can be formulated in terms of joint distributions. Any joint distribution can be used.

Expectation Value

One of the main reasons for introducing joint distribution functions is that one wishes to calculate quantum expectation values by phase space integration,

$$\iint g(q,p) \, F(q,p) \, dq dp = \int \psi^*(q) \, G(Q,P) \, \psi(q) \, dq \qquad (4.1)$$

where $g(q,p)$ is the classical function and $G(Q,P)$ is the quantum operator corresponding to it. To obtain the relation between them consider

$$\langle g(q,p) \rangle = \frac{1}{4\pi^2} \int g(q,p) \, e^{-i\theta q - i\tau p + i\theta u} f(\theta,\tau) \psi^*(u - \tfrac{1}{2}\tau\hbar) \psi(u + \tfrac{1}{2}\tau\hbar) \, du d\theta d\tau dq dp$$

$$= \iiint \gamma(\theta,\tau) \, f(\theta,\tau) \, e^{+i\theta u} \psi^*(u - \tfrac{1}{2}\tau\hbar) \psi(u + \tfrac{1}{2}\tau\hbar) du d\tau d\theta \qquad (4.2)$$

where we have set $\gamma(\theta,\tau)$ equal to the Fourier transform of $g(q,p)$,

$$\gamma(\theta,\tau) = \frac{1}{4\pi^2} \iint g(q,p) \, e^{-i\theta q - i\tau p} \, dq dp \ . \qquad (4.3)$$

Using

$$\int e^{+i\theta u} \psi^*(u - \tfrac{1}{2}\tau\hbar) \psi(u + \tfrac{1}{2}\tau\hbar) du = \int \psi^*(q) \, e^{+i\theta Q + i\tau P} \psi(q) \, dq \ , \qquad (4.4)$$

we have

$$\langle g(q,p) \rangle = \iiint \gamma(\theta,\tau) \, f(\theta,\tau) \, \psi^*(q) \, e^{i\theta Q + i\tau P} \psi(q) d\tau d\theta dq \ , \qquad (4.5)$$

and therefore we can take

$$G(Q,P) = \iint \gamma(\theta,\tau)\ f(\theta,\tau)\ e^{i\theta Q+i\tau P}\ d\tau d\theta \tag{4.6}$$

$$= \iint e^{+i\theta\tau\hbar/2}\ \gamma(\theta,\tau)\ f(\theta,\tau)\ e^{i\theta Q}\ e^{i\tau P}\ d\tau d\theta \tag{4.7}$$

$$= \iint e^{-i\theta\tau\hbar/2}\ \gamma(\theta,\tau)\ f(\theta,\tau)\ e^{i\theta P}\ e^{i\tau Q}\ d\tau d\theta\ . \tag{4.8}$$

If we define $G_Q(Q,P)$ to be $G(Q,P)$ after all the Q factors are made to precede the P factors then

$$G_Q(q,p) = \iint e^{i\theta\tau\hbar/2}\ \gamma(\theta,\tau)\ f(\theta,\tau)\ e^{i\theta q}\ e^{i\tau p}\ d\tau d\theta$$

$$= \frac{1}{4\pi^2} \iiiint f(\theta,\tau)\ e^{i\theta\tau\hbar/2}\ e^{+i\theta(q-q')+i\tau(p-p')}\ g(q',p')\ d\theta d\tau dq'dp' \tag{4.9}$$

which can be written in the operator form by using Eq. (3.35),

$$G_Q(q,p) = f\left(\frac{1}{i}\frac{\partial}{\partial q}, \frac{1}{i}\frac{\partial}{\partial p}\right) \exp\ (-i\hbar/2\ \partial^2/\partial q\partial p)\ g(q,p)\ . \tag{4.10}$$

In the case where g is a function of only q or p or the sum of such functions

$$G(Q) = g(Q) \longleftrightarrow g(q) \tag{4.11}$$

$$G(P) = g(P) \longleftrightarrow g(p) \tag{4.12}$$

$$G_1(Q) + G_2(P) = g_1(Q) + g_2(P) \longleftrightarrow g_1(q) + g_2(p) \tag{4.13}$$

where the double arrows signifies correspondence according to Eq. (4.6).

Conversely, if we have the quantum operator and wish to obtain a classical function which will assure Eq. (4.1) we must invert Eq. (4.9).

$$\gamma(\theta,\tau) = \frac{1}{4\pi^2}\frac{e^{-i\theta\tau\hbar/2}}{f(\theta,\tau)} \iint G_Q(q,p)\ e^{-i\theta q-i\tau p}\ dqdp\ , \tag{4.14}$$

or

$$g(q,p) = \frac{1}{4\pi^2} \iiiint \frac{e^{-i\theta\tau\hbar/2}}{f(\theta,\tau)} G_Q(q',p')\ e^{i\theta(q-q')+i\tau(p-p')}\ d\theta d\tau dq'dp'. \tag{4.15}$$

Using the theorem given by Eq. (3.35) this can be written as

$$g(q,p) = \frac{\exp\ (+i\hbar/2\ \partial^2/\partial q\partial p)}{f\left(\frac{1}{i}\frac{\partial}{\partial q}, \frac{1}{i}\frac{\partial}{\partial p}\right)}\ G_Q(q,p)\ . \tag{4.16}$$

For the case $f = 1$

$$g(q,p) = \exp\ (+i\hbar/2\ \partial^2/\partial q\partial p)\ G_Q(q,p) \tag{4.17}$$

For the same quantum operator two classical functions corresponding to two different distributions can be related to each other. From Eq. (4.6) the operator will be the same if

$$\gamma_1(\theta,\tau)\ f_1(\theta,\tau) = \gamma_2(\theta,\tau)\ f_2(\theta,\tau) \tag{4.18}$$

or

$$g_1(q,p) = \frac{1}{4\pi^2} \iiiint \frac{f_2(\theta,\tau)}{f_1(\theta,\tau)}\, g_2(q'p')\ e^{i\theta(q-q')+i\tau(p-p')}\ d\theta d\tau dq' dp' \tag{4.19}$$

$$= \frac{f_1(i\partial/\partial q, i\partial/\partial p)}{f_2(i\partial/\partial q, i\partial/\partial p)}\, g_2(q,p)\ . \tag{4.20}$$

Correspondence Rules

From the beginnings of quantum mechanics there has been the attempt to write quantum mechanical operators from their classical counterparts.[22-26] For functions of position or momentum only no difficulty arises but for mixed functions one can not do so straightforwardly since the choice is not unique. Methods which generate a quantum operator from a classical function are called correspondence rules or rules of association. Over the years there have been many rules proposed. Eq. (4.6) generates all possible such rules and as we shall see below particular choices of f yield the ones that have been previously investigated.

We work out the correspondence for

$$g(q,p) = e^{i\theta' q+i\tau' p} \tag{4.21}$$

where θ' and τ' are treated as parameters. For this case

$$\gamma(\theta,\tau) = \delta(\theta-\theta')\ \delta(\tau-\tau') \tag{4.22}$$

and G(Q,P) is therefore

$$G(Q,P) = \iint \delta(\theta-\theta')\ \delta(\tau-\tau')\ f(\theta,\tau)\ e^{i\theta Q+i\tau P}\ d\tau d\theta\ . \tag{4.23}$$

Hence

$$f(\theta',\tau')\ e^{i\theta' Q+i\tau' P} \longleftrightarrow e^{i\theta' q+i\tau' p}\ . \tag{4.24}$$

For the case

$$g(q,p) = q^n\ p^m \tag{4.25}$$

we have

$$\gamma(\theta,\tau) = \frac{1}{4\pi^2} \iint q^n p^m\ e^{-i\theta q-i\tau p}\ dq dp \tag{4.26}$$

$$= \left(i\,\frac{\partial}{\partial q}\right)^n \left(i\,\frac{\partial}{\partial p}\right)^m\ \delta(\theta)\ \delta(\tau)\ . \tag{4.27}$$

Substituting this into Eq. (4.7) we have

$$q^n\ p^m \longleftrightarrow \frac{1}{i^{n+m}}\ \frac{\partial^{n+m}}{\partial\theta^n\ \partial\tau^m}\, f(\theta,\tau)\ e^{i\theta\tau\hbar/2}\ e^{i\theta Q}\ e^{i\tau P}\ \Big|_{\theta,\tau=0} \tag{4.28}$$

For $f = 1$, $\sin(\theta\tau\hbar/2)/(\theta\tau\hbar/2)$, $\cos(\theta\tau\hbar/2)$, Eq. (4.28) evaluates to.

$$q^n\, p^m \longleftrightarrow \frac{1}{2^n}\sum_{k=0}^{n} \binom{n}{k}\, Q^{\,n-k}\, P^{\,m}\, Q^{\,k} \tag{4.29}$$

$$q^n\, p^m \longleftrightarrow \frac{1}{n+1}\sum_{k=0}^{n} Q^{\,n-k}\, P^{\,m}\, Q^{\,k} \tag{4.30}$$

$$q^n\, p^m \longleftrightarrow (\, Q^n\, P^m + P^m\, Q^n\,)/2\ . \tag{4.31}$$

The first two rules were suggested by Weyl[24-25] and by Born and Jordan[26] respectively.

For the case where $g = qp$ we have that

$$qp = (QP+PQ)/2 + \partial^2 f/\partial\theta\partial\tau\ \Big|_{\theta,\tau=0}\ . \tag{4.32}$$

Hence all rules for which

$$\partial^2 f/\partial\theta\partial\tau\ \Big|_{\theta,\tau=0} = 0 \tag{4.33}$$

yield symmetrization for qp. [27-28]

<u>Correspondence Rules and Distribution Functions</u>

Each correspondence rule will give a particular distribution function because each rule associates the classical moments with a quantum operator whose value can be calculated in the usual quantum manner. Once the moments are calculated the characteristic function can be obtained and the distribution calculated. To illustrate, we carry this out for the rule of Born and Jordan. Using (4.30)

$$\langle\, q^n\, p^m\, \rangle = \frac{1}{n+1}\sum_{k=0}^{n} \int \psi^*(q)\, q^{n-k}\left(\frac{\hbar}{i}\frac{\partial}{\partial q}\right)^m q^k\, \psi(q)\, dq \tag{4.34}$$

The characteristic function is

$$M(\theta,\tau) = \sum_{nm}^{n} \frac{(i\theta)^n\,(i\tau)^m}{(n+1)\,n!\,m!} \sum_{k=0} \int \psi^*(q)\, q^{n-k}\left(\frac{\hbar}{i}\frac{\partial}{\partial q}\right)^m q^k\, \psi(q)\, dq \tag{4.35}$$

$$= \sum_{n=0}^{\infty}\ \sum_{k=0}^{n} \frac{(i\theta)^n}{(n+1)\,n!} \int \psi^*(q)\, q^{n-k}\, (q+\tau\hbar)^k\, \psi(q+\tau\hbar)\, dq \tag{4.36}$$

but

$$\sum_{k=0}^{n} q^{-k}\,(q+\tau\hbar)^k = \frac{(1+\tau\hbar/q)^{n+1} - 1}{\tau\hbar/q} \tag{4.37}$$

$$\sum_{k=0}^{n} \frac{q^n\,(i\theta)^n}{(n+1)!}\ \frac{(1+\tau\hbar/q)^{n+1} - 1}{\tau\hbar/q} = e^{i\theta\tau\hbar/2 + i\theta q}\ \frac{\sin\theta\tau\hbar/2}{\theta\tau\hbar/2}\ . \tag{4.38}$$

and therefore

$$M(\theta,\tau) = \frac{\sin(\theta\tau\hbar/2)}{(\theta\tau\hbar/2)} \int e^{+i\theta u}\, \psi^*(u - \tfrac{1}{2}\tau\hbar)\, \psi(u + \tfrac{1}{2}\tau\hbar)\, du \tag{4.39}$$

which is the characteristic function of the distribution given by (3.14).

Algebra of Classical Functions

Previously we showed the connection between quantum operators and classical functions. We show here how classical functions are to be multiplied if they are to give the same expectation value as when their corresponding operators are multiplied. Ordinary multiplication will not do since it is commutative while operator multiplication is not. Suppose

$$A(Q,P) \longleftrightarrow a(q,p) \qquad B(Q,P) \longleftrightarrow b(q,p)\ . \tag{4.40}$$

What we seek is a classical function $c(q,p)$ which corresponds to the operator

$$A(Q,P)\, B(Q,P) \longleftrightarrow c(q,p)\ . \tag{4.41}$$

The derivation of $c(q,p)$ is straightforward and we present only the final result.[7]

$$c(q,p) = \iiiint e^{-i\hbar(\theta'\tau-\tau'\theta)/2}\, \frac{f(\theta,\tau)f(\theta',\tau')}{f(\theta+\theta',\tau+\tau')}\, \gamma_a(\theta,\tau)\, \gamma_b(\theta',\tau')$$

$$e^{iq(\theta+\theta')+ip(\tau+\tau')}\, d\theta d\tau d\theta' d\tau'. \tag{4.42}$$

This can be put in an operational form

$$c(q,p) = \exp\left(-i\frac{\hbar}{2}\left[\frac{\partial}{\partial q_b}\frac{\partial}{\partial p_a} - \frac{\partial}{\partial q_a}\frac{\partial}{\partial p_b}\right]\right) L(a,b)\; a(q,p)\; b(q,p) \tag{4.43}$$

where we have defined

$$L(a,b) = \frac{f\left(\dfrac{1}{i}\dfrac{\partial}{\partial q_a}\ ,\ \dfrac{1}{i}\dfrac{\partial}{\partial p_a}\right) f\left(\dfrac{1}{i}\dfrac{\partial}{\partial q_b}\ ,\ \dfrac{1}{i}\dfrac{\partial}{\partial p_b}\right)}{f\left(\dfrac{1}{i}\dfrac{\partial}{\partial q_a} + \dfrac{1}{i}\dfrac{\partial}{\partial q_b}\ ,\ \dfrac{1}{i}\dfrac{\partial}{\partial p_a} + \dfrac{1}{i}\dfrac{\partial}{\partial p_b}\right)} \tag{4.44}$$

and where $\partial/\partial q_a$ operates on a only and $\partial/\partial q_b$ operates on b only. The commutators can be expressed as such:

$$AB-BA \longleftrightarrow 2i\,\sin\frac{\hbar}{2}\left[\frac{\partial}{\partial p_a}\frac{\partial}{\partial q_b} - \frac{\partial}{\partial q_a}\frac{\partial}{\partial p_b}\right] L(a,b)\; a(q,p)\; b(q,p) \tag{4.45}$$

$$AB+BA \longleftrightarrow 2\,\cos\frac{\hbar}{2}\left[\frac{\partial}{\partial p_a}\frac{\partial}{\partial q_b} - \frac{\partial}{\partial q_a}\frac{\partial}{\partial p_b}\right] L(a,b)\; a(q,p)\; b(q,p)\ . \tag{4.46}$$

We note that if A and B commute it does not necessarily follow that the right hand side of Eq. (4.45) is zero.

Quantization

We first work out the quantization equation for the Wigner case. Starting with the eigenvalue equation

$$H(Q,P)\ \psi(q) = E\ \psi(q) \tag{4.47}$$

and using Eq. (4.7) for H(Q,P) we have

$$\iint e^{i\theta\tau\hbar/2}\ \gamma(\theta,\tau)\ f(\theta,\tau)\ e^{i\theta Q}\ \psi(q+\tau\hbar)\ d\tau d\theta = E\ \psi(q)\ . \tag{4.48}$$

Letting q go into q+τ'h/2, multiplying by

$$e^{-i\tau' p}\ \psi(q-\tau'\hbar/2) \tag{4.49}$$

and integrating with respect to τ', we have

$$\iint \gamma(\theta,\tau)\ e^{+i\theta q+i\tau p}\ F_W(q+\tfrac{1}{2}\tau\hbar,\ p-\tfrac{1}{2}\theta\hbar)\ d\tau d\theta$$

$$= \frac{1}{4\pi^2}\iiiint g(q',p')\ e^{i\theta(q-q')+i\tau(p-p')}\ H(q+\tfrac{1}{2}\tau\hbar, p-\tfrac{1}{2}\theta\hbar)\ d\theta d\tau dq'dp'$$

$$= E\ F_W(q,p) \tag{4.50}$$

where we have used Eq. (3.27). H(q,p) is the classical Hamiltonian. Using Eq. (3.38) this can be written as

$$H\left(q+\tfrac{1}{2}i\hbar\frac{\partial}{\partial p}, p-\tfrac{1}{2}i\hbar\frac{\partial}{\partial q}\right) F_W(q,p) = E\ F_W(q,p) \tag{4.51}$$

For the general case the quantization equation is,

$$\iint S(q,p,x,y)\ F(x,y)\ dxdy = E\ F(q,p) \tag{4.52}$$

where

$$S(q,p,x,y) = \frac{1}{4\pi^2}\iiiint \gamma(\theta,\tau)\ \frac{f(\theta,\tau)f(\theta',\tau')}{f(\theta+\theta',\tau+\tau')}$$
$$e^{ix(\theta+\theta')+iy(\tau+\tau')-i\theta'(q+\tau\hbar/2)-i\tau'(p-\theta\hbar/2)}\ d\theta d\tau d\theta' d\tau' \tag{4.53}$$

To obtain the equation in an operator form we use (3.41) and (4.51)

$$f\left(i\frac{\partial}{\partial q}, i\frac{\partial}{\partial p}\right) H\left(q+\tfrac{1}{2}i\hbar\frac{\partial}{\partial p}, p-\tfrac{1}{2}i\hbar\frac{\partial}{\partial q}\right) f^{-1}\left(i\frac{\partial}{\partial q}, i\frac{\partial}{\partial p}\right) F(q,p)$$
$$= E\ F(q,p)\ . \tag{4.54}$$

Dynamics

By differentiating F with respect to time the equation of motion for the distribution can be obtained. A straightforward but lengthy calculation gives

$$\frac{\partial F(q,p,t)}{\partial t} + \frac{p}{m}\frac{\partial}{\partial q}F(q,p,t) = \iint (J_1+J_2)\ F(q',p')\ dq'dp' \tag{4.55}$$

where

$$J_1 = \iint \frac{e^{i\theta(q'-q)+i\tau(p'-p)}}{f(\theta,\tau)} \left(\frac{\partial f}{\partial t} - \frac{\theta}{m}\frac{\partial f}{\partial \tau}\right) d\theta d\tau \quad , \tag{4.56}$$

$$J_2 = \frac{1}{4\pi^2}\,\frac{2}{h} \iiint \frac{f(\theta,\tau)}{f(\theta+x,\tau)} e^{i\theta(q'-q)+i\tau(p'-p)+ixq'} \overline{V}(x)\sin(x\tau\hbar/2)\; dx d\theta d\tau \tag{4.57}$$

$$\overline{V}(x) = \frac{1}{2\pi}\int V(q)\; e^{-ixq}\; dq \tag{4.58}$$

and V(q) is the potential. For the Wigner distribution

$$J_1 = 0 \tag{4.59}$$

$$J_2 = \frac{2}{\hbar}\frac{1}{2\pi\hbar}\,\delta(q-q') \int V(q+u/2)\, \sin(\, u(p'-p)/\hbar\,)\; du \quad . \tag{4.60}$$

The equation of motion can be transformed into the following form which makes clear the classical limit,

$$\frac{\partial F}{\partial t} = \frac{\dfrac{\partial f}{\partial t}\left(\dfrac{1}{i}\dfrac{\partial}{\partial q}\,,\,\dfrac{1}{i}\dfrac{\partial}{\partial p}\right)}{f\left(\dfrac{1}{i}\dfrac{\partial}{\partial q}\,,\,\dfrac{1}{i}\dfrac{\partial}{\partial q}\right)}\; F(q,p)$$

$$+\; f\left(\frac{1}{i}\frac{\partial}{\partial q_H},\frac{1}{i}\frac{\partial}{\partial p_H}\right) f\left(\frac{1}{i}\frac{\partial}{\partial q_F} + \frac{1}{i}\frac{\partial}{\partial q_H},\frac{1}{i}\frac{\partial}{\partial p_F} + \frac{1}{i}\frac{\partial}{\partial p_H}\right) f^{-1}\left(\frac{1}{i}\frac{\partial}{\partial q_F},\frac{1}{i}\frac{\partial}{\partial q_F}\right)$$

$$+\; \frac{2}{\hbar}\sin\; \frac{\hbar}{2}\left[\frac{\partial}{\partial p_F}\frac{\partial}{\partial q_H} - \frac{\partial}{\partial p_H}\frac{\partial}{\partial q_F}\right] H(q,p)\; F(q,p). \tag{4.61}$$

If we choose an f such that

$$\lim_{\hbar\to 0}\; f \longrightarrow 1 \qquad\qquad \lim_{\hbar\to 0}\; \partial f/\partial t \longrightarrow 1 \tag{4.62}$$

then (4.61) becomes, in the limit h approaching zero,

$$\frac{\partial F}{\partial t} = \frac{\partial H}{\partial q}\frac{\partial F}{\partial p} - \frac{\partial H}{\partial p}\frac{\partial F}{\partial q} \tag{4.63}$$

which is Liouville's equation.

The equation of motion for a dynamical variable a(q,p,t) is

$$\frac{da}{dt} = \frac{\partial a}{\partial t} + \frac{2}{\hbar}\sin\,\frac{\hbar}{2}\left[\frac{\partial}{\partial p_H}\frac{\partial}{\partial q_F} - \frac{\partial}{\partial p_F}\frac{\partial}{\partial q_H}\right] L(H,a)\; H(q,p)\; a(q,p) \tag{4.64}$$

Also, the equation of motion for the characteristic function is given by

$$\frac{\partial M(q,p,t)}{\partial t} = \frac{\theta f(\theta,\tau)}{m}\frac{\partial}{\partial \tau}\frac{M(\theta,\tau)}{f(\theta,\tau)} + \frac{M(\theta,\tau)}{f(\theta,\tau)}\frac{\partial f}{\partial t}$$

$$+\; \frac{2}{\hbar} f(\theta,\tau) \int \overline{V}(x)\, \sin(x\tau\hbar/2)\, \frac{M(\theta+x,\tau)}{f(\theta+x,\tau)}\, dx \quad . \tag{4.65}$$

Representability

An arbitrary normalized function of q,p is not necessarily a joint distribution since there may not exist a wave function which generates it. Distributions for which a wave function does exist, we shall call ψ representable.[11] To test whether a distribution is ψ representable we calculate the right hand side of the Eq. (3.28) and examine whether any function, ψ, exists for which it can be written in the form given by the left hand side.

Variational Principle

For a classical Hamiltonian H(q,p), we expect that

$$\iint H(q,p)\ F(q,p)\ dqdp \geq E_0 \tag{4.66}$$

where E_0 the lowest eigenvalue. For an arbitrary trial F(q,p) one can obtain a lower value then E_0 for the left had side of (4.66). Only if the trial F is representable will the variational principle hold.[11] This is so since if F is representable there exists a ψ which will make the left hand side equal to

$$\int \psi^*(q)\ H(Q,P)\ \psi(q)\ dq \tag{4.67}$$

and which of course must be greater than the lowest eigenvalue. Hence, before using the variational principle one must make certain that the trial functions are representable.

Mixtures

For the case of mixtures we write the general distribution as

$$F(q,p) = \frac{1}{4\pi^2} \iiint e^{-i\theta q - i\tau p + i\theta u}\ f(\theta,\tau)\ \rho(u - \tfrac{1}{2}\tau\hbar,\ u + \tfrac{1}{2}\tau\hbar)\ du d\tau d\theta \tag{4.68}$$

where ρ is the density matrix. Necessary and sufficient conditions for F to represent a pure case can be obtained by considering

$$F(q,p) = \iiiint F(q',p')\ F(q'',p'')\ g(q'-q,q''-q,p'-p,p''-p)\ dq'dq''dp'dp'' \tag{4.69}$$

where

$$g = \frac{\hbar}{8\pi^3} \iiiint e^{i\theta(q''-q)+i\theta'(q'-q)+i\tau(p''-p)+i\tau'(p'-p)} e^{+i\hbar(\theta'\tau-\tau'\theta)/2}\ f^*(\theta,\tau)\ f^*(\theta',\tau')\ f(\theta+\theta',\tau+\tau')\ d\theta d\tau d\theta' d\tau' \tag{4.70}$$

If Eq. (4.69) is satisfied then F(q,p) represents a pure case. For the Wigner distribution,

$$g = \frac{2}{\pi\hbar}\ e^{-2i(q'-q)(p''-p)/\hbar + 2i(q''-q)(p'-p)/\hbar}\ . \tag{4.71}$$

5. POSITIVE DISTRIBUTIONS II

Just as the bilinear distributions were characterized by the kernel $f(\theta,\tau)$ the positive distributions may be similarly characterized and

indeed $f(\theta,\tau)$ itself may be used to do so. More conveniently the kernel may be re defined to better accommodate the positive distributions. We define α, β, and μ by

$$F(q,p) = \int \mu(q-q',p-p')\ F_0(q,p)\ dq'dp' \tag{5.1}$$

and

$$\alpha(\theta,\tau) = 1 + \beta(\theta,\tau) = \int \mu(q,p)\ e^{+i\theta q+i\tau p} \tag{5.2}$$

where Fo is the correlationless distribution

$$F_0(q,p) = |\psi(q)|^2\ |\varphi(p)|^2\ . \tag{5.3}$$

If F is to satisfy the marginals we must have

$$\int \mu(q,p)dq = \delta(p)\ , \tag{5.4}$$

$$\int \mu(q,p)dp = \delta(q)\ , \tag{5.5}$$

$$\alpha(0,\tau) = \alpha(\theta,0) = 1\ , \qquad \beta(0,\tau) = \beta(\theta,0) = 0 \tag{5.6}$$

The positive distributions are characterized by taking

$$\beta(\theta,\tau) = \frac{c \iint F_0(q,p)\ \rho(u,v)\ e^{+i\theta q+i\tau p}\ dqdp}{M_q(\theta)\ M_p(\tau)} \tag{5.7}$$

where

$$M_q(\theta) = \int |\psi(q)|^2\ e^{i\theta q}\ dq \tag{5.8}$$

$$M_p(\tau) = \int |\varphi(p)|^2\ e^{i\tau p}\ dp\ . \tag{5.9}$$

The characteristic function of the distribution is then

$$M(\theta,\tau) = M_q(\theta)\ M_p(\tau)\ \alpha(\theta,\tau) = M_q(\theta)\ M_p(\tau)\ (\ 1 + \beta(\theta,\tau)\ )\ . \tag{5.10}$$

For two different distributions characterized by two different kernels the characteristic functions are related by

$$\frac{M_1(\theta,\tau)}{\alpha_1(\theta,\tau)} = \frac{M_2(\theta,\tau)}{\alpha_2(\theta,\tau)} \tag{5.11}$$

and the distributions are related by

$$F_1(q,p) = \frac{1}{4\pi^2} \iiiint \frac{\alpha_1(\theta,\tau)}{\alpha_2(\theta,\tau)}\ e^{i\theta(q'-q)+i\tau(p'-p)}\ F_2(q',p')\ d\theta d\tau dq'dp'. \tag{5.12}$$

If we take F_2 to be the correlationless case then

$$F(q,p) = \frac{1}{4\pi^2} \iiiint \alpha(\theta,\tau)\ e^{ia\theta(q-q')+ib\tau(p-p')}\ F_0(q',p')\ d\theta d\tau dq'dp' \tag{5.13}$$

or

$$F(q,p) = \alpha(i\partial/\partial q, i\partial/\partial p)\; F_0(q,p) \ . \tag{5.14}$$

The positive distributions can be obtained from Eq. (3.1) by choosing

$$f(\theta,\tau) = \frac{\alpha(\theta,\tau)\; M_q(\theta)\; M_p(\tau)}{\int e^{+i\theta u}\, \psi^*(u - \tfrac{1}{2}\tau\hbar)\; \psi(u + \tfrac{1}{2}\tau\hbar)\; du d\tau d\theta} \ . \tag{5.15}$$

as can be verified by direct substitution. The $\alpha(\theta,\tau)$ in (5.15) is obtained from Eq. (5.2) and (5.7). We note that this f is a functional of the state.

Choices for the Kernel

There are many classes of ρ's which satisfy Eq. (2.10) - (2.13). A particularly convenient case is where h(u,v) is factorable

$$h(u,v) = h_1(u)\; h_2(v) \ . \tag{5.16}$$

This leads to

$$\rho(u,v) = (\, h_1(u) - 1 \,)\; (\, h_2(v) - 1 \,) \ . \tag{5.17}$$

We illustrate with two examples. If we take

$$h_1(u) = (n+1)\; u^n \tag{5.18}$$

$$h_2(v) = (m+1)\; u^m \tag{5.19}$$

then

$$\rho(u,v) = \left[(n+1)u^n - 1\right] \quad \left[(m+1)v^m - 1\right] \tag{5.20}$$

and c can be chosen to be any number in the range

$$-\frac{1}{n\,m} \le c \le \frac{1}{\max(n,m)} \ . \tag{5.21}$$

Another class is

$$h_1(u) = (2n+1)\; (u-1)^{2n} \tag{5.22}$$

$$h_1(v) = (2m+1)\; (v-1)^{2m} \tag{5.23}$$

which gives

$$\rho(u,v) = \left[(2n+1)\; (u-1)^{2n} - 1\right] \quad \left[(2m+1)\; (v-1)^{2m} - 1\right] \tag{5.24}$$

and

$$-\frac{1}{4nm} \le c \le \frac{1}{2\max(n,m)} \ . \tag{5.25}$$

Many other choices are possible. Which h(u,v) to choose depends on the correlations we wish to impose.

Constraints on the Kernel

In this section we assume that the dependence of $\rho(u,v)$ on the state is only through u and v. As mentioned before the dependence can be more general. We use the same notation and definitions as in the corresponding section on the bilinear distributions. We have already shown how to assure that the marginals are satisfied and how to choose c so that the distribution is positive. For reality it is clear that ρ has to be real.

Galilei invariance. Letting $\psi_b(q) = \psi_a(q+d)$ we have that

$$u_b(q) = \int_{-\infty}^{q} |\psi_a(q' + d)|^2 \, dq' = \int_{-\infty}^{q+d} |\psi_a(q')|^2 \, dq' = u_a(q+d) \tag{5.26}$$

and similarly for v(p). Hence, any $\rho(u,v)$ will make the distribution invariant with respect to translations in position or momentum

Space reflection. In this case we take $\psi_b(q) = \psi_a(-q)$

$$u_b(q) = \int_{-\infty}^{q} |\psi_a(-q')|^2 \, dq' = - \int_{+\infty}^{-q} |\psi_a(q')|^2 \, dq' = \int_{-q}^{\infty} |\psi_a(q')|^2 \, dq'$$

$$= 1 - u_a(-q) \tag{5.27}$$

and similarly for $v_b(p)$. Hence, for space reflection invariance we must take a ρ such that

$$\rho(u,v) = \rho(1-u,1-v) \; . \tag{5.29}$$

Time reflection. We take $\psi_b(q) = \psi_a^*(q)$ and we have

$$u_b(q) = u_a(q) \; , \quad v_b(p) = 1 - v_a(-p) \tag{5.30}$$

and therefore we must have

$$\rho(u,v) = \rho(u,1-v) \; . \tag{5.31}$$

6. CONCLUSION

We have presented here some of the properties of the positive distributions and further development is currently under investigation. What needs to be done is to examine to what extent quantum mechanics can be written in terms of them in the same manner as was done for the bilinear distributions.

The general problem of finding joint distributions for given marginals is a particularly difficult one which has a long history.[29] The marginals in themselves do not determine uniquely the joint distribution or the correlations involved. Clearly, without further information any correlation is possible and any theory of joint distributions should reflect that fact. In the method presented here the wide choice available is reflected through $\rho(u,v)$. What correlations to impose, that is, which h to choose will require considerable further study and investigation.

As mentioned in the Introduction, the uncertainty principle is the most commonly stated reason why positive distributions cannot exist. The positive distributions given above yield the uncertainty principle because they have the correct marginals and that is all that is required to prove

the uncertainty principle. The uncertainty principle relates the indivi dual standard deviations in position and momentum but does not mix the two variables. The uncertainty relation is reflected in the relationship between the margins. It imposes no further conditions on the possible joint distribution and has no bearing on the existence of joint distributions.

Acknowledgement

This research was supported in part by a grant from the City University of New York Research Award Program.

References

1. L. Cohen and Y. I. Zaparovanny, J. Math. Phys. 21, 794, (1980).
2. L. Cohen, J. Chem. Phys. 80, 4277 (1984).
3. L. Cohen, ICASSP 84, 41B.1.1 (1984).
4. L. Cohen, J. Math. Phys., 25, 2402 (1984).
5. L. Cohen and T. Posch, Positive Time-Frequency Distribution Functions, IEEE Trans. on Acoustics, Speech and Signal Processing, (to appear).
6. P. D. Finch and R. Groblicki, Found. of Physics, 14, 549 (1984).
7. L. Cohen, J. Math. Phys. 7, 781 (1966).
8. L. Cohen, Phil. of Science, 33, 317 (1966).
9. H. Margenau and L. Cohen, in: Quantum Theory and Reality, ed. Mario Bunge p. 71, Springer-Verlag, New York (1967).
10. L. Cohen, in: Foundations and Philosophy of Quantum Mechanics, ed. C.A. Hooker, p.69 D. Reidel, New York (1973).
11. L. Cohen, J. Math. Phys. 17, 1863 (1976).
12. L. Cohen, J. Chem. Phys. 70, 788 (1978).
13. E. Wigner, Phys. Rev. 40, 749 (1932).
14. E. Wigner, in: Perspectives in Quantum Theory, eds. W. Yourgrau and A. van der Merwe, p.25 M.I.T., Cambridge, (1971).
15. J. E. Moyal, Proc. Cambridge Phil. Soc. 45, 99 (1949).
16. For a general review of the Wigner distribution function see the forthcoming article: M. Hillery, R.F. O'Connell, M.O. Scully and E.P. Wigner, Distribution Functions in Physics: Fundamentals, Physics Reports (to appear).
17. T. Takabayasi, Progr. Theoret. Phys. 11, 341 (1954).
18. H. Margenau and R. N. Hill, Prog. Theoret. Phys. 26, 722 (1961).
19. C. L. Mehta, J. Math. Phys. 5, 677 (1964).
20. O. von Roos, Phys. Rev. 119, 1174 (1960).
21. G. J. Ruggeri, Prog. Theoret. Phys. 46, 1703 (1971).
22. J. R. Shewell, Am. J. Phys. 27, 16 (1959).
23. H. J. Groenewold, Physica 12, 405 (1946).
24. H. Weyl, The Theory of Groups and Quantum Mechanics, E. P. Dutton and Co., New York, p.275 (1931).
25. N. H. McCoy, Proc. Natl. Acad. Sci. U.S.A. 18, 674 (1932).
26. M. Born and P. Jordan, Z. Phys. 34, 873 (1925).
27. N. Cartwright, Found. of Phys. 4, 127 (1974).
28. L. Cohen, Found. of Phys. 6, 739 (1976).
29. See, for example, the following papers and the references therein: D. Morgenstern, Mitt. Math. Stat. 8, 234 (1956); E.J. Gumbel, Rev. Fac. Ci. Univ. Lisboa Ser. 2A CI. Mat.7, 179 (1959); D.J.G. Farlie, Biometrica 50, 499 (1963).

the uncertainty principles. The uncertainty principle relating the [illegible] and [illegible] that [illegible] and [illegible] that the [illegible] [illegible] in the relation between the [illegible] joint distributions [illegible] on the existence of joint distributions [illegible].

Acknowledgement

This research was supported [illegible] grant from the City University of New York [illegible] Research Award Program.

References

1. L. Cohen and [illegible], J. Math. Phys. 21, 794 (1980).
2. L. Cohen, J. Math. Phys. [illegible] (1984).
3. L. Cohen, [illegible] (1986).
4. L. Cohen, J. Math. Phys. [illegible] (1966).
5. L. Cohen and C. Posch, [illegible] IEEE Trans. on Acoustics, Speech and Signal Processing [illegible].
6. P. D. Finch and R. Groblicki, Found. of Physics 14, 549 (1984).
7. L. Cohen, J. Math. Phys. 7, 781 (1966).
8. L. Cohen, Phil. of Science 33, 317 (1966).
9. H. Margenau and [illegible] Early [illegible], [illegible] (1961).
10. L. Cohen, [illegible] Foundations and Philosophy of Quantum Mechanics [illegible], C. A. Hooker, ed. (D. Reidel, New York, 1973).
11. L. Cohen, J. Math. Phys. 17, [illegible] (1976).
12. L. Cohen, J. Chem. Phys. 70, 788 (1979).
13. E. Wigner, Phys. Rev. 40, 749 (1932).
14. E. Wigner, [illegible] Perspectives in Quantum Theory, [illegible] (MIT Press, Cambridge, 1971).
15. [illegible]
16. For a general review of the [illegible] see [illegible] and L. Cohen, "Distribution Functions in [illegible]" Physics Reports (to appear).
17. [illegible]
18. [illegible] Prog. Theoret. Phys. [illegible]
19. [illegible] J. Math. Phys. 7, [illegible] (1966).
20. [illegible] (1950).
21. [illegible]
22. [illegible]
23. [illegible]
24. [illegible] New York [illegible]
25. [illegible]
26. [illegible]
27. [illegible]
28. [illegible] Found. of Phys. [illegible]
29. [illegible]

THREE LECTURES ON THE FOUNDATIONS OF QUANTUM THEORY AND QUANTUM ELECTRODYNAMICS

A. O. Barut

Department of Physics
University of Colorado
Boulder, CO. 80309, U.S.A.

Contents:

The common feature of these three lectures is to explore the borderline region between classical physics and quantum physics. I do not think that one kind of physical laws stop being applicable sharply and another regime of laws begin. Rather the intrinsic properties of electrons and photons are more complicated than the classical physics assumes them to be and that we have to explore how we approximate these internal structures in classical theory and in quantum theory.

LECTURE 1. Models for EPR-Problem

(i) Spin Correlation Experiments

We present here a theoretical analysis to the experiments discussed by Clauser[1] and Aspect[2] at this meeting.

We compare the correlation functions defined for quantum spins with those for classical spins, and come to the surprising conclusion that they are exactly the same. This leads us to discuss the assumptions underlying Bell inequalities, and we show that they are not as plausible as usually or tacitly assumed. Consequently, some of the general conclusions drawn from Bell's inequalities, such as the nonexistence of hidden variables, or nonexistence of an "objective reality",[1] are premature or incorrect.

We do not question the observed quantum mechanical correlations, nor the mathematical derivation of Bell's inequalities, but we question the physical meaning of the additional assumptions made and the physical conclusions drawn from these inequalities.

The classical spin correlation experiment is the following.[3] An object of total angular momentum zero splits into two parts, and we mea-

sure the components of two classical spin vectors $\vec{S}_1$ and $\vec{S}_2 = -\vec{S}_1$ relative to two vectors $\vec{a}$ and $\vec{b}$, respectively. Thus, the observables are $A = \vec{S}_1 \cdot \vec{a}$ and $B = \vec{S}_2 \cdot \vec{b}$. We repeat this experiment many times, assuming that the spins are isotropically distributed over all directions in space, and calculate expectation values and correlations of these two random variables A and B. Clearly $\langle A\rangle = \langle B\rangle = 0$, and we find

$$\langle A^2\rangle = \frac{1}{3} S_1 a \quad , \quad \langle B^2\rangle = \frac{1}{3} S_1 b$$

The correlation function of the two random variables A and B is given by

$$E_{\text{classical}}(\vec{a},\vec{b}) = \frac{\langle|AB|\rangle}{\sqrt{\langle A^2\rangle}\ \sqrt{\langle B^2\rangle}} = \frac{\langle|\ \vec{S}_1 \cdot \vec{a}\ S_2 \cdot \vec{b}|\rangle}{\langle A^2\rangle^{1/2}\ \langle B^2\rangle^{1/2}} \tag{1}$$

We shall compare this with the quantum mechanical correlation function in a singlet state ψ_s of two spin 1/2 particles in the EPR-experiment

$$E_Q(\vec{a},\vec{b}) = \frac{\langle\psi_s|\vec{\sigma}_1 \cdot \vec{a}\ \vec{\sigma}_2 \cdot \vec{b}|\psi_s\rangle}{\langle\psi_s|\psi_s\rangle} = \frac{\langle AB\rangle}{\sqrt{\langle A^2\rangle\langle B^2\rangle}},\ A = \tfrac{1}{2}\vec{\sigma}_1\cdot\vec{a} \tag{2}$$

in order "to make a picture how the observed strong correlation can come about."[2] It is important to emphasize that we must calculate (1) entirely classically, and (2) entirely quantum mechanically. Since classical spins (or angular momenta) can have any magnitudes we must divide in (1) the random variable A,B by the expected magnitudes $\sqrt{\langle A^2\rangle}$, $\sqrt{\langle B^2\rangle}$. In our experiment the common property carried by the two spins, or the supplementary hidden variable λ, is the set of angles of emissions of the two spins.

Now the surprising result is that the two correlation functions (1) and (2) are exactly the same:[3]

$$E_{cl}(\vec{a},\vec{b}) = E_Q(\vec{a},\vec{b}) = -\cos(\hat{a}\cdot\hat{b}) \tag{3}$$

This is not supposed to happen according to considerations of Bell.[4] We have "mimicked the quantum mechanical correlation by a classical reasonable looking model". So what is going on here? Bell's considerations are based on two further theoretical assumptions: The first assumption is

(1) "locality", i.e.

$$E(\vec{a},\vec{b}) = \int d\lambda\ A(a,\lambda)\ B(b,\lambda) \tag{4}$$

which is satisfied by our model, for eq. (1) reads explicitly

$$E_{cl}(\vec{a},\vec{b}) = 3\int d\Omega\ \cos(\hat{S}_1\cdot\hat{a})\ \cos(\hat{S}_2\cdot\hat{b}) \tag{5}$$

where $d\Omega$ is the solid angle.

(2) The second assumption is that in eq. (4) the unobserved densities $A(a,\lambda)$ and $B(b,\lambda)$ are bounded in absolute value by 1 (for photons)

or by 1/2 for spin 1/2-particles. Clearly this is a quantummechanical assumption. It is certainly true in quantum theory. The components of spin in any direction cannot be bigger than 1/2. But this assumption should not be invoked in a classical model. If we wish to see whether or not there exists a classical model, we must calculate the classical model entirely classically, as we have emphasized.

Because this second assumption is the central point of our disagreement with Bell's inequalities, we shall discuss it in several other forms.

Experimentally the correlation function E(a,b) is measured as discrete counts in the following form

$$E(a,b) = P_{++}(a,b) + P_{--}(a,b) - P_{+-}(a,b) - P_{-+}(a,b) \quad (6)$$

where P_{++} refers to spin component of S_1 in the direction $\vec{a}$ to be "up" and of S_2 in the direction $\vec{b}$ also "up", etc. Now both classical and quantum correlation functions can be written in the form.[3] For each of the P-functions we have in both cases the locality property

$$P_{ij}(\vec{a},\vec{b}) = \int d\lambda\, P_i(a,\lambda)\, P_j(b,\lambda); \quad i,j = \pm 1 \quad (7)$$

and in both cases the same observed forms

$$P_i(a) = \int d\lambda\, P_i(a,\lambda) = 1/2 \;, \quad i = \pm \quad (8)$$

and

$$P_{++}(a,b) = P_{--}(a,b) = \frac{1}{4}(1 - \cos(\hat{a} \cdot \hat{b}))$$

$$P_{+-}(a,b) = P_{-+}(a,b) = \frac{1}{4}(1 + \cos(\hat{a} \cdot \hat{b})). \quad (9)$$

But again the second assumption that the unobserved densities $p_i(a,\lambda)$ are positive and bounded by one is not satisfied in our model.

We conclude that the Bell's inequalities do not eliminate the purely classical spin model, but they eliminate mixed models with classical hidden variables λ, but quantum mechanical spin-values.

It is emphasized that observations are discrete "yes" or "no" experiments.[1,2] Each event contributes to one of the quantities $P_{ij}(a,b)$. Our model deals with a continuous values of the spin components and each event contributes a bit to all the $P_{ij}(a,b)$. Thus, although the final results of the two experiments are the same, individual counts are different. We will now show how we can discretize our classical experiment in order to obtain "yes" or "no" counts.[3] This will further give us some insight as to how spin quantization might come about, and about the notion of the collaps of the wave function thus going beyond the correlation functions.

We still perform our classical experiment, measure the components of spins along $\vec{a}$ and $\vec{b}$, but declare all spins in the upper hemisphere of a sphere with $\vec{a}$ pointing along the nordpole as "spin up", and in the lower hemisphere as "spin down". We have to normalize or weight the events so discretized into two groups and we do this by the length of projection of

the spin along $\vec{a}$, namely by cos $(\phi - \phi_a)$. It is sufficient to integrate over the azimuthal angles relative to $\vec{a}$ and $\vec{b}$, and we have

$$P_+(a) = \frac{1}{4} \int_{\phi_a - \pi/2}^{\phi_a + \pi/2} d\phi \left| \cos(\phi - \phi_a) \right| = 1/2 \qquad (10)$$

This just counts the number of events in the upper hemisphere relative to $\vec{a}$. Similarly, for $P_-(a)$, $P_+(b)$, $P_-(b)$. The correlation functions $P_{ij}(a,b)$ are now simply the number of events weighted by cos $(\phi - \phi_a)$ in the overlapp of two hemispheres relative to a and b. For example

$$P_{+-}(a,b) = \frac{1}{4} \int_{\phi_b - \pi/2}^{\phi_a + \pi/2} d\phi \left| \cos(\phi_a - \phi) \right| = \frac{1}{4} (1 + \cos(\hat{a} \cdot \hat{b})) \qquad (11)$$

This is precisely the overlap of the upper hemisphere with $\vec{a}$ at the nordpole with another hemisphere with $\vec{b}$ at the nordpole. [The weight factor cos $(\phi - \phi_b)$ gives the same result]. Again we see that in this discretized form, we get the same experimental correlation functions now in the "yes" or "no" form, but in the evaluation of the integrals for $P_{ij}(a,b)$ the limits of integrations are different and we have one single weight factor or measure cos $(\phi - \phi_a)$.

Quite intuitively this is how we would calculate the discrete spin correlation functions, namely just by counting the number of events in the overlapp region of the two hemispheres, and this is different than the assumptions underlying Bell's inequalities.

Our final conclusion is that Bell's inequalities have not eliminated the purely classical interpretation of spin correlation experiments, and our considerations further show how the discrete quantum spin correlations are obtained by classical spins. As to the last point, we might imagine a dynamical process by which all spins in the upper half-plane are attracted to the nordpole, and all spins in the lower plane to the southpole. There exist indeed a nonlinear spin procession equation for which this exactly happens.[5]

(ii) On EPR-Problem

Spin correlations and Bell's inequalities are only one aspect of the whole EPR-"paradox". The EPR-problem refers to the difference between the coherent state (e.g. singlet spin states) and the statistical mixture that seem to result by making a measurement on one of the particles, and further, to the seemingly paradoxical situation of being able to measure noncommuting observables. I believe these problems are resolved if we view the emission of two particles from a coherent state not in the language of second quantized theory as the creation of <u>two</u> particles, but as a dynamical process of a certain duration, and to be a <u>single</u> wave function, or a single quantum state. We should delete the word "two" from the beginning in the discussion of the EPR-problem. For this purpose I should like to exhibit a wave equation whose solution describes a <u>single</u> wave functions which as a function of time looks like the emission of two particles.

The wave equation, in one-dimension for simplicity, is the nonlinear equation

$$\frac{\partial}{\partial t} \psi - \frac{\partial}{\partial x} \psi = - \sin \psi(x,t) \quad .$$

The quantity $u = \sin^2 \psi$, has the solution

$$u = \cosh^{-2} [f(t - x) + f(t + x)] \quad .$$

Taking f to be a quadratic function, we can arrange[6] the solution such that it represents a wave-shape growing from zero, then splitting, and then the two parts going away from each other in opposite directions looking more and more like two distinct wave peaks. Yet the whole solution is a single wave. Clearly, the two peaks of the wave, are strongly correlated.

LECTURE 2. On Recent Advances in the Theory of the Electron

The fundamental reasons we are interested in the structure of the electron are:

(i) The electron is the first and most fundamental of all elementary particles. It is absolutely stable (as far as we can tell), carrier of the quantized charge, source of the electromagnetic field.

(ii) The charge is pointlike, as the latest high energy electron scattering experiments show, up to a distance of $R \approx 10^{-16}$ cm; yet the heavy leptons μ, τ may be interpreted in some sense as the "excited states" of a peculiar internal dynamics of the electron.

(iii) The behavior of the charge in the theory of the electron gives us an intuitive picture on the origin of spin as the orbital angular motion of the charge around its center of mass.

(iv) The internal dynamics of the electron may shed some new light on quantum behavior itself, and give a possible dynamical mechanism to calculate Planck's constant h or the fine structure constant α.

(v) The electron together with neutrino as the only stable remnant after the decay of all unstable particles, can be taken to be the building blocks of all matter. This immediately answers the conservation of charge as well as quantization of charge. If so, we must understand in a nonperturbative way the interaction of electrons at very short distances, which the perturbation theory of QED cannot give us.

After these motivational points I turn now briefly to our changing conception about the nature of the electron. Since its discovery the electron has gone through the following periods of evolution:

(1) The mechanical electron according to Lorentz
(2) The radiating electron according to Abraham, Lorentz, and finally to Dirac.
(3) The electron according to Heisenberg and Schrödinger
(4) The Dirac electron
(5) Relativistic classical electron with Zitterbewegung
(6) Electron in QED
(7) Dirac electron with self-field

Perhaps by 1992, the 100th anniversary of the discovery of the electron we might have a new picture of the electron.

I have recently reviewed the corresponding equations to these different notions of the electrons.[7] Here I shall only mention the salient features.

In the Lorentz-Dirac equation (written in proper-time relativistically)

$$m \ddot{Z}_\mu = e F^{ext}_{\mu\nu} \dot{Z}^\nu + \frac{2}{3} e^2 [\dddot{Z}_\mu + (\ddot{Z})^2 \dot{Z}_\mu] \tag{1}$$

where $Z_\mu(s)$ is the position of the electron (c = 1), the last terms comes from self-energy and has the follwing consequences:

(i) the radiative effects are all included nonperturbatively

(ii) velocity $\dot{Z}_\mu = u_\mu$ and momentum defined by $p_\mu = m\dot{Z}_\mu - \frac{2}{3} e^2 \ddot{Z}_\mu + eA_\mu$, are linearly independent. The phase space of the radiating electron has three sets of variables (Z_μ, $\dot{Z}_\mu$, p_μ).

(iii) The new degree of freedom due to self-field can be also interpreted as "spin", because angular momentum alone is no longer conserved.

(iv) The radiating electron has an integral of motion in proper time $\mathcal{H} = u^\mu(p_\mu - eA_\mu)$.

(v) A linearized form of (1) with a variable mass $\dot{m} = -\frac{2}{3} e^2 \dot{u}^2$ is

$$\frac{d}{ds} (m \dot{Z}_\mu) = e F_{\mu\nu} \dot{Z}^\nu + \frac{2}{3} e^2 \dddot{Z}_\mu \tag{2}$$

(vi) The characteristic length for the self energy effects are of the order of

$$r_{s.e.} \approx \frac{2}{3} \frac{e^2}{mc^2} \approx 10^{-13} \text{ cm.} \tag{3}$$

I mention these properties because they exactly parallel those of the Dirac electron. The quantum Dirac electron can namely be written in the Heisenberg representation as

$$\mathcal{H} \ddot{Z}_\mu = e F^{ext}_{\mu\nu} \dot{Z}^\nu + i \frac{\hbar}{2} \dddot{Z}_\mu \tag{4}$$

with $\mathcal{H} = \gamma^\mu (p_\mu - eA_\mu)$. Here the velocity operator $\dot{Z}_\mu$ is

$$\dot{Z}_\mu = i [\mathcal{H}, Z_\mu] = \frac{\partial \mathcal{H}}{\partial p^\mu} = \gamma_\mu \quad , \tag{5}$$

the Dirac matrices. Again the independent dynamical variables are (Z_μ, γ_μ, p_μ). The correspondance between (2) and (4) is

$$\mathcal{H} \leftrightarrow m \ , \quad \dot{Z}_\mu \leftrightarrow \gamma_\mu \ , \quad \frac{2}{3} e^2 \leftrightarrow i \hbar/2 \tag{6}$$

The characteristic quantum length is 1/2 of Compton wave-length:

$$r \approx \frac{\hbar}{2mc} \approx 10^{-11} \text{ cm.} \tag{7}$$

Thus, quantum effects come into play before the self-energy effects as we probe shorter and shorter distances. That is why the Lorentz-Dirac equation (1) has not been studied much — in fact it came much after the advent of quantum theory.

For Dirac electron also we can express the new degree of freedom as spin. But self-energy effects are not yet included in the Dirac equation. Note also the appearance of (i) in the correspondance (6), or in eq. (4) which converts runaway solutions of classical theory into oscillatory solutions of quantum theory. In fact the solution of the Heisenberg equations for the Dirac electron can be written as

$$Z_\mu(s) = X_\mu(s) + Q_\mu(s) \tag{8}$$

where $X_\mu(s) = \mathcal{H}^{-1} p_\mu s$ + const. is the motion of a relativistic free particle with velocity $\mathcal{H}^{-1} p_\mu$, interpreted as the center of mass motion, and Q_μ represent an oscillatory motion with amplitude and frequency of the order of

$$A \approx \frac{\hbar}{2mc} = 10^{-11} \text{ cm}, \qquad \omega \approx \frac{2mc^2}{\hbar^2} = 6 \times 10^{21} \text{ sec}^{-1}$$

In terms of the set of dynamical variables (X_μ, p_μ, Q_μ, $P_\mu = m\dot{Q}_\mu$) the Dirac equation[4] can be written as a dynamical system

$$\dot{X}_\mu = \mathcal{H}^{-1} p_\mu$$

$$\dot{p}_\mu = 0$$

$$\dot{Q}_\mu = \frac{1}{m} P_\mu$$

$$\dot{P}_\mu = -4 m^3 Q_\mu \tag{9}$$

where we see clearly in the last two lines the equation of the oscillator.

Thus the Dirac electron is not the quantization of a relativistic charge, nor that of the radiating electron. Recently a classical model of the electron has been given[8] which contains the oscillatory Zitterbewegung already at the classical level, and which when quantized gives precisely the Dirac electron. This classical model has additional internal degrees of freedom and is defined by the Lagrangian

$$L = i\frac{\lambda}{2}(\bar{\eta}\dot{\eta} - \dot{\bar{\eta}}\eta) + (p_\mu - eA_\mu)(\bar{\eta}\gamma^\mu\eta) - \dot{Z}^\mu P_\mu \tag{10}$$

Here λ is a constant of the dimension of action, η a <u>classical</u> spinor variable. Eq. (10) in effect says that the velocity of the particle is represented by $v_\mu = \dot{Z}_\mu = \bar{\eta}\gamma_\mu\eta$, (so that η is roughly the square root of the velocity). The solution for a free particle in proper time is

$$v_\mu = \frac{1}{m} p_\mu + (v(0)_\mu - \frac{p_\mu}{m}) \cos 2m\,s + \frac{\dot{v}(0)_\mu}{2m} \sin 2m\,s \tag{11}$$

and we recognize besides the velocity of the center of mass, $\frac{p_\mu}{m}$, the oscillatory motion.

The quantization of (10) proceeds by the representation of the elaborate Poisson-Bracket algebra of the classical theory by quantum commutators,[8] or by the path integral quantization. In fact the latter method solves the longstanding problem of representing discrete quantum spin as a path integral over a classical action with continuous variables.[12]

In Lecture 3 we shall complete the electron theories in QED and in nonperturbative QED.

LECTURE 3: The Development of Quantum Electrodynamics Based on Self-Energy

In Lecture 2 we saw the following description of the electron and its interaction with the electromagnetic field:

Classical spinless charge (Lorentz force) → Radiating spinless charge (Lorentz force + self-energy force)

Classical charge with Zitterbewegung → Radiating charge with Zbwg.
↓ ↓

Quantum Dirac Electron with Ztbwg. → Radiating quantum Electron (with self-energy)

In this lecture I shall discuss the last corner, framed, namely the quantum Dirac electron with its self-energy, possible in the form of a closed nonperturbative equation as in Lorentz-Dirac equation.

Quantum electrodynamics (QED) gives us an answer how to add radiative processes to the bare Dirac electron. In QED we throw away the self-field at the beginning, consider bare electrons, atoms or 2-level systems. We then introduce the self-field back photon by photon. This is done by decomposing the self-field into its modes, and introducing creation and annihilation operators for these modes.

In contrast, in classical electrodynamics we keep the full self-field and obtain the radiative effects non-perturbatively. The Lorentz-equation with self-field is

$$m_o \ddot{Z}_\mu = e F^{ext}_{\mu\nu} \dot{Z}^\nu + e F^{self}_{\mu\nu} \dot{Z}^\nu \tag{1}$$

The second term on the right hand side is dropped in ordinary electrodynamics although it is infinite on the surface of a point electron. The reason we do quite well so is that this term can be decomposed as follows:

$$e F^{self}_{\mu\nu} \dot{Z}^\nu = -\delta m \ddot{Z}_\mu + \frac{2}{3} e^2 (\dddot{Z}_\mu + (\ddot{Z})^2 \dot{Z}_\mu) \tag{2}$$

the large (in fact infinite) first term is of the form of an inertial force, we bring it to the left hand side and renormalize the mass $m = m_o + \delta m$, and obtain eq. (1) of Lecture 2. The finite and small second term in (2) contains, we believe, all the radiative effects in a non-perturbative way. Thus, classical electrodynamics would be a complete closed theory if the mass renormalization term would be finite.

Can we now do the same procedure of obtaining self-energy effects non-perturbatively in quantum theory? This approach I shall call "Non-perturbative QED".

In analogy with (1) I start from the coupled Maxwell-Dirac equations including self-field:

$$[\gamma^\mu(i\partial_\mu - eA_\mu^{ext} - e\, A_\mu^{self}) - m_o]\, \psi(x) = 0$$

$$\Box\, A_\mu = e\, \overline{\psi}\, \gamma_\mu\, \psi \tag{3}$$

we can eliminate A_μ from these equations to obtain

$$[\gamma^\mu(i\partial_\mu - e_o\, A_\mu^{ext}) - m_o]\psi = e^2\, \gamma^\mu\, \psi \int dy\, D(x-y)\, \overline{\psi}(y)\, \gamma_\mu\, \psi(y) \tag{4}$$

This equation correspond in classical theory to the evaluation of self-force in (1) using Lienard-Wiechert potentials.

The main problem is now to extract from the right hand side of (4), i.e. the self-potential, those terms which renormalize m_o and e_o, the only parameters of the theory. The remaining terms would be new small observable potentials, as the Larmor term of the classical theory.

This has been done, in the Dirac equation,[9] or directly in terms of the action,[10] and in the nonrelativistic form.[11] I shall here indicate only the steps and the final results.

If one makes a Fourier expansion in energy of the wave function

$$\psi(x) = \sum\!\!\!\!\!\!\int_n\, \psi_n(\vec{x})\, e^{-iE_n t} \tag{5}$$

and insert this in the nonlinear equation (4), or in the corresponding Schrödinger equation with self-field, separates the term $\psi_n(x)$ on the left, the right hand self energy expression contains the following terms:

$$\int \psi_n^* \left(E_n - \frac{p^2}{2m} - \frac{Ze^2}{r}\right) \psi_n\, d\vec{x} =$$

$$= \frac{2}{3}\,\alpha\, \frac{1}{\pi m^2} \sum_m (E_n - E_m) \int_0^\infty \frac{dk}{E_n - E_m - k}\, T^j_{mn}(\vec{k})\, T^j_{nm}(-\vec{k})$$

$$- \frac{2}{3}\, i\alpha\, \frac{1}{m^2} \sum_m \int_0^\infty dk\, \delta(E_n - E_m - k)\, T^j_{mn}(\vec{k})\, T^j_{nm}(-\vec{k})$$

$$- \frac{\alpha}{\pi} \sum_m \int_0^\infty dk\, T^o_{mm}(\vec{k})\, T^o_{nn}(-\vec{k})$$

$$- \frac{2}{3} \frac{\alpha}{\pi m^2} \sum_m \int_0^\infty dk \; T^j_{mn}(\vec{k}) \; T^j_{nm}(-\vec{k})$$

$$- \frac{\alpha}{\pi} \sum_m \int_0^\infty dk \; T^o_{mn}(\vec{k}) \; T^o_{nm}(-\vec{k}) \tag{6}$$

Here $\sum_m$ means a sum over both discrete and continuous spectrum, T_{mn}'s are form factors which in non-relativistic version are given by

$$T^j_{mn}(\vec{k}) = \int d\vec{x}' \, e^{i\vec{k}\cdot\vec{x}'} \, \psi^*_m(\vec{x}') \, \frac{\overleftrightarrow{\nabla}^j}{i} \, \psi_n(x') \quad ; \quad j = 1, 2, 3$$

$$T^o_{mn}(\vec{k}) = \int d\vec{x}' \, e^{i\vec{k} \cdot \vec{x}'} \, \psi^*_m(\vec{x}') \, \psi_n(\vec{x}')$$

The first term on the RHS of (6) represents the generalized Bethe expression for the Lamb-shift; the second term is the spontaneous emission. The third term is the vacuum polarization and charge renormalization term. The last two terms are pure renormalizaton terms, because they can be written as $\int \psi^*_n \, (- \frac{2}{3\pi} \frac{\alpha}{m^2} \Lambda \, \vec{p}^{\,2}) \, \psi_n d\vec{x}$ and as $\int \psi^*_n \, (- \frac{\alpha}{\pi} \Lambda) \, \psi_n$, respectively, where

$$\Lambda = \sum_m \int_0^\infty dk \, d\vec{x}' \, e^{i\vec{k} \cdot (\vec{x} - \vec{x}')} \, \psi_m(x') \, \psi_m(x) \quad .$$

These terms can be brought to LHS of (6) and use to renormalize mass and charge. All integrals in (6) are finite. The only infinities come from summing over infinitely many bound and scattering states. This can be remedied, if we consider that actual physical states have finite widths. We have used the expansion (5) in order to establish contact with the existing methods.

References

1. J. Clauser, These proceedings.
2. A. Aspect, These proceedings.
3. A. O. Barut and P. Meystre, Phys. Lett. 105A, 458 (1984)
4. J. S. Bell, Physics 1, 195 (1965).
5. For this discussion see A. O. Barut, in Proceedings of the Conference on the Foundations of Quantum Theory, SUNY, Albany, N.Y. 1984.
6. A. O. Barut, Phys. Lett. 67A, 257 (1978).
7. A. O. Barut, in Proceedings of the Intern. Symposium on the Foundations of Quantum Mechanics, Japan Phys. Soc. 1984; p. 321. See also, "Quantum Optics, Experimental Gravitation and Measurement Theory", edit. by P. Meystre et al. (Plenum 1983); p. 155.
8. A. O. Barut and N. Zanghi, Phys. Rev. Lett. 52, 2009 (1984).
9. A. O. Barut and J. Kraus, Found. of Phys. 13, 189 (1983).
10. A. O. Barut, in Quantum Electrodynamics and Quantum Optics, (Plenum Press) 1984.
11. A. O. Barut and J. F. van Huele, Helvet. Physica Acta.
12. A. O. Barut and I. H. Duru, Phys. Rev. Lett.

THE INTERFACE BETWEEN QUANTUM MECHANICS AND STATISTICAL MECHANICS

Indeterminism in classical statistical mechanics merely reflects our ignorance of initial conditions. In quantum statistical mechanics indeterminism seems to be unavoidable even in principle. The inequalities discovered by Bell, which are satisfied in a very wide class of hidden-variable theories, are not valid in quantum theory. Their violation has also been observed experimentally, as discussed in a paper here by Professor Aspect.

We introduce this section with two reprinted papers which explore the possibility for useful hidden-variable descriptions of reality which are in accordance with quantum mechanics. These considerations are particularly germane to the subject of this conference, in that they use conventional quantum-mechanical techniques, but in a formulation which is not as common as the usual wave-function or state-vector formulation. In particular, the use of quantum distribution function techniques for spin 1/2 systems is applied to the EPR problem. These techniques represent an extention of the more usual joint distribution function for position and momentum, as developed by Wigner. In particular, it is possible to develop a quantum-mechanical formulation along these lines of problems previously treated via hidden-variable theories (for example, by Bellefonte). In this way one learns by specific calculations how it is that quantum mechanics is similar to and different from various notions of what a hidden variable theory "might be". We see once again how strange hidden variables theories are. The Physical Review paper by Dr. Scully shows that quantum mechanics is much more appealing than the hidden variable theory.

These papers are followed by two papers of Dr. Zurek dealing with the measurement process in quantum mechanics. Lastly come two papers of Dr. Aspect, one just mentioned, the other commenting on the classical model of the EPR Gedanken experiment proposed in the reprint by Drs. Barut and Meystre.

HOW TO MAKE QUANTUM MECHANICS LOOK LIKE A HIDDEN-VARIABLE THEORY AND VICE VERSA

Marlan O. Scully

Max-Planck-Institut für Quantenoptik
D-8046 Garching bei München, Federal Republic of Germany
and Center for Advanced Studies, University of New Mexico
Albuquerque, New Mexico 87131

I. INTRODUCTION

A. Background and motivation

Conventional quantum mechanics is a superb calculational tool. It has successfully solved mysteries ranging from macroscopic superconductivity[1(a)] to the microscopic theory of the electron[1(b)] and has provided deeper insight into the nature of the vacuum[1(c)] on the one hand and the description of the nucleon[1(d)] on the other. Whole new fields[2(a)–2(f)] such as quantum optics and quantum electronics owe their very existence to this body of knowledge.

Nevertheless certain aspects, concerning the foundations and interpretation of the theory, are regarded by many[3] as incomplete, unsatisfactory, or contradictory. Thus, various alternative theories to quantum mechanics have been considered over the years, hidden-variable (HV) theories[4] being prominent among these.

In this latter HV context the inequalities of Bell[5] have been instrumental in rendering various philosophical arguments susceptible to experimental[6] test. Such experiments are important from at least two points of view: first they provide a sharp focus on the essential differences between quantum and HV theories and, of course, help us choose between theories. More important, perhaps, is the ability of such studies to assist us in gaining a better understanding and appreciation for what quantum theory is not. For example, in the current HV vernacular it is not a "local-realistic"[7] theory. It is in this latter context that the present author finds these studies most interesting.

It is thus not a desire to replace quantum theory, but rather a desire to better understand its foundations that stimulates this worker. With this motivation in mind let us consider the framework in which this paper fits.

The stimulus for the present study was provided by Belinfante's[4] chapter on "Some Examples of Hidden-Variables Theories of the Second Kind for 'Explaining' Polarization Correlations." In that chapter it is argued that certain "polarization hidden-variables" (α), as discussed in Sec. II of this paper, are interesting in a HV context, but do not exist in quantum theory. While it is true that conventional quantum mechanics involves "no such a thing as α," we show in Sec. III that a completely acceptable quantum theory can be built around this α variable via an associated quasiclassical quantum distribution function[8] $P(\alpha)$ [e.g., Eqs. (3.14)]. With this distribution function in hand, we are naturally motivated to consider a class of closely related HV distribution functions $\tilde{P}(\alpha)$ [e.g., Eqs. (3.15)]. Following along these lines the present hidden-variables theory reproduces the quantum two-particle correlation.

To sum up, in this paper we show that

(1) various polarization "hidden variables" α have a proper quantum-mechanical interpretation which suggests a

(2) hidden-variable theory, in agreement[9,10] with quantum theory insofar as the current two-particle correlation experiments are concerned.

No consideration as to the "philosophical" pros and cons of the present HV calculational strategy is included here, although an essential difference between the epistemology of this work and quantum theory is developed in Sec. VI. Various extensions and further discussion of the ways in which our HV theory agrees and disagrees with quantum theory will be presented elsewhere.

B. A Pico review of the HV concept, EPRB, and orthodox quantum theory

Just as statistical mechanics provides a deeper "hidden-variable" description of thermodynamical phenomena, so, many people would like to see quantum mechanics supplanted by a deeper hidden-variable theory. According to this point of view the various parameters of interest (e.g., *all three* components of a particle's angular momentum) are knowable but just not known to us.[11] For example, a spin-$\frac{1}{2}$ system might be thought of as having a kind of hidden spin parameter ϕ as in Fig. 1, such that the measured spin projection is given by

$$S_z=\frac{\hbar}{2}[\theta(\pi-\phi)-\theta(\phi-\pi)] .$$

This article is reprinted with permission from Physical Review D, 28, 10, 2477-2484, copyright © by The American Physical Society.

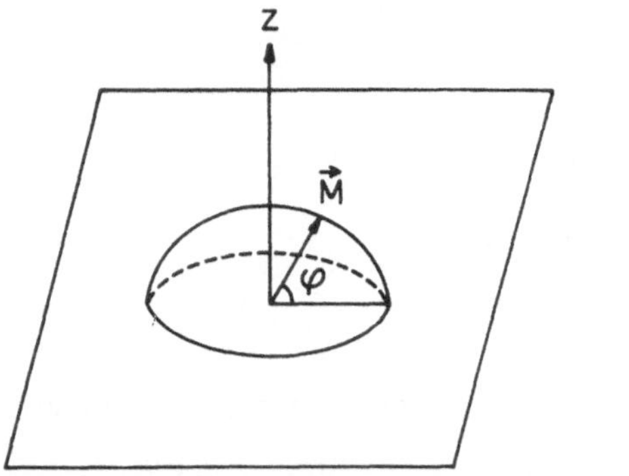

FIG. 1. Hidden-variable description of a spin-$\frac{1}{2}$ particle in which the hidden variable ϕ lying between 0 and π corresponds to spin-up while angle ϕ in the lower half plane corresponds to spin-down.

At the heart of most modern discussions of "hidden variables" in quantum mechanics lies a *Gedankenexperiment* of the Einstein-Podolsky-Rosen-Bohm[12] variety. As sketched in Fig. 2 the essence of this "experiment" involves the passage (or blockage) of particles 1 and 2 through appropriate Stern-Gerlach apparati (SGA).

In order to set the stage for our considerations, let us first review what orthodox quantum mechanics says about the problem of Fig. 2. To calculate the probability that particle 1 is passed by a SGA oriented at an angle ϕ to the vertical $(+z)$ direction and that particle 2 is passed by a SGA oriented at an angle θ to the vertical, we must evaluate the correlation function

$$\mathscr{P}(\theta,\phi)=\langle\psi\,|\,\hat{\pi}_\theta^{(2)}\hat{\pi}_\phi^{(1)}\,|\,\psi\rangle\ . \tag{1.1}$$

In Eq. (1.1), the singlet state may be written as

$$|\psi\rangle=\frac{1}{\sqrt{2}}(\,|+_1-_2\rangle-|-_1+_2\rangle)\ , \tag{1.2}$$

where, for example,

$$|+_i\rangle=|+\phi_i\rangle=e^{-i\phi_i\hat{\sigma}_y/2}\,|\uparrow_i\rangle\ ,\quad i=1,2\ .$$

The projection operators are given by

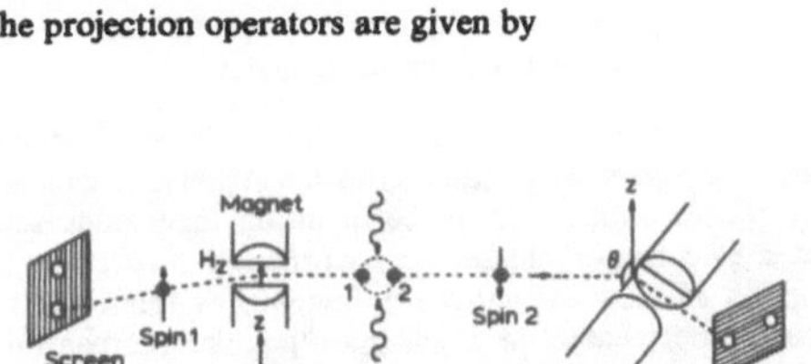

FIG. 2. Singlet-spin system, such as orthohydrogen, is split by external radiation field and the corresponding spin-$\frac{1}{2}$ particles (protons) proceed to opposite ends of laboratory where they are passed through SG apparati oriented along the z axis in the case of particle 1, and at an angle θ to the z axis in the case of spin 2.

$$\begin{aligned}\hat{\pi}_\phi^{(1)}&=|\phi_1\rangle\langle\phi_1|\\&=e^{-i\phi\hat{\sigma}_y/2}\,|\uparrow_1\rangle\langle\uparrow_1|\,e^{i\hat{\sigma}_y\phi/2}\\&=\tfrac{1}{2}(1+\hat{\sigma}_z^{(1)}\cos\phi+\hat{\sigma}_x^{(1)}\sin\phi)\end{aligned} \tag{1.3a}$$

and

$$\hat{\pi}_\theta^{(2)}=\tfrac{1}{2}(1+\hat{\sigma}_z^{(2)}\cos\theta+\hat{\sigma}_x^{(2)}\sin\theta)\ . \tag{1.3b}$$

The Pauli spin operators appearing in Eqs. (1.3) are defined as

$$\hat{\sigma}_x=\begin{bmatrix}0&1\\1&0\end{bmatrix},\quad \hat{\sigma}_y=\begin{bmatrix}0&-i\\i&0\end{bmatrix},$$

and

$$\hat{\sigma}_z=\begin{bmatrix}0&1\\1&0\end{bmatrix}.$$

Inserting Eqs. (1.2), (1.3a), and (1.3b) into Eq. (1.1) we find the usual result

$$\mathscr{P}(\theta,\phi)=\tfrac{1}{4}[1-\cos(\theta-\phi)]\ . \tag{1.4}$$

II. A HIDDEN-VARIABLE TREATMENT OF SPIN CORRELATIONS FOLLOWING BELINFANTE[13]

A. Spin polarization as a possible hidden variable

The leading idea in hidden-variable theories involves the assumption that the two spins will have some properties which will have correlated values. These values will help the spin decide, upon reaching a SGA, whether or not it should pass through that apparatus.

Following the treatment of Belinfante, we require that in order to give such theories an air of possibility we want to invent them in such a way that at least in the simplest cases they yield the same results as quantum mechanics. For example, in an unpolarized beam only $\frac{1}{2}$ the particles should pass a given SGA. Further the probability of passing a second SGA placed behind (and at an angle θ) relative to the previous (vertical) SGA should be given by

$$\langle\uparrow|\,\pi_\theta\,|\uparrow\rangle\ , \tag{2.1}$$

which, using Eq. (1.3b), is

$$\tfrac{1}{2}(1+\cos\theta)\ . \tag{2.2}$$

Or, more generally, if a spin emerges from a SGA oriented at an angle α and then passes into a SGA tipped through an angle θ relative to the vertical, then the likelihood that the particle will emerge from the second SGA is given by

$$\tfrac{1}{2}[1+\cos(\theta-\alpha)]\ . \tag{2.3}$$

Thus we might say that a "hidden variable" α determined whether the spin passed through the apparatus whose angle θ is determined by the experimenter.

With this in mind we define the hidden-variable probability function

$$\bar{\pi}_\theta(\alpha) \equiv \tfrac{1}{2}[1+\cos(\theta-\alpha)] \tag{2.4}$$

as giving the "simultaneous passage" through the α and θ oriented SGA.

B. Correlation between singlet spins in hidden-variables theory

Consider the case where the two spins of our singlet system of Fig. 2 (having polarization angles α and β for spins 1 and 2) are correlated such that

$$I(\alpha,\beta)d\alpha\, d\beta \tag{2.5}$$

is the probability that the spins carry polarizations α and β, while all other hidden variables can be randomly distributed. For maximum polarization correlation, we might then think that

$$I(\alpha,\beta)=\delta(\alpha-\beta-\pi)\frac{1}{2\pi} \; . \tag{2.6}$$

Let us next ask: what is the probability of simultaneous passage of spins 1 and 2 through the double SG system of Fig. 1? That is, what is $\mathscr{P}(\theta,\phi)$ in hidden-variable theory? Following the above discussion and in view of Eqs. (2.4) and (2.5) we would expect

$$\bar{\mathscr{P}}(\theta,\phi)=\int d\alpha\, d\beta\, I(\alpha,\beta)\bar{\pi}_\theta^{(2)}(\beta)\bar{\pi}_\phi^{(1)}(\alpha) \; , \tag{2.7}$$

and from Eqs. (2.6) and (2.4) this implies

$$\bar{\mathscr{P}}(\theta,\phi)=\int d\alpha\, d\beta\, \delta(\alpha-\beta-\pi) \times\frac{1}{2\pi}\{\tfrac{1}{2}[1+\cos(\theta-\beta)]\tfrac{1}{2}[1+\cos(\phi-\alpha)]\} \; . \tag{2.8}$$

Carrying out the simple integrations in Eq. (2.8) we find

$$\bar{\mathscr{P}}(\theta,\phi)=\tfrac{1}{4}[1-\tfrac{1}{2}\cos(\theta-\phi)] \; . \tag{2.9}$$

It is precisely the difference between the quantum correlation (1.4) and the hidden-variable result (2.9) which concerns us here. How should we react to this difference? From a pragmatic, calculational point of view quantum theory is a superb tool, hence Eq. (1.4) has never really been seriously doubted (although such results should be and have been verified experimentally).

What then should we think of the hidden-variable prediction Eq. (2.9)? It is not all that different from the quantum prediction. Might there not be a germ of "truth" hidden in Eq. (2.9) and, more to the point, the arguments leading to it?

In this context we quote Belinfante[9]: "The polarization [spin] hidden-variable here introduced is, of course, a quantity which does not exist in quantum theory. ... in quantum theory no such a thing as α even exists."

In the next section we shall argue that a rigorous quantum mechanical treatment of the present problem can in fact be couched in terms of the angular variable α. In so doing we shall be led to reconsider the correlation function [Eq. (2.7)] and by a simple extension of Eq. (2.6) regain the quantum result [Eq. (1.4)] via a "hidden-variable" calculation.

III. QUANTUM DISTRIBUTION THEORY AND A CORRESPONDING HIDDEN-VARIABLE TREATMENT OF SPIN-$\frac{1}{2}$ SYSTEMS

A. Distribution functions

Some 50 years ago Eugene Wigner succeeded in developing a phase-space distribution[8] description of quantum mechanics in terms of what is now called the Wigner distribution $W(p,q)$. This distribution has many nice properties and has been fruitfully applied to a variety of problems in conventional quantum theory. Practicing quantum opticians have such notions bred into their bones at an early age. For example, the probability density function,[14] $P(E)$, for the quantized electromagnetic field on the one hand and the phase-space ("Bloch" type) description of two-level systems[15] on the other.

Proceeding toward an "α" description of our spin singlet problem, consider first the expectation value of an operator $\hat{Q}(\vec{\sigma})$ given in terms of the density matrix ρ:

$$\langle \hat{Q} \rangle = \mathrm{Tr}[\rho(t)\hat{Q}(\sigma)] \; .$$

Introducing the operator δ function

$$\delta(\gamma-\hat{c}) \equiv \int \frac{d\xi}{2\pi} \exp[-i(\gamma-\hat{c})] \; , \tag{3.1}$$

where $\hat{c}$ is an operator (e.g., $\hat{\sigma}_z$) and γ is the associated classical variable (e.g., m_z); we may then write $\hat{Q}(\vec{\sigma})$ as

$$\hat{Q}(\vec{\sigma})=\int d^3m\, Q(m_x,m_y,m_z)\delta(m_x-\hat{\sigma}_x) \times\delta(m_y-\hat{\sigma}_y)\delta(m_z-\hat{\sigma}_z) \; . \tag{3.2}$$

Inserting Eq. (3.2) into Eq. (3.1) we then find the expectation value for $\hat{Q}$ to be given by

$$\langle \hat{Q} \rangle = \int d^3m\, P(\vec{m},t)Q(\vec{m}) \; , \tag{3.3}$$

where

$$P(\vec{m},t) \equiv \mathrm{Tr}[\rho(t)\delta(m_x-\hat{\sigma}_x)\delta(m_y-\hat{\sigma}_y)\delta(m_z-\hat{\sigma}_z)] \; . \tag{3.4}$$

Recalling, however, from the discussion of Secs. I and II that we shall only be interested in operators Q involving $\hat{\sigma}_x$ and $\hat{\sigma}_z$, e.g., $\hat{\pi}_\theta$ of Eq. (1.3b) we may rigorously restrict our treatment to operator expansions of the form

$$\hat{Q}(\hat{\sigma}_x,\hat{\sigma}_z)=\int Q(m_x,m_z)\delta(m_x-\hat{\sigma}_x)\delta(m_z-\hat{\sigma}_z) \times dm_x dm_z \; ,$$

and the associated quantum distribution function

$$P(m_x,m_z,t)=\mathrm{Tr}[\rho(t)\delta(m_x-\hat{\sigma}_x)\delta(m_z-\hat{\sigma}_z)] \; . \tag{3.5}$$

The distribution function Eq. (3.5) is the vehicle by which we shall realize a quantum treatment of the present spin-$\frac{1}{2}$ problem in terms of the angle α. In the next few paragraphs we treat the simple problem of a single "spin up" particle.

For the "spin up" case the density matrix is, of course,

$$\rho=|\uparrow\rangle\langle\uparrow| \tag{3.6}$$

and the associated distribution function is

$$P_\uparrow(m_x,m_z)=\mathrm{Tr}[\,|\uparrow\rangle\langle\uparrow|\,\delta(m_x-\hat{\sigma}_x)\delta(m_z-\hat{\sigma}_z)]$$
$$=\langle\uparrow|\,\delta(m_x-\hat{\sigma}_x)\delta(m_z-\hat{\delta}_z)\,|\uparrow\rangle\ , \qquad (3.7)$$

which by Eq. (3.1) becomes

$$P_\uparrow(m_x,m_z)=\int\frac{d\xi}{2\pi}\int\frac{d\eta}{2\pi}\langle\uparrow|\,e^{i\xi\hat{\sigma}_x}e^{i\eta\hat{\sigma}_z}\,|\uparrow\rangle e^{-im_x\xi}e^{-im_z\eta}$$
$$=\int\frac{d\xi}{2\pi}\int\frac{d\eta}{2\pi}\cos\xi e^{i\eta}e^{-im_x\xi}e^{-im_z\eta}$$
$$=\tfrac{1}{2}[\delta(m_x+1)\delta(m_z-1)+\delta(m_x-1)\delta(m_z-1)]\ . \qquad (3.8)$$

Consulting Fig. 3 we see that the quantum distribution function for the state $|\uparrow\rangle$ corresponds to equal admixtures of "probability" at $\pm\pi/4$.

Define the angle which m makes with the vertical $(+z)$ direction as α, see Fig. 3, so that

$$m_x=m\sin\alpha \qquad (3.9a)$$

and

$$m_z=m\cos\alpha\ . \qquad (3.9b)$$

In terms of this angle α, it is clear that Eq. (3.8) now becomes

$$P_\uparrow(\alpha)=\tfrac{1}{2}\left[\delta\left[\alpha-\frac{\pi}{4}\right]+\delta\left[\alpha+\frac{\pi}{4}\right]\right]\delta(m-\sqrt{2})\ . \qquad (3.10a)$$

In like manner it is easy to show that the probability density for "spin down" is given by

$$P_\downarrow(\alpha)=\tfrac{1}{2}\left[\delta\left[\alpha-\frac{3\pi}{4}\right]+\delta\left[\alpha+\frac{3\pi}{4}\right]\right]\delta(m-\sqrt{2})\ . \qquad (3.10b)$$

Let us now rewrite the operator $\hat{\pi}_\theta$ of Eq. (1.3) in its associated c number representation, that is,

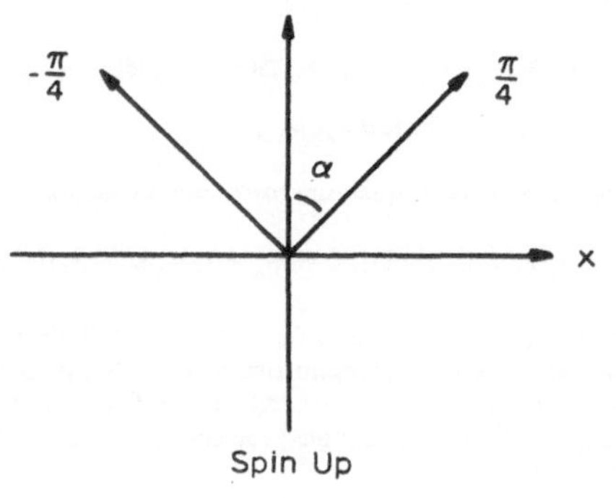

FIG. 3. Figure depicting spin-up particle in quantum distribution theory.

$$\pi_\theta=\tfrac{1}{2}(1+m_z\cos\theta+m_x\sin\theta)\ , \qquad (3.11)$$

and, using Eq. (3.9), this becomes

$$\pi_\theta=\tfrac{1}{2}[1+m\cos(\theta-\alpha)]\ . \qquad (3.12)$$

We may check the vector length m by requiring that $\langle\uparrow|\hat{\pi}_0|\uparrow\rangle$ be unity so that from Eqs. (3.10) and Eq. (3.11b) we have

$$1=\int d\alpha\,\tfrac{1}{2}\left[\delta\left[\alpha-\frac{\pi}{4}\right]+\delta\left[\alpha+\frac{\pi}{4}\right]\right]\tfrac{1}{2}(1+m\cos\alpha)$$
$$=\tfrac{1}{2}\left[1+m\cos\frac{\pi}{4}\right]\ . \qquad (3.13)$$

This then fixes the value of m, as noted in (3.10), to be

$$m=\sqrt{2}\ .$$

In conclusion we note that the probability density calculated for states corresponding to an arbitrary angle ϕ are given by

$$P_{+\phi}(\alpha)=\tfrac{1}{2}\left[\delta\left[\alpha-\phi-\frac{\pi}{4}\right]+\delta\left[\alpha-\phi+\frac{\pi}{4}\right]\right] \qquad (3.14a)$$

and

$$p_{-\phi}(\alpha)=\tfrac{1}{2}\left[\delta\left[\alpha-\phi-\frac{3\pi}{4}\right]+\delta\left[\alpha-\phi+\frac{3\pi}{4}\right]\right]\ . \qquad (3.14b)$$

B. Connection with a revised hidden-variable scheme

Thus we see from the calculations of Sec. III A that the angle α "exists" in quantum theory. Furthermore we note that the probability/projection functions in hidden-variable and quantum theory, i.e., $\tilde{\pi}_\theta(\alpha)$ from Eq. (2.4) and $\pi_\theta(\alpha)$ as given by Eq. (3.11b) are quite similar. Indeed the quantum probability densities given by Eqs. (3.14a) and (3.14b) suggest that we introduce similar probability densities into our hidden-variable considerations as follows:

$$\tilde{P}_{+\phi}(\alpha)=\delta(\alpha-\phi) \qquad (3.15a)$$

and

$$\tilde{P}_{-\phi}(\alpha)=\delta(\alpha-\phi-\pi)\ . \qquad (3.15b)$$

Belinfante, in his scholarly treatment of the subject, divides HV theories into two main groups:

1. Hidden-variable theories of the first kind (HV_1) in which the theory will make exactly the quantum predictions when the hidden variables are in "equilibrium." Deviations from quantum mechanics would occur within the context of HV_1 only for a nonequilibrium situation which is supposed to exist for very short times in any given experiment.

2. Hidden-variable theories of the second kind (HV_2) in which the theory disagrees with quantum mechanics even when the supposed hidden variables are in a steady state or equilibrium condition. The current experimental stud-

TABLE I. Tabulation emphasizing close similarity between the single-particle descriptions in quantum mechanics and HV_3.

	QUANTUM	HV_3
PASSAGE "PROBABILITY"	$\pi_\theta(\alpha) = \frac{1}{2}[1 + m\cos(\theta - \alpha)]$ $m = \sqrt{2}$	$\tilde{\pi}_\theta(\alpha) = \frac{1}{2}[1 + \cos(\theta - \alpha)]$
DISTRIBUTION FOR SPIN UP	$P_\uparrow(\alpha) = \frac{1}{2}[\delta(\alpha - \frac{\pi}{4}) + \delta(\alpha + \frac{\pi}{4})]$ 	$\tilde{P}_\uparrow(\alpha) = \delta(\alpha)$ 
EXPECTATION OF PASSING $\theta = 0$ SGA FOR ↑ SPIN	$\mathcal{P}_\uparrow = \int P_\uparrow(\alpha)\pi_0(\alpha)d\alpha$ $= 1$	$\tilde{\mathcal{P}}_\uparrow = \int \tilde{P}_\uparrow(\alpha)\tilde{\pi}_0(\alpha)d\alpha$ $= 1$

ies, stimulated by the inequalities of Bell, support quantum theory in favor of HV_2.

In this section, as well as the next, we are lead to consider a hidden-variable theory which is a kind of mixture of HV_1 and HV_2, wherein correlation functions (associated with the above-mentioned type of experiments) as calculated from quantum theory and our HV theory are identical. In fact the present HV theory is a close cousin to and directly suggested by the quantum distribution treatment of the problem. In this sense there is a definite HV_1 flavor to our work although no mention of equilbrium (or nonequilibrium) hidden variables is made here. On the other hand, while this work is even closer to an HV_2 theory, it does not fall into this category either since to quote Belinfante: "... this (second) kind of hidden-variables theory may be expected to contradict quantum theory." To be sure, the HV theory here presented will be seen to deviate from quantum theory, but is not immediately contradictory in so far as the correlation functions of current interest are concerned.

For those reasons, and for want of a better term, we call the present HV theory a hidden-variable theory of the third kind (HV_3). A summary of the comparison between the quantum and hidden-variable descriptions for the present example is given in Table I.

IV. QUANTUM DISTRIBUTION AND HV_3 TREATMENT OF THE SINGLET SPIN PROBLEM

A. Quantum and hidden-variable correlation functions

Corresponding to the anticorrelated (singlet) state for a pair of spin-$\frac{1}{2}$ particles

$$|\psi\rangle = \frac{1}{\sqrt{2}}[\,|+_1-_2\rangle - |-_1+_2\rangle\,] \tag{4.1}$$

we have the associated probability distribution

$$P(\vec{m}_1,\vec{m}_2) = \mathrm{Tr}[\rho\,\delta(m_x^{(1)} - \hat{\sigma}_x^{(1)})\delta(m_z^{(1)} - \hat{\sigma}_z^{(1)}) \times \delta(m_x^{(2)} - \hat{\sigma}_x^{(2)})\delta(m_z^{(2)} - \hat{\sigma}_z^{(2)})]\,. \tag{4.2}$$

In Eq. (4.2) above,

$$\rho = |\psi\rangle\langle\psi| \text{ with } |\psi\rangle \text{ given by Eq. (4.1)}\,, \tag{4.3a}$$

$\hat{\sigma}_x^{(i)}$ and $\hat{\sigma}_z^{(i)}$ are the usual Pauli operators

$$\text{for the } i\text{th particle } (i = 1,2), \tag{4.3b}$$

and $m_x^{(i)}$ and $m_z^{(i)}$ are the associated quasiclassical variables.

We must now carry out a calculation for $P(\vec{m}^{(1)},\vec{m}^{(2)})$ similar to the calculation of $P(\vec{m})$ given in Sec. III B; after some algebra, we find

$$P_\phi(\alpha,\beta) = \delta(\alpha - \beta - \pi)\tfrac{1}{2}\left\{\tfrac{1}{2}\left[\delta\left(\alpha - \phi - \frac{\pi}{4}\right) + \delta\left(\alpha - \phi + \frac{\pi}{4}\right)\right] + \tfrac{1}{2}\left[\delta\left(\alpha - \phi - \frac{3\pi}{4}\right) + \delta\left(\alpha - \phi + \frac{3\pi}{4}\right)\right]\right\} \tag{4.4}$$

where α and β are the angular variables corresponding to particles 1 and 2. The physical meaning of the various terms in Eq. (4.4) is discussed in Fig. 4 for the case $\phi = 0$.

We note immediately that the "spin anticorrelation" factor occurring in Eq. (4.4), i.e., the $\delta(\alpha - \beta - \pi)$ term, is

$$P(\alpha,\beta) = \delta(\alpha - \beta - \pi)\,\tfrac{1}{2}\left\{\tfrac{1}{2}\left[\delta(\alpha - \tfrac{\pi}{4}) + \delta(\alpha + \tfrac{\pi}{4})\right] + \tfrac{1}{2}\left[\delta(\alpha - \tfrac{3\pi}{4}) + \delta(\alpha + \tfrac{3\pi}{4})\right]\right\}$$

anticorrelation of spins

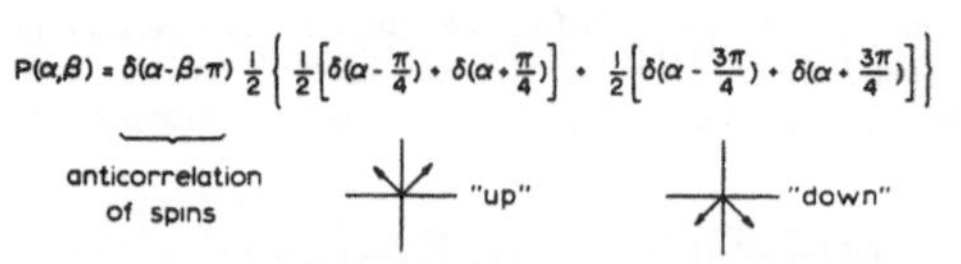

FIG. 4. Physical interpretation of the joint probability density for spin-spin correlation functions.

TABLE II. Table summarizing correspondence between two-particle spin-correlation functions in HV_2, quantum mechanics, and HV_3.

	$P(\alpha,\beta)$ TWO PARTICLE DISTRIBUTION	$\mathscr{P}(\theta,\phi)$ JOINT PROBABILITY
HV_2	$\delta(\alpha-\beta-\pi)\frac{1}{2\pi}$	$\frac{1}{4}(1-\frac{1}{2}\cos(\theta-\phi))$
QM	$\delta(\alpha-\beta-\pi)\frac{1}{2}\left\{\frac{1}{2}\left[\delta(\alpha-\phi-\frac{\pi}{4})+\delta(\alpha-\phi+\frac{\pi}{4})\right]+\frac{1}{2}\left[\delta(\alpha-\phi-\frac{3\pi}{4})+\delta(\alpha-\phi+\frac{3\pi}{4})\right]\right\}$	$\frac{1}{4}(1-\cos(\theta-\phi))$
HV_3	$\delta(\alpha-\beta-\pi)\frac{1}{2}\left\{\delta(\alpha-\phi)+\delta(\alpha-\phi-\pi)\right\}$	$\frac{1}{4}(1-\cos(\theta-\phi))$

identical with that contained in Belinfante's $I(\alpha,\beta)$ given by Eq. (2.6). The difference, of course, is that the present result follows directly from quantum theory. Furthermore the curly bracketed factor corresponds to the sum of single-particle probability density functions for spin along $+\phi$ and $-\phi$.

In view of Eqs. (2.5), (4.4), (3.14), and (3.15), we are motivated to consider the "hidden-variable" probability function

$$\tilde{P}_\phi(\alpha,\beta)=\delta(\alpha-\beta-\pi)\tfrac{1}{2}[\delta(\alpha-\phi)+\delta(\alpha-\phi-\pi)] \ . \tag{4.5}$$

The comparison between Eq. (4.5), Belinfante's equation (2.6), and the quantum-mechanical result [Eq. (4.4)] is summarized in Table II.

As we shall see in the next section $\tilde{P}_\phi(\alpha,\beta)$ allows us to regain the quantum-mechanical result Eq. (1.4).

B. Joint count probabilities in HV_3

The joint count probabilities such as Eqs. (1.4) and (2.9) can now be easily calculated via the general expression

$$\mathscr{P}(\theta,\phi)=\int \tilde{P}_\phi(\alpha,\beta)\tilde{\pi}_\theta(\beta)\tilde{\pi}_\phi(\alpha)d\alpha\, d\beta \ . \tag{4.6}$$

The present hidden-variables "'theory" involves inserting Eqs. (2.4) and (4.5) into Eq. (4.6) to obtain

$$\mathscr{P}(\theta,\phi)=\int \delta(\alpha-\beta-\pi)\tfrac{1}{2}[\delta(\alpha-\phi)+\delta(\alpha-\phi-\pi)]\tfrac{1}{2}[1+\cos(\theta-\beta)]\tfrac{1}{2}[1+\cos(\phi-\alpha)]d\alpha\, d\beta \tag{4.7}$$

and, upon carrying out a couple of simple integrations, this yields

$$\mathscr{P}(\theta,\phi)=\tfrac{1}{4}[1-\cos(\theta-\phi)] \ . \tag{4.8}$$

The present HV_3 result (4.8), agrees with the quantum-mechanical prediction [Eq. (1.4)]. The results of the various theories are summarized in Table II.

V. THE DIFFERENCE BETWEEN HV_3 AND QUANTUM MECHANICS

In this section we treat a somewhat subtle but important point. If one calculates $\mathscr{P}(\theta,\phi)$ via $P_0(\alpha,\beta)$ from Eq. (4.4) with $\phi=0$ we find

$$\begin{aligned}\mathscr{P}(\theta,\phi)&=\int \delta(\alpha-\beta-\pi)\tfrac{1}{2}\left\{\tfrac{1}{2}\left[\delta\left[\alpha-\frac{\pi}{4}\right]+\delta\left[\alpha+\frac{\pi}{4}\right]\right]+\tfrac{1}{2}\left[\delta\left[\alpha-\frac{3\pi}{4}\right]+\delta\left[\alpha+\frac{3\pi}{4}\right]\right]\right\}\\&\quad\times\tfrac{1}{2}[1+m\cos(\theta-\beta)]\tfrac{1}{2}[1+m\cos(\phi-\alpha)]d\alpha\, d\beta\\&=\tfrac{1}{4}[1-\cos(\theta-\phi)] \ ,\end{aligned} \tag{5.1}$$

which is the same as before, i.e., the same as that obtained when $P_\phi(\alpha,\beta)$ from Eq. (4.4) was used. This is as would be expected since use of either spherically symmetric state

$$|\psi\rangle=\frac{1}{\sqrt{2}}[\,|\uparrow_1\downarrow_2\rangle-|\downarrow_1\uparrow_2\rangle]\rightarrow P_0(\alpha,\beta) \tag{5.2a}$$

or

$$|\psi\rangle=\frac{1}{\sqrt{2}}[\,|+_1-_2\rangle-|-_1+_2\rangle]\rightarrow P_\phi(\alpha,\beta) \tag{5.2b}$$

should give the same results.

Note, however, that the operator associated quasiclassical expression $\pi_\phi(\alpha)$, corresponds to and suggests that the particular choice $P_\phi(\alpha,\beta)$ is to be preferred. That is, in any consistent treatment of quantum theory from an asso-

ciated quasiclassical distribution point of view the counterparts to operators $\hat{Q}(\hat{c})$, i.e., $Q(\gamma)$ are chosen to correspond to the operator ordering, etc., associated with $P(\gamma)$.

However, if, within HV_3, we carry out the calculation of $\tilde{\mathscr{P}}(\theta,\phi)$, using $\tilde{P}_0(\alpha,\beta)$ as given in Fig. 4 we would have

$$\tilde{\mathscr{P}}(\theta,\phi)=\int \delta(\alpha-\beta-\pi)\tfrac{1}{2}[\delta(\alpha)+\delta(\alpha-\pi)] \times\tfrac{1}{2}[1+\cos(\theta-\beta)] \times\tfrac{1}{2}[1+\cos(\phi-\alpha)]d\alpha\, d\beta \tag{5.3}$$

which yields

$$\tilde{\mathscr{P}}(\theta,\phi)=\tfrac{1}{4}[1+\tfrac{1}{2}\cos(\theta-\phi)-\tfrac{1}{2}\cos(\theta+\phi)]\ . \tag{5.4}$$

This result is clearly in agreement with the quantum prediction only if $\phi=0$.

The physical content and interpretation of passage (projection) functions, e.g., $\pi_\theta(\alpha)$ and the conditional probability distribution $P_\phi(\alpha,\beta)$ in quantum theory and HV_3 differ, and the roles of state preparation and experimental parametrization enter into the theories in different ways.

In quantum theory state preparation involves choosing $|\psi\rangle$, i.e., $P_\phi(\alpha,\beta)$ and then the experiment in question tells us what operator, e.g., $\hat{\pi}_0$, we should calculate. However, if we change from $P_\phi(\alpha,\beta)$ to $P_0(\alpha,\beta)$ the calculation should remain unchanged since we are considering a spherically symmetric state.

In HV_3 on the other hand the choice of $\tilde{P}_\phi(\alpha,\beta)$, for the present spherically symmetric problem, must "match" the choice of $\pi_\phi(\alpha)$ as the calculations of Eqs. (5.3), (5.4), and (4.8) indicate. Both $\tilde{P}_\phi(\alpha,\beta)$ and $\tilde{\pi}_\phi(\alpha)$ involve state preparation and experimental specification. Thus while HV_3 reproduces the quantum spin-spin correlation function, it differs from quantum theory in this important aspect. As will be developed in more detail elsewhere this implies that HV_3 is a nonlocal theory.

VI. SUMMARY AND COMMENTS

By way of summary, the major points of this paper are as follows:

A. The variable α has a satisfactory and rigorous quantum-mechanical interpretation via quantum distribution theory.

B. The spin analogy of "Malus's cosine law," i.e.,

$$\tilde{\pi}_\theta(\alpha)=\tfrac{1}{2}[1+\cos(\theta-\alpha)] \tag{6.1}$$

is always positive and has a near transcription in quantum distribution theory, namely,

$$\pi_\theta(\alpha)=\tfrac{1}{2}[1+m\cos(\theta-\alpha)]\ , \tag{6.2}$$

where $m=\sqrt{2}$, and $\pi_\theta(\alpha)$ is therefore not positive definite.

C. The quantum mechanical probability density for a spin-up particle is given by

$$P_0(\alpha)=\frac{1}{2}\left[\delta\left[\alpha-\frac{\pi}{4}\right]+\delta\left[\alpha+\frac{\pi}{4}\right]\right]\ , \tag{6.3}$$

and the corresponding HV_3 quantity is

$$\tilde{P}_0(\alpha)=\delta(\alpha)\ , \tag{6.4}$$

which is always positive.

D. The correlation $I(\alpha,\beta)$ implied by Ref. 1,

$$I(\alpha,\beta)\propto(\alpha-\beta-\pi)\ , \tag{6.5}$$

is contained in the quantum treatment as indicated in Eq. (4.4).

E. The HV_3 probability density

$$\tilde{P}_\phi(\alpha,\beta)=\delta(\alpha-\beta-\pi)\tfrac{1}{2}[\delta(\alpha-\phi)+\delta(\alpha-\phi-\pi)] \tag{6.6}$$

taken together with Eq. (2.4) implies

$$\tilde{\mathscr{P}}(\theta,\phi)=\int \tilde{P}_\phi(\alpha,\beta)\tilde{\pi}_\phi(\alpha)\tilde{\pi}_\theta(\beta)d\alpha\, d\beta =\tfrac{1}{4}[1-\cos(\theta-\phi)] \tag{6.7}$$

which reproduces the quantum joint probability distribution for the passage of the first spin through a SGA oriented at ϕ and the second spin through a corresponding apparatus tipped through an angle θ.

F. In quantum mechanics

$$\int P_\phi(\alpha,\beta)\pi_\phi(\alpha)\pi_\theta(\beta)d\alpha\, d\beta =\int P_0(\alpha,\beta)\pi_\phi(\alpha)\pi_\theta(\beta)d\alpha\, d\beta \tag{6.8}$$

and is therefore a "local" theory. However in HV_3

$$\int \tilde{P}_\phi(\alpha,\beta)\tilde{\pi}_\phi(\alpha)\tilde{\pi}_\theta(\beta)d\alpha\, d\beta \neq \int \tilde{P}_0(\alpha,\beta)\tilde{\pi}_\phi(\alpha)\tilde{\pi}_\theta(\beta)d\alpha\, d\beta \tag{6.9}$$

and is, in this sense, "nonlocal."

A central theme of this paper is the observation that quantum mechanics (in the α,β representation) is rather closer to our "HV intuition" than might have been expected. As mentioned earlier, other facets of this work will be presented elsewhere.[16]

ACKNOWLEDGMENTS

The author wishes to thank A. Aspect, I. Białynicki-Birula, A. Barut, J. Clauser, D. Leiter, P. Meystre, and S. Stenholm for their helpful suggestions concerning the manuscript. The author also wishes to express his appreciation to the institutions which supported and encouraged this work, in particular: The present work was stimulated by discussions with A. Barut and P. Meystre at the Max-Planck Institute for Quantum Optics, the calculation was then carried out while a lecturer at the 1982 International School of Physics (Islamabad, Pakistan) and extended at the University of New Mexico Center for Advanced Studies where this paper was written.

[1](a) J. Bardeen, L. Cooper, and J. Schrieffer, Phys. Rev. 108, 1175 (1957); (b) P. Dirac, *Quantum Mechanics* (Oxford University, London, 1935); (c) W. Lamb and R. Retherford, Phys. Rev. 72, 241 (1947); (d) see, for example, K. Huang, *Quarks, Leptons and Gauge Fields* (World Scientific, Singapore, 1982).

[2](a) *Quantum Optics and Electronics,* 1964 Les Houches Lectures, edited by C. DeWitt, A. Blandin, and C. Cohen-Tannoudji (Gordon and Breach, New York, 1965); (b) M. Lax, in *Brandeis University Summer Lectures,* edited by M. Chretien, S. Deser, and E. Gross (Gordon and Breach, New York, 1966); (c) H. Haken, *Handbuch Der Physik* (Springer, Berlin, 1920), Vol. 25/2C; (d) W. H. Louisell, *Quantum Statistical Properties of Radiation* (Wiley, New York, 1973); (e) R. Loudon, *The Quantum Theory of Light* (Oxford University, London, 1973); (f) M. Sargent, M. Scully, and W. Lamb, *Laser Physics* (Addison-Wesley, Reading, Mass., 1974).

[3]A. Einstein, B. Podolsky, and N. Rosen, Phys. Rev. 47, 777 (1935). The present spin-$\frac{1}{2}$ example is due to D. Bohm. In this regard B. Hiley quotes Dirac and Feynman [New Sci. 6 (1981)] as follows:

"Paul Dirac, in referring to the apparent non-local effects that arise in quantum mechanics, wrote, 'It is against the spirit of relativity, but it is the best we can do . . . We cannot be content with such a theory.'"

"Richard Feynman recently wrote, 'We have always had a great deal of difficulty in understanding the world view that quantum mechanics represents . . . It has not yet become obvious to me that there's no real problem.'"

We wish to thank J. Cresser for bringing to our attention the article by E. Jaynes, in *Foundations of Radiation—Theory and Quantum Electrodynamics,* edited by A. Barut (Plenum, New York, 1980), in which he says

"Quantum theory not only does not use—it does not even dare to mention—the notion of a 'real physical situation.' Defenders of the theory say that this notion is philosophically naive, a throwback to outmoded ways of thinking, and that recognition of this constitutes deep new wisdom about the nature of human knowledge. I say that it constitutes a violent irrationality, that somewhere in this theory the distinction between reality and our knowledge of reality has become lost, and the result has more the character of medieval necromancy than of science. It has been my hope that quantum optics, with its vast new technological capability, might be able to provide the experimental clue that will show us how to resolve these contradictions."

The present author's point of view on this subject is to be found in an article by M. Scully, R. Shea, and T. McCullen, Phys. Rep. 43C, 487 (1978).

[4]F. Belinfante, *A Survey of Hidden-Variable Theories* (Pergamon, New York, 1973); J. Clauser and A. Shimony, Rep. Prog. Phys. 41, 1881 (1978).

[5]J. Bell, Rev. Mod. Phys. 38, 447 (1966).

[6]A good summary of experimental tests up to 1978 is given by Clauser and Shimony in Ref. 4. The recent work of Aspect is especially interesting in this regard, see A. Aspect, P. H. Grangier and G. Roger, Phys. Rev. Lett. 49, 91 (1982).

[7]B. Hiley, New Sci. 6, 17 (1983).

[8]See, for example, the review article by M. Hillery, R. O'Connell, M. Scully, and E. Wigner (unpublished).

[9]F. Belinfante, in Ref. 4, p. 13

[10]F. Belinfante, in Ref. 4, p. 279.

[11]In this context see especially E. Wigner, Am. J. Phys. 38, 1005 (1970).

[12]See, for example, C. Cantrell and M. Scully, Phys. Rep. 43C, 499 (1978).

[13]The work of Belinfante referred to here actually involves photon polarization correlations rather than the spin correlations which we deal with. However, the two problems are holomorphic and we present here the arguments couched in spin-$\frac{1}{2}$ language. We emphasize however the intellectual content of the arguments is credited to Belinfante. In fact we have even tried to use his phraseology where appropriate in order to remain faithful to his logic.

[14]See especially R. Glauber, in Ref. 2(a) and L. Mandel and E. Wolf, Rev. Mod. Phys. 37, 231 (1965).

[15]For a nice overview of the subject, see E. Hahn in *NMR Grundlagen und Fortschritte* (Springer, Berlin), Vol. 13.

[16]A. Barut, P. Meystre, and M. Scully (unpublished).

A CLASSICAL MODEL OF EPR EXPERIMENT WITH QUANTUM MECHANICAL CORRELATIONS AND BELL INEQUALITIES

A.O. Barut[1] and P. Meystre

Max-Planck Institute für Quantenoptik
D-8046 Garching
Federal Republic of Germany

A simple model of a classical break-up process is given in which the correlation $E(a, b)$ of the components A and B of the spins of the two subsystems along directions $\boldsymbol{a}$ and $\boldsymbol{b}$ gives precisely the quantum mechanical result $-\cos(\boldsymbol{a}\cdot\boldsymbol{b})$. The model is "local", but the normalization procedure of correlation functions in terms of "hidden variables" is different from that used in deriving Bell's inequalities. A discretization procedure of the classical spins is then given which reproduces fully the dichotomous quantum mechanical results both for probabilities and for correlation functions. This procedure illustrates particularly clearly the difference between quantum and classical spins and provides a possible intuitive picture for the notion of the "reduction of the wave function".

1. Introduction. Despite considerable amounts of literature in recent years, one of the difficulties with the interpretation of Bell's inequalities has been the lack of explicit models of hidden variables with which to compare classical and quantum correlations and to test the assumptions underlying their derivation. It is now generally believed that the so-called "local-realistic" hidden-variable (HV) theories are ruled out by experiment [1,2]. In this paper, we present a simple classical model of correlated events which can in principle be realized experimentally, and which exhibits the same correlations as those obtained in spin-1/2 correlation experiments. Our model is local in the sense of Bell, but is not in conflict with Bell's inequalities, because as we shall see the normalization of the correlation function is performed in different ways in our model and in the derivation of these inequalities.

We then use a technique formally similar to quantum projection operators to construct "joint probabilities" from the correlation function $E(a, b)$. Again, the observed probabilities are the same as in the quantum case, but the corresponding quantities at the "hidden variables" level are *not* positive definite, in contrast to Bell's requirement, and cannot be interpreted as probabilities. This situation can however be remedied by a discretization procedure leading to quantities that can be interpreted as probabilities also at the HV level. We then recover all results of quantum mechanics within a classical theory, but have to use a form of integration measure over the space of "hidden variables" different from that proposed by Bell. This simple example, and the three distinct ways to approach it, illustrate particularly simply the limitations of the hypotheses underlying the derivation of Bell's theorem.

2. Correlation functions. Consider a classical system initially at rest and which splits into two parts with opposite classical spins. The experiment consists in measuring the projections of the momenta (or equivalently of the angular momenta or spins) of the two separated subsystems along two directions $\boldsymbol{a}$ and $\boldsymbol{b}$, respectively (fig. 1). The measured quantities are thus $\boldsymbol{p}\cdot\boldsymbol{a}$ and $-\boldsymbol{p}\cdot\boldsymbol{b}$. We indicate the relevant vectors in a 3-dimensional coordinate system shown in fig. 2.

We have

$$\boldsymbol{p}\cdot\boldsymbol{a} = pa\cos\mu\,, \quad -\boldsymbol{p}\cdot\boldsymbol{b} = -pb\cos\lambda\,, \tag{1}$$

[1] Permanent address: Department of Physics, University of Colorado, Boulder, CO 80309, USA.

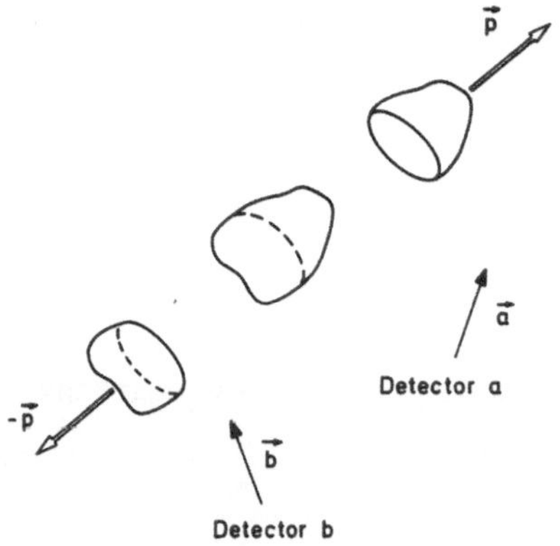

Fig. 1. The break-up process.

where p, a, and b are the magnitudes of the vectors $\boldsymbol{p}$, $\boldsymbol{a}$, and $\boldsymbol{b}$. With respect to the axes as chosen in fig. 2,

$$A \equiv \boldsymbol{p}\cdot\boldsymbol{a} = pa \sin\theta \cos\varphi ,$$
$$B \equiv -\boldsymbol{p}\cdot\boldsymbol{b} = -pb(\cos\theta_0 \sin\theta \sin\varphi + \sin\theta_0 \sin\theta \sin\varphi) , \quad (2)$$

where θ_0 is the angle between $\boldsymbol{a}$ and $\boldsymbol{b}$, $\cos\theta_0 = \boldsymbol{a}\cdot\boldsymbol{b}/ab$. We repeat the experiment a large number of times, and assume that the directions of the subsystems after splitting are randomly distributed in all directions. After averaging, we then obtain $\langle A\rangle = \langle B\rangle = 0$,

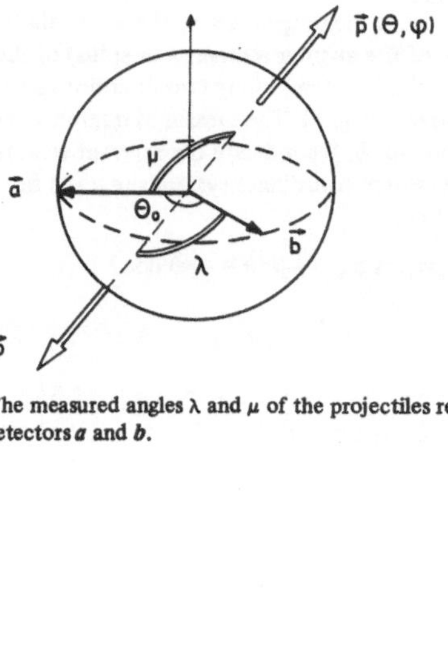

Fig. 2. The measured angles λ and μ of the projectiles relative to the detectors $\boldsymbol{a}$ and $\boldsymbol{b}$.

$$\langle A^2\rangle = \langle\langle(\boldsymbol{p}\cdot\boldsymbol{a})^2\rangle\rangle = \frac{(pa)^2}{4\pi}\int_0^{2\pi} d\varphi \int_0^{\pi} d\theta \sin\theta \cos^2\mu = (pa)^2/3 , \quad (3)$$

and similarly

$$\langle B^2\rangle = (pb)^2/3 . \quad (4)$$

The correlation $\langle AB\rangle = \langle\langle(\boldsymbol{p}\cdot\boldsymbol{a})(-\boldsymbol{p}\cdot\boldsymbol{b})\rangle\rangle$ is found to be

$$\langle AB\rangle = -\frac{1}{4\pi}\iint d\theta\, d\varphi\, papb \sin\theta \cos\lambda \cos\mu = -\tfrac{1}{3}p^2ab \cos\theta_0 . \quad (5)$$

The averages (3) to (5) still depend on the magnitudes of the various vectors. This is to be expected, since a classical spin can have any magnitude and therefore any value for a given component. If we are only interested in information about the direction of $\boldsymbol{p}$, the relevant quantity is the normalized correlation function

$$E_c(a,b) = \frac{\langle AB\rangle}{\langle A^2\rangle^{1/2}\langle B^2\rangle^{1/2}} = \left\langle \frac{A}{\langle A^2\rangle^{1/2}}\,\frac{B}{\langle B^2\rangle^{1/2}}\right\rangle \equiv \widetilde{A}\widetilde{B} = -\cos\theta_0 . \quad (6)$$

This is identical with the correlation function obtained in the measurement of the spin components of a quantum mechanical spin-0 coherent state splitting into two spin-1/2 states:

$$E_Q(a,b) = \langle\psi|\tau_1\boldsymbol{a}\otimes\sigma_2\boldsymbol{b}|\psi\rangle , \quad (7)$$

where $|\psi\rangle = 2^{-1/2}[\uparrow\downarrow - \downarrow\uparrow]$.

In Bell's proof of the inequality [2]

$$-2 \leqslant E(a,b) - E(a,b') + E(a',b) + E(a',b') \leqslant 2 , \quad (8)$$

a crucial step is to write $E(a,b) = \int d\Lambda\, A_a(\Lambda)B_b(\Lambda)$, where Λ is a complete set of HV, and to assume that the expectation values of the unobserved $A_a(\Lambda)$ and $B_b(\Lambda)$ are bounded by one. This is an assumption, dictated by reasons to which we will return later on. In our case, we normalize the correlation function correctly by dividing it by the "magnitudes":

$$\widetilde{A}(\mu) = pa\cos\mu/\langle\langle(\boldsymbol{p}\cdot\boldsymbol{a})^2\rangle\rangle^{1/2} = \sqrt{3}\cos\mu ,$$
$$\widetilde{B}(\lambda) = -pb\cos\lambda/\langle\langle(\boldsymbol{p}\cdot\boldsymbol{b})^2\rangle\rangle^{1/2} = -\sqrt{3}\cos\lambda . \quad (9)$$

It is easily seen that in this case, Bell's proof does not

follow. It is important to emphasize, however, that in our classical example, it is the procedure of normalization of the correlation function, rather than "nonlocality", that leads to a violation of Bell's inequalities.

We see, then, that the measured correlation functions alone do not tell too much about the details of the break-up process and can be reproduced by a simple classical picture. They merely describe the kinematics of the process, nothing more. In the next section, we show that joint probabilities already give a better indication of the difference between classical and quantum mechanics.

3. Probabilities. In experiments of the type performed e.g. by Aspect et al. [1], one does not measure directly the correlation function $E(a, b)$, but rather, the joint probabilities $P_{++}(a, b)$, $P_{+-}(a, b)$, etc. for the a- and b-component of the first and second spin, respectively, to be $+\hbar/2$ and $\pm\hbar/2, \dots$. These are obtained by decomposing the operator $\sigma_1 \boldsymbol{a} \otimes \sigma_2 \boldsymbol{b}$ in terms of projection operators as

$$4\sigma_1 \boldsymbol{a} \otimes \sigma_2 \boldsymbol{b} = (1 + \sigma_1 \boldsymbol{a})(1 + \sigma_2 \boldsymbol{b}) + (1 - \sigma_1 \boldsymbol{a})(1 - \sigma_2 \boldsymbol{b}) - (1 - \sigma_1 \boldsymbol{a})(1 + \sigma_2 \boldsymbol{b}) - (1 + \sigma_1 \boldsymbol{a})(1 - \sigma_2 \boldsymbol{b}) . \quad (10)$$

This yields

$$E_Q(a, b) = -\cos\theta_0 = \sin^2(\tfrac{1}{2}\theta_0) + \tfrac{1}{2}\sin^2(\tfrac{1}{2}\theta_0) - \tfrac{1}{2}\cos^2(\tfrac{1}{2}\theta_0) - \tfrac{1}{2}\cos^2(\tfrac{1}{2}\theta_0) \equiv P_{++}(a, b) + P_{--}(a, b) - P_{-+}(a, b) - P_{+-}(a, b) . \quad (11)$$

In order to find a counterpart to this decomposition in our model, we use the identity

$$4\cos\lambda\cos\mu = (\alpha + \cos\lambda)(\beta + \cos\mu) + (\alpha - \cos\lambda)(\beta - \cos\mu) - (\alpha - \cos\lambda)(\beta + \cos\mu) - (\alpha + \cos\lambda)(\beta - \cos\mu) , \quad (12)$$

valid for arbitrary constants α and β.

By an appropriate choice of α and β we can find the observables whose correlation functions give the quantum mechanical ones $P_{ij}(a, b)$, $(i, j = \pm)$, namely

$$X_\pm(a, \Lambda) = \tfrac{1}{2}(1 \pm \widetilde{A}) , \quad Y_\pm(b, \Lambda) = \tfrac{1}{2}(1 \pm \widetilde{B}) . \quad (13)$$

In our case $\Lambda = [\lambda, \mu]$, and $X_\pm = \frac{1}{2}(1 \pm \sqrt{3}\cos\mu)$ and $Y_\pm = \frac{1}{2}(1 \mp \sqrt{3}\cos\lambda)$ are a measure of the correlation of the original observables A and B relative to their mean square deviations. Indeed, we find

$$P_{ij}(a, b) = \langle X_i Y_j \rangle = \tfrac{1}{4}(1 \pm \cos\theta_0) \quad (14)$$

(i.e., $\frac{1}{4}(1 - \cos\theta_0)$ for P_{++} and P_{--}, and $\frac{1}{4}(1 + \cos\theta_0)$ for P_{+-} and P_{-+}). We can now see why Bell's inequalities are inoperative here. If we write

$$P_{ij}(a, b) = \frac{1}{4\pi}\int d\Lambda\, P_{ij}(a, b, \Lambda) = \frac{1}{4\pi}\int d\Lambda\, P_i(a, \Lambda) P_j(b, \Lambda) , \quad (15)$$

then, our model is local in the sense of Bell, and all the final observed quantities (integrated over HV) are the same as quantum mechanically, but the densities $P_i(a, \Lambda) \equiv X_i(a, \Lambda) = \frac{1}{2}[1 \pm \widetilde{A}(a, \lambda)]$ are not positive definite. (Note that they take the same form in 3 dimensions as the "probabilities" obtained by Scully when describing the quantum mechanical problem with a Wigner distribution function [3].) Thus, they cannot be interpreted as probabilities.

In experiments aimed at testing Bell's theorem, the system may conceptually be regarded as a blackbox connected to four lamps, two on each side, which always light up in pairs. One measures then the joint probability that a given pair of lamps light up by accumulating the results of single experiments [‡1]. Thus the experiment is reduced to counting. This is why one insists on having positive definite densities $P_i(a, \Lambda)$ and $P_j(b, \Lambda)$ for each single event. (It is a similar argumentation which leads to the requirement that $A_a(\Lambda)$ and $B_b(\Lambda)$ be bounded by 1.) We now show how this can be done in our model.

4. Discretization. Although all final correlations, probabilities, and joint probabilities are the same for both the classical and quantum mechanical cases, and apart from the difficulty with negative densities, there is clearly a fundamental difference between a quantized and a classical spin. In the quantum case, each event contributes to exactly one count in one of the $P_{ij}(a, b)$, say $P_{++}(a, b)$. In the classical case, each event contributes, in general, a certain amount to all of the $P_{ij}(a, b)$. This is of course related to another feature of the quantum theory of measurement, namely the collapse of the wave function. We do not wish to erase this difference. However, we can operationally discretize our continuous measurement so that the final recording of events will be exactly like the quantum mechanical "yes" or "no" experiment, with the same probabilities and correlation functions.

For this purpose, we construct a detector that first measures both observables $X_+(a, \Lambda)$ and $X_-(a, \Lambda)$, and compares them. The system is then instructed

[‡1] For a tutorial discussion of this point, see ref. [4].

to associate the region where $X_+(a, \Lambda) > X_-(a, \Lambda)$ with spin "up", or "+", relative to a. Assuming, without loss of generality, a and b to be in the equatorial plane, as shown in fig. 2, and generalizing slightly from our preceding discussion to label a by the angle φ_a (instead of zero), the region of angles for which $X_+ > X_-$ is given by

$$-\pi/2 \leqslant \varphi - \varphi_a \leqslant \pi/2 . \tag{16}$$

This is the upper hemisphere with the vector a as north pole. The lower hemisphere will then be spin "down". Similarly the region for which $Y_- > Y_+$ is given by

$$-\pi/2 \leqslant \varphi - \varphi_b \leqslant \pi/2 , \tag{17}$$

i.e. the upper hemisphere with vector b as north pole. (Note that the spin S_2 of the second particle is opposite to the spin S_1 of the first one, $S_2 = -S_1$, so that the component of S_2 in the upper half-plane is negative.)

The joint probability of, say, X_+ and Y_- occurs then in the intersection of the two regions (16) and (17). Similar considerations hold for X_+Y_+, X_-Y_- and X_-Y_+.

Recording each event at an angle φ with a weight factor $N|\cos(\varphi - \varphi_a)|$, respectively $N|\cos(\varphi - \varphi_b)|$, where $N = 1/4$ is a normalization constant yields for the single probabilities

$$P_+(a) = \frac{1}{4} \int_0^{2\pi} d\varphi |\cos(\varphi - \varphi_a)|$$
$$\times [H(\varphi - (\varphi_a - \pi/2)) - H(\varphi - (\varphi_a + \pi/2))]$$
$$= \frac{1}{4} \int_{\varphi_a - \pi/2}^{\varphi_a + \pi/2} d\varphi |\cos(\varphi - \varphi_a)| = 1/2 , \tag{18a}$$

where $H(x)$ is the Heaviside function. This is just the sum of all spins in the upper hemisphere with respect to a, weighted by the factor

$$\Pi(\varphi, \varphi_a) = \tfrac{1}{4} |\cos(\varphi - \varphi_a)| \tag{19}$$

i.e., the sum of the projections of the spin components along the detector axis a. Similarly, $P_-(a) = 1/2$, and

$$P_-(b) = \frac{1}{4} \int_0^{2\pi} d\varphi |\cos(\varphi - \varphi_b)|$$
$$\times [H(\varphi - (\varphi_b - \pi/2)) - H(\varphi - (\varphi_b + \pi/2))]$$
$$= \frac{1}{4} \int_{\varphi_b - \pi/2}^{\varphi_b + \pi/2} d\varphi |\cos(\varphi - \varphi_b)| = 1/2 . \tag{18b}$$

Note the difference between this and the averaging procedure used in the derivation of Bell's inequalities. In this last case, one uses the same measure to integrate $P_i(a, \Lambda)$, $P_j(b, \Lambda)$, and $P_{ij}(a, b, \Lambda)$ over the hidden-variables space:

$$P_i(a) = \int d\Lambda\, \rho(\Lambda) P_i(a, \Lambda) , \tag{20a}$$

$$P_j(b) = \int d\Lambda\, \rho(\Lambda) P_j(b, \Lambda) , \tag{20b}$$

$$P_{ij}(a, b) = \int d\Lambda\, \rho(\Lambda) P_i(a, \Lambda) P_j(b, \Lambda) , \tag{20c}$$

while in our case, the measure is replaced by a "projector" on the detectors axes a or b.

In our detection scheme, the joint probabilities are given by the intersection of the two relevant hemispheres, with a weight function that may be chosen as either that of the first detector, or of the second detector (or the average of both). For example:

$$P_{+-}(a, b) = \int_0^{2\pi} d\varphi\, \Pi(\varphi, \varphi_a)$$
$$\times [H(\varphi - (\varphi_a - \pi/2)) - H(\varphi - (\varphi_a + \pi/2))]$$
$$\times [H(\varphi - (\varphi_b - \pi/2)) - H(\varphi - (\varphi_b + \pi/2))]$$
$$= \int_0^{2\pi} d\varphi\, \Pi(\varphi, \varphi_b)$$
$$\times [H(\varphi - (\varphi_a - \pi/2)) - H(\varphi - (\varphi_a + \pi/2))]$$
$$\times [H(\varphi - (\varphi_b - \pi/2)) - H(\varphi - (\varphi_b + \pi/2))]$$
$$= \int_{\varphi_b - \pi/2}^{\varphi_a + \pi/2} d\varphi\, \Pi(\varphi, \varphi_a) = \tfrac{1}{4}(1 + \cos\theta_0) . \tag{18c}$$

This definition of the joint probability $P_{+-}(a, b)$, while quite different from the prescription of local-realistic HV theories, presents interesting features. Most notably, in the evaluation of joint probabilities one applies only one of the "projectors" $\Pi(\varphi, \varphi_a)$ or $\Pi(\varphi, \varphi_b)$. This is reminiscent of the quantum mechanical reduction of the wave packet: in that last case, once one of the spins is measured, then, the value of the other one is known with certainty. Our prescription does essentially the same thing: once one of the spins has been "projected" on its detection axis, no further projection is necessary. Thus, we can simulate the results of quantum mechanics by a completely classical descretization procedure.

Is this procedure local? The single probabilities (18a) and (18b) do not depend on the setting of the other detector. But the way $P_{ij}(a, b)$ is constructed, although independent of the relative setting of the two detectors, does depend on the fact that both detectors are present. This, of course, is also true in local-realistic HV theories. But the difference is that in that last case, one multiplies the single probabilities [eq. (20c)] while in our model $P_{ij}(a, b)$ is determined by the overlap region shown in fig. 3. In this region, the events at $\boldsymbol{a}$ are weighted by the "projector" (19), and the events at $\boldsymbol{b}$ with probability one (or vice versa) which introduces explicitly the fact that these are absolutely correlated events. (If the component of the first spin along $\boldsymbol{a}$ is known, we know for sure the value of the component of the second spin along $\boldsymbol{b}$.) Whether our system can be called local, in some generalized sense, is then probably a question of taste.

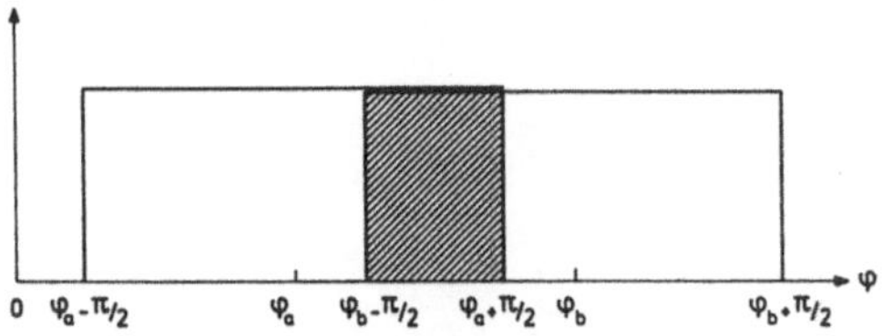

Fig. 3. Overlap region of spin up for the first particle and down for the second one.

5. Conclusions. We have compared the quantum mechanical spin correlation function (7) with the corresponding classical spin correlation (6) calculated completely classically, and found the same result in both cases. We also computed joint probabilities, and again found the same final observable results. Bell's inequalities are based on supplementary assumptions about the probability densities in the hidden variables space. There are no additional assumptions in our model.

Since quantum spins are discrete and the experiments yield discrete "yes" or "no" counts, but classical spins are continuous, we have introduced a natural discretization process counting all spins in the upper hemisphere relative to the observer as "up", and all spins in the lower hemisphere as down, together with a suitable measure. With this, our experiment also reproduces the discrete quantum mechanical correlations. We therefore conclude that a classical experiment can reproduce exactly observed quantum mechanical correlations, but that some of the assumptions underlying Bell's inequalities have to be removed. Our model further sheds some light on the differences and similarities between the classical and quantum mechanical spins and the corresponding spin correlation observables.

References

[1] A. Aspect, J. Dalibard and G. Roger, Phys. Rev. Lett. 49 (1982) 1804.
[2] J.F. Clauser and A. Shimony, Rep. Progs. Phys. 41 (1978) 1881.
[3] M.O. Scully, Phys. Rev. D28 (1983) 2477.
[4] N.D. Mermin, Am. J. Phys. 49 (1981) 940.

REDUCTION OF THE WAVEPACKET: HOW LONG DOES IT TAKE?

W.H. Zurek

Theoretical Astrophysics
Los Alamos National Laboratory
Los Alamos, New Mexico 87545
and
Institute for Theoretical Physics
University of California
Santa Barbara, California 93106

ABSTRACT

We show that the "reduction of the wavepacket" caused by the interaction with the environment occurs on a timescale which is typically many orders of magnitude shorter than the relaxation timescale τ. In particular, we show that in a system interacting with a "canonical" heat bath of harmonic oscillators decorrelation timescale of two pieces of the wavepacket separated by N thermal de Broglie wavelengths is approximately τ/N^2. Therefore, in the classical limit $\hbar \to 0$ dynamical reversibility ($\tau \to \infty$) is compatible with "instantaneous" coherence loss.

I. INTRODUCTION

It is sometimes argued that observables of macroscopic objects which obey, to a good approximation, reversible classical dynamics -- i.e. their relaxation timescale τ is, for all practical purposes, infinite -- could not have lost coherence and become "classical" due to the interaction with the environment through environment-induced superselection.[1-5] For, the reasoning goes, relaxation rate is the measure of the strength of the coupling with the environment. In particular, when $\tau \to \infty$ one can neglect dissipation of energy. Consequently, one should be equally justified in neglecting any influence of the environment. We show that this argument is fallacious in an example of a free particle interacting with the environment of quantum oscillators in the high-temperature weak coupling limit. In particular, we show that the coherence between two pieces of the wavepacket Δx apart is lost on a decorrelation timescale θ which is typically

$$\theta = \tau[(\hbar/\sqrt{4mkT})/\Delta x]^2. \tag{1}$$

Here, m is the mass of the particle, k is Boltzmann's constant, and T is temperature. For "canonical" classical systems ($m \sim 1g$, $T \sim 300^{\circ}K$) and standard "macroscopic" separations $\Delta x \sim 1cm$, $\theta/\tau \sim 10^{-40}$. Moreover, in the classical limit, $\hbar \to 0$, $\theta/\tau \to 0$. This enormous disparity between the two timescales can be regarded as the explanation of the apparent

"instantaneous" collapse of the state vector of macroscopic objects, including distinguishable (i.e. separated by many de Broglie wavelength* $\lambda_{dB} = \hbar/\sqrt{4mkT}$) outcomes of measurements performed by a classical apparatus on a quantum system.

II. DECORRELATION OF A "FREE PARTICLE"

Consider an otherwise free particle of mass m interacting with the environment of many harmonic oscillators via the Hamiltonian:

$$H_{INT} = x \sum c_i q_i \; . \tag{2}$$

Above, x is the coordinate of the free particle while q_i are the coordinates of harmonic oscillators. This interaction Hamiltonian was used extensively in many earlier discussions of relaxation,[7,8] and, more recently, it is being used in calculations of dephasing in a harmonic oscillator.[9]

We note that H_{INT}, Eq. (1), commutes with the position observable of the free particle:

$$[H_{INT}, x] = 0 \; . \tag{3}$$

Therefore, position can be regarded as pointer observable,[1,3,5] measured continuously by the environment of harmonic oscillators. In the absence of the self-Hamiltonian:

$$H_0 = -(\hbar^2/2m)(\partial/\partial x)^2 \tag{4}$$

x would be a constant of motion. One would then expect combined system-environment state vector to evolve from an initial, uncorrelated state

$$|\Phi_0\rangle = |\psi\rangle|\varepsilon\rangle$$

into the time dependent, correlated state:

$$|\Phi_t\rangle \propto \int dx \; |\psi(x)\rangle|\varepsilon_x\rangle \quad .$$

Tracing out an environment after it has performed idealized, perfect "measurement" -- i.e. after states of the environment are correlated with different positions become orthogonal, $\langle\varepsilon_x|\varepsilon_y\rangle \sim \delta(x-y)$ -- yields, for the system, the density matrix:

$$\rho \propto \int dx \; |\psi(x)|^2 \; |x\rangle\langle x| \tag{5}$$

This density matrix is diagonal in the pointer basis $\{|x\rangle\}$.

In the more realistic case of finite H_0, the density matrix ρ will not achieve perfect diagonalization, Eq. (5). Rather, it will have a finite correlation length $\sim \lambda_{dB}$. Moreover, the distribution will become uniform, $\langle x|\rho|x\rangle$ = const., on a relaxation timescale. The estimate of the timescales of these two processes can be obtained from the effective master equation for the free particle. We shall use it in the form given by Caldeira and Leggett[7]. Its three consecutive terms correspond to the von Neumann's equation for the density matrix of a free particle, to the dissipation with viscosity $\eta = 2m\gamma$, and to the fluctuating force responsible for Brownian motion:

---

*The more popular definition of thermal de Broglie wavelength is $\lambda_T^2 = h^2/2\pi mkT$. It differs by a factor $\sqrt{\pi/2}$ ($\lambda_{dB}^2 = (2/\pi)\lambda_T^2$) from the de Broglie wavelength λ_{dB} we shall use here.

$$\dot{\rho} = \{(i\hbar/2m)(\partial^2/\partial x^2 - \partial^2/\partial y^2) - \gamma(x-y)\left(\frac{\partial}{\partial x} - \frac{\partial}{\partial y}\right)$$
$$- (2m\pi kT/\hbar^2)(x-y)^2\}\rho \ . \tag{6}$$

To compare relaxation and decorrelation timescales we consider an initial wavepacket of half-width δ. As we shall argue in the next section, this half-width will be typically of the order of the de Broglie wavelength. We now suppose that the initial wavepacket has been "split," coherently, into two pieces, $|\alpha\rangle$ and $|\beta\rangle$, so that the free particle is described by the wave function:

$$|\psi\rangle = (|\alpha\rangle + |\beta\rangle)/\sqrt{2} \ . \tag{7}$$

Here we assume for simplicity:

$$\langle x|\alpha\rangle = (2\pi\delta^2)^{-1/4} \exp[-(x - \Delta x/2)^2/4\delta^2] \ , \tag{8a}$$

$$\langle x|\beta\rangle = (2\pi\delta^2)^{-1/4} \exp[-(x - \Delta x/2)^2/4\delta^2] \ . \tag{8b}$$

The resulting initial density matrix

$$\rho = |\psi\rangle\langle\psi| \tag{9}$$

has, in the position representation, four extrema. Two of them occur on the diagonal: (1) $x = y = \Delta x/2$; (2) $x = y = -\Delta x/2$. They are the maxima of $|\langle x|\alpha\rangle|^2$ and $|\langle x|\beta\rangle|^2$. In addition, there are off-diagonal maxima of $\langle x|\alpha\rangle\langle\beta|y\rangle$ and of its Hermitian conjugate which lie at: (3) $x = -y = \Delta x/2$; (4) $x = -y = -\Delta x/2$. The size of these off-diagonal maxima provides a measure of the coherence between $|\alpha\rangle$ and $|\beta\rangle$.

The rate of change of the diagonal terms due to the interaction with the environment can be estimated by calculating, from Eq. (6),

$$\tau^{-1} = \langle\alpha_t|\dot{\rho}|\alpha_t\rangle \cong -(\gamma/2)\langle\alpha_t|(x-y)^2|\alpha_t\rangle(1/\delta^2 + 1/\lambda_{dB}^2) \tag{10a}$$

Here $|\alpha_t\rangle = \exp(-iH_0t/\hbar)|\alpha$ was used to separate out the evolution due to the environment from the evolution induced by the self-Hamiltonian H_0. Similarily, the rate of change of the off-diagonal term is:

$$\theta^{-1} = \langle\alpha_t|\dot{\rho}|\beta_t\rangle \cong -(\gamma/2)\langle\alpha_t|(x-y)^2|\beta_t\rangle(1/\delta^2 + 1/\delta_{dB}^2) \tag{10b}$$

The key and only difference between the two rates is then the size of the factor $\langle|(x-y)^2|\rangle$. For the diagonal terms it is given by

$$\langle\alpha_t|(x-y)^2|\alpha_t\rangle = \delta^2 \sim \lambda_{dB}^2 \ . \tag{11a}$$

For the off-diagonal elements, it is, on the other hand

$$\langle\alpha_t|(x-y)^2|\beta_t\rangle = \Delta x^2 \ . \tag{11b}$$

Therefore, the ratio of the two rates is

$$\tau/\theta = (\Delta x/\delta)^2 \sim (\Delta x/\lambda_{dB})^2 \tag{12}$$

in accord with Eq. (1). For "macroscopic" values of Δx, m, and T, this ratio is enormous and enforces environment-induced superselection. It is worth pointing out that qualitative conclusions of our discussion are in accord with more elaborate path integral treatment of the harmonic

oscillator, given recently by Caldeira and Leggett.[9]

III. DISCUSSION: A CLASSICAL LIMIT?

In the previous section we have shown that when $\delta \sim \lambda_{dB}$, decorrelation is much more rapid than relaxation. The purpose of this section is to justify why, in the practical context, the assumption $\delta \sim \lambda_{dB}$ is natural. Moreover, we shall briefly point out consequences of the disparity between the two timescales for the interpretation of quantum mechanics.

Let us first consider a classic example of measurement, patterned after the one discussed by von Neumann.[10] We couple the measured system, initially in a state $|\phi\rangle$, with the free particle measuring apparatus, so that their total Hamiltonian is

$$H = H_{SYSTEM} + H_0 - i\hbar\, \Delta x\, \delta(t-t_0) P(\partial/\partial x) \ . \tag{13}$$

Here P is the measured operator which we assume has 0 and 1 as the eigenvalues, while x is the position of the free particle which will record the outcome of the measurement.

Just before the observation the state of the apparatus must be determined with the accuracy better than Δx. If the free particle apparatus is already in contact with the heat bath of temperature T, as discussed previously, then the measurement of its position with some accuracy σ, $\Delta x \gg \sigma \gg \lambda_{dB}$, will be a typical, sufficient preparation. Therefore, the apparatus will be left in an incoherent mixture of $n = \sigma/\lambda_{dB}$ wavelets. Such inexhaustive measurements may be not only "realistic," but also advantageous, as the resulting mixture will spread slower than the pure wavepacket of comparable width.[11] In the course of the interaction at $t = t_0$, each of the de Broglie-sized wavepackets will be split into an "unmoved" $|\alpha\rangle$ portion, and into the shifted one: $\exp(-i\Delta x \partial/\partial x)|\alpha\rangle = |\beta\rangle$. Therefore, immediately after the observation, the state of the combination (system-free particle apparatus) is, in effect, described by a mixture of terms of the form:

$$|T\rangle = |\alpha\rangle(1-P)|\phi\rangle + |\beta\rangle P|\phi\rangle \tag{14}$$

with all the $|\alpha\rangle$ contained within σ. Now the analysis of the decay of the pure state $|T\rangle$ into the density matrix of the form

$$\rho = |\alpha\rangle\langle\alpha|(1-P)|\phi\rangle\langle\phi|(1-P) + |\beta\rangle\langle\beta|P|\phi\rangle\langle\phi|P \tag{15}$$

can be conducted in accord with the discussion of the previous section. In particular, $\delta \sim \lambda_{dB}$ will apply as long as the resolution σ of the measurement which prepares free-particle apparatus is worse than λ_{dB}. Moreover, even if $\sigma < \lambda_{dB}$, our qualitative conclusions still hold, as in that case decorrelation will be even more rapid.

The most intriguing corollary of our discussion is, perhaps, the possibility that in the classical limit of $\hbar \to 0$ the relaxation timescale may approach infinity,

$$\tau \to \infty \tag{14}$$

which allows the system to follow reversible, Newtonian dynamics, and yet the decorrelation timescale will remain arbitrarily short, or, indeed, it may approach zero:

$$\theta \to 0 \ . \tag{15}$$

We regard this limit as a true classical limit: Not only does it allow classical Newton's equations of motion, but it also prevents long-range quantum correlations, by imposing the environment-induced superselection.[1,2]

It is worth stressing that the loss of coherence and the accompanying "irreversibility" is a consequence of the deliberate tracing out of the environment, which disposes of the mutual information,[5] and not of the approximations involved in the derivation of Eq. (6). This is particularly clearly demonstrated by the analogous results obtained by path-integral methods for harmonic oscillators.[9]

I would like to thank Amir Caldeira and Dan Walls for discussions on the subject of this paper. This research was supported by the National Science Foundation under Grant No. PHY77-27084, supplemented by funds from the National Aeronautics and Space Administration.

REFERENCES

1. W.H. Zurek, "Pointer Basis of Quantum Apparatus: Into What Mixture Does the Wavepacket Collapse?," Phys. Rev. D 24:1516 (1981).
2. W.H. Zurek, "Environment-Induced Superselection Rules," Phys. Rev. D 26:1862 (1982).
3. E.P. Wigner, "Review of Quantum Mechanical Measurement Problem," p. 43 of Ref. 6.
4. H.D. Zeh, "On the Irreversibility of Time and Observation in Quantum Theory," in Foundations of Quantum Mechanics, B. d'Espagnat, ed., Academic Press, New York (1971).
5. W.H. Zurek, "Information Transfer in Quantum Measurements: Irreversibility and Amplification," p. 87 in Ref. 6.
6. P. Meystre and M.O. Scully, eds., Quantum Optics, Experimental Gravitation and Measurement Theory, NATO ASI Series, Plenum Press, New York (1983).
7. A.O. Caldeira and A.J. Leggett, "Path Integral Approach to Quantum Brownian Motion," Physica 121A:587 (1983), and references therein.
8. A.J. Leggett, "The Superposition Principle in a Macroscopic System," p. 74 in Proceedings of the International Symposium on Foundation of Quantum Mechanics, S. Kamefuchi et al., eds., Physical Soc. of Japan, Tokyo (1983).
9. A.O. Caldeira and A.J. Leggett, "Influence of Damping on Quantum Interference: An Exactly Soluble Model," Phys. Rev. A, submitted.
10. J. von Neumann, Mathematical Foundations of Quantum Mechanics, Princeton University Press, Princeton (1955).
11. N.S. Krylov, Works on Foundations of Statistical Physics, Princeton University Press, Princeton (1979).

MAXWELL'S DEMON, SZILARD'S ENGINE AND QUANTUM MEASUREMENTS

W.H. Zurek

Theoretical Astrophysics
Los Alamos National Laboratory
Los Alamos, New Mexico 87545

and

Institute for Theoretical Physics
University of California
Santa Barbara, California 93106

ABSTRACT

We propose and analyze a quantum version of Szilard's "one-molecule engine." In particular, we recover, in the quantum context, Szilard's conclusion concerning the free energy "cost" of measurements: $\Delta F \geqslant k_B T \ln 2$ per bit of information.

I. INTRODUCTION

In 1929 Leo Szilard wrote a path-breaking paper entitled "On the Decrease of Entropy in a Thermodynamic System by the Intervention of Intelligent Beings."[1] There, on the basis of a thermodynamic "gedanken experiment" involving "Maxwell's demon," he argued that an observer, in order to learn, through a measurement, which of the two equally probable alternatives is realized, must use up at least

$$\Delta F = k_B T \ln 2 \tag{1}$$

of free energy. Szilard's paper not only correctly defines the quantity known today as information, which has found a wide use in the work of Claude Shannon and others in the field of communication science.[3] It also formulates physical connection between thermodynamic entropy and information-theoretic entropy by establishing "Szilard's limit," the least price which must be paid in terms of free energy for the information gain.

The purpose of this paper is to translate Szilard's classical thought experiment into a quantum one, and to explore its consequences for quantum theory of measurements. A "one-molecule gas" is a "microscopic" system and one may wonder whether conclusions of Szilard's classical analysis remain valid in the quantum domain. In particular, one may argue, following Jauch and Baron,[4] that Szilard's analysis is inconsistent, because it employs two different, incompatible classical idealizations of the one-

molecule gas--dynamical and thermodynamical--to arrive at Eq. (1). We shall show that the apparent inconsistency pointed out by Jauch and Baron is removed by quantum treatment. This is not too surprising, for, after all, thermodynamic entropy which is central in this discussion is incompatible with classical mechanics, as it becomes infinite in the limit $\hbar \to 0$. Indeed, information--theoretic analysis of the operation of Szilard's engine allows one to understand, in a very natural way, his thermodynamical conclusion, Eq.(1), as a consequence of the conservation of information in a closed quantum system.

The quantum version of Szilard's engine will be considered in the following section. Implications of Szilard's reasoning for quantum theory and for thermodynamics will be explored in Sec. III--where we shall assume that "Maxwell's Demon" is classical, and in Sec. IV, where it will be a quantum system.

II. QUANTUM VERSION OF SZILARD'S ENGINE

A complete cycle of Szilard's classical engine is presented in Fig. 1. The work it can perform in the course of one cycle is

$$\Delta W = \int_{V/2}^{V} p(v)dv = k_B T \int_{V/2}^{V} dv/v = k_B T \ln 2 \qquad (2)$$

Above, we have used the law of Gay-Lussac, $p=kT/V$, for one-molecule gas. This work gain per cycle can be maintained in spite of the fact that the whole system is at the same constant temperature T. If Szilard's model engine could indeed generate useful work, in the way described above at no additional expense of free energy, it would constitute a perpetuum mobile, as it delivers mechanical work from an (infinite) heat reservoir with no apparent temperature difference. To fend off this threat to the thermodynamic irreversibility, Szilard has noted that, "If we do not wish to admit that the Second Law has been violated, we must conclude that...the measurement...must be accompanied by a production of entropy." Szilard's conclusion has far-reaching consequences, which have not yet been fully explored. If it is indeed correct, it can provide an operational link between the concepts of "entropy" and "information." Moreover, it forces one to admit that a measuring apparatus can be used to gain information only if measurements are essentially irreversible.

Before accepting Szilard's conclusion one must realize that it is based on a very idealized model. In particular, two of the key issues have not been explored in the original paper. The first, obvious one concerns fluctuations. One may argue that the one-molecule engine cannot be analyzed by means of thermodynamics, because it is nowhere near the thermodynamic limit. This objection is overruled by noting that arbitrarily many "Szilard's engines" can be linked together to get a "many-cylinder" version of the original design. This will cut down fluctuations and allow one to apply thermodynamic concepts without difficulty.

A more subtle objection against the one-molecule engine has been advanced by Jauch and Baron.[4] They note that "Even the single-molecule gas is admissible as long as it satisfies the gas laws. However, at the exact moment when the piston is in the middle of the cylinder and the opening is closed, the gas violates the law of Gay-Lussac because gas is compressed to half its volume without expenditure of energy." Jauch and Baron "...therefore conclude that the idealizations in Szilard's experiment are inadmissible in their actual context..." This objection is not easy to refute for the classical one-molecule gas. Its molecule should behave as a billiard ball. Therefore, it is difficult to maintain that

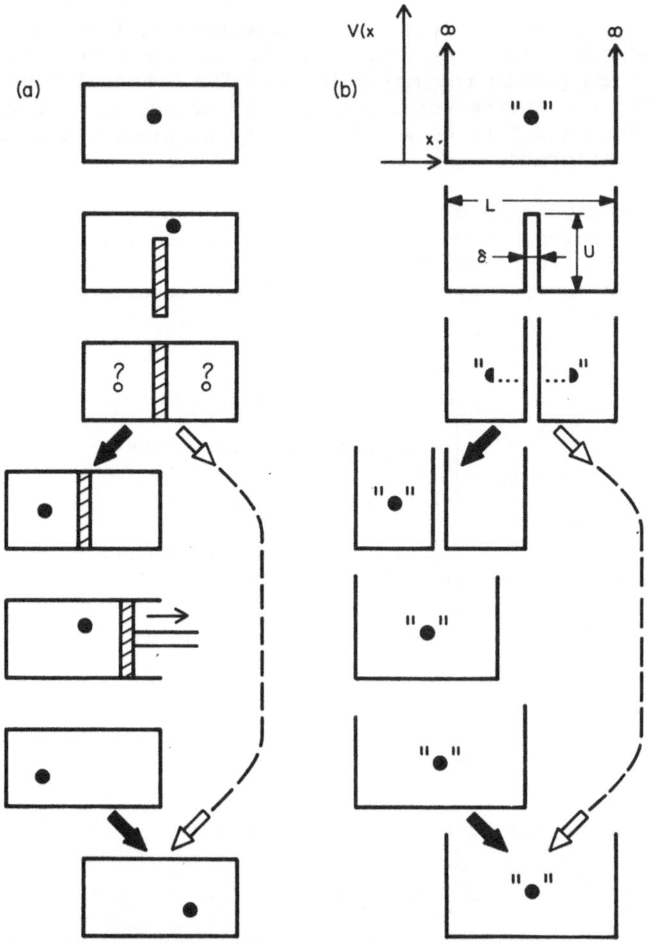

Figure 1. A cycle of Szilard's engine.
a) Original, classical version
b) Quantum version discussed here. The measurement of the location of the molecule is essential in the process of extracting work in both classical and quantum design.

after the piston has been inserted, the gas molecule still occupies the whole volume of the container. More defensible would be a claim that while the molecule is on a definite side of the partition, we do not know on which side, and this prevents extraction of the promised $k_B T \ln 2$ of energy. (This was more or less Szilard's position.) The suspicious feature of such argument is its subjectivity: Our (classical) intuition tells us that the gas molecule is on a definite side of a partition. Moreover, what we (or Maxwell's demon) know should have nothing to do with the objective state of the gas. It is this objective state of the gas which should allow one to extract energy. And, objectively, the gas has been compressed to half its volume by the insertion of the partition no matter on which side of the piston the molecule is. The eventual "observation" may help in making the engine work, but one may argue, as do Jauch and Baron, that the "potential to do work" seems to be present even before such a measurement is performed.

To re-examine arguments of Jauch and Baron, consider the quantum version of Szilard's engine, shown in Fig. 1. The role of the container is now played by the square potential well. A single gas molecule of mass m is described by the (nonrelativistic) Schrödinger equation with the boundary conditions $\psi(-L/2)=\psi(L/2)=0$. Energy levels and eigenfunctions are given by;

$$E_n = n^2\pi^2\hbar^2/(2mL^2) = \varepsilon n^2 \tag{2a}$$

$$\langle x|\psi_n\rangle = \begin{cases} (2/L)^{1/2}\cos 2\pi n x/L \text{ for } n=2k+1 & (2b) \\ (2/L)^{1/2}\sin 2\pi n x/L \text{ for } n=2k & (2c) \end{cases}$$

At a finite temperature $T=\beta^{-1}k_B^{-1}$ the equilibrium state of the system will be completely described by the density matrix:

$$\rho = Z^{-1}\sum_n \exp(-\beta E_n)|\psi_n\rangle\langle\psi_n|. \tag{3}$$

Above, Z is the usual partition function:

$$Z = \sum_{n=1}^{\infty}\exp(-\beta\varepsilon n^2) = \sum_{n=1}^{\infty}\zeta^{n^2} \tag{4}$$

For $1/2<\zeta\leq 1$, Z can be adequately approximated by:

$$Z = \frac{1}{2}\left(\sqrt{\pi/|\ln\zeta|}-1\right). \tag{5}$$

For our purposes a still simpler, high-temperature approximation

$$Z = (\pi/\varepsilon\beta)^{1/2}/2 = L(h^2/2\, mk_BT)^{1/2}\ . \tag{6}$$

valid for $\varepsilon \ll k_B T$, will be sufficient most of the time. This is, of course, the familiar Boltzmann gas partition function. It can be readily generalized to the three-dimensional box

$$Z = L_xL_yL_z/(h^2/2\pi mk_BT)^{3/2}\ , \tag{7}$$

as well as to the case when there are N "classically" indistinguishable particles. The point of this elementary calculation is to demonstrate that, in the further analysis, we can rely on classical estimates of pressure, internal energy, entropy, etc...for one molecule gas which were used by Szilard[1]: A partition function completely determines thermodynamic behavior of the system.

Consider now a "piston," slowly inserted in the middle of the potential well "box". This can be done either (1) while the engine is coupled with the reservoir, or (2) when it is isolated. In either case, it must be done slowly, so that the process remains either (1) thermodynamically or (2) adiabatically reversible. We shall imagine the piston as a potential barrier of width $d \ll L$ and height that eventually attains $U \gg k_B T$. The presence of this barrier will alter the structure of the energy levels: these associated with even n will be shifted "upwards" so that the new eigenvalues are:

$$E'_{2k} = \varepsilon'(2k)^2 + \Delta_k = E_k + \Delta_k \tag{8}$$

where

$$\varepsilon' = \varepsilon\, L^2/(L-d)^2 \tag{9}$$

and

$$\Delta_k \cong (4\varepsilon'/\pi)\exp(-d\sqrt{2m(U-E_k)}/\hbar). \tag{10}$$

Energy levels corresponding to the odd values of n are shifted upwards by $\Delta E_n \sim (2n+1)\varepsilon'$ so that

$$E'_{2k-1} = \varepsilon'(2k)^2 - \Delta_k = E_k - \Delta_k \tag{11}$$

A pair of the eigenvalues E'_{2k}, E'_{2k-1} can be alternatively regarded as the kth doubly degenerate eigenstate of the newly created two-well potential with degeneracy broken for finite values of U.

For $U \to \infty$ exact eigenfunctions can be given for each well separately. For finite U, for these levels where $\Delta_k \ll E_k$, eigenfunctions of the complete potential can be reconstructed from the kth eigenfuctions of the left ($|L_k\rangle$) and right ($|R_k\rangle$) wells:

$$E_k + \Delta_k \quad \leftrightarrow |\psi_k^+\rangle = (|L_k\rangle - |R_k\rangle)/\sqrt{2}$$

$$E_k - \Delta_k \quad \leftrightarrow |\psi_k^-\rangle = (|L_k\rangle + |R_k\rangle)/\sqrt{2}$$

Alternatively, eigenfunctions of the left and right wells can be expressed in terms of energy eigenfunctions of the complete Hamiltonian

$$|L_k\rangle = (|\psi_k^+\rangle + |\psi_k^-\rangle)/\sqrt{2} \tag{12c}$$

$$|R_k\rangle = (|\psi_k^-\rangle - |\psi_k^+\rangle)/\sqrt{2} \tag{12d}$$

III. MEASUREMENTS BY THE CLASSICAL MAXWELL'S DEMON

Consider a measuring apparatus which, when inserted into Szilard's engine, determines on which side of the partition the molecule is. Formally, this can be accomplished by the measurement of the observable

$$\hat{\Pi} = \lambda(|L\rangle\langle L| - |R\rangle\langle R|). \tag{13}$$

Here $\lambda > 0$ is an arbitrary eigenvalue, while

$$|L\rangle = \left(\sum_{k=1}^{N} |L_k\rangle\right)/N^{1/2} \tag{13a}$$

$$|R\rangle=\left(\sum_{k=1}^{N}|R_k\rangle\right)/N^{1/2} \tag{13b}$$

and N is sufficiently large, $N^2\varepsilon\beta>>1$.

The density matrix before the measurement, but after piston is inserted, is given by

$$\tilde{\rho}=\tilde{Z}^{-1}\sum_{k=1}^{\infty}\exp(-\beta E_k)\{\exp(-\beta\Delta_k)|\psi_k^+\rangle\langle\psi_k^+|+\exp(\beta\Delta_k)|\psi_k^-\rangle\langle\psi_k^-|\} \tag{14}$$

$$=\tilde{Z}^{-1}\sum_{k=1}^{\infty}\exp(-\beta E_k)\{\cosh\beta\Delta_k(|L_k\rangle\langle L_k|+|R_k\rangle\langle R_k|)+\sinh\beta\Delta_k(|L_k\rangle\langle R_k|+|R_k\rangle\langle L_k|)\}$$

Depending on the outcome of the observation, the density matrix becomes either ρ_L or ρ_R where;

$$\rho_L=Z_L^{-1}\sum_{k=1}^{\infty}\exp(-\beta E_k)\cosh\beta\Delta_k|L_k\rangle\langle L_k|, \tag{15a}$$

$$\rho_R=Z_R^{-1}\sum_{k=1}^{\infty}\exp(-\beta E_k)\cosh\beta\Delta_k|R_k\rangle\langle R_k|. \tag{15b}$$

Both of these options are chosen with the same probability. The classical "demon" "knows" the outcome of the measurement. This information can be used to extract energy in the way described by Szilard.

We are now in a position to return to the objection of Jauch and Baron. The "potential to do work" is measured by the free energy A(T,L) which can be readily calculated from the partition function

$$A(T,L) = -k_BT\ln Z(T,L) .$$

For the one-molecule engine this free energy is simply

$$A= -k_BT\ln[L/(h^2/2\pi mk_BT)^{1/2}] \tag{16}$$

before the partition is inserted. It becomes

$$\tilde{A}= -k_BT\ln[(L-d)/(h^2/2\pi mk_BT)^{1/2}] \tag{17}$$

after the insertion of the partition. Finally, as a result of the measurement, free energy increases regardless of the outcome

$$A_L=A_R= -k_BT\ln[((L-d)/2)/h^2/2\pi mk_BT)^{1/2}]$$
$$= -k_BT(\tilde{A}-\ln 2) \tag{18}$$

Let us also note that the change of A as a result of the insertion of the partition is negligible:

$$\tilde{A}-A=k_BT\ln(L/(L-d))\sim O(d/L). \tag{19}$$

However, the change of the free energy because of the measurement is precisely such as to account for the $W=k_BT\ln 2$ during the subsequent expansion of the piston:

$$A_L-\tilde{A}=k_BT\ln 2. \tag{20}$$

It is now difficult to argue against the conclusion of Szilard. The formalism of quantum mechanics confirms the key role of the measurement in

the operation of the one-molecule engine. And the objection of Jauch and Baron, based on a classical intuition, is overruled.

The classical gas molecule, considered by Szilard, as well as by Jauch and Baron, may be on the unknown side of the piston, but cannot be on "both" sides of the piston. Therefore, intuitive arguments concerning the potential to do useful work could not be unambiguously settled in the context of classical dynamics and thermodynamics. Quantum molecule, on the other hand, can be on "both" sides of the potential barrier, even if its energy is far below the energy of the barrier top, and it will "collapse" to one of the two potential wells only if $\hat{\Pi}$ is "measured."

It is perhaps worth pointing out that the density matrix $\tilde{\rho}$, Eq. (15), has a form which is consistent with the "classical" statement that " a molecule is on a definite, but unknown side of the piston" almost equally well as with the statement that "the molecule in in a thermal mixture of the energy eigenstates of the two-well potential." This second statement is rigorously true in the limit of weak coupling with the resevoir: gas is in contact with the heat bath and therefore is in thermal equilibrium. On the other hand, the off-diagonal terms of the very same density matrix in the $|L_k>, |R_k>$ representation are negligible ($\sim \sinh\beta\Delta_k$). Therefore, one can almost equally well maintain that this density matrix describes a molecule which is on an "unknown, but definite" side of the partition. The phase between $|R_n>$ and $|L_n>$ is lost.

The above discussion leads us to conclude that the key element needed to extract useful work is the correlation between the state of the gas and the state of the demon. This point can be analyzed further if we allow, in the next section, the "demon" to be a quantum system.

IV. MEASUREMENT BY "QUANTUM MAXWELL'S DEMON"

Analysis of the measurement process in the previous section was very schematic. Measuring appartus, which has played the role of the "demon" simply acquired the information about the location of the gas molecule, and this was enough to describe the molecule by ρ_L or ρ_R. The second law demanded the entropy of the apparatus to increase in the process of measurement, but, in the absence of a more concrete model for the apparatus it was hard to tell why this entropy increase was essential and how did it come about. The purpose of this section is to introduce a more detailed model of the apparatus, which makes such an analysis possible.

We shall use as an apparatus a two-state quantum system. We assume that it is initially prepared by some external agency -- we shall refer to it below as an "external observer" -- in the "ready to measure" state $|D_0\rangle$. The first stage of the measurement will be accomplished by the transition:

$$|L_n>|D_0> \rightarrow |L_n> |D_L> \qquad (21a)$$

$$|R_n>|D_0> \rightarrow |R_n>|D_R> \qquad (21b)$$

for all levels n. Here, $|D_R>$ and $|D_L>$ must be orthogonal if the measurement is to be reliable. This will be the case when the gas and the demon interact via an appropriate coupling Hamiltonian, e.g.:

$$H_{int}=i\delta(|L_n><L_n|-|R_n><R_n|)(|D_L><D_R|-|D_R><D_L|) \qquad (22)$$

for the time interval $\Delta t=\Pi\hbar/(4\delta)$ [5,6], and the initial state of the demon is:

$$|D_0\rangle=(|D_L\rangle+|D_R\rangle)/\sqrt{2} \tag{23}$$

For, in this case the complete density matrix becomes:

$$P=\exp(-iH_{int}\Delta t/\hbar)\tilde{\rho}|D_0\rangle\langle D_0|=$$

$$\tilde{=}(\rho_L|D_L\rangle\langle D_L|+\rho_R|D_R\rangle\langle D_R|)/2 \tag{24}$$

Here and below we have omitted small off-diagonal terms ($\sim\beta\Delta_n$) present in $\tilde{\rho}$, Eq. (14) and, by the same token, we shall drop corrections ($\sim\beta^2\Delta_n^2$) from the diagonal of ρ_L and ρ_R. As it was pointed out at the end of the last section, for the equilibrium there is no need for the "reduction of the state vector." The measured system -- one molecule gas -- is already in the mixture of being on the left and right-hand side of the divided well.

To study the relation of the information gain and entropy increase in course of the measurement, we employ the usual definition of entropy in terms of the density matrix[3, 6, 7, 8]

$$S(\rho)=-k_B \mathrm{Tr}\,\rho\ln\rho \tag{25}$$

The information is then:

$$I(\rho)=\ln(\mathrm{Dim}(\mathcal{H}))-S(\rho)/k_B, \tag{26}$$

where $\mathcal{H}$ is the Hilbert space of the system in question. Essential in our further discussion will be the mutual information $I_\mu(P_{AB})$ defined for two subsystems, A and B, which are described jointly by the density matrix P_{AB}, while their individual density matrices are ρ_A and ρ_B:

$$I_\mu(P_{AB})=I(P_{AB})-(I(\rho_A)+I(\rho_B)). \tag{27}$$

In words, mutual information is the difference between the information needed to specify A and B are not correlated, $P_{AB}=\rho_A\rho_B$, then $I_\mu(P_{AB})=0$.

The readoff of the location of the gas molecule by the demon will be described by an external observer not aware of the outcome by the transition:

$$\tilde{\rho}|D_0\rangle\langle D_0|\rightarrow(\rho_L|D_L\rangle\langle D_L|+\rho_R|D_R\rangle\langle D_R|)/2 \tag{28}$$

The density matrix of the gas is then:

$$\tilde{\rho}=\tilde{Z}^{-1}\sum_{n=1}\exp(-\beta\varepsilon n^2)(|L_n\rangle\langle L_n|+|R_n\rangle\langle R_n|)=(\rho_L+\rho_R)/2 \tag{29}$$

Thus, even after the measurement by the demon, the density matrix of the gas, ρ_G will be still :

$$\rho_G=\langle D_L|P|D_L\rangle+\langle D_R|P|D_R\rangle\tilde{=}\tilde{\rho} \tag{30}$$

The state of the demon will, on the other hand, change from the initial pure state $|D_0\rangle\langle D_0|$ into a mixture:

$$\rho_D=\sum_n(\langle L_n|P|L_n\rangle+\langle R_n|P|R_n\rangle)=(|D_L\rangle\langle D_L|+|D_R\rangle\langle D_R|)/2 \tag{31}$$

Entropy of the gas viewed by the external observer remains constant:

$$S(\rho_G)=S(\tilde{\rho})= \partial(\beta \ln\tilde{Z})/\partial\beta \qquad (32)$$

Entropy of the demon has however <u>increased</u>:

$$S(\rho_D)-S(|D_0\rangle\langle D_0|)=k_B \ln 2 \qquad (33)$$

Nevertheless, the combined entropy of the gas-demon system could not have changed: In our model evolution during the read-off was dynamical, and the gas-demon system was isolated. Yet, the sum of the entropies of the two subsystems -- the gas and the demon -- has increased by $k_B \ln 2$. The obvious question is then: where is the "lost information" $\Delta I=I(P)-(I(\tilde{\rho})+I(\rho_D))$? The very form of this expression and its similarity to the right-hand side of Eq. (27) suggests the answer: The loss of the information by the demon is compensated for by an equal increase of the mutual information:

$$\Delta I_\mu=I_\mu(P)-I_\mu(\tilde{\rho}|D_0\rangle\langle D_0|). \qquad (34)$$

Mutual infromation can be regarded as the information gained by the demon. From the conservation of entropy during dynamical evolutions it follows that an increase of the mutual information must be compensated for by an equal increase of the entropy.

$$\Delta I_\mu-\Delta S/k_B=0 \qquad (35)$$

This last equation is the basis of Szilard's formula, Eq. (1).

At this stage readoff of the location of the gas molecule is reversible. To undo it, one can apply the inverse of the unitary operation which has produced the correlation. This would allow one to erase the increase of entropy of the demon, but only at a price of the loss of mutual information.

One can now easily picture further stages of the operation of Szilard's engine. The state of the demon can be used to decide which well is "empty" and can be "discarded."[6] The molecule in the other well contains twice as much energy as it would if the well were <u>adiabatically</u> (i.e. after decoupling it from the heat reservoir) expanded from its present size $\sim L/2$ to L. Adiabatic expansion conserves entropy. Therefore, one can immediately gain $\Delta W=\frac{1}{2}\langle E\rangle=k_B T/4$ without any additional entropy increases. One can also gain $\Delta W=k_B T\ln 2$, Eq. (2), by allowing adiabatic expansion to occur in small increments, inbetween which one molecule gas is reheated by the re-established contact with the reservoir. In the limit of infinitesimally small increments this leads, of course, to the isothermal expansion. At the end the state of the gas is precisely the same as it was at the beginning of the cycle, and we are ΔW "richer" in energy. If Szilard's engine could repeat this cycle, this would be a violation of the second law of thermodynamics. We would have a working model of a <u>perpetuum mobile</u>, which extracts energy in a cyclic process from an infinite reservoir <u>without</u> the temperature difference.

Fortunately for the second law, there is an essential gap in our above design: The demon is still in the mixed state, ρ_D, Eq. (31). To perform the next cycle, the observer must "reset" the state of the demon to the initial $|D_0\rangle$. This means that the entropy $dS=k_B \ln 2$ must be somehow removed from the system. If we were to leave demon in the mixed state, coupling it with the gas through H_{int} will not result in the increase of their mutual information: The one-bit memory of the demon is still filled by the outcome of the past measurement of the gas. Moreover, this measurement cannot any longer be reversed by allowing the gas and the demon to interact. For, even though the density matrix of each of these two systems has remained the same, their joint

density matrix is now very different:

$$\tilde{\rho}\rho_D = \{(\rho_L|D_L\rangle\langle D_L| + \rho_R|D_R\rangle\langle D_R|)/2 + (\rho_L|D_R\rangle\langle D_R| + \rho_R|D_L\rangle\langle D_L|)/2\}/2 \quad (36)$$

One could, presumably, still accomplish the reversal using the work gained in course of the cycle. This would, however, defy the purpose of the engine. The only other way to get the demon into the working order must be executed by an "external observer" or by its automated equivalent: The demon must be reset to the "ready-to-measure" state $|D_0\rangle$.

As it was pointed out by Bennett in his classic analysis, this operation of memory erasure is the true reason for irreversibility, and the ultimate reason why the free energy, Eq. (1), must be expended.[9] Short of reversing the cycle, any procedure which resets the state of the demon to $|D_0\rangle$ must involve a measurement. For instance, a possible algorithm could begin with the measurement whether the demon is in the state $|D_L\rangle$ or in the orthogonal state $|D_R\rangle$. Once this is established then, depending on the outcome, the demon may be rotated either "rightward" from $|D_L\rangle$, or "leftward" from $|D_R\rangle$, so that its state returns to $|D_0\rangle$. The measurement by some external agency -- some "environment" -- is an essential and unavoidable ingredient of any resetting algorithm. In its course entropy is "passed on" from the demon to the environment.

V. SUMMARY

The prupose of our discussion was to analyze a process which converts thermodynamic energy into useful work. Our analysis was quantum, which removed ambiguities pointed out by Jauch and Baron in the original discussion by Szilard. We have concluded, that validity of the second law of thermodynamics can be satisfied only if the process of measurement is accompanied by the increase of entropy of the measuring apparatus by the amount no less than the amount of gained information. Our analysis confirms therefore conclusions of Szilard (and earlier discussion of Smoluchowski [10], which has inspired Szilard's paper). Moreover, we show that the ultimate reason for the entropy increase can be traced back to the necessity to "reset" the state of the measuring apparatus[9], which, in turn, must involve a measurement. This necessity of the "readoff" of the outcome by some external agency -- the "environment" of the measuring apparatus has been already invoked to settle some of the problems of quantum theory of measurements.[11] Its role in the context of the above example is to determine what was the outcome of the measurement, so that the apparatus can be reset. In the context of quantum measurements its role is very similar: it forces one of the outcomes of the measurement to be definite, and, therefore, causes the "collapse of the wavepacket."

In view of the title of Szilard's paper, it is perhaps worthwhile to point out that we did not have to invoke "intelligence" at any stage of our analysis: all the systems could be presumably unanimated. However, one can imagine that the ability to measure, and make the environment "pay" the entropic cost of the measurement is also one of the essential attributes of animated beings.

ACKNOWLEDGMENTS

I would like to thank Charles Bennett and John Archibald Wheeler for many enjoyable and stimulating discussions on the subject of this paper. The research was supported in part by the National Science Foundation under Grant No. PHY77-27084, supplemented by funds from the National Aeronautics and Space Administration.

REFERENCES

1. L. Szilard, "On the Decrease of Entropy in a Thermodynamic System by the Intervention of Intelligent Beings," Z. Phys. 53:840 (1929); English translation reprinted in Behavioral Science 9:301 (1964), as well as in Ref. 2, p. 539.
2. J.A. Wheeler and W.H. Zurek, eds., Quantum Theory and Measurement (Princeton University Press, Princeton, 1983).
3. C.E. Shannon and W. Weaver, The Mathematical Theory of Communication, (University of Illinois Press, Urbana, 1949).
4. J.M. Jauch and J.G. Baron, "Entropy, Information, and Szilard's Paradox," Helv. Phys. Acta, 45:220 (1972).
5. W.H. Zurek, "Pointer Basis of Quantum Apparatus: Into What Mixture Does the Wavepacket Collapse?" Phys. Rev. D24:1516 (1981).
6. W.H. Zurek, "Information Transfer in Quantum Measurements: Irreversibility and Amplification," Quantum Optics, Experimental Gravitation and Measurement Theory, eds. P. Meystre and M.O. Scully (Plenum Press, New York, 1983).
7. J.R. Pierce and E.C. Rosner, Introduction to Communication Science and Systems (Plenum Press, New York, 1980).
8. H. Everett III, Dissertation, reprinted in B.S. DeWitt and N. Graham, eds., The Many -- Worlds Interpretation of Quantum Mechanics (Princeton University Press, Princeton, 1973).
9. C.H. Bennett, "The Thermodynamics of Computation" Int. J. Theor. Phys. 21:305 (1982).
10. M. Smoluchowski, "Vorträge über die Kinetische Theorie der Materie und Elktrizitat," Leipzig 1914.
11. W.H. Zurek, "Environment-Induced Superselection Rules," Phys. Rev. D26:]862 (1982).

EXPERIMENTAL TESTS OF BELL'S INEQUALITIES WITH PAIRS OF LOW ENERGY CORRELATED PHOTONS

Alain Aspect

Institut d'Optique Théorique et Appliquée
Bâtiment 503 - Centre Universitaire d'Orsay - BP 43
91406 Orsay Cedex - France

INTRODUCTION

Early in the development of quantum mechanics[1], the following question was raised : is it possible (is it necessary) to understand the probabilistic nature of the predictions of quantum mechanics by invoking a more precise description of the world, at a deeper level ? Such a description would complete quantum mechanics, like statistical mechanics complete thermodynamics by invoking the motions of the molecules. Reasoning on a Gedanken-experiment, Einstein Podolsky and Rosen[2] concluded to the necessity of completing quantum mechanics. On the other hand, Bohr disagreed with this conclusion[3], and one could think that the commitment to either position was only a matter of taste or of philosophical position. This situation changed dramatically when John Bell[4] discovered that the two points of view lead to different predictions for the Bohm's version[5] of the E.P.R. Gedanken-experiment. Bell's paper opened a route towards real experiments. The closest realization of that Gedanken-experiment uses pairs of low energy photons correlated in polarization, as suggested in the late sixties by Clauser, Horne, Shimony and Holt[6]. The second part of this paper will be devoted to the description of these experiments.

In the first part of the paper, I intend to spend some time to recalling the essence of Bell's reasoning. I will try to convince the reader that a very natural way of understanding the E.P.R. correlations is to complete quantum mechanics in the way considered by Bell. What is surprising is the fact that such supplementary parameters theory conflict with the Quantum mechanical predictions. This result is the essence of Bell's theorem. This first part is intended to be tutorial, adressed to non specialists, and I will not go into the many details of the subtle discussions of this field. The interested reader will find more in excellent reviews[7,8].

THE EINSTEIN-PODOLSKY-ROSEN CORRELATIONS

Let us consider the optical variant of the E.P.R. Gedanken-experiment modified by Bohm[5]. A source S emits a pair of photons with

different energies, $h\nu_1$ and $h\nu_2$, counterpropagating along $\pm\vec{Oz}$ (Fig. 1). Suppose that the polarization part of their state vector is

$$|\psi(\nu_1, \nu_2)\rangle = (1/\sqrt{2}) \; \{|x,x\rangle + |y,y\rangle\} \tag{1}$$

where $|x\rangle$ and $|y\rangle$ are linear polarizations states.

We perform on these photons linear polarization measurements. The analyser I in orientation $\vec{a}$, followed by two detectors, gives + 1 or - 1 result, corresponding to a linear polarization found parallel or perpendicular to $\vec{a}$. Similarly acts analyzer II, in orientation $\vec{b}$.*

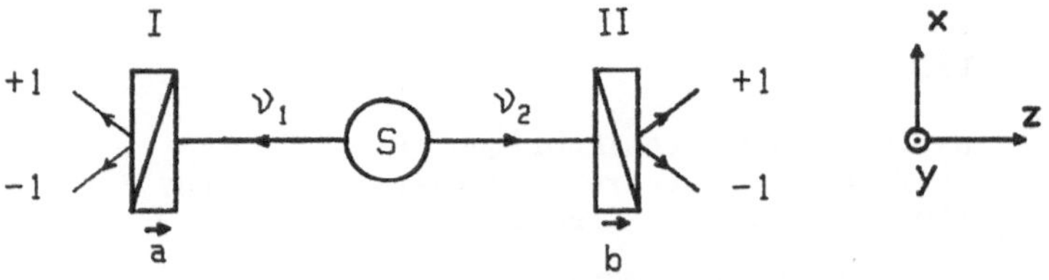

Fig. 1. Einstein-Podolsky-Rosen-Bohm Gedanken-experiment with photons. The two photons ν_1 and ν_2, emitted in the state (1), are analyzed by linear polarizers in orientations $\vec{a}$ and $\vec{b}$. One can measure the probalities of single or joint detection after the polarizers.

It is easy to derive the Quantum Mechanical predictions for these measurements, single or in coincidence.

Let $P_\pm(\vec{a})$ be the probability of getting the result $\pm$ 1 for ν_1 ; similarly $P_\pm(\vec{b})$ is related to ν_2. Quantum Mechanics predicts

$$\begin{aligned} P_+(\vec{a}) &= P_-(\vec{a}) = 1/2 \\ P_+(\vec{b}) &= P_-(\vec{b}) = 1/2 \end{aligned} \tag{2}$$

Let $P_{\pm\pm}(\vec{a},\vec{b})$ be the probability of joint detection of ν_1 in channel $\pm$ 1 of I (in orientation $\vec{a}$), and of ν_2 in channel $\pm$ 1 of II($\vec{b}$). Quantum Mechanics predicts

$$\begin{aligned} P_{++}(\vec{a},\vec{b}) &= P_{--}(\vec{a},\vec{b}) = \frac{1}{2}\cos^2(\vec{a},\vec{b}) \\ P_{+-}(\vec{a},\vec{b}) &= P_{-+}(\vec{a},\vec{b}) = \frac{1}{2}\sin^2(\vec{a},\vec{b}) \end{aligned} \tag{3}$$

It is easy to see that the results of the measurements on ν_1 and ν_2 are strongly correlated. Consider for instance the special case $(\vec{a},\vec{b}) = 0$, for which formula (3) yields

$$P_{++}(\vec{a},\vec{b}) = P_{--}(\vec{a},\vec{b}) = 1/2$$

$$P_{+-}(\vec{a},\vec{b}) = P_{-+}(\vec{a},\vec{b}) = 1/2$$

*There is a one to one correspondance with the Gedanken-experiment dealing with a pair of 1/2 spin particles in a singlet state, and analyzed by two Stern-Gerlach filters.[5]

This means that if ν_1 is found in the channel + 1 of polarizer I (the probability of which is 50 % according to formula (2)), we are sure to find ν_2 in the channel + 1 of polarizer II. Similarly, if ν_1 is found in the − 1 channel then ν_2 is for sure in − 1. The measurement results are perfectly correlated.

More generally, we define the polarization correlation coefficient, as the usual correlation coefficient of the results (+ 1 or − 1) of measurements at I and II, which is here

$$E(\vec{a},\vec{b}) = P_{++}(\vec{a},\vec{b}) + P_{--}(\vec{a},\vec{b}) - P_{+-}(\vec{a},\vec{b}) - P_{-+}(\vec{a},\vec{b}) \quad (4)$$

Using formula (3), we find the coefficient predicted by quantum mechanics

$$E_{MQ}(\vec{a},\vec{b}) = \cos 2(\vec{a},\vec{b}) \quad (5)$$

Which reaches the values + 1 or − 1, indicating a complete correlation.

Supplementary parameters

Correlations between distant measurements on two systems that have separated may be easily understood in terms of some common properties of the two systems. Let us consider again the correlations of polarization measurements in the case $(\vec{a}, \vec{b}) = 0$. When we find + 1 for ν_1, we are sure to find + 1 for ν_2. We are thus led to admit that there is some property (Einstein said "an element of physical reality") pertaining to this particular pair, and determining the result ++. For another pair the results will be --, and the invoked property is different. Such properties, differing from one pair to another one, are not taken into account by the Quantum Mechanical state vector $|\psi(1,2)\rangle$ which is the same for all pairs. This is why Einstein concluded that Quantum Mechanics is not complete. And this is why such properties are referred to as "supplementary parameters" (sometimes calles "hidden-variables").

As a conclusion, one can hope to "understand" the E.P.R. correlations by such a classical-looking picture, involving supplementary parameters differing from one pair to another one. It can be hoped to recover the Quantum Mechanical predictions when averaging over the supplementary parameters. It seems that so was Einstein's position[9]. At this stage, a commitment to this view-point is just a matter of taste.

Remark. Since Einstein spoke of "an element of the physical reality", some authors call "Realistic Theories" these theories invoking supplementary parameters.[7]

Bell could translate into mathematics the preceding discussion, by introducting explicit supplementary parameters, denoted λ. Their distribution on an ensemble of emitted pairs is specified by a probability distribution $\rho(\lambda)$, such that

$$\rho(\lambda) \geq 0 \qquad \text{and} \qquad \int d\lambda\, \rho(\lambda) = 1 \quad (6)$$

For a given pair, characterized by a given λ, the results of measurement will be

$$A(\lambda,\vec{a}) = \begin{cases} +1 \\ \text{or} \\ -1 \end{cases} \quad \text{at analyzer I}$$

$$B(\lambda,\vec{b}) = \begin{cases} +1 \\ \text{or} \\ -1 \end{cases} \quad \text{at analyzer II} \qquad (6')$$

A particular theory must be able to afford explicitly the functions $\rho(\lambda)$, $A(\lambda,\vec{a})$ and $B(\lambda,\vec{b})$. It is then easy to express the probabilities of the various results.

For instance $\quad P_+(\vec{a}) = \frac{1}{2}\int d\lambda\ \rho(\lambda)\ \{A(\lambda,\vec{a}) + 1\}$ $\quad$ etc...

In particular, we will use the correlation function

$$E(\vec{a},\vec{b}) = \int d\lambda\ \rho(\lambda)\ A(\lambda,\vec{a})\ B(\lambda,\vec{b})$$

A (naive) exemple

Let us suppose that the two photons of a pair are emitted "with the same linear polarization", defined by its angle λ with $O\vec{x}$ (Fig. 2).

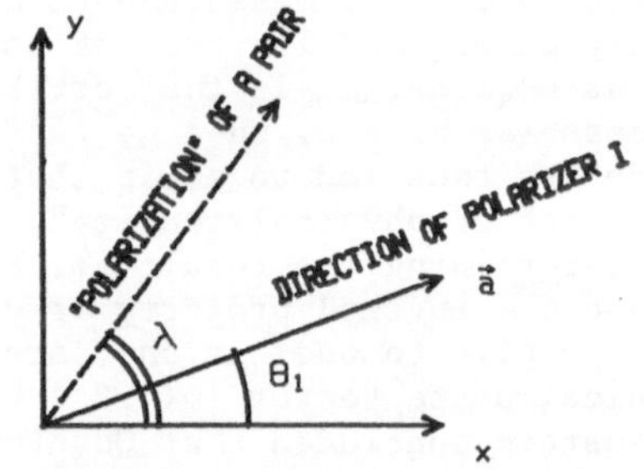

Fig. 2. Our example. Each pair has a "direction of polarization", defined by λ.

The probability distribution is taken isotropic

$$\rho(\lambda) = 1/2\pi$$

As a simple model for the polarizer I we assume that we get the result + 1 if

$$|\theta_I - \lambda| \leq \frac{\pi}{4} \qquad \text{or} \qquad |\theta_I - \lambda| \geq 3\,\frac{\pi}{4}$$

The result − 1 is obtained for

$$\frac{\pi}{4} < |\theta_I - \lambda| < 3\pi/4$$

The response of the polarizer I can thus be written

$$A(\lambda,\vec{a}) = \frac{\cos 2(\Theta_I-\lambda)}{|\cos 2(\Theta_I-\lambda)|}$$

Similarly, the response function of polarizer II is

$$B(\lambda,\vec{b}) = \frac{\cos 2(\Theta_I-\lambda)}{|\cos 2(\Theta_I-\lambda)|}$$

With this model, we find

$$P_+(\vec{a}) = P_-(\vec{a}) = P_+(\vec{b}) = P_+(\vec{b}) = 1/2$$

which is identical to the Quantum Mechanical result. As correlation function, our model yields

$$E(\vec{a},\vec{b}) = 1 - 4\,\frac{|\Theta_I-\Theta_{II}|}{\pi} = 1 - 4\,\frac{|(\vec{a},\vec{b})|}{\pi}$$

Like the Quantum Mechanical result, $E(\vec{a},\vec{b})$ depends only on the relative angle $(\vec{a},\vec{b})$. Fig. 3 shows a comparison between this result and the Quantum Mechanical prediction.

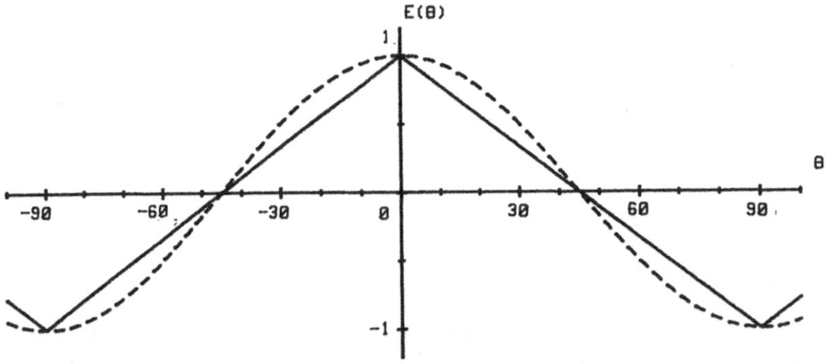

Fig. 3. Polarization correlation coefficient, as a function of the relative orientation of the polarizers.

---- : Calculated by Quantum Mechanics ;

—— : Given by our simple model.

The agreement is not bad. It might be hoped that some more complicated model would be able to reproduce exactly the Quantum Mechanical predictions.

Bell's discovery is that the search for such a model is hopeless.

BELL'S THEOREM

The reasoning will not hold on a particular model, but rather on any model in agreement with the formalism defined in equations (6) and (6'). The first step is to derive the so.called "Bell's inequalities", which are some constraints on the possible values of the polarization correlation coefficient at various angles. A convenient form of these inequalities is the one found by Clauser, Horne, Shimony and Holt[6].

$$-2 \leqq S(\vec{a},\vec{a}',\vec{b},\vec{b}') \leqq 2$$

with (7)

$$S = E(\vec{a},\vec{b}) - E(\vec{a},\vec{b}') + E(\vec{a}',\vec{b}) + E(\vec{a}',\vec{b}')$$

These inequalities bear upon a combination of four polarization correlation coefficients, measured in four orientations of the polarizers. S is thus a measurable quantity.

The demonstration is straight forward by considering the quantity

$$s = A(\lambda,\vec{a}).B(\lambda,\vec{b}) - A(\lambda,\vec{a}).B(\lambda,\vec{b}') + A(\lambda,\vec{a}').B(\lambda,\vec{b}) + A(\lambda,\vec{a}').B(\lambda,\vec{b}')$$

$$= A(\lambda,\vec{a})\left\{B(\lambda,\vec{b}) - B(\lambda,\vec{b}')\right\} + A(\lambda,\vec{a}')\left\{B(\lambda,\vec{b}) + B(\lambda,\vec{b}')\right\}$$

Remembering that the four numbers, A, B take only the values ± 1, we find that

$$s(\lambda,\vec{a},\vec{a}',\vec{b},\vec{b}') = \pm 2 \tag{8}$$

The average of s over λ is therefore included between + 2 and - 2, i.e.

$$-2 \leqq \int d\lambda\ \rho(\lambda).\ s(\lambda,\vec{a},\vec{a}',\vec{b},\vec{b}') \leqq 2$$

which is the inequalities (7).

The next step is to show that some quantum mechanical predictions violate the Bell's inequalities. As an example let us take the particular set of orientations of Fig. 4a. Replacing the E's by their Quantum Mechanical values (5) for pairs in state (1), we obtain

$$S_{MQ} = 2\sqrt{2}$$

This quantum mechanical prediction strongly violates the upper limit of inequalities (7). We thus find it impossible to reconcile the formalism defined in (6) and (6') with the predictions of Quantum Mechanics for the particular E.P.R.-type state (1).

More generally, we can look for the greatest conflict between quantum mechanics and the inequalities (7). We derivate s by respect to the three angles $(\vec{a},\vec{b})$, $(\vec{b},\vec{a}')$ and $(\vec{a}',\vec{b}')$ (which are independant). S_{MQ}

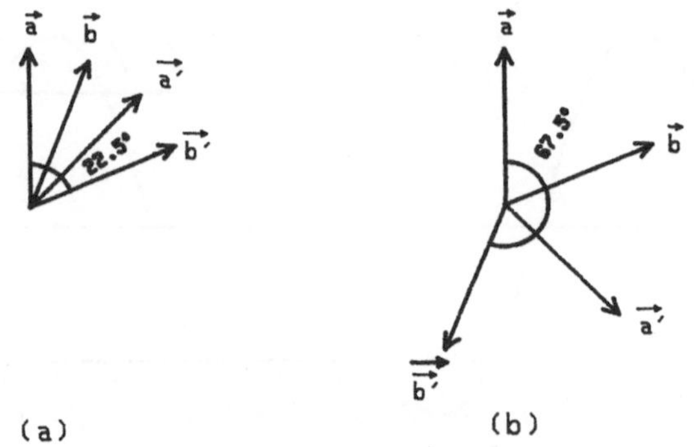

Fig. 4. Orientations yielding the largest conflict between Bell's Inequalities and Quantum Mechanics.

is extremum if

$$(\vec{a},\vec{b}) = (\vec{b},\vec{a}') = (\vec{a}',\vec{b}') = \Theta$$

and it takes the value (9)

$$S_{MQ}(\Theta) = 3\cos 2\Theta - \cos 6\Theta$$

Derivating now by respect to Θ, we obtain the maximum and minimum values of S_{MQ}

$$S_{MQ}^{Max} = 2\sqrt{2} \quad \text{for} \quad \Theta = \pi/8$$

(10)

$$S_{MQ}^{Min} = -2\sqrt{2} \quad \text{for} \quad \Theta = 3\pi/8$$

The corresponding orientations are the ones displayed on Fig. 4.

Figure 5 shows the variations of $S_{MQ}(\Theta)$, and the limits given by B.C.H.S.H. inequalities. One sees that the conflict is serious.

In order to try to understand which part of the formalism causes this conflict, let us point out the hypotheses implied by formalism (6) and (6'). First, we have introduced supplementary parameters, for explaining the E.P.R. correlations by some common properties of the two photons. This point has already been discussed. We can also remark that the used formalism is deterministic. When λ is fixed, then the results of measurements $A(\lambda,\vec{a})$ and $B(\lambda,\vec{b})$ are certain, i.e. λ determines the result. It might be thought that it is the reason for the conflict with Quantum Mechanics. But Bell[4], and Clauser and Horne,[10] have exhibited Stochastic Supplementary Parameters Theories that are not deterministic,

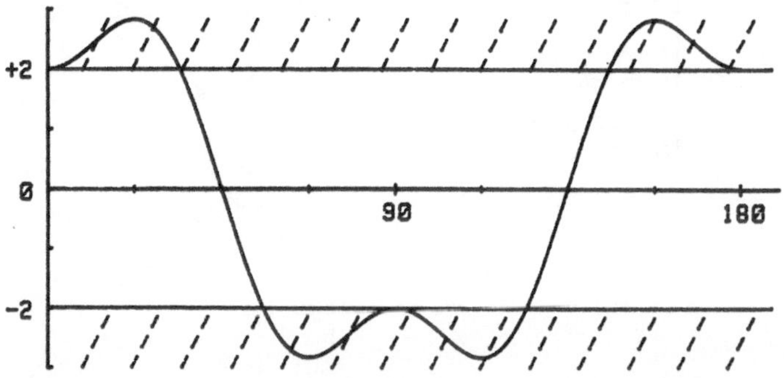

Fig. 5. $S(\Theta)$ as predicted by Quantum Mechanics for pairs in state (1). The conflict arises in the /// zone.

and which nevertheless lead to Bell's Inequalities. The deterministic character does not seem sufficient to lead to a conflict.*

The main point, stressed by Bell, is that the formalism follows a Locality Condition. The result of measurement at I, $A(\lambda,\vec{a})$, does not depend on the orientation $\vec{b}$ of the remote polarizer II, and vice-versa, nor depends $\rho(\lambda)$ (i.e. the way in which pairs are emitted) on the orientations $\vec{a}$ et $\vec{b}$. Bell's Inequalities no longer hold if we don't make the locality assumption (It is easy to see that the demonstration fails with quantities such as $A(\lambda,\vec{a},\vec{b})$ or $\rho(\lambda,\vec{a},\vec{b})$).

As an abstract of this discussion, we can say that Bell's theorem states a conflict between Local Supplementary Parameters Theories and certain Quantum Mechanical predictions. It yields a quantitative criterion for this conflict, that will allow us to design sensitive experiments. Before describing these experiments, we want to comment further on the locality condition. We can remark that in static experiments, in which the polarizers are held fixed for the whole duration of a run, the Locality Condition must be stated as an assumption. Although highly reasonable, it is not prescribed by any fundamental physical law. To quote J. Bell "the settings of the instruments are made sufficiently in advance to allow them to reach some mutual rapport by exchange of signals with velocity less than or equal to that of light". If such interactions existed, the Locality Condition would no longer hold for static experiments, nor would Bell's Inequalities. Bell thus insisted upon the importance of "experiments of the type proposed by Bohm and Aharonov[9], in which the settings are changed during the flight of the particles". In such a timing-experiment, the locality condition would become a consequence of Einstein's Causality that prevents any faster-than-light influence.

* This conclusion is not shared by all authors. For instance A. FINE[11] argues that the stochastic theories of Bell or of Clauser and Horne achieve no further generality, since they can be mimicked by a deterministic theory. We think that this argument rather proves that the determinism is not the issue of this discussion.

As shown in our 1975 proposal[12], it is sufficient to switch each polarizer's orientation between two particular settings ($\vec{a}$ and $\vec{a}'$ for I, $\vec{b}$ and $\vec{b}'$ for II). It then becomes possible to test experimentally a larger class of Supplementary Parameters Theories, those obeying Einstein's Causality. In such theories, the response of polarizer I at time t, is allowed to depend on the orientation $\vec{b}$ (or $\vec{b}'$) of II at time t - L/C (L being the distance between the polarizers). A similar retarded dependance is considered for the way in which pairs are emitted at the source (characterized by the supplementary parameters distribution). For random switching times, with both sides uncorrelated, the predictions of these more general theories are constrained by generalized Bell's Inequalities[12].

On the other hand, it is easy to show that the polarization correlations predicted by Quantum Mechanics depend only on the orientations $\vec{a}$ or $\vec{a}'$ and $\vec{b}$ or $\vec{b}'$ at the very time of the measurements, and do not involve any retardations terms such as L/C. For a suitable choice of the set of orientations $(\vec{a},\vec{a}',\vec{b},\vec{b}')$ - for instance the sets displayed in Fig. 4 - the Quantum Mechanical predictions still conflict with generalized Bell's Inequalities. Such a timing-experiment with variable analyzers would thus provide a test of Supplementary-Parameters-Theories obeying Einstein's Causality, versus Quantum Mechanics.

All the preceding discussion is based on Einstein's ideas about the physical world, and we can summarize Bell's theorem by the following sentence : *some predictions of quantum mechanics (for instance in E.P.R. situations) cannot be mimicked by a "reasonable-classical-looking model", in the spirit of Einstein's ideas.*

At his point, the question is "How does nature behave ?", and the answer belongs to the experimentalists.

THE ROUTE TO THE EXPERIMENTS

Quantum Mechanics has been so much upheld in a great variety of experiments that Bell's theorem might appear as an impossibility proof of supplementary parameters. However, situations in which this conflict arises (sensitive situations) are rare ; in 1965 none had been realized. The following remarks will help to understand this surprizing fact.

First let us notice that Bell's Inequalities obviously constrain the whole classical physics, i.e. classical mechanics and classical electrodynamics, which can be expressed according to the formalism (6) and (6'). (For instance, in classical mechanics, we can take as λ the initial positions and velocities. In classical electrodynamics, the λ would be the charges and currents of the sources).

Moreover, in a situation involving two correlated measurement onto two separated subsystems, Quantum Mechanics will very seldom predict a violation of Bell's Inequalities. Without being exhaustive, we can point out two important necessary conditions for a sensitive experiment (according to Quantum Mechanics) :

(i) the two subsystems must be in a non-factorizing state, such as a singlet state for two spin 1/2 particles, or the similar state (1) for two photons ;

(ii) for each subsystem, it must possible to choose the measured quantity among at least two non-commuting observables (such as polarization measurements along directions $\vec{a}$ and $\vec{a}'$ neither parallel nor perpendicular).

As a matter of fact, these are stringent conditions.

Only four years after the publications of Bell's theorem, C.H.S.H.[6] pointed out that pairs of photons emitted in suitable atomic radiative cascades are good candidate for a sensitive test. Consider for instance a$(J = 0) \rightarrow (J = 1) \rightarrow (J = 0)$ cascade, in the singlet states of an alkaline earth (Fig. 6). Suppose that we select, with the use of wavelengths filters and collimators, two plane waves of frequencies ν_1 and ν_2 propagating along $- O\vec{z}$ and $+ O\vec{z}$ (Fig. 7).

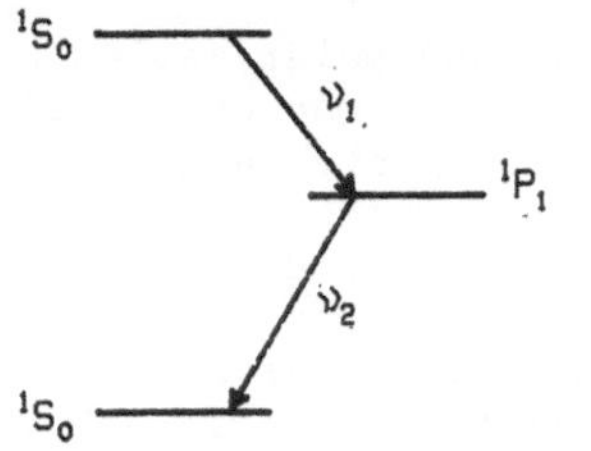

Fig. 6. Radiative cascade emitting pairs of photons correlated in polarization.

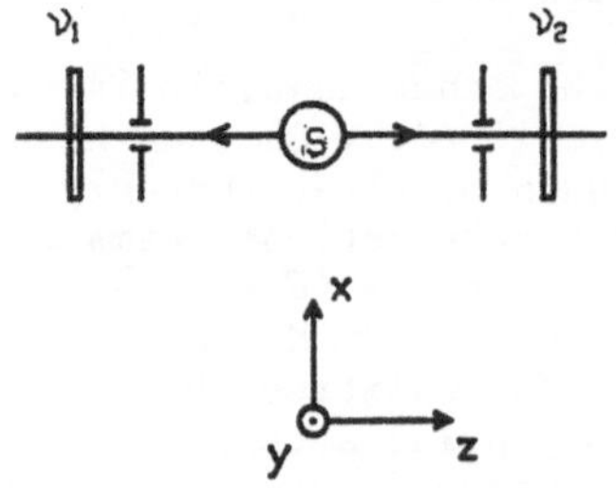

Fig. 7. Ideal configuration (infinitely small solid angles).

It is easy to show, by invoking parity and angular momentum conservation , that the polarization part of the state vector discribing the pair (ν_1, ν_2) writes :

$$(1/\sqrt{2}) \ \{|R,R\rangle + |L,L\rangle\} \tag{11}$$

where R and L are circularly polarized states. By expressing $|R\rangle$ and $|L\rangle$ on a linear polarization basis, we obtain the state (1)

$$|\psi(\nu_1,\nu_2)\rangle = (1/\sqrt{2}) \ \{|x,x\rangle + |y,y\rangle\}$$

We know that such a pair is a good candidate for a sensitive experiment, since corresponding quantum mechanical predictions violate Bell's Inequalities.

A real experiment differs from the ideal one in several respects. For instance, the light should be collected in finite solid angles, as large as possible (Fig. 8). One can show[13] that the visibility of the correlation function then decreases, since (5) is replaced by :

$$E_{MQ}(\vec{a},\vec{b}) = F(\vec{u}) \, . \, \cos 2(\vec{a},\vec{b}) \tag{12}$$

where

$$F(\vec{u}) \leq 1$$

Fig. (9) displays F(u) for a $0 \rightarrow 1 \rightarrow 0$ alkaline-earth cascade (with no hyperfine structure). Fortunately, one can use large angles without a great harm. For $u = 32°$ (our experiments), $F(u) = 0.984$.

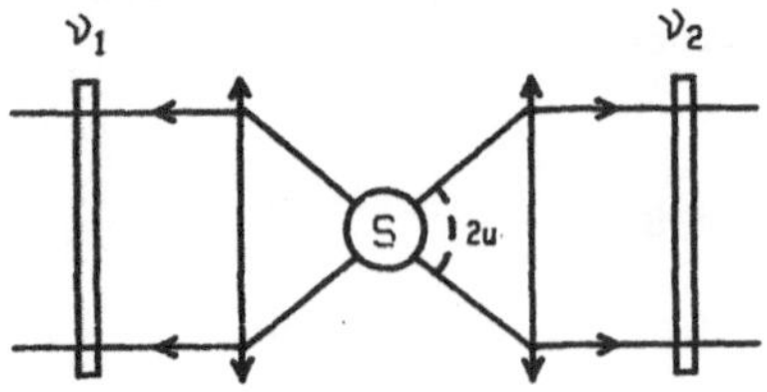

Fig. 8. Realistic configuration, with finite solid angles.

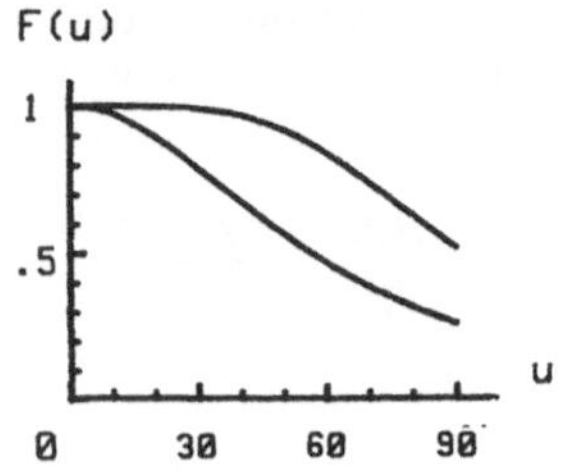

Fig. 9. F(u) for a 0-1-0 cascade.

All other inefficiencies - polarizers defects, accidental birefringences etc... - will similarly lead to a decrease of the correlation function $E(\vec{a},\vec{b})$. The function $S_{MQ}(\theta)$ (Fig. 5) is then multiplied by a factor less than 1, and the conflict with Bell's Inequalities decreases, or even vanishes.

Therefore, an actual experiment must be carefully designed and every auxiliary effect has to be evaluated. Everything must be perfectly controlled, since one can assume that a forgotten effect would similarly lead to a decrease of the conflict (one knows for instance that hyperfine structure dramatically decreases F(u), so that only even istopes can be used[13]).

After the publication of the C.H.S.H. paper in 1969, two groups began to design an experiment. Following the C.H.S.H. proposition, they used a simpler experimental scheme, involving one channel polarizers. In this simplified experimental scheme, one uses polarizers that transmit light polarized parallel to $\vec{a}$ (or $\vec{b}$), and block the orthogonal one. One thus only detects the + 1 results, and the coincidence measurements only yield $N_{++}(\vec{a},\vec{b})$.

Auxilliary runs are performed with one or both polarizers removed (we denote ∞ the "orientation" of a removed polarizers). We can write relations such as :

$$N(\infty,\infty) = N_{++}(\vec{a},\vec{b}) + N_{+-}(\vec{a},\vec{b}) + N_{-+}(\vec{a},\vec{b}) + N_{--}(\vec{a},\vec{b})$$

$$N_{++}(a,\infty) = N_{++}(\vec{a},\vec{b}) + N_{+-}(\vec{a},\vec{b})$$

By substitution into inequalities (7), one gets new B.C.H.S.H. inequalities

$$-1 \leq S' \leq 0 \qquad (13)$$

with

$$S' = (1/N(\infty,\infty)) \left\{ N(\vec{a},\vec{b}) - N(\vec{a},\vec{b}') + N(\vec{a}',\vec{b}) + N(\vec{a}',\vec{b}') - N(\vec{a}',\infty) - N(\infty,\vec{b}) \right\}$$

(we omitted the subscripts ++)

For the orientations sets already considered (Fig. 4), the Quantum Mechanical predictions violate ineq. (13), since one finds

$$S'^{Max}_{MQ} = \frac{\sqrt{2}-1}{2} \qquad \text{for} \qquad \Theta = \pi/8$$

$$S'^{Min}_{MQ} = \frac{-\sqrt{2}-1}{2} \qquad \text{for} \qquad \Theta = 3\pi/8 \tag{14}$$

The derivation of inequ. (13) requires a supplementary assumption. Since the detection efficiencies are low (due to small angular acceptance and low photomultipliers efficiencies), the probabilities involved in the E(a,b) (Eq. (4)) must be redefined on the ensemble of pairs that would be detected with polarizers removed. This procedure is valid only if one assumes a reasonable hypothesis about the detectors. The C.H.S.H. assumption states that, "given that a pair of photons emerges from the polarizers, the probability of their joint detection is independent of the polarizer orientations" (or of their removal)[6]. Clauser and Horne have exhibited another assumption[10], leading to the same inequalities.*

Surprisingly, the two first experiments yielded conflicting results. In the Berkeley experiment (Clauser and Freedman[14]), the $4p^2\ {}^1S_0 - 4s4p\,{}^1P_1 - 4S^2\ {}^1S_0$ cascade of Calcium was excited by ultraviolet absorption towards a ${}^1P_1$ upper state. Since the signal was weak, and spurious cascades occured, it took more than 200 hours of measurement for a significant result. The experiment upheld Quantum Mechanics, and violated inequalities (13) by several standard deviations.

At the same time, in Harvard, Holt and Pipkin[15] found a result in disagreement with Quantum Mechanical predictions, and in agreement will Bell's Inequalities. They excited the $9^1P_1 \rightarrow 7^3P_1 \rightarrow 6^3P_0$ cascade in Mercury 200 by an electron beam. The data accumulation lasted 150 hours.

Clauser[16] repeated their experiment in Mercury 202. He found an agreement with Quantum Mechanics, and a violation of Bell's Inequalities.

In 1976, at Texas A & M University, Fry and Thompson[17] used the $7^3S_1 \rightarrow 6^3P_1 \rightarrow 6^3S_0$ cascade in Mercury 200. Their selective excitation involved a C.W. single-line-laser. The signal was several order of magnitude larger than in previous experiments, allowing them to collect the data in a period of 80 minutes. Their result was in excellent agreement with Quantum Mechanics and violated generalized Bell's inequalities by 4 standard deviations.

---

* Although these assumptions are reasonable, let us mention that there exist supplementary-parameters theories that do not obey them. From the view point of supplementary-parameters theories, there is no way for experimentally testing these assumptions[7].

LATER EXPERIMENTS

By the end of the seventies, we felt that the technological progress (mostly in lasers) were large enough to allow a new generation of experiments. Our goal was to perform experiments closer to the ideal schemes on which the theoretical discussions bore. The first possible progress was to design an experiment using two channel-polarizers, in order to reproduce the theoretical scheme of Fig. 1. The other proposed improvement was to fulfil some requirements on the detection times, in order to have the locality condition derived from Einstein's causality (see the discussion in the section on Bell's theorem). This requirements can be split into two conditions

(i) the measurements onto the 2 subsystems are space-like separated ;

(ii) the choices of the quantities measured on each subsystem are made at random, and are space-like separated from the measurement on the opposite side. It is obviously much more difficult to fulfil the second condition.

Experimental set-up

Since our aim was to use more sophisticated experimental schemes, we had first to build a high-efficiency and very stable and well controlled source. This was carried out (Fig. 10) by a two-photon-excitation of the $4p^2\ ^1S_0$ - $4s4p\ ^1P_1$ - $4s^2\ ^1S_0$ cascade of calcium[18]. This cascade is very well suited to coincidence counting experiments since the lifetime τ_r of the intermediate level is rather short (5ns). If one can reach an excitation rate about $1/\tau_r$, then an optimum signal-to-noise ratio for this cascade is attained.

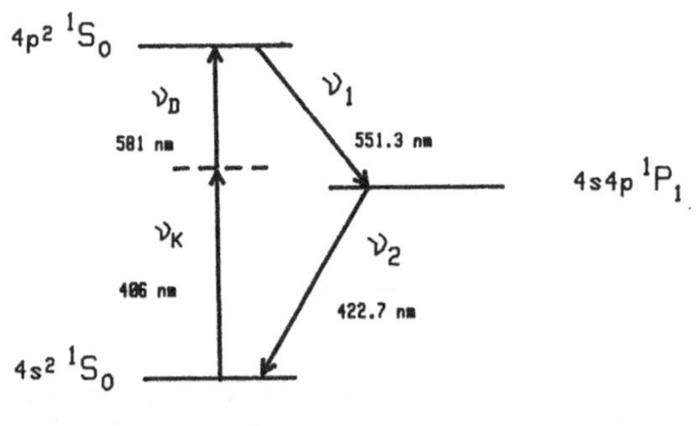

Fig. 10. Two-photon excitation of the chosen cascade in Calcium.

We have achieved this optimum rate with the use of a Krypton laser (λ_K = 406.7 nm) and a dye laser (λ_D = 581 nm) tuned to resonance for the two-photon process. Both lasers are single-mode operated, and they have parallel polarizations. They are focused onto a Calcium atomic beam (laser beam waists about 50 μm). Two feedback loops provide the required

stability of the source (better than 0.5 % for several hours) : the first loop controls the wavelength of the tunable laser to ensure the maximum fluorescence signal ; a second loop controls the power of one laser and compensates all the fluctuations at low frequency.

With a few tens of milliwatts from each laser, the cascade rate is about $N = 4 \times 10^7\ s^{-1}$. An increase beyond this rate would not significantly improve the signal-to-noise ratio for coincidence counting, since the accidental coincidence rate increases as N^2, while the true coincidence rate increases as N.

The fluorescence photons ν_1 and ν_2 are collected by large-aperture aspherical lenses, followed by interference filters, lenses, the polarizers, and photomultipliers.

The photomultipliers feed the coincidence-counting electronics, that includes a time-to-amplitude converter and a multichannel analyzer, yielding the time-delay spectrum of the two-photon detections (Fig. 11). This spectrum involves a flat background due to accidental coincidences (i.e. between photons emitted by different atoms). True coincidences yield a peak around the null-delay, with an exponential decrease (time constant τ_r). The true-coincidence signal is thus taken as the signal in the peak.

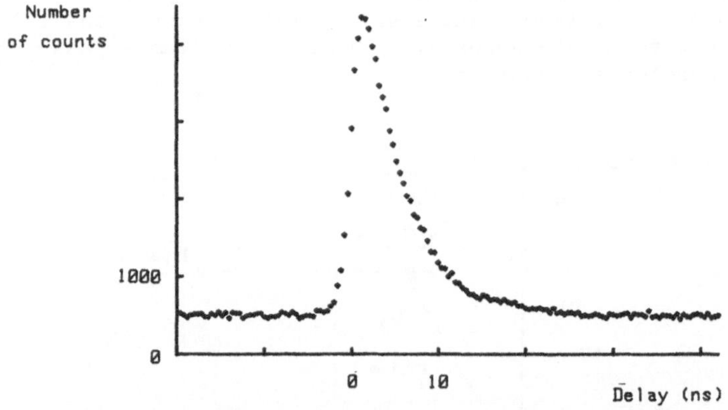

Fig. 11. Time-delay spectrum. Number of detected pairs as a function of the delay between the detections of two photons.

Additionally, a standard coincidence circuit with a 19 ns coincidence window monitors the rate of coincidences around null delay, while a delayed-coincidence channel monitors the accidental rate. It is then possible to check that the true coincidence rate obtained by substraction is equal to the signal in the peak of the time-delay spectrum.

In the second and third experiments, we have used a fourfold coindidence system, involving a fourfold multichannel analyzer and four double-coincidence circuits. The data were automatically gathered and processed by a computer.

Experiment with one-channel polarizers[19]

Our first experiment was carried out using one-channel-pile-of plates polarizers, made of ten glass-plates at Brewster angle.

Thanks to our high-efficiency source, the statistical accuracy was better than 2 % in a 100 s run (with polarizers removed).This allowed us to perform various checks.

The test of Bell's inequalities at $\Theta = \frac{\pi}{8}$ has yielded

$$S'_{exp} = 0.126 \pm 0.014 \tag{15}$$

violating inequalities (13) by 9 standard deviations, and in good agreement with the Quantum Mechanical predictions (for our polarizers and solid angles) :

$$S_{QM} = 0.118 \pm 0.005$$

(this error accounts for uncertainly in the measurements of the polarizers efficiencies).

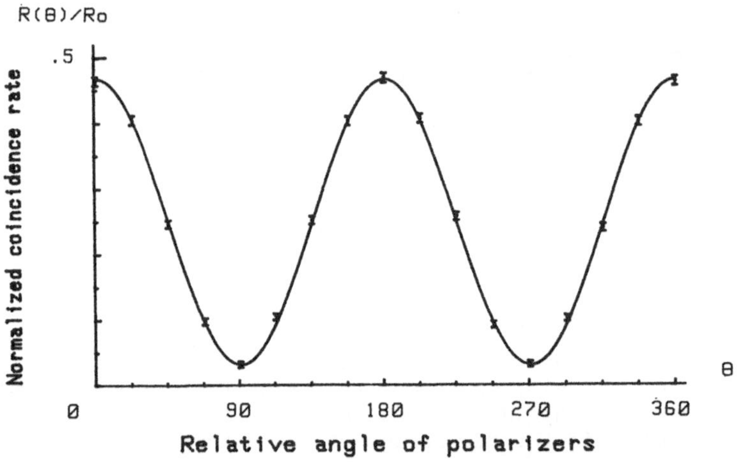

Fig. 12. Experiment with one channel polarizers : Normalized coincidence rate as a function of the relative polarizers orientation. Indicated errors are ± 1 standard deviation. The solid curve is not a fit to the data but the prediction by Quantum Mechanics.

The agreement between the experimental data and the Quantum Mechanical predictions has been checked in a full 360° range of orientations (Fig. 12).

In order to fulfil the first time-condition we have repeated these measurements with the polarizers at 6.5 m from the source. At such a distance (four cohenrence-lengths of the wave packet associated with the lifetime τ_r the detection) events are space-like separated. No modification of the experimental results was observed.

Experiment with two-channel analyzers[20]

With single-channel polarizers, the measurements of polarization are inherently incomplete. When a pair has been emitted, if no count is obtained at one of the photomultipliers, there is no way to know if "it has been missed" by the detector or if it has been blocked by the polarizer (only the later case corresponds to a result - 1 for the measurement). This is why one has to resort to auxilliary experiments, and indirect reasoning, in order to test Bell's inequalities.

With the use of two-channel polarizers, we have performed an experiment following much more closely the ideal scheme of Fig. 1.* Our polarizers were polarizing cubes transmitting one polarization (parallel to $\vec{a}$, or respectively to $\vec{b}$) and reflecting the orthogonal one. Such a polarisation splitter, and the two corresponding photomultipliers, are mounted in a rotatable mechanism. This device (polarimeter) yields + 1 and - 1 results for linear polarization measurements along $\vec{a}$ (respectively $\vec{b}$). It is an optical analog of a Stern-Gerlach filter for spin 1/2 particles.

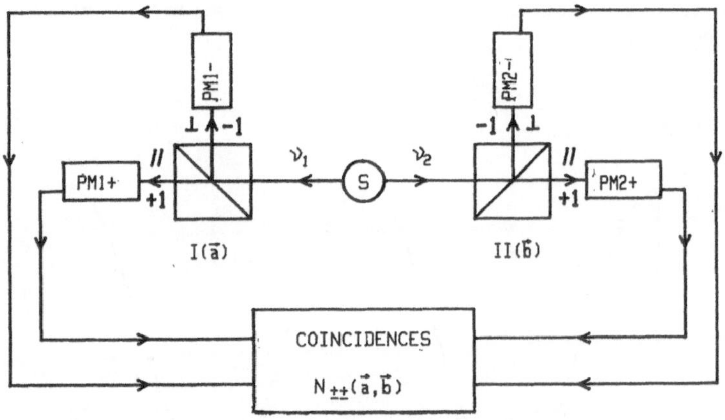

Fig. 13. Experiment with two-channel polarizers. Each polarization splitter is followed by two photomultipliers, feeding a fourfold coincidence circuit, which monitors simultaneously the four coincidence rates $N_{\pm\pm}(\vec{a},\vec{b})$. Each analyzer can rotate around the beam axis for the choice of the orientation.

With polarimeters I and II in orientations $\vec{a}$ and $\vec{b}$, and the fourfold coincidence counting system, we are able to measure in a single run the four coincidence rates $R_{\pm\pm}(\vec{a},\vec{b})$. We then get directly the correlation coefficient for the measurement along $\vec{a}$ and $\vec{b}$.

$$E(\vec{a},\vec{b}) = \frac{R_{++}(\vec{a},\vec{b}) + R_{--}(\vec{a},\vec{b}) - R_{+-}(\vec{a},\vec{b}) - R_{-+}(\vec{a},\vec{b})}{R_{++}(\vec{a},\vec{b}) + R_{--}(\vec{a},\vec{b}) + R_{+-}(\vec{a},\vec{b}) + R_{--}(\vec{a},\vec{b})} \quad (16)$$

* A similar experiment, using calcite polarizers, has been undertaken at the University of Catania, Italy[21].

This procedure is sound if the measured values (16) of the correlation coefficients can be taken equal to the definition (4), i.e. if we assume that the ensemble of actually detected pairs is a faithfull sample of all emitted pairs. This assumption is very reasonable with our symetrical scheme, where the two measurements + 1 and - 1 are treated in the same way (the detection efficiencies in both channels of a polarimeter are equal). Moreover, we have checked that the sum of the four coincidence rates $R_{\pm\pm}(\vec{a},\vec{b})$ is constant, when changing the orientations, although each rate strongly varies. The size of the selected sample of pairs is thus found constant.

We have performed measurements of $E(\vec{a},\vec{b})$ at various angles, allowing a direct comparison with he quantum mechanical predictions[20]. In particular, measurements at sets of orientations as defined in equations (9) yield directly the quantity $S_{exp}(\Theta)$, that can be tested versus the B.C.H.S.H. inequalities (7). The results are displayed on Fig. 14, with the prediction of quantum mechanics for the actual experiment, and the limits of Bell's inequalities.

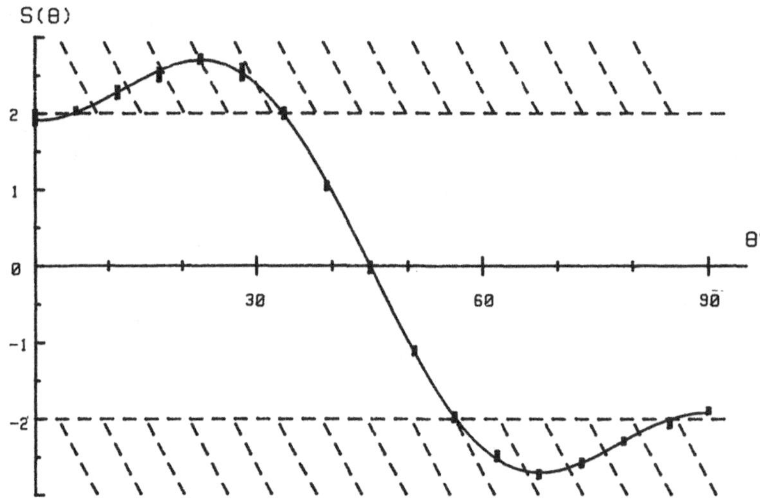

Fig. 14. Relevant combination of correlation coefficients for various sets of orientations. According to Bell's inequalities, S would not be in the hatched region. The errors are ±2 standard deviations. The solid curve is the prediction by quantum mechanics.

The agreement of the data with quantum mechanics, and the violation of Bell's inequalities, are clearly shown. At $\Theta = 22.5°$ (orientations of Fig. 4 a) we have found

$$S_{exp} = 2.697 \pm 0.015 \tag{17}$$

that is a violation of inequality (7) by 40 standard deviations, and a good agreement with the quantum mechanical prediction

$$S_{QM} = 2.70 \pm 0.05$$

Experiment with switched polarizers[22]

As emphasized previously, it would be worth doing an experiment in which the orientations of the polarizers "are changed during the flight of the photons". More precisely, such a thought experiment would need random changes with an auto-correlation time shorter than L/c[15]) (L is the distance between the two polarizers, and c the speed of light).

Following our proposal,[12] we have made a step towards such a thought experiment by replacing each polarizer by an optical switch followed by two polarizers in different orientations (Fig. 15). Each setup is therefore equivalent to a single polarizer, the orientation of which is switched from one direction to another one.

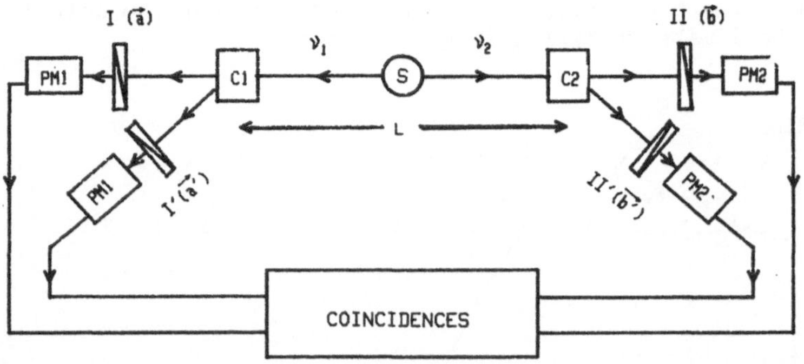

Fig. 15. Experiment with time-varying analyzers. Each optical switch followed by two polarizers is equivalent to a single polarizer fastly switched between two orientations. A switching occurs each 10 ns, while L/c is 40 ns.

The switching of the light is effected by acousto-optical interaction of the light with an ultrasonic standing wave at 25 MHz, providing a commutation at 50 MHz, i.e., a change of orientation each 10 ns. This time is short compared to L/c (40 ns), but unfortunately it is not possible with these devices to achieve a random switching. In this respect, the experiment is far from the thought experiment. Nevertheless, let us mention that the two switches on both sides were driven by independent generators drifting separately.

In this experiment we had to reduce the size of the beams and as a consequence the coincidence signal decreased by a factor 40. We had thus to collect the data for longer periods, and there were drift problems. This is why our results are far less accurate than in the first two experiments. In about 10 hours of experiment, we obtained data violating the Bell's inequalities (suited to this experiment) by 5 standard deviations. These data also allow a comparison with quantum mechanics, with which they show a reasonable agreement in the limit of our accuracy (Fig. 16).

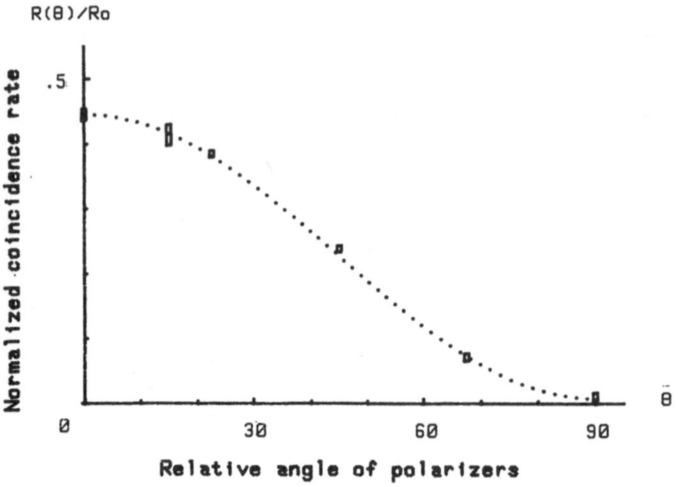

Fig. 16. Experiment with optical switches. Indicated errors are ± 1 standard deviation. The dashed curve is the prediction by quantum mechanics.

CONCLUSION

As previous ones, our experiments have still some imperfections that leave some loopholes open for the advocates of hidden variable theories obeying Einstein's causality. Improved experiments can be devised,[23] but the agreement of existing data with quantum mechanics is already impressive.

We are thus led to the conclusion that the predictions of quantum mechanics in E.P.R.- type situations are vindicated by the experiments. Bell's theorem shows us that it is not possible to understand these correlations in a "classical-looking" fashion. As often, the situation is difficult to understand classically when we have interferences between terms to which we assign a macroscopic scale (each term $|x_1, x_2\rangle$ and $|y_1, y_2\rangle$ in formula (1) describes a pair of photons separated by 12 meters in our experiments).

We can also notice that this impossibility of describing the situation classically is related to the appearance of negative values in Wigner-type distributions of probabilities. It is indeed possible to reconcile the quantum mechanical predictions (that is to say the experimental results) and Bell's formalism (equations (6) or their generalization[4,10]) by allowing some "probabilities" to assume negative values[24].

As usual, the Wigner transform is a powerful tool for comparing quantum mechanics and classical physics. However, as an experimentalist, I can hardly understand the corresponding picture of the world[25], since the realization of an event with a negative probability would mean that an event that previously happened would be erased from the world, and in particular from the computer memories where it was already registered!

REFERENCES

1. J. Von Neumann, Mathematical Foundations of Quantum Mechanics, Princeton University Press, Princeton, N.J. (1955).

2. A. Einstein, P. Podolsky and N. Rosen, Can Quantum Mechanical description of physical reality be considered complete ?, Phys. Review 47 : 777 (1935).

3. N. Bohr, Can Quantum-Mechanical description of physical reality be considered complete ?, Phys. Rev. 48 : 696 (1935).

4. J.S. Bell, On the Einstein-Podolsky-Rosen Paradox, Physics 1 195 (1964).
J.S. Bell, Introduction to the Hidden-Variable Question, in ; "Foundations of Quantum Mechanics", B. d'Espagnat ed., Academic, N.Y. (1972).

5. D. Bohm, "Quantum Theory", Prentice-Hall, Englewoods Cliffs, N.J. (1951).

6. J. F. Clauser, M.A. Horne, A. Shimony and R.A. Holt, Proposed experiment to test local hidden-variable theories, Phys. Rev. Lett. 23 : 880 (1969).

7. J.F. Clauser and A. Shimony, Bell's Theorem : Experimental Tests and Implications, Rep. Progr. Phys. 41 : 1881 (1978).

8. F. Selleri and G. Tarrozzi, Quantum Mechanics Reality and Separability Riv. Nuovo Cimento 4 : 1 (1981).

9. D. Bohm and Y. Aharonov, Discussion of Experimental Proof for the paradox of Einstein, Rosen and Podolsky, Phys. Rev. 108 : 1070 (1957).

10. J.F. Clauser, and M.A. Horne, Experimental consequences of objective local theories, Phys. Rev. D 10 : 526 (1974).

11. A. Fine, Hidden Variables, Joint Probability, and the Bell Inequalities, Phys. Rev. Lett. 48 : 291 (1982).

12. A. Aspect, Proposed Experiment to Test Separable Hidden-Variable Theories, Phys. Lett. 54 A : 117 (1975).
A. Aspect, Proposed Experiment to test the nonseparability of Quantum Mechanics, Phys. Rev. D 14 : 1944 (1976).

13. E.S. Fry, Two-photons Correlations in Atomic Transitions, Phys. Rev. A 8 : 1219 (1973).

14. S. J. Freedman and J.F. Clauser, Experimental test of local hidden-variable theories, Phys. Rev. Lett. 28 : 938 (1972).

15. F. M. Pipkin, Atomic Physics Tests of the Basic Concepts in Quantum Mechanics, in : "Advances in Atomic and Molecular Physics", D.R. Bates and B. Bederson, ed., Academic (1978).

16. J.F. Clauser, Experimental investigation of Polarization Correlation Anomaly, Phys. Rev. Lett. 36 : 1223 (1976).

17. E. S. Fry, and R.C. Thompson, Experimental Test of Local Hidden-Variable Theories, Phys. Rev. Lett. 37 : 465 (1976).

18. A. Aspect, C. Imbert, and G. Roger, Absolute Measurement of an Atomic Cascade Rate Using a Two Photon coincidence Technique. Application to the $4p^{2\,1}S_0$ - $4s4p\ ^1P_1$ - $4s^{2\,1}S_0$ Cascade of Calcium excited by a Two Photon Absorption, Opt. Comm. 34 : 46 (1980).

19. A. Aspect, P. Grangier and G. Roger, Experimental Tests of Realistic Local Theories via Bell's Theorem, Phys. Rev. Lett. 47 : 460 (1981)

20. A. Aspect, P. Grangier and G. Roger, Experimental Realization of Einstein-Podolsky-Rosen-Bohm Gedankenexperiment : A New Violation of Bell's Inequalities, Phys. Rev. Lett. 49 : 91 (1982).

21. A. Garuccio and V.A. Rapisarda, Bell's Inequalities and the Four-Coincidence Experiment, Nuovo Cimento 65 A : 269 (1981).

V.A. Rapisarda, On the measurement by Dichotomic Analyzers of the Polarization Correlation of Optical Photons Emitted in Atomic Cascade, Lett. Nuovo Cimento 33 : 437 (1982).

22. A. Aspect, J. Dalibard and G. Roger, Experimental Test of Bell's Inequalities Using Variable Analyzers, Phys. Rev. Lett. 49, 1804 (1982).

23. T. K. Lo, and A. Shimony, Proposed Molecular Test of Local Hidden-Variables Theories, Phys. Rev. A 23 : 3003 (1981)

24. P. Grangier, Correlation de polarisation de photons émis dans la cascade $4p^{2\,1}S_0$ - $4s4p\ ^1P_1$ - $4s^{2\,1}S_0$ du Calcium : test des inégalités de Bell.
Thèse de 3ème Cycle, Paris (1982)

M. O. Scully, How to make quantum mechanics look like a hidden-variable theory and vice versa, Phys. Rev. D 10 : 2477 (1983).

Although it is not fully pointed out by the author, this last model can be imbedded in a generalized formalism of local supplementary parameters theories[4,10], with quantities - interpreted as probabilities - assuming values greater than 1 or negative.

25. W. Mückenheim, A resolution of the Einstein-Podolsky-Rosen Paradox, Lett. Nuovo Cim. 35 : 300 (1981).

R. P. Feynman, Negative Probabilities, Preprint, California Institute of Technology, Pasadena (1984).

COMMENT ON "A CLASSICAL MODEL OF EPR EXPERIMENT

WITH QUANTUM MECHANICAL CORRELATIONS AND BELL INEQUALITIES"*

Alain Aspect

Institut d'Optique Théorique et Appliquée
Bâtiment 503 - Centre Universitaire d'Orsay - BP 43
91406 ORSAY Cédex

In my talk, I have claimed that the following statement brings the essence of Bell's theorem : *"It is impossible to mimick the quantum mechanical predictions for the EPR correlations, with a reasonable classical-looking model, in the spirit of Einstein's ideas"*. Then Azim Barut disputed this statement. He did so in what would be the most decisive way, that is to say by exhibiting a counter-example. I must say that I would be completely convinced that I am wrong if somebody could make a classical model (i.e. following the laws of classical physics) mimicking *all* the quantum mechanical predictions for the E.P.R. correlations. But I intend to show here that it is not the case for Barut's model, for the following reasons :

1) the first version of his model is classical, but doesn't mimick at all an E.P.R. type experiment ;

2) by reinterpretation we can get a model that *does mimick* the experiment, but this model is no longer "reasonable classical looking" since it involves negative probabilities.

As a matter of fact, I find this model very interesting. It is an excellent example of how one can modify Bell's formalism (in a non-classical way) in order to make it agree with quantum mechanics.

The Challenge

For the sake of clarity, I will put my claim in the form of a challenge. I challenge anybody to reproduce with a classical model all the features of an E.P.R. type experiment (Figure (1)), which I recall now.

At each measuring apparatus, one registers for each pair the result of the measurement (+ 1 or - 1) and the setting of the instrument, allowing to derive the relative frequencies (or probabilities) of single detections ($P_\pm(\vec{a})$ and $P_\pm(\vec{b})$) and of joint detections ($P_{\pm\pm}(\vec{a},\vec{b})$). It is convenient to consider the results of the measurements as random variables $A(\vec{a})$ and $B(\vec{b})$ with values + 1 or - 1. The measurement results

* See this volume

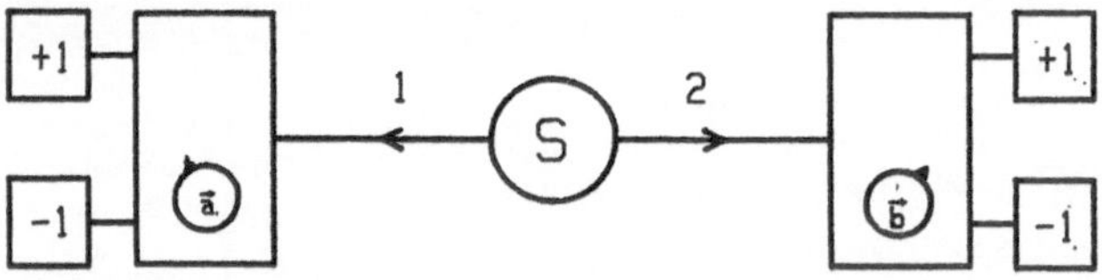

Fig. C-1. An idealized E.P.R. type experiment. A source emits pairs of objects, on which some measuring apparatus perform dichotomic measurements (only two-possible out comes, + 1 or − 1). The setting of each instruments ($\vec{a}$ or $\vec{b}$) can be selected at will. The experiment is repeated on a great number of pairs.

yield eventually the average values :

$$\overline{A(\vec{a})} \qquad \overline{B(\vec{b})} \qquad \text{and} \qquad \overline{A(\vec{a}).\,B(\vec{b})}$$

For an E.P.R. type experiment following the quantum mechanical predictions, one will find :

$$\begin{aligned} \overline{A(\vec{a})} &= p_+(\vec{a}) - p_-(\vec{a}) = 0 \\ \overline{B(\vec{b})} &= p_+(\vec{b}) - p_-(\vec{b}) = 0 \end{aligned} \qquad \text{(C-1)}$$

and

$$E(\vec{a},\vec{b}) = \overline{A(\vec{a})\; B(\vec{b})} = p_{++}(\vec{a},\vec{b}) + p_{--}(\vec{a},\vec{b}) - p_{+-}(\vec{a},\vec{b}) - p_{-+}(\vec{a},\vec{b})$$

This last quantity is equal to the usual correlation function ($\overline{A(\vec{a})} = 0$ and $\overline{A^2(\vec{a})} = 1$)

$$\frac{\overline{A(\vec{a}).B(\vec{b})} - \overline{A(\vec{a})}.\overline{B(\vec{b})}}{(\overline{A^2(\vec{a})}.\overline{B^2(\vec{b})})^{1/2}}$$

The quantum mechanical predictions are

$$E(\vec{a},\vec{b}) = \cos 2(\vec{a},\vec{b}) \qquad \text{(C-2')}$$

for photon polarizations (see my talk). For spin 1/2 particles in a singlet state, as considered by Barut, quantum mechanics predict :

$$E(\vec{a},\vec{b}) = -\cos(\vec{a},\vec{b}) \qquad \text{(C-2")}$$

Let us summarize the experimental facts that should be mimicked in order to disprove my claim :

(i) two valued measurements on each particle, allowing to define $A(\vec{a})$ and $B(\vec{b})$ with only two possible values (+ 1 or − 1) ;

(ii) $\overline{A(\vec{a})} = \overline{B(\vec{b})} = 0$

(iii) $E(\vec{a},\vec{b}) = \overline{A(\vec{a}).B(\vec{b})}$ identical to the quantum mechanical predictions (C-2') or (C-2").

Barut's model

This model follows the laws of classical mechanics. Two systems produced by a break up mechanism fly apart, bringing some exactly opposite vectorial quantities $\vec{p}$ and $\vec{p}'$. Many conservation laws (such as momentum or angular momentum conservation) can provide with such a scheme. The angular distribution of $\vec{p}$ over the ensemble of all emitted pairs is taken isotropic.

The model then considers measurements yielding *the projections* $A = \vec{p}.\vec{a}$ and $B = -\vec{p}.\vec{b}$ of $\vec{p}$ on the orientations $\vec{a}$ and $\vec{b}$ of the two measuring apparatus. These measurement results clearly depend on the initial condition $\vec{p}$ and on the orientation $\vec{a}$ or $\vec{b}$. By average on the isotropic initial conditions, one easily gets :

$$\overline{A(\vec{a})} = \overline{B(\vec{b})} = 0$$

and (C-3)

$$E(\vec{a},\vec{b}) = \frac{\overline{A.B.}}{(A^2.B^2)^{1/2}} = -\cos(\vec{a},\vec{b})$$

These results clearly agree with the quantum mechanical predictions, and the model fulfills the points (ii) and (iii).

But the model *does not* agree with point (i), since the measurements do not yield two-valued results, but they rather yield results in a continuous range. In this respect the model does not mimick an E.P.R. type experiment as described in figure C-1, and as performed in one of our experiments (ref. 20 of my talk).

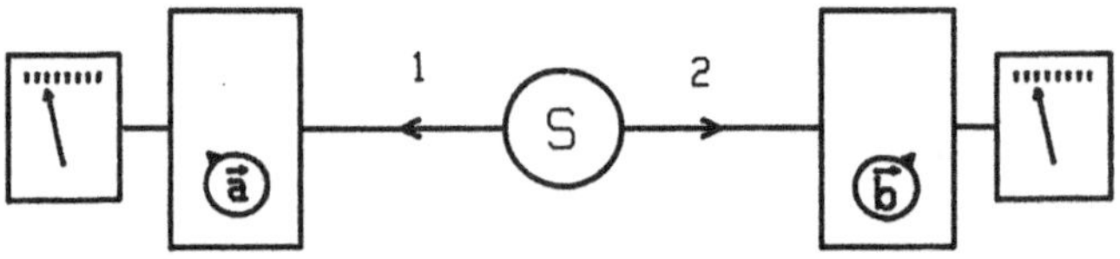

Fig. C-2. The model under discussion differs from an EPR type experiment since it yields outputs in a continuous range, instead of two-valued outputs.

Barut was certainly aware of this fact, and the discussion in the second part of his paper can be understood as adressed to this point. As a matter of fact, all his discussion about probabilities can be interpreted in the framework of a model modified to give values + 1 or -1 as results of measurements. I will discuss this point now.

Reinterpretation of the model

In order to get a model reproducing the features (i), one has just to introduce in the measuring device an extra part, that *discretizes* the results of measurements. In other words, given a result in a continuous range (between $-|\vec{p}|$ and $+|\vec{p}|$, it is easy to imagine a device returning a discrete value (either + 1 or - 1). We have all heard for instance of analogic to digital converters.

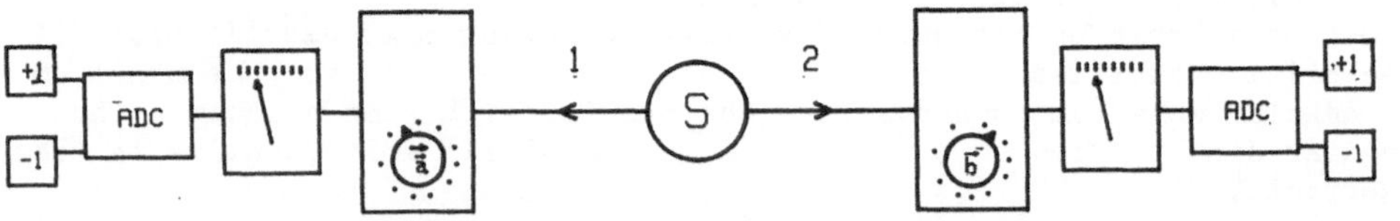

Fig. C-3. The model under discussion can be reinterpreted by adding a chip converting the continuous outputs into two-valued outputs.

The simplest system that one can think of would be a device measuring the sign of $\vec{p}.\vec{a}$ (respectively $-\vec{p}.\vec{b}$) and yielding + 1 or - 1 according to this sign. I have actually considered such a model in my talk (the so-called "naive example"), and we have found that a model of this type fulfills points (i) and (ii), but not point (iii) (the predicted correlation function is not the correct one).

Barut's model thus needs a more subtle device, namely a chip involving some randomness. For a given projection $\vec{p}.\vec{a}$ (or respectively $-\vec{p}.\vec{b}$), such a chip would return the values + 1 or - 1 according to some law of probability

$$\pi_+(\vec{p}.\vec{a}) \qquad \text{and} \qquad \pi_-(\vec{p}.\vec{a}) \tag{C-4}$$

that will of course obey the equation

$$\pi_+(\vec{p}.\vec{a}) + \pi_-(\vec{p}.\vec{a}) = 1 \tag{C-5}$$

For a given initial value $\vec{p}$, one can thus define an averaged output of the chip (averaged on the internal random system of the chip)

$$\tilde{A}(\vec{p},\vec{a}) = \pi_+(\vec{p}.\vec{a}) - \pi_-(\vec{p}.\vec{a}) \tag{C-6}$$

A similar quantity $\tilde{B}(\vec{p},\vec{b})$ is defined for the second measuring apparatus. Following Barut, we take

$$\tilde{A}(\vec{p},\vec{a}) = \sqrt{3}\cos(\vec{p},\vec{a})$$

and (C-7)

$$\tilde{B}(\vec{p},\vec{b}) = -\sqrt{3}\cos(\vec{p},\vec{b})$$

If now we average on all the emitted pairs, we obtain values identical to the quantum mechanical predictions (C-1) and (C-2"). Thus, we have a model reproducing all the features of an E.P.R. type experiment, that is to say fulfilling the points (i), (ii) and (iii). The only remaining question is to know if it is a reasonable classical looking model.

There is obviously no problem neither with the break-up process, nor with the measurement of the projections $\vec{p}.\vec{a}$ or $\vec{p}.\vec{b}$. The remaining part is the analogic to digital converter, that effects the discretization of the results. According to formula (C-4) to (C-7), this converter should follow the law of probabilities:

$$\begin{aligned}\pi_+(\vec{p}.\vec{a}) &= \frac{1}{2}(1+\sqrt{3}\cos(\vec{p},\vec{a}))\\ \pi_-(\vec{p}.\vec{a}) &= \frac{1}{2}(1-\sqrt{3}\cos(\vec{p},\vec{a}))\end{aligned} \tag{C-8}$$

These probabilities can assume values negative or larger than one for some angles $(\vec{p},\vec{a})$. To my knowledge, there is no chip able to realize such a function. For a breakup process producing $\vec{p}$ parallel to $\vec{a}$, the chip should "fire the output + 1 with a probability 1.35, and the output - 1 with a probability - 0.35". I don't ever understand what these words mean*.

To paraphrase the claim under discussion, *"It is impossible to afford such a chip from a reasonable classical-looking company"* and Barut's model, reinterpreted in order to fulfill points (i) (ii) and (iii), is not a classical model.

---

* Conversely, the condition that the quantities $\pi_\pm$ have values between 0 and 1 justifies the constraints on $|\tilde{A}(\lambda,a)|$ and $|\tilde{B}(\lambda,b)|$ that are used in the derivation of Bell's inequalities for stochastic, local, supplementary-parameters theories.

FLUCTUATIONS, IRREVERSIBILITY, AND CHAOS

This section begins with a pair of papers by Professor Beretta presenting a novel dynamical theory of irreversibility and the approach to stable equilibrium. This is a quite different (and no doubt controversial) approach to the problem treated by Professor Lamb in the opening chapter. The paper by Dr. Kleber explores the use of quantum trajectories as a technique for solving time-dependent problems in quantum statistical physics. Next come papers by Dr. Liboff and Dr. Ambegaokar which treat statistical properties of plasmas and quantum Brownian motion in fluids. Dr. Banyai's paper studies the scaling relations which relate the microscopic and macroscopic descriptions of physical systems. Finally, the paper by Drs. Aguirregabiria and Bel explores the role of time retardation in promoting the onset of chaotic behavior.

A GENERAL NONLINEAR EVOLUTION EQUATION FOR IRREVERSIBLE CONSERVATIVE APPROACH TO STABLE EQUILIBRIUM

Gian Paolo Beretta

Massachusetts Institute of Technology, Cambridge, MA 02139
and Politecnico di Milano, 20133 Milano, Italy

INTRODUCTION

The problem of understanding entropy and irreversibility has been tackled by thousands of physicists during the past century. Schools of thought have formed and flourished around different perspectives of the problem. But a definitive solution has yet to be found.[1]

We address a mathematical problem very relevant to the question of nonequilibrium and irreversibility, namely, that of "designing" a general evolution equation capable of describing irreversible but conservative relaxation towards equilibrium. Our objective is to present an interesting mathematical solution to this "design" problem, namely, a new nonlinear evolution equation that satisfies a set of very stringent relevant requirements.

In this lecture, we do not claim any physical meaning for the proposed nonlinear evolution equation. Indeed, we define three essentially different frameworks within which the new equation could be adopted, with entirely different interpretations. We purposely devoid this presentation of our school's quite unorthodox perspective on the physical meaning of entropy and irreversibility, because we feel that the proposed nonlinear equation constitutes an important advance in itself, independently of the physical context for which it was designed and developed.[2-4]

The lecture is organized as follows. First, we define three familiar mathematical frameworks within which the subsequent results may be interpreted. Then, we list the "design specifications" that we intend to impose on the desired evolution equation. We review

some useful well-known mathematics involving Gram determinants and, finally, present our nonlinear evolution equation which meets the stringent design specifications.

Our views and hypotheses on entropy, nonequilibrium and irreversibility will be discussed in our second lecture in this volume.

FRAMEWORK A: QUANTUM STATISTICAL MECHANICS AND QUANTUM THERMODYNAMICS

Let $\mathcal{H}$ be a Hilbert space ($\dim \mathcal{H} \leq \infty$[5]), and $\mathcal{L}$ the set of all linear operators A, B, ... on $\mathcal{H}$, equipped with the real inner product $(\cdot|\cdot)$ defined by

$$(A|B) = \frac{1}{2}\mathrm{Tr}(A^{\dagger}B + B^{\dagger}A) \tag{1a}$$

where $A^{\dagger}$ denotes the adjoint of operator A and Tr the trace functional. We denote by $\mathcal{P}$ the subset of all self-adjoint, nonnegative-definite, unit-trace operators ρ in $\mathcal{L}$, i.e.,

$$\mathcal{P} = \{\rho \text{ in } \mathcal{L} \,|\, \rho^{\dagger} = \rho,\ \rho \geq 0,\ \mathrm{Tr}\rho = 1\} \tag{2a}$$

We will then consider a set

$$\{H, N_1, \ldots, N_r\} \tag{3a}$$

of self-adjoint operators in $\mathcal{L}$, where each N_i commutes with H, i.e., is such that $HN_i = N_iH$, for $i = 1, \ldots, r$.

In Quantum Statistical Mechanics,[1] ρ is the von Neumann statistical or density operator which represents the index of statistics from a generally heterogeneous ensemble of identical systems (with associated Hilbert space $\mathcal{H}$) distributed over a range of possible quantum mechanical states.

In Quantum Thermodynamics,[3] ρ is the state operator which represents the individual quantum state of a strictly isolated system with associated Hilbert space $\mathcal{H}$, in the same sense as in Quantum Mechanics a vector ψ in $\mathcal{H}$ represents the individual quantum state.

In both theories, H is the Hamiltonian operator, and operator N_i, for $i = 1, \ldots, r$, is the number operator for particles of type i in the system (if the system has a fixed number n_i of particles of type i, then $N_i = n_iI$, where I is the identity operator on $\mathcal{H}$).

FRAMEWORK B: CLASSICAL STATISTICAL MECHANICS

Let Ω be a phase space, and $\mathcal{L}$ the set of real, square-integrable functions A, B, ... on Ω, equipped with the inner product $(\cdot|\cdot)$ defined by

$$(A|B) = \mathrm{Tr}AB = \int_\Omega AB d\Omega \tag{1b}$$

where Tr in this framework denotes $\int_\Omega d\Omega$. We denote by $\mathcal{P}$ the subset of all nonnegative-definite, normalized functions ρ in $\mathcal{L}$, i.e.,

$$\mathcal{P} = \{\rho \text{ in } \mathcal{L} \,|\, \rho \geq 0, \int_\Omega \rho d\Omega = 1\} \tag{2b}$$

We will then consider a set

$$\{H, N_1, \ldots, N_r\} \tag{3b}$$

of functions in $\mathcal{L}$.

In Classical Statistical Mechanics, ρ is the Gibbs density-of-phase function which represents the index of statistics from a generally heterogeneous ensemble of identical systems (with associated phase space Ω) distributed over a range of possible classical mechanical states. H is the Hamiltonian function, and N_i the number function for particles of type i.

FRAMEWORK C: INFORMATION THEORY

Let $\mathcal{L}$ be the set of all n × n real, diagonal matrixes $A = \mathrm{diag}(a_j)$, $B = \mathrm{diag}(b_j)$, ... ($n \leq \infty$), equipped with the inner product $(\cdot|\cdot)$ defined by

$$(A|B) = \mathrm{Tr}AB = \sum_{j=1}^{n} a_j b_j \tag{1c}$$

We denote by $\mathcal{P}$ the subset of all nonnegative-definite, unit-trace matrixes ρ in $\mathcal{L}$, i.e.,

$$\mathcal{P} = \{\rho = \mathrm{diag}(p_j) \,|\, p_j \geq 0, \mathrm{Tr}\rho = \sum_{j=1}^{n} p_j = 1\} \tag{2c}$$

We will then consider a set

$$\{H, N_1, \ldots, N_r\} \tag{3c}$$

of diagonal matrixes $H = \mathrm{diag}(e_j)$, $N_1 = \mathrm{diag}(n_{1j})$, ..., $N_r = \mathrm{diag}(n_{rj})$ in $\mathcal{L}$.

In Information Theory,[6] $\rho = \mathrm{diag}(p_j)$ represents the probability assignment to a set of n events, p_j being the probability of occurrence of the j-th event. H, N_1, ..., N_r are characteristic features of the events in the set, taking on the values e_j, n_{1j}, ..., n_{rj}, respectively, for the j-th event.

MEAN VALUE FUNCTIONALS AND S-FUNCTIONAL

From here on, our notation allows us to treat at once the three frameworks just defined. For reasons to become apparent below, we call the elements H, N_1, ..., N_r in Equations 3 the generators of the motion. We will assume that such sets always contain at least element H, that we call the Hamiltonian generator of the motion.

For each generator of the motion, we then define a mean value functional on $\mathcal{P}$ as follows

$$
\begin{aligned}
m(\rho;H) &= \mathrm{Tr}\,\rho H = (\sqrt{\rho}|\sqrt{\rho}H) \\
m(\rho;N_i) &= \mathrm{Tr}\,\rho N_i = (\sqrt{\rho}|\sqrt{\rho}N_i)
\end{aligned}
\tag{4}
$$

Moreover, we define the S-functional on $\mathcal{P}$ as

$$S(\rho) = -k\mathrm{Tr}\,\rho\ln\rho = -k(\sqrt{\rho}|\sqrt{\rho}\ln\rho) \tag{5}$$

Depending on the context, the S-functional represents the thermodynamic entropy, the statistical uncertainty as to the actual state of a system, or the information carried by the occurrence of one of the possible events.

For each given set of values $\langle H\rangle$, $\langle N_1\rangle$, ..., $\langle N_r\rangle$, in the range of the mean value functionals (Equations 4) corresponding to the generators of the motion, we consider the subset of all elements ρ in $\mathcal{P}$ that share the given mean values, i.e.,

$$
\begin{aligned}
&\mathcal{P}(\langle H\rangle,\langle N_1\rangle,\ldots,\langle N_r\rangle) = \\
&\{\rho \text{ in } \mathcal{P}\,|\,m(\rho;H) = \langle H\rangle,\ m(\rho;N_i) = \langle N_i\rangle \text{ for } i = 1,\ldots,r\}
\end{aligned}
\tag{6}
$$

On each such subset, i.e., for fixed mean values $\langle H\rangle$, $\langle N_1\rangle$, ..., $\langle N_r\rangle$ of the generators of the motion, the S-functional (Equation 5) achieves a unique maximum at the point

$$\rho = e^{-\alpha}\exp(-\beta H + \sum_{i=1}^{r}\nu_i N_i) \tag{7}$$

where

$$\alpha = \ln \mathrm{Tr} \exp(-\beta H + \sum_{i=1}^{r} \nu_i N_i) \tag{8}$$

$$\beta = \beta(\langle H\rangle, \langle N_1\rangle, \ldots, \langle N_r\rangle) \tag{9}$$

$$\nu_i = \nu_i(\langle H\rangle, \langle N_1\rangle, \ldots, \langle N_r\rangle) \tag{10}$$

It is noteworthy that the maximum-S points satisfy the condition

$$\sqrt{\rho}\ln\rho = -\alpha\sqrt{\rho} - \beta\sqrt{\rho}H + \sum_{i=1}^{r} \nu_i \sqrt{\rho}N_i \tag{11}$$

where α, β and ν_i, for i = 1, ..., r, are real numbers. In words, the maximum-S element ρ is such that $\sqrt{\rho}\ln\rho$ lies on the linear manifold generated by elements $\sqrt{\rho}$, $\sqrt{\rho}H$, $\sqrt{\rho}N_1$, ..., $\sqrt{\rho}N_r$. This observation will prove useful in what follows. In particular, it is noteworthy that Condition 11 is satisfied not only by the maximum-S elements given by Equation 7, but also by the elements given by

$$\rho = e^{-a} B \exp(-bH + \sum_{i=1}^{r} c_i N_i) \tag{12}$$

where

$$a = \ln \mathrm{Tr}\, B \exp(-bH + \sum_{i=1}^{r} c_i N_i) \tag{13}$$

$$b = b(B; \langle H\rangle, \langle N_1\rangle, \ldots, \langle N_r\rangle) \tag{14}$$

$$c_i = c_i(B; \langle H\rangle, \langle N_1\rangle, \ldots, \langle N_r\rangle) \tag{15}$$

and B is any idempotent element in $\mathcal{L}$ (i.e., $B^2 = B$) commuting with ρ (this last condition is trivally satisfied within Frameworks b and c). An element ρ satisfying Equation 12 maximizes S on $\mathcal{P}(\langle H\rangle, \langle N_1\rangle, \ldots, \langle N_r\rangle)$ and satisfies Equation 7 only if B = I (I = identity operator on $\mathcal{H}$ in Framework a; I = constant function equal to 1 on the whole Ω in Framework b; I = diag(1) in Framework c).

DESIGN SPECIFICATIONS

Our "architectural" problem is to design a function $F(\cdot)$ such that every solution $\rho(t)$ of the autonomous differential equation

$$\frac{d}{dt}\rho(t) = F(\rho(t)) \tag{16}$$

with $\rho(0)$ in $\mathcal{D}(F) \subset \mathcal{P}$ [7] satisfies the following conditions for all $t \geq 0$:

(i) $\rho(t)$ lies entirely in $\mathcal{D}(F)$;

(ii) $m(\rho(t);H) = m(\rho(0);H)$, and $m(\rho(t);N_i) = m(\rho(0);N_i)$ for $i = 1,\ldots,r$;

(iii) $S(\rho(t+u)) \geq S(\rho(t))$ for all $u > 0$;

(iv) among all the elements ρ in $\mathcal{P}$ with given mean values $\langle H\rangle$, $\langle N_1\rangle$, ..., $\langle N_r\rangle$ of the generators of the motion, the unique maximum-S element $\rho = e^{-\alpha} \exp(-\beta H + \sum_{i=1}^{r} \nu_i N_i)$ (Equations 7 to 10) is the only equilibrium solution that is stable according to Liapunoff.[8]

It is noteworthy that requirement (iv) is most restrictive. For example, within Framework a, it rules out the von Neumann evolution equation ($F(\rho) = -i(H\rho-\rho H)/\hbar$) because all the equilibrium solutions (ρ such that $H\rho = \rho H$) are stable according to Liapunoff and, in general, there are many such solutions for each given set of mean values of the generators of the motion.

SOME NECESSARY MATHEMATICAL BACKGROUND

Given a subset of elements A, B, ..., Z in $\mathcal{L}$, we denote by M(A,B,...,Z) the Gram matrix

$$M(A,B,\ldots,Z) = \begin{bmatrix} (A|A) & (A|B) & \cdots & (A|Z) \\ (B|A) & (B|B) & \cdots & (B|Z) \\ \vdots & \vdots & \ddots & \vdots \\ (Z|A) & (Z|B) & \cdots & (Z|Z) \end{bmatrix} \tag{17}$$

where $(\cdot|\cdot)$ is the real symmetric inner product defined on $\mathcal{L}$. We denote by G(A,B,...,Z) the Gram determinant of A,B,...,Z with respect to the inner product $(\cdot|\cdot)$, i.e., G(A,B,...,Z) = det[M(A,B,...,Z)]. Matrix M(A,B,...,Z) is nonnegative definite and, therefore, its determinant G(A,B,...,Z) is also nonnegative. A necessary and sufficient condition for elements A,B,...,Z to be linearly independent is that their Gram determinant G(A,B,...,Z) be nonzero and, hence, strictly positive.

Given a subset of elements A, B, ..., Z in $\mathcal{L}$, we denote by L(A,B,...,Z) the linear manifold spanned by all linear combinations with real coefficients of the elements A, B, ..., Z. With respect to the inner product $(\cdot|\cdot)$ defined on $\mathcal{L}$, we denote the projection of a given element V in $\mathcal{L}$ onto a linear manifold L by the symbol $(V)_L$. $(V)_L$ is the unique element in L such that $((V)_L|X) = (V|X)$ for all X in L.

The theory of Gram determinants, very seldom used in the physics literature, offers a useful explicit way of writing the projection $(V)_L$ of V onto a given linear manifold L.

Let a linear manifold L be given, e.g., by specifying it as the set of all real linear combinations of given elements A, B, ..., Z in $\mathcal{L}$, not necessarily linearly independent, i.e., L = L(A,B,...,Z). Let us select any subset of linearly independent elements E_1, E_2, ..., E_m spanning L, i.e., such that $G(E_1,E_2,\ldots,E_m) > 0$ and $L(E_1,E_2,\ldots,E_m) = L$. By the definition of $(V)_L$, $((V)_L|E_j) = (V|E_j)$ for every j = 1, 2, ..., m, and $(V)_L = \sum_{i=1}^{m} v_i E_i$, where v_i are real scalars. Thus,

$$\sum_{i=1}^{m} v_i (E_i|E_j) = (V|E_j) \text{ for } j = 1, 2, \ldots, m \tag{18}$$

Because $(E_i|E_j) = [M(E_1,E_2,\ldots,E_m)]_{ij}$ and the elements E_1, E_2, ..., E_m are linearly independent, Equations 18 are linearly independent and can be solved for the v_i's to yield

$$v_i = \sum_{j=1}^{m} (V|E_j)[M(E_1,E_2,\ldots,E_m)^{-1}]_{ji} \qquad \text{for } i=1,2,\ldots,m \tag{19}$$

and, therefore,

$$(V)_L = \sum_{i=1}^{m} \sum_{j=1}^{m} (V|E_j)[M(E_1,E_2,\ldots,E_m)^{-1}]_{ji} E_i \tag{20}$$

An alternative, completely equivalent, but more elegant expression for $(V)_L$ is given by

$$(V)_L = -\frac{1}{G(E_1,E_2,\ldots,E_m)} \det \begin{bmatrix} 0 & E_1 & E_2 & \cdots & E_m \\ (E_1|V) & (E_1|E_1) & (E_1|E_2) & \cdots & (E_1|E_m) \\ (E_2|V) & (E_2|E_1) & (E_2|E_2) & \cdots & (E_2|E_m) \\ \vdots & \vdots & \vdots & \ddots & \vdots \\ (E_m|V) & (E_m|E_1) & (E_m|E_2) & \cdots & (E_m|E_m) \end{bmatrix} \tag{21}$$

We next consider an important example that we shall use directly in what follows. For a given element ρ in the set $\mathcal{P}$, we will need to consider the projection of $\sqrt{\rho}\ln\rho$ onto the linear manifold $L(\sqrt{\rho},\sqrt{\rho}H,\sqrt{\rho}N_1,\ldots,\sqrt{\rho}N_r)$ where $\sqrt{\rho}$, $\sqrt{\rho}H$, $\sqrt{\rho}N_1$, ..., $\sqrt{\rho}N_r$ are not necessarily linearly independent. Using Equation 21 and Definitions 1 of the inner product $(\cdot|\cdot)$, we find

$$(\sqrt{\rho}\ln\rho)_{L(\sqrt{\rho}R_0,\sqrt{\rho}R_1,\ldots,\sqrt{\rho}R_z)} = -\frac{1}{G(\sqrt{\rho}R_0,\sqrt{\rho}R_1,\ldots,\sqrt{\rho}R_z)}\,\cdot$$

$$\cdot\det \begin{bmatrix} 0 & \sqrt{\rho}R_0 & \sqrt{\rho}R_1 & \cdots & \sqrt{\rho}R_z \\ \mathrm{Tr}\rho R_0\ln\rho & \mathrm{Tr}\rho R_0^2 & \frac{1}{2}\mathrm{Tr}\rho\{R_0,R_1\} & \cdots & \frac{1}{2}\mathrm{Tr}\rho\{R_0,R_z\} \\ \mathrm{Tr}\rho R_1\ln\rho & \frac{1}{2}\mathrm{Tr}\rho\{R_1,R_0\} & \mathrm{Tr}\rho R_1 & \cdots & \frac{1}{2}\mathrm{Tr}\rho\{R_1,R_z\} \\ \vdots & \vdots & \vdots & \ddots & \vdots \\ \mathrm{Tr}\rho R_z\ln\rho & \frac{1}{2}\mathrm{Tr}\rho\{R_z,R_0\} & \frac{1}{2}\mathrm{Tr}\rho\{R_z,R_1\} & \cdots & \mathrm{Tr}\rho R_z^2 \end{bmatrix} \tag{22}$$

where $\{A,B\} = AB + BA$, and the subset of elements R_0, R_1, ..., R_z in $\mathcal{L}$ is such that $L(\sqrt{\rho}R_0,\sqrt{\rho}R_1,\ldots,\sqrt{\rho}R_z) = L(\sqrt{\rho},\sqrt{\rho}H,\sqrt{\rho}N_1,\ldots,\sqrt{\rho}N_r)$ and $G(\sqrt{\rho}R_0,\sqrt{\rho}R_1,\ldots,\sqrt{\rho}R_z) > 0$.

With this background, we can present in a quite compact form an evolution equation meeting our very restrictive design specifications.

NEW NONLINEAR EVOLUTION EQUATION

The author designed and proposed the following evolution equation:[2,3]

$$\frac{d}{dt}\rho(t) = F(\rho(t)) \tag{23a}$$

$$F(\rho) = -\frac{i}{\hbar}[H,\rho] - \frac{1}{\tau}\frac{1}{2}(\sqrt{\rho}D(\rho) + D^{\dagger}(\rho)\sqrt{\rho}) \tag{23b}$$

$$D(\rho) = \sqrt{\rho}\ln\rho - (\sqrt{\rho}\ln\rho)_{L(\sqrt{\rho},\sqrt{\rho}H,\sqrt{\rho}N_1,\ldots,\sqrt{\rho}N_r)} \tag{23c}$$

where $[H,\rho] = H\rho - \rho H$ (= 0 within Frameworks b and c), $\hbar$ is the reduced Planck constant (playing a role only within Framework a), τ is a characteristic time constant, H, N_1, ..., N_r are fixed generators of the motion and, for each ρ, $L(\sqrt{\rho},\sqrt{\rho}H,\sqrt{\rho}N_1,\ldots,\sqrt{\rho}N_r)$ is the linear manifold generated by elements $\sqrt{\rho}$, $\sqrt{\rho}H$, $\sqrt{\rho}N_1$, ..., $\sqrt{\rho}N_r$. Function $F(\rho)$ is clearly nonlinear in ρ.

It is noteworthy that along any solution $\rho(t)$, $D(\rho(t))$ is the "instantaneous distance" element between element $\sqrt{\rho}(t)\ln\rho(t)$ and the linear manifold spanned by $\sqrt{\rho}(t)$, $\sqrt{\rho}(t)H$, $\sqrt{\rho}(t)N_1$, ..., $\sqrt{\rho}(t)N_r$. At each instant in time along a solution, the inner product of $i[H,\rho]$ and $\sqrt{\rho}D(\rho) + D^{\dagger}(\rho)\sqrt{\rho}$ is equal to zero and, therefore, the unitary contribution to $d\rho/dt$ due to the first linear term in Equation 23b is orthogonal to the nonunitary contribution to $d\rho/dt$ due to the second nonlinear term in Equation 23b.

The original motivation for the design of this new evolution equation, together with the physical justification of the design specifications, is discussed in our second lecture of this series, and in References 2 to 4. Here we are mainly concerned with the mathematical properties of the new evolution equation and its solutions. The properties that we list below emerge from our analysis of Equation 23 in References 3 and 9. In Reference 3, we adopt Equation 23 as the general quantum thermodynamic equation of motion for a single strictly isolated constituent of matter. Much technical mathematical work remains to be done on our conjectures in Reference 3. In Reference 9, by specializing the analysis to a single strictly isolated two-level system, we prove rigorously all the properties of the equation, including the existence and uniqueness of solutions to the initial value problem both in forward and in backward time.

For all t in the range $(-\infty,+\infty)$, i.e., both in forward and backward time, if $\rho(0)$ is in $\mathcal{D}(F) \subset \mathcal{P}$ then:

(i) $\rho(t)$ remains in $\mathcal{D}(F)$, i.e., there are no forward nor backward escape times;

(ii) the mean value of any linear combination of I, H, N_1, ..., N_r is time-invariant, i.e., for every set of real scalars a, b, c_1, ..., c_r,

$$\mathrm{Tr}\,\rho(t)[aI + bH + \sum_{i=1}^{r} c_i N_i] = \text{const.} \tag{24}$$

(iii) $S(\rho(t+u)) \geq S(\rho(t))$ for all $u > 0$, because

$$\frac{d}{dt}S(\rho) = (D(\rho)|D(\rho)) = \frac{G(\sqrt{\rho}\ln\rho, \sqrt{\rho}R_0, \sqrt{\rho}R_1, \ldots, \sqrt{\rho}R_z)}{G(\sqrt{\rho}R_0, \sqrt{\rho}R_1, \ldots, \sqrt{\rho}R_z)} \geq 0 \tag{25}$$

Moreover, the strict equality applies, i.e., $dS(\rho)/dt = 0$, if and only if ρ satisfies the condition

$$\sqrt{\rho}\ln\rho = -a\sqrt{\rho} - b\sqrt{\rho}H + \sum_{i=1}^{r} c_i\sqrt{\rho}N_i \tag{26}$$

for some real scalars a, b, c_1, ..., c_r, in which case ρ is said to be nondissipative, and can be written according to Equation 12;

(iv) every nondissipative ρ is either an equilibrium solution of Equation 23 or it belongs to a limit cycle. An equilibrium solution is unstable if ρ can be written according to Equation 12 with $B \neq I$, whereas it is stable if ρ can be written according to Equation 7.

In Reference 10, by specializing the analysis to a single strictly isolated N level system, we prove that Equation 23 implies a generalization of Onsager's reciprocal relations valid for all nonequilibrium states, close and far from stable equilibrium. Onsager's coefficients emerge as well-defined nonlinear functionals of the quantum thermodynamic state operator.

ADDITIONAL DESIGN SPECIFICATIONS

A generalization of Equation 23, consistent with additional design specifications required for the description of composite systems is proposed in References 1 and 11. There, we conjecture that the nonlinear term in the new evolution equation implies an intrinsic mechanism of loss of correlations between component subsystems.

A SIMPLEST EXAMPLE

As an illustrative example, let us select Framework c, with $n = 2$, $\rho = \mathrm{diag}(x,1-x)$, $0 \leq x \leq 1$, $H = \mathrm{diag}(1,1)$ and $r = 0$. We discuss only the mathematical aspects of this simplest example, not its possible information-theoretic applications. Equation 23 becomes

$$\dot{x} = \begin{cases} 0 & \text{if } x = 0 \text{ or } x = 1 \\ -\dfrac{1}{\tau}\, x(1-x)\ln\dfrac{x}{1-x} & \text{if } 0 < x < 1 \end{cases} \tag{27}$$

There are three equilibrium solutions, $x = 0$, $x = 1$, and $x = 1/2$. The nonequilibrium solutions are

$$x(t) = \frac{[x_o/(1-x_o)]^{\exp(-t/\tau)}}{1 + [x_o/(1-x_o)]^{\exp(-t/\tau)}} \tag{28a}$$

or, equivalently,

$$\ln\frac{x(t)}{1-x(t)} = e^{-t/\tau}\ln\frac{x(0)}{1-x(0)} \tag{28b}$$

We easily verify that $x(t)$ is either always greater or always smaller than 1/2, for $-\infty < t < +\infty$. The solutions are defined both in forward and backward time, with no finite escape times. If $x_o > 1/2$, then $x(t) \to 1$ as $t \to -\infty$. If $x_o < 1/2$, then $x(t) \to 0$ as $t \to -\infty$. In either case, $x(t) \to 1/2$ as $t \to +\infty$. The only equilibrium solution which is stable is the maximum-S $x = 1/2$, whereas equilibrium solutions $x = 0$ and $x = 1$ are unstable.

CONCLUSIONS

We presented a nonlinear evolution equation new to physics. The new equation is applicable within the frameworks of Quantum Thermodynamics, Classical and Quantum Statistical Mechanics, and Information Theory. The nonlinear equation (and its generalization discussed in Reference 11) yields a unique deterministic description of irreversible, but conservative relaxation from nonequilibrium towards equilibrium, and satisfies a very restrictive stability requirement.

We believe that the new evolution equation will constitute an important mathematical "tool" for several, radically different nonequilibrium problems. Different schools of thought, with contrasting perspectives on the physical meaning of entropy and irreversibility, could all extract important insights from the richness of structure implied by this relevant nonlinear equation. Applications could also be found beyond the domain of physics, e.g., to time-dependent problems in Information Theory.

REFERENCES

1. For recent reviews of the problem, see: R. Jancel, Foundations of Classical and Quantum Statistical Mechanics, Pergamon Press, Oxford, 1969; A. Wehrl, Rev. Mod. Phys., Vol. 50, 221 (1978); O. Penrose, Rep. Mod. Phys., Vol. 42, 129 (1979); J.L. Park and R.F. Simmons, Jr., in Old and New Questions in Physics, Cosmology, Philosophy, and Theoretical Biology, ed. A. van der Merwe, Plenum Press, 1983.
2. G.P. Beretta, Thesis, MIT, 1981, unpublished.
3. G.P. Beretta, E.P. Gyftopoulos, J.L. Park and G.N. Hatsopoulos, Nuovo Cimento B, Vol. 82, 169 (1984).
4. G.N. Hatsopoulos and E.P. Gyftopoulos, Found. Phys., Vol. 6, 15, 127, 439, 561 (1976).
5. Throughout this lecture we proceed heuristically and disregard all questions of purely technical mathematical nature.
6. E.T. Jaynes, Phys. Rev., Vol. 106, 620 (1957); Vol. 108, 171 (1957).
7. Here $\mathcal{D}(F)$ is the domain of definition of function F, which does not necessarily coincide with the set $\mathcal{P}$.
8. Equilibrium solution ρ_e is stable according to Liapunoff if and only if for every $\varepsilon > 0$ there is a $\delta(\varepsilon) > 0$ such that any solution $\rho(t)$ with $||\rho(0) - \rho_e|| < \delta(\varepsilon)$ remains with $||\rho(t) - \rho_e|| < \varepsilon$ for every $t > 0$, where $||\cdot||$ denotes the norm on $\mathcal{L}$ defined by $||A|| = (A|A)$.
9. G.P. Beretta, Int. J. Theor. Phys., Vol. 24, 119 (1985).
10. G.P. Beretta, to be published.
11. G.P. Beretta, E.P. Gyftopoulos and J.L. Park, Nuovo Cimento B, in press.

INTRINSIC ENTROPY AND INTRINSIC IRREVERSIBILITY FOR A SINGLE ISOLATED CONSTITUENT OF MATTER: BROADER KINEMATICS AND GENERALIZED NONLINEAR DYNAMICS

Gian Paolo Beretta

Massachusetts Institute of Technology, Cambridge, MA 02139
and Politecnico di Milano, 20133 Milano, Italy

INTRODUCTION

What if entropy, rather than a statistical, information theoretic, macroscopic or phenomenological concept, were an intrinsic property of matter in the same sense as energy is universally understood to be an intrinsic property of matter? What if irreversibility were an intrinsic feature of the fundamental dynamical laws obeyed by all physical objects, macroscopic and microscopic, complex and simple, large and small? What if the second law of thermodynamics, in the hierarchy of physical laws, were at the same level as the fundamental laws of mechanics, such as the great conservation principles? Is it inevitable that the gap between mechanics and thermodynamics be bridged by resorting to the usual statistical, phenomenological, or information-theoretic reasoning, and by hinging on the hardly definable distinction between microscopic and macroscopic reality? Is it inevitable that irreversibility be explained by designing ad hoc mechanisms of coupling with some heat bath, reservoir or environment, and ad hoc mechanisms of loss of correlation? What if, instead, mechanics and thermodynamics were both special cases of a more general unified fundamental physical theory valid for all systems, including a single strictly isolated particle, such as a single isolated harmonic oscillator or a single isolated two-level spin system?

In spite of a century of inquiries, scrutinies, claims, proposals and conjectures, we can still venture no definitive answers to these profound unresolved fundamental physical questions. However, we can show that it is possible to construct a logically consistent, mathematically sound and definite, physically intriguing and appealing, nonstatistical unified quantum theory encompassing

both quantum mechanics and thermodynamics as special cases. In this unified quantum theory, that we call Quantum Thermodynamics, entropy emerges as an intrinsic property of matter, irreversibility emerges as an intrinsic feature of the individual internal dynamics of each constituent of matter, and the second law of thermodynamics emerges as a fundamental theorem on the stability of the equilibrium states. The new theory is not a statistical theory, it requires no layer of statistical or information-theoretic reasoning, it requires no underlying distinction between microscopic and macroscopic description, and it is not a merely phenomenological theory. It is a fundamental physical theory valid for all the systems for which quantum mechanics is usually understood to apply.

Our objective in this lecture is not to try to convince you that our new unified quantum theory is physically correct. Rather, we wish to underline that the new theory and its underlying physical hypotheses offer an entirely new perspective on the meaning of entropy, nonequilibrium, irreversibility and thermodynamics. The new perspective is indeed very unorthodox, unconventional and "revolutionary," but these are no sufficient reasons to dismiss it as "unphysical." Only a very careful scrutiny of the logical consistency and experimental verifiability of the new hypotheses could resolve the question of physical validity.

The lecture is organized as follows. First, we review the essential differences between mechanics and thermodynamics, and underline a common premise of the many proposals to rationalize such differences. Then, we propose an entirely new rationalization based on a new physical hypothesis.

MECHANICS VERSUS THERMODYNAMICS

The sciences of mechanics and thermodynamics developed quite independently and for different applications. Both sciences are physical theories of nature. Indeed, the objective of both mechanics and thermodynamics is to describe the properties of material systems and the laws governing their changes. However, the two theories give conflicting descriptions of the same physical reality. The dilemma created by such contrasts has been the subject of intensive inquiry throughout the past century.

A striking way to underline the essential distinctions between quantum mechanics and thermodynamics is to contrast their conflicting answers to the following three questions, that refer to the description of an isolated system with fixed number of particles of each type and Hamiltonian operator H with eigenvalues e_k and degeneracy D_k:

1. Among the states with given mean value $\langle H \rangle$ of the energy, how many are stable equilibrium?

2. Are there states from which it is not possible to extract all the energy (above the ground state) adiabatically, i.e., by means of a time dependent Hamiltonian $H(t)$?

3. Are irreversible processes conceivable?

The answers within quantum mechanics are:

1. None, if the given mean value $\langle H \rangle$ is not equal to one of the eigenvalues e_k. One, if $\langle H \rangle = e_k$ for some k and $D_k = 1$. Many, i.e., a (D_k-1)-fold infinity, if $\langle H \rangle = e_k$ and $D_k > 1$.
These results follow from a straightforward analysis of the equilibrium solutions of the time-dependent Schroedinger equation $\hbar d\psi/dt = -iH\psi$, which are all stable according to Liapunoff.

2. No. By a well-known theorem proved by von Neumann,[1] for each given initial state ψ, it is always conceivable to find a time-dependent Hamiltonian $H(t)$ such that $H(0) = H$, $H(T) = H$ and the solution of the Schroedinger equation with $\psi(0) = \psi$ passes through the lowest energy ground state $\psi(T) = \varepsilon_0$ at time T, thus describing an adiabatic extraction of all the energy of the initial state in excess to the ground state energy.

3. No. By the same von Neumann theorem, given any process $\psi_1 \rightarrow \psi_2$ of the isolated system, it is always possible to find a time-dependent Hamiltonian $H(t)$ such that $H(0) = H$, $H(T) = H$ and the solution of the Schroedinger equation with $\psi(0) = \psi_2$ passes through $\psi(T) = \psi_1$ at time T, thus restoring the initial state with no net external effects. Hence, any process $\psi_1 \rightarrow \psi_2$ is reversible.

The answers within thermodynamics are:

1. One and only one, for each given value of $\langle H \rangle$. Indeed, as shown by Hatsopoulos and Keenan,[2] the statement that for each given value of the energy (and of the number of particles and the external parameters, which are fixed in our context) there exists one and only one stable equilibrium state, is a general statement of the second law of thermodynamics, which implies all the other known statements.

2. Yes. In particular, no energy can be extracted adiabatically from the stable thermodynamic equilibrium states.

3. Yes. Irreversible processes are conceivable within thermodynamics, although their existence is not required by the second law.

Mechanics and thermodynamics, therefore, seem to describe physical reality in sharply conflicting ways. This century-old dilemma is invariably resolved by stating that the true fundamental description of physical reality is given by quantum mechanics, whereas thermodynamics copes with the phenomenological description of systems whose fundamental mechanical description would be overwhelmingly complicated, hardly reproducible and, therefore, scarcely useful. Many rationalizations of the essential differences between mechanics and thermodynamics have been proposed during the past century.[3] They all share a common underlying premise, namely, that mechanics and thermodynamics occupy different levels in the hierarchy of physical theories.

A NEW HYPOTHESIS

Instead of regarding mechanics and thermodynamics as two different theories, belonging to two different levels of fundamentality, we wish to consider a new hypothesis, first conceived by Hatsopoulos and Gyftopoulos,[4] namely, that mechanics and thermodynamics are two particular aspects of a more general fundamental theory, within which the three key physical questions we just considered admit the mechanical answers for a special class of states and the thermodynamical answers in general. This can only be achieved by means of the Hatsopoulos-Gyftopoulos hypothesis that, in addition to the individual states contemplated by mechanics, there exists a new class of individual states, unconceivable within conventional mechanics, for which the mechanical answers do not necessarily apply.

We have formulated a unified quantum theory of mechanics and thermodynamics based on the new hypothesis.[5,6] We call the new theory Quantum Thermodynamics. In this lecture, we outline the main premises, postulates and theorems of the new nonstatistical theory. Quantum Thermodynamics applies to all the systems to which conventional Quantum Mechanics applies, including, for example, a single isolated two-level quantum system.

Quantum Thermodynamics contains Quantum Mechanics entirely as a special case. We present it in two stages, as is natural for any theory of nature, namely, we first postulate a kinematics, i.e., we structure the fundamental description of the individual states of a physical system, and then we postulate a dynamics, i.e., we structure the fundamental causal description of the time evolution of the individual states.

The quantum mechanical description of the individual states of a strictly isolated, i.e., noninteracting and uncorrelated, system with associated Hilbert space $\mathcal{H}$ is in terms of one-dimensional orthogonal projection operators P on $\mathcal{H}$. We denote by $\mathcal{P}_M$ the set of all the quantum mechanical individual states, i.e.,

$$\mathcal{P}_M = \{P \text{ on } \mathcal{H} \,|\, P^\dagger = P,\ P \geq 0,\ \mathrm{Tr}P = 1,\ P^2 = P\} \tag{1}$$

Following the pioneering work of Hatsopoulos and Gyftopoulos,[4] our hypothesis is that, in addition to the quantum mechanical states in $\mathcal{P}_M$, a strictly isolated system has also access to a new class of nonmechanical individual states that must be described by self-adjoint, nonnegative-definite, unit-trace operators ρ on $\mathcal{H}$ that are not idempotent, i.e., such that $\rho^2 \neq \rho$. We denote by $\mathcal{P}$ the set of all quantum thermodynamic individual states, i.e.,

$$\mathcal{P} = \{\rho \text{ on } \mathcal{H} \,|\, \rho^\dagger = \rho,\ \rho \geq 0,\ \mathrm{Tr}\rho = 1\} \tag{2}$$

Clearly, set $\mathcal{P}$ contains set $\mathcal{P}_M$ entirely. Indeed, set $\mathcal{P}_M$ contains the extreme points of convex set $\mathcal{P}$.

Operators ρ in $\mathcal{P}$ have the same defining <u>mathematical</u> properties as the von Neumann statistical or density operators of quantum statistical mechanics, but in Quantum Thermodynamics their <u>physical</u> meaning is entirely different. They do <u>not</u> represent the index of statistics from a generally heterogeneous ensemble of identical systems possibly distributed over a range of different quantum mechanical individual states. Operators ρ represent the quantum thermodynamic individual states in exactly the same sense in which state vectors ψ represent the quantum mechanical individual states. To stress this extremely important point, we call operators ρ the <u>state</u> operators.

Some physical observables are represented by linear functionals on $\mathcal{P}$. Thus, the linear functional $a(\cdot)$ associated with a physical observable A, e.g., the magnetic moment in a given direction, is given by

$$a(\cdot) = \mathrm{Tr}A\cdot \tag{3}$$

and the value of the observable for a system in individual state ρ is given by $a(\rho) = \mathrm{Tr}A\rho$. The set of all linear and nonlinear functionals on $\mathcal{P}$ represents the set of all intrinsic <u>state properties</u> of the system, because the individual state operator ρ fixes uniquely their values. As shown by Hatsopoulos and Gyftopoulos,[4] and by Ochs[7] in a more rigorous and technical mathematical framework, the only functional $s(\cdot)$ on $\mathcal{P}$ that could

represent the intrinsic entropy of a strictly isolated system in every quantum thermodynamic individual state is

$$s(\rho) = -kTr\rho\ln\rho \tag{4}$$

In particular, $s(\cdot)$ is the only functional[4,7] which:
(i) is invariant under all unitary processes on $\mathcal{P}$;
(ii) is additive for independent systems; and
(iii) is equal to the same constant (zero) for all mechanical states.

NONLINEAR FUNDAMENTAL DYNAMICS

The quantum mechanical causal description of the time evolution of the individual states of a strictly isolated system is given by the time-dependent Schroedinger equation of motion, which in terms of the projection operators P in $\mathcal{P}_M$ becomes

$$\frac{d}{dt}P = -\frac{i}{\hbar}[H,P] \tag{5}$$

Because in Quantum Thermodynamics we postulate the existence of a broader class of individual states, we must also postulate a law of causal evolution valid in general for both the mechanical and the new nonmechanical individual states. For a single strictly isolated constituent of matter, i.e., a single particle or a single field, with Hamiltonian operator H and number operators N_1, ..., N_r for each type of particle in the field ($[N_i,H] = 0$ for $i = 1, \ldots, r$), we postulate that the fundamental causal description of the time evolution of all the quantum thermodynamic individual states is given by the equation of motion proposed by the author,[8] i.e.,

$$\frac{d}{dt}\rho = -\frac{i}{\hbar}[H,\rho] - \frac{1}{\tau}\frac{1}{2}(\sqrt{\rho}D(\rho) + D^{\dagger}(\rho)\sqrt{\rho}) \tag{6}$$

where τ is an intrinsic relaxation time constant whose value can only be determined experimentally, and operator $D(\rho)$ is a nonlinear function of the state operator ρ given by

$$D(\rho) = \sqrt{\rho}\ln\rho - (\sqrt{\rho}\ln\rho)_{L(\sqrt{\rho},\sqrt{\rho}H,\sqrt{\rho}N_1,\ldots,\sqrt{\rho}N_r)} \tag{7}$$

The notation and explicit forms for $D(\rho)$ are discussed in our first lecture in this volume. The generalization of the equation of motion for systems composed of more than one constituent of matter is given in Reference 6.

For mechanical states in $\mathcal{P}_M$, i.e., for $\rho = \rho^2 = P$, Equation 6 reduces to the Schroedinger Equation 5 and, hence, Quantum Thermodynamics contains indeed the whole of Quantum Mechanics, i.e., it contains all the quantum mechanical individual states and all

their unitary time evolutions. Along every solution of Equation 6, the mean values of the energy and the number of particles of each type are time invariant, and the rate of change of the entropy functional is always nonnegative. Thus, the fundamental equation of motion postulated in quantum thermodynamics for a single isolated constituent of matter entails the theorems of energy conservation, number of particle conservation and entropy nondecrease. The equation implies an explicit formula for the rate of entropy production valid for all states, close and far from equilibrium, namely, $ds(\rho)/dt = kTrD^{\dagger}(\rho)D(\rho)/\tau$. It also implies a generalization to all nonequilibrium states of Onsager's reciprocal relations and gives explicit formulae for the Onsager coefficients which emerge as nonlinear functionals of the individual state operator.[9]

The equation entails a rich variety of equilibrium solutions and unitary limit cycles. However, all the limit cycles, including all the unitary evolutions of the quantum mechanical states, and all the equilibrium solutions, with the exception of the equilibrium states with maximum entropy among the states with same mean values of the energy and the number of particles, are unstable according to Liapunoff. Very interestingly, the quantum mechanical equilibrium states (stationary states), that are all stable within Quantum Mechanics because the stability analysis is limited to the set $\mathcal{P}_M$ of mechanical states obeying the Schroedinger equation of motion, become unstable within Quantum Thermodynamics, because the stability analysis must be broadened to the set $\mathcal{P}$ of individual states obeying our general nonlinear equation of motion. For every set of mean values of the energy and the number of particles, there is one and only one stable equilibrium state. This is a statement of the second law of thermodynamics which, therefore, is entailed as a fundamental theorem of our unified quantum theory.

CONCLUSIONS

We presented a unified quantum theory of mechanics and thermodynamics new to physics. The new theory, that we call Quantum Thermodynamics, contains both Quantum Mechanics and equilibrium thermodynamics as special cases. In addition, it entails a general, definite description of nonequilibrium states and their law of relaxation towards stable equilibrium. Quantum Thermodynamics is to be understood as a new fundamental theory, playing the same role that we are used to assign to Quantum Mechanics. It is not a statistical theory, nor a merely phenomenological theory, nor an information theory. Entropy emerges as an intrinsic property of matter in the same sense as energy emerges as an intrinsic property of matter. Irreversibility emerges as an intrinsic feature of the fundamental dynamical law obeyed by the individual elementary contituents of matter. The second law emerges as a theorem of the law of causal

evolution and, therefore, acquires the same level of fundamentality as the great conservation principles.

We believe that the new unified and generalized nonstatistical quantum theory is logically consistent, mathematically sound and physically appealing. The theory and, in particular, its equation of motion is definite enough to call for further careful scrutiny of its conceptual consequences, and experimentally verifiable implications. The theory does not contradict any of the successes of quantum mechanics, but contains a richer variety of physical implications. For example, it gives explicit and general formulas for the rate of entropy production and for the relaxation of individual nonmechanical nonequilibrium states towards equilibrium. Conceptually, Quantum Thermodynamics shows that the key premises of the conventional statistical rationalizations of thermodynamics, such as the key distinction between microscopic and macroscipic reality, and the key role played by heat baths and reservoirs, are not inevitable. Thus, the questions with which we started our lecture remain short of definitive answers and call for further theoretical and experimental efforts of clarification.

REFERENCES

1. J. von Neumann, Mathematical Foundations of Quantum Mechanics, English translation of the 1932 German edition, Princeton Univ. Press, 1955.
2. G.N. Hatsopoulos and J.H. Keenan, Principles of General Thermodynamics, Wiley, 1965. J.H. Keenan, G.N. Hatsopoulos and E.P. Gyftopoulos, Principles of Thermodynamics, article in the Encyclopaedia Britannica, Chicago, 1972.
3. For a recent review, see J.L. Park and R.F. Simmons, Jr., in Old and New Questions in Physics, Cosmology, Philosophy, and Theoretical Biology, ed. A. van der Merwe, Plenum Press, New York, 1983.
4. G.N. Hatsopoulos and E.P. Gyftopoulos, Found. Phys., Vol. 6, 15, 127, 439, 561 (1976).
5. G.P. Beretta, E.P. Gyftopoulos, J.L. Park and G.N. Hatsopoulos, Nuovo Cimento B, Vol. 82, 169 (1984).
6. G.P. Beretta, E.P. Gyftopoulos and J.L. Park, to be published.
7. W. Ochs, Rep. Math. Phys., Vol. 8, 109 (1975).
8. G.P. Beretta, Thesis, MIT (1981), unpublished. The new equation for a single constituent of matter is presented in Reference 5. See also the first lecture in this volume.
9. G.P. Beretta, to be published.

USE OF QUANTUM TRAJECTORIES FOR TIME-DEPENDENT PROBLEMS

M. Kleber

Institute for Theoretical Sciences
University of New Mexico
Albuquerque, NM 87131, USA
and
Physik Department
T. U. München
D-8046 Garching bei München
West Germany

INTRODUCTION

Ehrenfest's principle states the correspondence between a classical trajectory and the expectation values of the corresponding quantum operator. In most cases the equations of motion for the average values of momentum, position, etc. are not closed and, therefore, cannot be solved without further assumptions. It is then useful to rewrite the Schrödinger equation as a time-dependent eigenvalue equation. In the following we shall outline: (1) how to solve the eigenvalue problem within variational perturbation theory, (2) how to calculate quantum trajectories, and (3) how to determine transition probabilities from trajectories. The method presented here will be illustrated by two examples from scattering theory.

TIME-DEPENDENT EIGENVALUE EQUATION

The time-dependent Schrödinger equation

$$(H - i\hbar\partial_t)|\psi\rangle = 0 \tag{2.1}$$

is equivalent to a time-dependent eigenvalue problem. The equivalence can be seen by writing $|\psi\rangle$ in the following way:

$$|\psi\rangle = N(t)|\phi\rangle \quad . \tag{2.2}$$

If we insert (2.2) in (2.1) we obtain the eigenvalue equation

$$(H - i\hbar\partial_t)|\phi\rangle = E(t)|\phi\rangle \quad , \tag{2.3}$$

with

$$E(t) = \langle\phi|H - i\hbar\partial_t|\phi\rangle/\langle\phi|\phi\rangle \tag{2.4}$$

the time-dependent eigenvalue, and

$$N(t) = N(t_0) \exp[-i/\hbar \int_{t_0}^{t} dt'E(t')] \tag{2.5}$$

a time-dependent amplitude.

There are only very few time-dependent Hamiltonians which allow for an exact analytical solution. The most important example is the problem of a harmonic oscillator, driven by a variable force, $-f(t)$, and also subjected to a time-dependent spring constant, $\Omega(t)$:

$$H_{os}(t) = -\frac{\hbar^2}{2m}\frac{d^2}{dx^2} + \frac{1}{2}m\Omega^2(t)x^2 + f(t)x \quad . \tag{2.6}$$

In Eq. (2.6), m denotes the mass of the linear oscillator. We note that Eq. (2.6) is quadratic in the coordinate x. Therefore the ansatz

$$\langle x|\phi(t)\rangle = \langle x|\kappa,\lambda\rangle/[\langle\kappa,\lambda|\kappa,\lambda\rangle]^{\frac{1}{2}}$$

$$\langle x|\kappa,\lambda\rangle = (\tfrac{\omega}{\pi})^{\frac{1}{4}} \exp[-\lambda \tfrac{\omega}{2} (x - \kappa)^2] \tag{2.7}$$

will solve the eigenvalue problem (2.3).

In Eq. (2.7), κ and λ are time-dependent intrinsic coordinates (parameters). Both κ and λ, and also the frequency ω are undefined at present. Since E(t) is independent of x, the linear and quadratic preexponential terms in x on the left hand side of Eq. (2.3) must vanish. These two conditions yield two equations of motion for $\kappa(t)$ and $\lambda(t)$:

$$i\dot{\lambda}\omega = (\lambda\omega)^2 - \Omega^2(t) \quad ,$$

$$i\dot{\kappa}\lambda\omega = \kappa\Omega^2 + f(t)/\hbar \quad . \tag{2.8}$$

The differential equations (2.8) describe the response of the oscillator to an external force and to a change of the spring constant. For given $f(t)$ and $\Omega(t)$, we can calculate $\kappa(t)$ and $\lambda(t)$ from Eq. (2.8) if we specify their initial values.

In the following we shall assume that the oscillator was in the ground state for $t = -\infty$. This means $f(-\infty) = 0$ and $\kappa(-\infty) = 0$. If we now choose $\omega = \Omega(-\infty)$, we find $\lambda(-\infty) = m/\hbar$. It is obvious from Eq. (2.8) that κ and λ will have in most cases both real and imaginary components. The meaning of complex, time-dependent parameters becomes clear when we express the expectation values of x and p in terms of κ and λ:

$$\langle x\rangle = (\kappa\lambda + \kappa^*\lambda^*)/(\lambda + \lambda^*) \quad ,$$

$$\langle p\rangle = - i\hbar\omega\lambda\lambda^*(\kappa - \kappa^*)/(\lambda + \lambda^*) \quad . \tag{2.9}$$

The fluctuations in coordinate and momentum are

$$\Delta x^2 = \langle x^2\rangle - \langle x\rangle^2 = 1/[\omega(\lambda + \lambda^*)],$$

$$\Delta p^2 = \langle p^2\rangle - \langle p\rangle^2 = \hbar^2\omega\lambda\lambda^*/(\lambda + \lambda^*) \ , \tag{2.10}$$

Equations (2.9) and (2.10) can be interpreted as a mapping of κ and λ onto $\langle x\rangle$, $\langle p\rangle$, Δx^2 and Δp^2. It is, therefore, not surprising that the Ehrenfest equations of motion for $\langle x\rangle$, $\langle p\rangle$, Δx^2 and Δp^2 are equivalent to Eq. (2.8).

THE OPIC APPROXIMATION

The eigenvalue problem (2.3) can be derived from the variational principle:

$$\delta I = \delta \int_{t_i}^{t_f} dt \langle\phi|H - i\hbar\partial_t|\phi\rangle/\langle\phi|\phi\rangle = 0 \ . \tag{3.1}$$

If we generalize the ansatz (2.7) to wave functions $\langle x|\rho\rangle$ which are assumed to depend on a set of N intrinsic parameters, $\underline{\rho} = \{\rho_1, \rho_2 ..., \rho_N\}$, then the variational principle (3.1) becomes [1,2]

$$\partial_{\rho_i^*}\left[\frac{\langle\rho|H|\rho\rangle}{\langle\rho|\rho\rangle} - i\hbar\left(\sum_{i=1}^{N} \rho_i\partial_{\rho_i}\right)\ln\langle\rho|\rho\rangle\right] = 0 \ . \tag{3.2}$$

$$i = 1, 2, \dots N$$

Eq. (3.2) determines the time dependence of the $\rho_i(t)$. In other words, the N variational Eqs. (3.2) enable us to calculate the optimum path of the intrinsic coordinates $\rho_i(t)$. The OPIC approximation consists in replacing the exact wave function by a wave function which depends on suitably chosen parameters $\rho_i(t)$ whose time dependence follows from Eq. (3.2).

As an example we take the Gaussian wave packet (2.7) and the Hamiltonian (2.6). In this case, $\rho_1 = \kappa$ and $\rho_2 = \lambda$, and it is obvious that Eq. (3.2) will lead to Eq. (2.8). If the reader has any doubt about this fact, he should carry out the missing steps in the calculation.

The question of how to choose the intrinsic coordinates cannot be generally answered. The answer must depend on the physical problem under consideration. In an actual calculation only very few intrinsic coordinates, i.e. variational parameters, will be used. Consider, for example the Landau-Teller model [3] for a collinear atom-molecule collision. The corresponding Hamiltonian is the sum of a harmonic molecular oscillator, H_0, and an exponential model interaction, H_1, between the atom and the diatomic molecule:

$$H_0/\hbar\omega = -\frac{1}{2}\frac{d^2}{d\xi^2} + \frac{1}{2}\xi^2 \quad , \tag{3.3}$$

$$H_1/\hbar\omega = \frac{E}{\hbar\omega}\,\mathrm{sech}^2\left[\alpha\left(\frac{E/\hbar\omega}{2m}\right)^{1/2}\tau\right]\ \exp(\alpha\xi)\ . \tag{3.4}$$

In Eqs. (3.3) and (3.4) dimensionless quantities have been introduced. $E/\hbar\omega$ is the energy of relative motion in units of the oscillator energy. The scaled time and space coordinates are denoted by $\tau = \omega t$ and $\xi = x/d$ where d is the oscillator length. The other constants are the reduced mass, m, and the relative width, α, of the molecular potential. In figures 1 and 2 we show [4] the reaction dynamics for values of α and m which represent approximately the collision HBr + He. Figure 1 depicts the deformation of the oscillator in configuration and momentum space as a function of time. Figure 2 shows the time dependence of the uncertainty Δx^2. Compared with quasiclassical methods [5], quantal trajectories are always smeared out because of the uncertainty principle. Note the transient squeezing of Δx^2 as a result of a temporary increase of the denominator in Eq. (2.10). The numerical results were obtained in OPIC approximation with a Gaussian trial wave function (2.7). By solving Eq. (3.2) for $\rho_1 = \kappa$ and $\rho_2 = \lambda$ we determined the trajectories in phase space from Eqs. (2.9) and (2.10). The ansatz (2.7) is, however, not the exact solution for the Landau-Teller model, and we are left with the question:

HOW GOOD IS OPIC?

In order to answer the question we have to estimate the error of OPIC. This can be done by means of the Schwarz inequality. For this purpose we define an error vector $|\varepsilon\rangle$ by writing

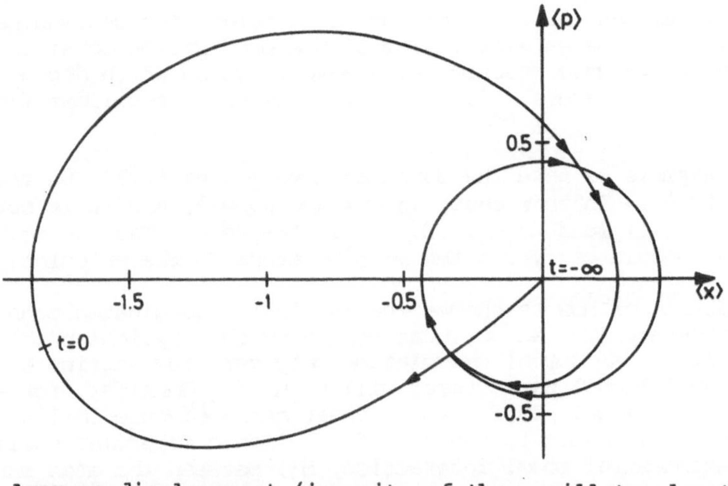

Fig. 1. Average displacement (in units of the oscillator length d) and momentum (in units of $\hbar/d$) of the harmonic oscillator in a collinear collision. Starting point $t = -\infty$, turning point $t = 0$. Arrows point in the direction of increasing time. $\alpha = 0.5584$, $m = 3.737$; scaled total energy $\tilde{E} = E/\hbar\omega + 0.5 = 6$.

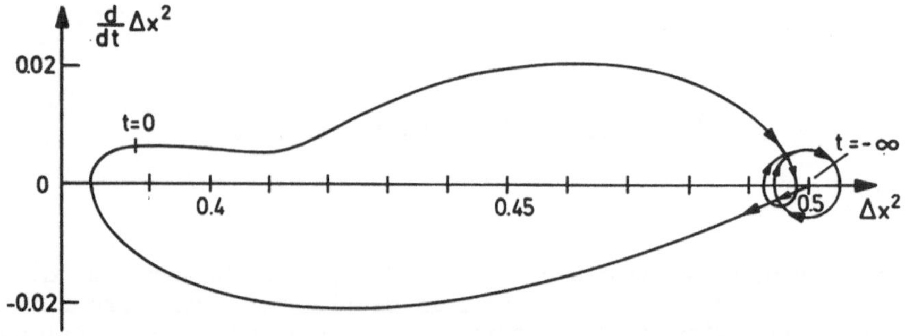

Fig. 2. Change of the collision induced fluctuation Δx^2. Length and time are in units of d and ω^{-1} respectively. Trajectory starts with $\Delta x^2 = 0.5$ at $t = -\infty$. All other details as in Fig. 1.

$$(H - i\hbar\partial_t)|\phi\rangle = E(t)|\phi\rangle + |\varepsilon\rangle \; , \tag{4.1}$$

where $|\phi\rangle$ is the approximate solution, and where E(t) is given by Eq. (2.4). Then one can show [6,7] that the square of the overlap between the exact and the approximate solution will satisfy the inequality

$$|\langle \text{exact}|\phi(t)\rangle|^2 \geqq 1 - \sin^2 \min \begin{Bmatrix} \pi/2 \\ \frac{1}{\hbar}\int_{-\infty}^{t} dt'(\langle\varepsilon(t')|\varepsilon(t')\rangle)^{1/2} \end{Bmatrix} \tag{4.2}$$

For the Landau-Teller model it turns out [4] that the lower bound expression on the r.h.s. of Eq. (4.2) is a rather pessimistic estimate for the actual overlap. In fact, by comparing with an (exact) coupled channel calculation for the Landau-Teller problem (Eqs. (3.3) and (3.4)), we find that the OPIC results are practically exact. But what if we are not so lucky? Can we improve OPIC? Yes, we can. One possible way is to consider [8] fluctuations of the intrinsic coordinates. But instead of improving the overall accuracy of the wave function, we should look for accurate transition probabilities which we can compare to experiment. This goal is achieved by combining OPIC and perturbation theory.

PERTURBATIVE VARIATIONAL METHOD

The idea of the method is to first carry out an OPIC calculation and then, to take care of the remaining error by perturbation theory. For this purpose we make use of the Demkov variational principle [9], which states that the transition amplitudes are stationary with respect to a variation of the corresponding trial wave functions: The variational approach to the transition amplitude $T(i \to f)$ is given by

$$T_{var}(i \to f) = \langle\psi^f_{tr}(\infty)|\psi^i_{tr}(\infty)\rangle$$

$$+ \frac{1}{i\hbar}\int_{-\infty}^{\infty} dt \langle\psi^f_{tr}(t)|H - i\hbar\partial_t|\psi^i_{tr}(t)\rangle \quad . \tag{5.1}$$

In other words: T is correct up to second order in the error of the trial wave functions - as long as the error is small. We do not vary T because it conserves unitarity only approximately; instead we insert variational wave functions, determined according to the OPIC method (3.2), into the r.h.s. of Eq. (5.1).

In order to illustrate the power of the perturbative variational method for complicated physical problems, we have calculated [8] the dynamics of Dirac electrons during a heavy-ion collision. For physical reasons the effective charge seen by the atomic electrons was chosen as the variational parameter. In Fig. 3 a phase-space trajectory for a 1s-electron is shown. When the corresponding variational wave functions are used in Eq. (5.1), the transition amplitudes can be calculated. Figure 4 depicts the positron production which results from a heavy-ion collision.

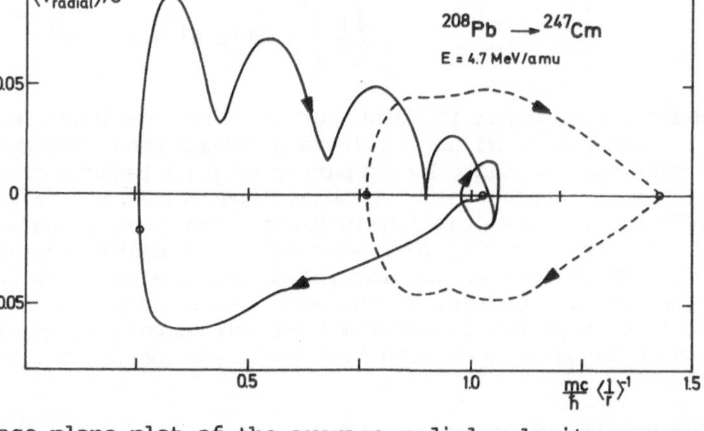

Fig. 3. Phase plane plot of the average radial velocity versus the average Bohr radius of a 1s-electron in Cm during a head-on collision between Pb and Cm. Starting point ($t = -\infty$) and point of closest internuclear distance are marked by small circles. Sense of circulation is clockwise. Dashed line is the result of a non-relativistic calculation. All values are for point-like nuclei.

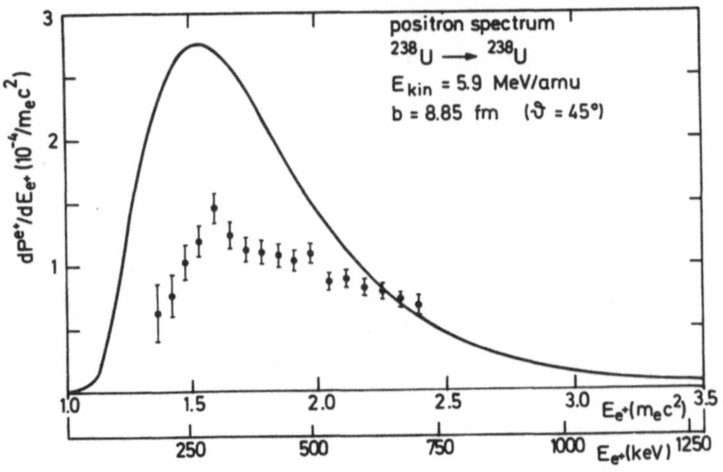

Fig. 4. Differential probabilitiy for positron production in a collision between U-ion and U-atom. Full curve: theoretical result [8] from the collision-induced excitation of the four 1s-electrons. The experimental data [10] are corrected for nuclear background. Bombarding energy, lab scattering angle and impact parameter as shown in the Figure.

RELATION TO OTHER NONPERTURBATIVE METHODS

Let us first consider the matrix element for the equation of motion of an arbitrary time-independent operator O. Then, by using Eq. (4.1) we find

$$i\hbar \frac{d}{dt} \langle\phi|O|\phi\rangle = \langle\phi|[O,H]|\phi\rangle + \langle\varepsilon|O|\phi\rangle - \langle\phi|O|\varepsilon\rangle \, , \tag{6.1}$$

which means that wave functions determined by the OPIC method will automatically satisfy Ehrenfest's principle, if the last two terms in Eq. (6.1) give a vanishing contribution. A detailed study can be found in Ref. [4]. Here, the message is that wave functions with parameters determined from the variational principle (3.2) can be equivalent to wave functions with the same parameters, but now determined from equations of motion for operators like x, p, etc. In case of discrepancies the OPIC method is less biased and should be preferred.

It is obvious that Wigner functions [11] can be constructed from OPIC wave functions. Let us consider the Wigner distribution function

$$W(x,p,t) = \frac{1}{2\pi\hbar} \int_{-\infty}^{\infty} dq\psi^*\left(x - \frac{q}{2}, t\right)\psi\left(x + \frac{q}{2}, t\right) \times \exp(-ipq/\hbar) \tag{6.2}$$

for the wave function (2.7):

$$W(x,p,t) = \frac{(2/(\lambda + \lambda^*))^{1/2}}{\pi\hbar\langle\kappa,\lambda|\kappa,\lambda\rangle} \exp\left\{-\frac{\omega}{2}[\lambda^*(x - \kappa^*)^2 + \lambda(x - \kappa)^2]\right\} \times \exp\left\{\frac{\omega}{2(\lambda + \lambda^*)}\left[\lambda(x - \kappa) - \lambda^*(x - \kappa^*) + \frac{2ip}{\hbar\omega}\right]^2\right\} \tag{6.3}$$

We note that within OPIC the dependence of W on time enters only through the variational parameters. Therefore, if we determine $\kappa(t)$ and $\lambda(t)$, then we are in a position to plot and to study Wigner trajectories.

REFERENCES

1. M. Kleber and J. Zwiegel, Phys. Rev. A19, 579 (1979).
2. K. Unterseer and M. Kleber, Nucl. Instr. and Meth. 192, 35 (1982).
3. M. S. Child, "Molecular Collision Theory", Academic Press, London and New York 1974.
4. W. Elberfeld and M. Kleber, J. Phys. B16, 3557 (1983).
5. H. W. Lee and M. O. Scully, J. Chem. Phys. 73, 2238 (1980).
6. L. Spruch in "Boulder Lectures in Theoretical Physics", Gordon and Breach, New York 1969, Vol. XI-C, p. 77.
7. L. Aspinall and J. C. Percival, J. Phys. B1, 589 (1968).
8. J. Krause and M. Kleber, to be published in Phys. Rev. A.
9. Yu. N. Demkov, Sov. Physics JETP 11 1351 (1960).
10. M. Clemente, et al., Phys. Lett. 137B, 41 (1984).
11. E. Wigner, Phys. Rev. 40, 749 (1932).

EQUATION OF MOTION FOR CORRELATION FUNCTION OF STRONGLY-COUPLED PLASMA

Richard L. Liboff

Schools of Applied Physics and
Electrical Engineering
Cornell University
Ithaca, NY 14853

I. INTRODUCTION

The spatial correlation function for a one-component plasma (OCP)[1,2,3,4] may be constructed within two different formalisms. These involve: (a) derivation from the partition function for the system[5,6,7] and (b) construction of an equation of motion for the correlation function[8,9,10,11] stemming from the BBKGY[12,13,14] sequence. In the present work, a new linear, second-order differential equation for strongly-coupled plasmas is constructed from a generic form obtained previously[11] within the scheme (b). The new equation incorporates fundamental statistical mechanical properties of the correlation function and is further augmented to give results in accord with previous numerical studies.[15,16]

Equations of motion for the correlation functions in the weakly-coupled[17,18] (WCP) and strongly-coupled[4,16] (SCP) limits are found to have identical structures in the domain of small interparticle displacement. The form of the general solution to this equation is obtained and an asymptotic series is developed which returns the canonical starting value of the correlation function. For larger interparticle displacement, the equation of motion for the SCP gives a decaying oscillating correlation function which, due to the construction of the equations, has characteristic values which agree with those of previous numerical studies.

Strongly coupled plasmas play an important role in the description of solid-state plasmas,[19] x-ray plasmas, laser-fusion devices,[20] and in the interiors of certain super-dense stars.[6]

II. ANALYSIS

Parameters, Correlation Functions and Macroscopic Variables

Two correlation functions are particularly relevant in equilibrium statistical mechanics.[21,22] These are the "total" and "radial" distributions, h(r) and g(r), respectively. These functions are related as

$$h(r) = g(r)-1$$

where r is interparticle displacement. In classical physics, $g(0) = 0$, $h(0) = -1$.

Two important parameters related to plasma physics are the so-called plasma parameter, γ, and Debye distance,[23,24,25] λ_d,

$$\gamma = \frac{1}{4\pi n\lambda_d^3}, \qquad \lambda_d^2 = \frac{k_B T}{4\pi n(Ze)^2} \tag{1}$$

These expressions are relevant to an aggregate of charged particles of density n, charge Ze at temperature T. An OCP includes a stationary transparent charge-neutralizing background. The plasma is said to be weakly coupled for $\gamma \ll 1$ and strongly coupled for $\gamma \gtrsim 1$.

The correlation functions $g(r)$ and $h(r)$ find important application in the structure of matter. Thus, for example,

$$\frac{E}{Nk_B T} = \frac{3}{2} + \frac{n}{2k_B T}\int u(r)g(r)4\pi r^2 dr$$

$$\frac{P}{nk_B T} = 1 - \frac{1}{6nk_B T}\int u'(r)g(r)4\pi r^3 dr$$

$$S(k) = 1 + n\int h(r)e^{i\underset{\sim}{k}\cdot\underset{\sim}{r}}4\pi r^2 dr$$

In these expressions $u(r)$ is the two-particle potential, E represents total energy, and P is pressure. The structure factor, $S(k)$, serves to relate data from scattering experiments to the correlation functions.

Weak-Coupling Limit

The equation of motion for $h(r)$ as first obtained by Frieman and Book[8] for a weakly-coupled plasma is given by

$$h'' + \left(\frac{2}{r} - \frac{\ell}{r^2}\right)h' - \frac{h}{\lambda_d^2} = 0 \tag{2}$$

where ℓ is the length

$$\ell = (Ze)^2/k_B T$$

This equation was subsequently analyzed by Lamb and Burdick[26] and rederived by Levitt, Richardson and Cohen.[9] Invoking expansion techniques it was concluded that $h(r)$ has the correct canonical form near the origin,

$$h(r) \simeq -1 + \exp -\left[\frac{(Ze)^2/r}{k_B T}\right], \qquad r \ll \lambda_d \tag{3}$$

Substitution appears to show that in fact (3) is not a solution to (2). The validity of this solution was argued on the basis of orders of magnitude.[8,9,26] In the present analysis we will show more formally that (3) is in fact a solution to (2) near the origin, irrespective of order-of-magnitude arguments.

It is important to note that the canonical form given by (3) is in keeping with exact classical statistical mechanical results concerning the universal composition of h(r) near the origin.[27]

For large r one obtains the well-established Debye-Hückel correlation function,[25]

$$h(r) \simeq \frac{(Ze)^2/r}{k_B T} e^{-r/\lambda_d}, \qquad r >> \lambda_d \tag{4}$$

The overall behavior given by (3,4) is consistent with an extensive numerical calculation by Brush, Sahlin and Teller[15] ("BST") who found that for $\gamma << 1$, h(r) varies exponentially with no oscillation. For purposes of discussion to follow, we note that (2) may be rewritten in the more concise form

$$h'' + \left(\frac{2}{y} - \frac{k}{y^2}\right) h' - h = 0 \tag{5}$$

$$y = r/\lambda_d, \qquad k = \gamma$$

Strong-Coupling Limit

In this same numerical study,[15] it was demonstrated that in the strong-coupling limit, $\gamma \gtrsim 1$, h(r) exhibits oscillation for interparticle displacement on the order of mean interparticle separation.

An equation of motion for h(r) relevant to this limit was recently obtained by Guillen and Liboff[11] from expansion of the BBKGY hierarchy about relatively large γ. Recent study indicates that it must be modified to completely satisfy certain well-established basic properties of h(y) appropriate to an OCP.

Such properties include:

a) Near the origin (y = 0), in addition to reducing to the canonical form (3), $h'(0_+) > 0$.

b) For $y \gtrsim 0$, h'(y) grows smaller with increasing Γ (defined below).

c) For $y \gtrsim 2$, BST data indicate that with varying Γ, h(y) curves have a common wavelength.

d) Each h(y) curve becomes oscillatory beyond a certain critical value of y, which we label $\bar{y}$. BST data indicate the manner in which $\bar{y} = \bar{y}(\Gamma)$.

These properties will be referred to in construction of the revised equation for h(y) to follow.

In the study of SCP, the appropriate nondimensional interparticle displacement is

$$y = r/a$$

The parameter a gives the volume per particle,

$$\frac{4}{3}\pi a^3 = 1/n$$

and Γ is an alternative plasma parameter. Note in particular the parallel construction,

$$\gamma = \frac{(Ze)^2/\lambda_d}{k_B T}$$

$$\Gamma = \frac{(Ze)^2/a}{k_B T}$$

so that

$$\gamma^2 = 3\Gamma^3$$

III. EQUATION OF MOTION

In the following analysis, it proves convenient to introduce the smooth 'switching' function

$$\Theta(y) = 1, \quad y \geq 1 + \varepsilon$$

$$\Theta(y) = 1/2, \quad y = 1$$

$$\Theta(y) = 0, \quad y \leq 1 - \varepsilon$$

where ε is an infinitesimal parameter. Our revised equation appears as

$$h'' + \left(\frac{2}{y} - \frac{k}{y^2}\right) h' + Q^2 h = 0 \tag{6}$$

where we have set

$$Q^2 = \Theta(2y)(q^2+1) - 1 \tag{7}$$

Note that $Q^2 \cong q^2$ for $y \geq 1/2$, and $Q^2 = -1$ for smaller y. As we will see below, this choice for $\tilde{Q}$ serves to satisfy properties (a) and (c) listed above. Specifically, matching with wavelength of h(y) from BST data gives $q \simeq 3.9$.

The transformation[28]

$$h(y) = \frac{1}{y} e^{-k/2y} v(y) \tag{8}$$

collapses (6) to the Schroedinger-like form[29]

$$v'' + \left[Q^2 - \frac{(k/2)^2}{y^4}\right] v = 0 \tag{9}$$

For $Q^2 < 0$, it is clear that solutions are exponential only. For $Q^2 > 0$ oscillatory solutions occur for

$$y^2 > \frac{k/2}{Q^2} \tag{10}$$

whereas the solution is exponential for smaller y.

As described in property (d), BST data suggest that h(y) curves for varying Γ become oscillatory for y greater than the transition values

$$\bar{y} \simeq A\left(\frac{\Gamma}{2+\Gamma}\right) \tag{11}$$

where A ≃ 1.3. To more fully match numerical results, in (10) we set Q ≃ 1, which with (11) gives

$$\sqrt{k/2} \simeq A\left(\frac{\Gamma}{\Gamma+2}\right) \tag{12}$$

As we will find in the asymptotic study of h(y) below, in order for (6) to return the canonical form (3) near the origin, we must have k ≃ Γ in this domain. Combining this constraint with (12) we write

$$k = [1-\Theta(2y)]\Gamma - \Theta(2y)2(A)^2\left(\frac{\Gamma}{\Gamma+2}\right)^2 \tag{13}$$

In this manner we find our revised equation of motion for h(y) is given by (6), with the coefficients Q^2 and k given by (7) and (13), respectively.

In the following sections we examine the structure of h(y) in its domains of oscillation and decay, respectively.

IV. STRUCTURE OF ASYMPTOTIC SOLUTIONS

Oscillating Domain

For y >> k/2, (6) yields spherical Bessel functions with the independent variable, qy. The transformed equation (9) gives a better estimate of this asymptotic structure.[30] Namely,

$$h(y) = -B(\Gamma)\frac{e^{-k/2y}}{y}[\sin(qy) + b\cos(qy)] \tag{14}$$

with k given by (12). The coefficients B(Γ) and b are determined by comparing the amplitude and zeros of (14) with those of BST, respectively. We find

$$B = B_o \log\Gamma, \qquad b = 2.36$$

The remaining positive coefficient B_o is fixed by matching solutions of different domains.

Starting Solution

In the domain near the origin, (6) reduces to

$$h'' + \left(\frac{2}{y} - \frac{k}{y^2}\right)h' - h = 0 \tag{15}$$

This equation is relevant to a strongly-coupled plasma. However we note that it is identical in form to (5) relevant to a weakly coupled plasma. We have used the same symbols (y,k) in both equations (5,15) to bring out this similarity. The actual values of these parameters in the small y domain are listed below.

	WCP	SCP
y	r/λ_d	r/a
k	γ	Γ
rk/y	$(Ze)^2/k_BT$	$(Ze)^2/k_BT$

Thus, the discussion to follow is relevant to both SCP and WCP.

General Solution

We first note that

$$\Phi(y) \equiv y^2 e^{k/y} \tag{16}$$

is an integrating factor for (15) which permits it to be rewritten

$$(h'\Phi)' - h\phi = 0 \tag{17}$$

Note in particular that

$$\tilde{h}' = \Phi^{-1}, \qquad (\tilde{h} = C\,e^{-k/y}) \tag{18}$$

gives $(\tilde{h}'\Phi)' = 0$. This property suggests a trial solution in the form

$$h = \tilde{h}f \tag{19}$$

Substitution into (15) gives

$$f'' + \left(\frac{2}{y} + \frac{k}{y^2}\right)f' - f = 0 \tag{20}$$

We may conclude that if

$$h(y) = f(y,k)$$

is a solution to (15) then

$$\bar{h}(y) = C\,e^{-k/y}\,f(y,-k)$$

is also a solution. Thus we find that the general solution to (15) is given by

$$h(y) = A\,f(y,k) + B\,e^{-k/y}\,f(y,-k) \tag{21}$$

Asymptotic Expansion Near the Origin

The description by Ince[31] concerning irregular singular points leads us to conclude that at most, one solution of (15) is regular at the origin. [This conclusion is evident from (21).] Suppose this equation has a regular solution about y = 0. Then such solution has a Taylor series expansion. Substitution of the expansion into (15) gives

$$f(y,k) = c_o \left[1 - \frac{y^3}{3k} - \frac{y^4}{k^2} - \frac{4y^5}{k^3} + \frac{1}{c_o} \sum_{n=6}^{\infty} c_n y^n \right] \tag{22}$$

where, for $n \geq 3$,

$$n\, c_n k = - c_{n-3} + n(n-1)c_{n-1} \tag{22a}$$

This recurrence relation implies that (22) is divergent for all y. Thus, with Ince's theorem we conclude that both solutions to (15) are irregular at the origin.

However, the series (22) is of value because it offers an approximate solution to (15) in the form of an asymptotic expansion for sufficiently small values of y. The series is asymptotic in the formal sense.[32,33] Namely, although divergent, any sum of a finite number of starting terms of (22) represents a good approximate solution to (15) for sufficiently small y. Thus, for example, keeping terms up to y^n in (22) and substituting into (15) leaves a remainder that includes only y^{n-2}, y^{n-1}, y^n terms which vanish rapidly for small y.

To recapture the canonical starting value of h(y), given by (3), one merely adds the leading terms in the two independent solutions contained in (21,22) with $A = -B = -1$, and $c_o = +1$. Thus, for y sufficiently small we find

$$h(y) \simeq -f(y,k) + e^{-k/y} f(y,-k) \tag{23}$$

which is seen to return the canonical starting value (3).

Differentiation of (23) gives

$$h'(y) \simeq -f'(y,k) + (e^{-k/y})' f(y,-k) + e^{-k/y} f'(y,-k)$$

Near $y = 0$ we find

$$h'(y) \simeq -f'(y,k) \simeq \frac{y^2}{k} > 0 \tag{24}$$

and we may conclude that

$$h'(0_+) > 0$$

in accord with property (a). Furthermore in that $k \sim \Gamma$ (for SCP), $f'(0_+)$ grows smaller with increasing Γ, in agreement with property (b).

Here is a brief recapitulation of asymptotic solutions for SCP:

$y^2 \gtrsim 0$, $k = \Gamma$

$$h(y) = - \left[1 - \frac{y^3}{3k} - \frac{y^4}{k^2} - \ldots \right] + e^{-k/y} \left[1 + \frac{y^3}{3k} - \frac{y^4}{k^2} + \ldots \right] \tag{25}$$

$y > k/2$, $\sqrt{k/2} \simeq qA(\Gamma/\Gamma+2)$

$$h(y) \simeq -B_o(\log \Gamma) \frac{e^{-k/2y}}{y} [\sin(qy) + b \cos(qy)] \tag{26}$$

V. DISCUSSION AND CONCLUSIONS

Comparison of WCP and SCP Equations

The oscillatory behavior of h(y) relevant to SCP stems from the positive sign of Q in (6). This value of sign may be attributed to an effective attractive component of the two-particle force due to the neutralizing background.[11] However, for sufficiently small y, such neutralizing effects are overcome by Coulomb repulsion, and Q must change sign in this domain. We may speculate that the related negative sign in the corresponding term in (5) for all values of y is due to temperature dominance over interaction which occurs for WCP and has the effect of reducing the influence of the background.

Concluding Remarks

A new linear, second-order differential equation for the total correlation function of a strongly-coupled OCP was constructed incorporating fundamental statistical mechanical properties and features from previous numerical studies. Comparison with the companion equation for a WCP reveals that these equations have identical structures in the domain of small y with an irregular singularity at the origin. The form of the general solution of this equation was demonstrated and asymptotic series were developed which were found to return the canonical starting value of the correlation function.

For larger y, the equation of motion for the SCP gives an oscillating correlation function with characteristic values in agreement with previous numerical studies.

ACKNOWLEDGMENTS

Fruitful discussions on these and related topics with my colleagues Brian Jones and Kenneth Gardner are gratefully acknowledged.

References

1. E. H. Lieb and H. Narnhofer, J. Stat. Phys. 12, 291 (1975).
2. C. Deutsch, Y. Forutani and M. Gombart, Phys. Rev. A 13, 2244 (1976).
3. R. L. Liboff, Am. J. Phys. 48, 532 (1980).
4. S. Ichimaru, Revs. Mod. Phys. 54, 1017 (1982).
5. D. D. Carley, Phys. Rev. 131, 1406 (1963).
6. D. L. Bowers and E. E. Salpeter, Phys. Rev. 119, 1180 (1960).
7. M. S. Cooper, Phys. Rev. A 7, 1 (1973).
8. E. A. Freeman and D. L. Book, Phys. Fluids 6, 1700 (1963).
9. L. C. Levitt, J. M. Richardson and E. R. Cohen, Phys. Fluid 10, 406 (1967).
10. R. Cauble and J. J. Duderstadt, Phys. Rev. A 23, 3182 (1981).
11. M. A. Guillen and R. L. Liboff, "Unified kinetic theory of plasma correlations," to be published: J. Plasma Phys. (1984).
12. R. L. Liboff, "Introduction to the Theory of Kinetic Equations," Krieger, Melbourne, FL (1979).
13. R. L. Liboff and G. E. Perona, J. Math. Phys. 8, 2001 (1967).
14. N. N. Bogoliubov, "Problemi Dynamitcheskij Theorie v Statistitcheskey Phisike," OGIS, Moscow (1946).
15. S. G. Brush, H. L. Sahlin and E. Teller, J. Chem. Phys. 45, 2102 (1966).
16. M. Baus and J. P. Hansen, Phys. Repts. 59, 1 (1980).
17. N. Rostoker and M. N. Rosenbluth, Phys. Fluids 3, 1 (1960).
18. L. J. Caroff and R. L. Liboff, J. Plasma Phys. 4, 83 (1970).

19. R. L. Liboff, Criteria for physical domains in laboratory and solid state plasmas," to be published: J. Appl. Phys. (1984).
20. D. M. Heffernan and R. L. Liboff, J. Plasma Phys. 27, 473 (1982).
21. D. L. Goodstein, "States of Matter," Prentice-Hall, Englewood Cliffs, NJ (1975)
22. D. A. McQuarrie, "Statistical Mechanics," Harper and Row, New York.
23. W. B. Kunkel, "Plasma Physics in Theory and Application," McGraw Hill, New York (1966).
24. N. A. Krall and A. W. Trivelpiece, "Principles of Plasma Physics," McGraw Hill, New York (1973)
25. P. Debye and E. Hückel, Phys. Z. 24, 185 (1923).
26. G. L. Lamb and B. Burdick, Phys. Fluids 7, 1087 (1964)
27. B. Widom, J. Chem. Phys. 39, 2808 (1963).
28. E. T. Whittaker and G. N. Watson, "A Course of Modern Analysis," Cambridge University, Cambridge (1952)
29. R. L. Liboff, "Introductory Quantum Mechanics," Holden Day, Oakland, CA (1981).
30. E. Kamke, "Differentialgleichungen," Akademische Verlagsgesellschaft, Leipzig (1943), Vol. I, 2nd Ed.
31. E. L. Ince, "Ordinary Differential Equations," Longmans, Green and Co., London (1927).
32. A. Erdélyi, "Asymptotic Expansions," Dover, New York (1956).
33. C. Bender and S. Orszag, "Advanced Mathematics for Scientists and Engineers," McGraw Hill, New York (1978).

QUANTUM BROWNIAN MOTION

Vinay Ambegaokar

Institute for Theoretical Physics
University of California
Santa Barbara, California 93106
and
Laboratory of Atomic & Solid State Physics[a]
Cornell University
Ithaca, New York 14853

ABSTRACT

A tutorial discussion is given of path integral methods for dealing with the fluctuations of a quantum variable ("particle") coupled to a large quantum system ("environment") that provides a friction force linearly proportional to the particle velocity. It is shown explicitly how the sum over paths can be reduced to an ordinary double integral for the time evolution of the density matrix describing the particle when the potential in which it moves has one of the following three simple forms: constant, linear, or quadratic. A brief summary is given of results that can be obtained by this method.

I. INTRODUCTION

The Feynman path integral formulation of quantum mechanics[1] replaces the classical notion of a trajectory by a coherent sum over trajectories, each weighted by the factor $\exp[iS/\hbar]$ where S is the classical action for the given path. For certain simple potentials -- constant, linear, or quadratic -- the sum over paths is particularly simple (a Gaussian functional integral). In these cases it turns out that all of the physics of quantum mechanics can be obtained from the extremum of the action, i.e. the classical action along the classical path.

In this paper, I point out that the same surprising technical simplification also occurs in the quantum mechanics of a particle coupled to an environment whose effect in the classical limit is to provide a damping force linear in the velocity of the particle. My aim is to give an introduction to these ideas, and, at the risk of seeming simple-minded, to be fairly explicit. Some parts of the presentation follow the suggestion of two graduate students who asked after my lecture that I show them how calculations involving path integrals are actually done.

2. PATH INTEGRALS IN QUANTUM MECHANICS

First, a very few words about path integrals in the quantum mechanics of one particle, and the special role of classical paths for simple potentials.

The transition amplitude to go from position x_i to position x_f in time t is, as Feynman[1] has shown,

$$\langle x_f | \exp[-iHt] | x_i \rangle = \int \mathcal{D} x \exp \frac{i}{\hbar} S[x] \tag{2.1}$$

whre S is the classical action

$$S = \int_0^t dt' \left[\frac{1}{2} m\dot{x}^2 - V(x) \right] \tag{2.2}$$

and $\mathcal{D}x$ means a sum over all paths such that $x(0) = x_i$ and $x(t) = x_f$. Meaning is given to the "sum" by dividing the time interval into finite slices and interpreting $\mathcal{D}x$ as the limit of a product of integrals over x at times separating the slices. The measure to be associated with the multiple integral is calculated by considering (2.1) for a free particle, whereupon it becomes clear that

$$\mathcal{D} x = \sqrt{\frac{m}{2\pi i\hbar\Delta t}} \prod_{j=1}^{N-1} dx_j \sqrt{\frac{m}{2\pi i\hbar\Delta t}} \tag{2.3}$$

where $N\Delta t = t$, corresponding to dividing the interval into N slices. Suppose one wants to evaluate (2.1) for a harmonic oscillator. Written as a Riemann sum (2.2) becomes a quadratic form. The integral one has to do can, in general, then be deduced from the formula

$$\int \mathcal{D}V \, e^{-\frac{1}{2} VMV} e^{iAV} = \frac{1}{\sqrt{\det M}} e^{-\frac{1}{2} AM^{-1}A} \tag{2.4}$$

where V and A are N-dimensional real vectors, M is a real symmetric non-singular matrix, and $\mathcal{D}V = \prod_j dV_j/\sqrt{2\pi}$. One proves (2.4) by completing the square, shifting the origin of the V integrals, and transforming to the eigenvectors of M. For the problem at hand, the inhomogeneous part AV comes from the x_f and x_i in (2.1). One can finesse the question of boundary conditions by expanding around the extremum value of S, i.e. by writing $x(t) = x_{cl}(t) + \delta x(t)$ where $x_{cl}(t)$ obeys the classical equation $m\ddot{x}_{cl} = - (\partial V/\partial x_{cl})$ and satisfies the boundary conditions. Now, since everything is quadratic and x_{cl} gives an extremum of S, we have $S = S_{cl} + \delta S^{(2)}$ where $\delta S^{(2)}$ is the same functional, Eq. (2.2), of δx as S was of x, but the boundary conditions are now $\delta x(0) = \delta x(t) = 0$. We thus have a homogeneous Gaussian integral to do, i.e. (2.4) with A = 0. Now, look at the matrix in the exponent in (2.1) for the harmonic oscillator potential $\frac{1}{2} m\omega_0^2 x^2$:

$$\delta S^{(2)} = \sum_{j=1}^{N} \frac{1}{2} \frac{m}{\Delta t} (\delta x_j - \delta x_{j-1})^2 - \frac{1}{2} m\omega_0^2 \Delta t \sum_{j=1}^{N-1} (\delta x_j)^2 . \tag{2.5}$$

[In the first term, the boundary conditions require $\delta x_0 = \delta x_N = 0$.] Taking into account the factors in (2.3), we see that the $(N-1)\times(N-1)$

matrix that would fit (2.1) into the mold of (2.4), with A = 0, is tridiagonal:

$$\begin{pmatrix} 2-\omega_0^2\Delta t^2 & -1 & 0 & 0 & . \\ -1 & 2-\omega_0^2\Delta t^2 & -1 & 0 & . \\ 0 & -1 & . & . & . \end{pmatrix} \tag{2.6}$$

We also see that because of the extra factor of $(\Delta t)^{-1/2}$ in (2.3) we need to calculate Δt times the determinant of (2.6). Let's call this product D_N. A simple recursion relation follows from (2.6):

$$D_N = (2 - \omega_0^2\Delta t^2)\, D_{N-1} - D_{N-2} \tag{2.7}$$

which in the limit $\Delta t \to 0$ is the differential equation

$$\frac{d^2D}{dt^2} + \omega_0^2 D = 0\ . \tag{2.8}$$

A little playing around with the Δt's shows that the initial conditions are $D(0) = 0$ $\dot{D}(0) = 1$, so that the required solution is $D(t) = \omega_0^{-1}\sin\omega_0 t$. Thus, the solution of (2.1) for the harmonic oscillator is

$$\langle x_f|e^{-iHt}|x_i\rangle = \sqrt{\frac{m\omega_0}{2\pi i\hbar\sin\omega_0 t}}\ e^{\frac{i}{\hbar}S_{cl}(x_f,x_i,t)}\ . \tag{2.9}$$

One obtains S_{cl} most easily by integrating (2.2) by parts and using the equation of motion to get $S_{cl} = (m/2)(\dot{x}_f x_f - \dot{x}_i x_i)$ where $\dot{x}_i$ and $\dot{x}_f$ are the initial and final classical velocities. A short calculation then yields

$$S_{cl}(x_f,x_i,t) = (m\omega_0/2)[\cot\omega_0 t(x_i^2+x_f^2) - 2x_i x_f \operatorname{cosec}\omega_0 t]\ . \tag{2.10}$$

Note that there is nothing semi-classical about (2.9); it contains all of the quantum mechanics of a harmonic oscillator. The only place where deviations from the classical path show up is in the prefactor, and, actually, this can be determined from the nonlinear requirement of consistency

$$\int dx_m\ \langle x_f|e^{-i(Ht/2)}|x_m\rangle\langle x_m|e^{-i(Ht/2)}|x_i\rangle = \langle x_f|e^{-iHt}|x_i\rangle \tag{2.11}$$

In particular, if one writes

$$\langle x|e^{-iHt}|x'\rangle = N(t)\ e^{\frac{i}{\hbar}S_{cl}(x,x't)} \tag{2.12}$$

with S_{cl} given by (2.10), (2.11) yields

$$\left(\frac{2\pi i\hbar}{m\omega_0}\right)^{1/2} \frac{N^2(t/2)}{\sqrt{2\cot(\omega_0 t/2)}} = N(t) \tag{2.13}$$

whose solution is

$$N(t) = \sqrt{\frac{m\omega_0}{2\pi i\hbar \sin\omega_0 t}} \quad . \tag{2.14}$$

Nothing deep has been done here. The point is merely that the quantum mechanics of a harmonic oscillator, which is technically simple any way you want to look at it, is also simple in the language of path integrals.

3. BROWNIAN MOTION AND PATH INTEGRALS

I now want to describe some recent developments in the treatment of Brownian motion in quantum mechanics. My interest in this subject has come from attempting to understand better the role of quantum fluctuations in superconducting tunnel junctions,[2,3] but I shall not be referring to that topic here.

To discuss friction microscopically, which is, I believe, the only correct way to treat the topic quantum-mechanically, it is necessary to include the environment in the starting Hamiltonian. Caldeira and Leggett,[4,5] building on work by Feynman and Vernon,[6] and others,[7] have shown how to model an environment as an infinite collection of harmonic oscillators in such a way as to produce in the classical limit a friction force linearly proportional to the velocity. I shall be using this model here, and what I have to say is implicit in one of the papers by Caldeira and Leggett,[4] though my methods are slightly different and, possibly, more transparent. The Hamiltonian introduced in Ref. 5 is

$$H = \frac{p^2}{2m} + V(q) + \sum_\alpha \left[\frac{p_\alpha^2}{2m_\alpha} + \frac{1}{2} m_\alpha \omega_\alpha^2 \left(x_\alpha - \frac{c_\alpha q}{m_\alpha \omega_\alpha^2} \right)^2 \right] \quad . \tag{3.1}$$

For the classical friction to be linear in the particle velocity one needs

$$J(\omega) \equiv \frac{\pi}{2} \sum_\alpha \frac{c_\alpha^2}{m_\alpha \omega_\alpha} \delta(\omega - \omega_\alpha) = m\gamma\omega\theta(\omega_c - \omega) \tag{3.2}$$

where ω_c is an upper frequency cut-off on which certain physical quantities do depend. Information about the particle without reference to the state of the environment is contained in the reduced density matrix

$$\rho(q_1 q_2) \equiv \mathrm{Tr}_\alpha \, \rho(q_1,\{x_\alpha\};\, q_2,\{x_\alpha\}) \quad . \tag{3.3}$$

A study of the time development of this density matrix, the method used in most investigations, will be described here. A treatment of the complete Hamiltonian (3.1) for a free particle $[V(q) = 0]$ in the usual language of quantum mechanics is contained in a forthcoming paper by V. Hakim and the present author.[8]

Feynman and Vernon[6] have shown how $\rho(q_1^f, q_2^f)$ is related by a double path integral to its value at t = 0, $\rho(q_1^i, q_2^i)$. For a non-interacting particle the two path integrals are given by the product of the transition amplitude (2.1) to carry q_1^i to q_1^f and its complex conjugate to carry q_2^i to q_2^f. The effect of the environment is to couple the two paths. To start the system off, one has to choose an initial condition. Following Feynman and Vernon, assume that at $t = 0$ the system and environment are uncoupled and that the latter is in thermal equilibrium at a temperature T. This temperature then, as we shall see, communicates itself to the particle. [More general initial conditions are possible, and are discussed

in Ref. 8.] It is convenient to use sum and difference variables $x \equiv (q_1+q_2)/2$ and $y = q_1-q_2$. These are the variables introduced by Wigner,[9] and described by him at this school. Then, the result of the procedure of Feynman and Vernon when applied to the system described by (3.1) and (3.2) is

$$\rho(x_f,y_f,t) = \int dx_i dy_i \mathcal{D}x \mathcal{D}y \; e^{\frac{i}{\hbar}[S_1+iS_2]} \rho(x_i,y_i,0) \tag{3.4}$$

$$S_1 = \int_0^t dt' \left[m\dot{x}\dot{y} - m\gamma\dot{x}y - V(x+\tfrac{y}{2}) + V(x-\tfrac{y}{2})\right] \tag{3.5}$$

$$S_2 = \frac{m\gamma k_B T}{\hbar} \int_0^t dt' \int_0^t dt' \; y(t') K(t',t'')y(t'') \tag{3.6}$$

with $K(t',t'')$ given by

$$K(t',t'') = \int_{-\omega_c}^{\omega_c} \frac{d\omega}{2\pi} \frac{\hbar\omega}{2k_BT} \coth \frac{\hbar\omega}{2k_BT} \exp[-i\omega(t'-t'')] \tag{3.7}$$

Above $\mathcal{D}x$, $\mathcal{D}y$ indicate sums over paths, with the meaning of Eq. (2.3), beginning at x_i,y_i and ending at x_f,y_f; T is the temperature of the environment; k_B is Boltzmann's constant and the other symbols are defined in Eqs. (3.1)-(3.2). The set of Eqs. (3.4)-(3.7) can also be obtained with other models for the environment.[2,3,10] The key assumption for obtaining these results is that the environment is large and therefore weakly perturbed by the particle, and that it provides a linear damping mechanism.

Three remarks are worth making about this scheme:

(1) For $\gamma \to 0$, single-particle quantum mechanics is recovered, in the form discussed below Eq. (3.3).

(2) In the classical limit ($\hbar \to 0$) the scheme yields the usual theory of classical Brownian motion. To see this, first note that in the classical limit $K(t'-t'') \to \delta(t'-t'')$. Then one can interpret $\exp-[S_2/h]$ as the average effect of a classical fluctuating force:

$$e^{-\frac{m\gamma k_BT}{\hbar^2}\int_0^t dt' y^2(t')} = \left\langle e^{\frac{i}{\hbar}\int_0^t dt\, f(t)y(t)} \right\rangle \tag{3.8}$$

where f(t) is a Gaussian random variable with the correlation function

$$\langle f(t)\, f(t')\rangle = 2m\gamma k_BT\, \delta(t-t') \, . \tag{3.9}$$

Substituting (3.8) in (3.4), we can go back to the variables q_1 and q_2, and in the limit $\hbar \to 0$ conclude that the dominant trajectory is given by the condition of least action:

$$m\ddot{q}_1 + m\gamma\dot{q}_1 + \frac{\partial V}{\partial q_1} = f \, . \tag{3.10}$$

This is the classical Langevin equation.

(3) When $V(q) = 0$ or $-\lambda q$ or $\frac{1}{2} m\omega_0^2 q^2$, (3.4) is a Gaussian functional integral. Simplifications should, and do, occur.

4. SOLUTIONS IN SIMPLE CASES

Let us consider the cases listed in remark (3) above. It turns out that the evaluation of the path integral is even simpler than its Gaussian form suggests. As in Sec. 2, one is tempted to expand the paths around a reference path. However, since the action is complex, and the sum is over real paths, there seems at first sight to be no obvious reference like the least action path x_{cl} for the non-interacting system. In fact, it turns out to be a useful tactic to expand around the extrema of S_1. Suppose one does this. Let $x_{cl}(t)$, $y_{cl}(t)$ be the solutions determined from $\delta S_1 = 0$ and the boundary conditions, and let $\delta x(t)$, $\delta y(t)$ be the deviations. Now write $S = S_1 + iS_2 = S_{cl} + R$. R contains terms linear in δy, so that although the boundary conditions have been taken care of by the expansion one still has an inhomogeneous Gaussian integral to do. Although Eq. (2.4) applies, it would seem to require the inversion of a functional matrix. However, there is a simplification. Consider the part of R quadratic in δx, δy. This plays the role of M in (2.4). Note that there are no terms of the type $\delta x_i \delta x_j$. Also, note that there is no term linear in δx. Thus, in the space spanned by (δx, δy) we have the structures

$$M = \begin{pmatrix} 0 & p \\ \tilde{p} & r \end{pmatrix} \qquad A = \begin{pmatrix} 0 \\ a \end{pmatrix} \tag{4.1}$$

It follows that M^{-1} has a zero in the lower right-hand corner, and that $AM^{-1}A = 0$!

There is a good physical reason for the zeros in (4.1), which A. Garg has very patiently explained to me.[11] The argument is very general. Paths for which $y \equiv 0$ are paths for which $q_1(t) = q_2(t)$. Now, in (3.4) the action, both technically and colloquially, comes from: (1) terms in the Hamiltonian which act on the particle coordinate alone; and (2) terms, proportional to γ, which involve the coupling of the particle and the environment. In terms of type (1), the action, in the technical sense now, involves $S_0(q_1) - S_0(q_2)$ which is clearly zero if $q_1 = q_2$. For the other terms, called the influence functional by Feynman and Vernon,[6] the argument is slightly more subtle. If $q_1 = q_2$, the particle coordinate enters the Hamiltonian for the environment exactly like a time-dependent external potential. We are integrating over all environment coordinates at times t and 0. The time development of the environment density matrix associated with unitary transformations, such as those arising from external potentials, cannot contribute anything because $\mathrm{Tr}\{U(t0)\rho(0)U(0t)\}=1$. Thus, everything in the exponent of (3.4) must, and does, go to zero for paths $y(t) \equiv 0$.

As a result of these simplifications, we can perform the functional integrals in (3.4) with hardly any work for the cases of a free particle, a particle in a constant force, or a particle in a harmonic potential. Using exactly the same tricks described in Sec. 2, we obtain

$$\rho(x_f,y_f,t) = \frac{m}{2\pi\hbar d(t)} \int dx_i dy_i \; e^{\frac{i}{\hbar} m(\dot{x}_f y_f - \dot{x}_i y_i)}$$

$$\times\, e^{-\frac{m\gamma k_B T}{\hbar^2} \int dt' dt'' y_{cl}(t') K(t't'') y_{cl}(t'')}$$

$$\times\, \rho(x_i,y_i,0) \quad . \tag{4.2}$$

Everything in the exponents in (4.2) comes from the paths which make S_1 an extremum. The fluctuations about this path produce the prefactor, which is proportional to the reciprocal of $\det M = (\det p)^2$ in the symbolic expression (4.1), and has been evaluated using the method exploited in Eqs. (2.6)-(2.8). Explicitly, for the potential $V(q) = -\lambda q + \frac{1}{2}m\omega_0^2 q^2$ one has:

$$\ddot{x}_{cl} + \gamma\dot{x}_{cl} + \omega_0^2 x_{cl} = (\lambda/m) \tag{4.3a}$$

$$\ddot{y}_{cl} - \gamma\dot{y}_{cl} + \omega_0^2 y_{cl} = 0 \tag{4.3b}$$

$$\ddot{d} + \gamma\dot{d} + \omega_0^2 d = 0 \; . \tag{4.3c}$$

The boundary conditions are $x_{cl}(0) = x_i$, $x_{cl}(t) = x_f$, $y_{cl}(0) = y_i$, $y_{cl}(t) = y_f$, $d(0) = 0$ $\dot{d}(0) = 1$. In (4.2), $\dot{x}_f$ and $\dot{x}_i$ have to be expressed in terms of x_f, x_i, and the other parameters in the problem via (4.3a) and the boundary conditions.

5. RESULTS

Results using this general scheme can be found in various places, and there is no point in repeating these exercises here. In Ref. 4, the time evolution of a damped harmonic oscillator is discussed. An interesting result, easily obtained from (4.2), is Eq. (6.35) of that reference where the mean square displacement at times much greater than $(1/\gamma)$ is given. It is found that for $\gamma \to 0$, this corresponds to thermal equilibrium for the oscillator. For finite γ, one can work out a formula valid at $T = 0$

$$\langle x^2(t\to\infty)\rangle = \frac{\hbar}{2m\omega}\left[1 - \frac{2}{\pi}\tan^{-1}\frac{\gamma}{2\omega}\right] \tag{5.1}$$

where $\omega^2 \equiv \omega_0^2 - (\gamma/2)^2$, and the notation is that of Eq. (4.3) above. [Note that my γ is twice that of Ref. 4.] This shows that the interaction with an environment of the form (3.1) contracts the width of the ground state of the particle -- an effect which is at the root of the prediction[5] that damping decreases the rate of tunneling from a metastable well.

Results for a free particle are given in Ref. 8. The main result for the motion of an arbitrary wave packet is that the normal quantum mechanical spreading, due to phase coherence between different plane waves making up the packet, is quenched on a time scale $\sim \gamma^{-1}$, and that the environment also induces a random walk or diffusion, with the Einstein diffusion constant, for times $t > (\hbar/k_B T)$, which is the period of the most rapid oscillator of (3.1) in classical thermal equilibrium. In the interval $\gamma^{-1} < t < (\hbar/k_B T)$, one finds a logarithmic diffusion: $\langle x^2\rangle = (2\hbar\gamma/m\pi)\ln(t\sqrt{\omega_c\gamma})$ + const.

It is interesting that the case of a constant force can be handled perfectly well by this scheme, because the environment stops the runaway acceleration and instead introduces a terminal velocity. From (4.2), and (4.3) with $\omega_0 = 0$, one finds that the mean position

$$<x>_t \equiv \int dx_f \, x_f \, \rho(x_f, y_f=0, t) \tag{5.2}$$

obeys the equation

$$<x>_t = <x>_0 + \frac{d(t)}{m} <p>_0 + \frac{\lambda}{m} \int_0^t dt' \, d(t') \tag{5.3}$$

where $d(t) = \frac{1}{\gamma}(1 - e^{-\gamma t})$. Here x_0 is the initial mean position given by a formula like (5.2) and $<p>_0$ is given by

$$<p>_0 = \int dx_i \frac{\hbar}{i} \frac{\partial}{\partial y_i} \rho(x_i, y_i, 0)_{y_i=0} \quad . \tag{5.4}$$

Equation (5.3) is exactly the motion of a classical damped particle in an external force. The fluctuations around the mean come out to be precisely those for a free quantum particle in the scheme described by Eqs. (4.2)-(4.3).

6. CONCLUSIONS

I have tried to show here that there is a conceptually simple and correct way of dealing with friction in quantum mechanics. The method is also mathematically tractable in simple situations. Unfortunately, this simplicity evaporates when one comes to more interesting situations, such as the escape from a metastable well. It is possible, however, that the simple solutions have not yet been sufficiently exploited. The very problems we have treated exactly here are the ones which, in ordinary quantum mechanics, yield the turning point formulas which make semiclassical methods useful. It is conceivable that the exact solutions given here might be similarly useful. As they stand, they do answer in an extremely explicit way questions one might have about how a pure state in quantum mechanics evolves into a mixed state. In an equation like (4.2), the environment is carrying away information from, and thereby increasing the entropy of, the system. While this does not resolve the global paradoxes of quantum mechanics, or make it any less strange, it does show how the wave function of a subsystem is "collapsed" by interaction with its environment.

ACKNOWLEDGMENTS

I thank Amir Caldeira, Anupam Garg, and Emil Mottola for particularly useful conversations on these matters, and the members of the Program on Quantum Noise in Macroscopic Systems, held at the Institute for Theoretical Physics during January - August 1984, for their collective wisdom. The John Simon Guggenheim Memorial Foundation is to thanked for the award of a fellowship. This work was supported in part by the National Science Foundation under Grant No. PHY77-27084, supplemented by funds from the National Aeronautics and Space Administration, and in part by the Office of Naval Research.

REFERENCES

a. Permanent address.
1. R.P. Feynman and A.R. Hibbs, "Quantum Mechanics and Path Integrals," McGraw-Hill, New York, 1965.
2. V. Ambegaokar, U. Eckern, and G. Schön, Phys. Rev. Lett. 48:1795 (1982); and NSF-ITP-84-83, to be published.
3. V. Ambegaokar, "Quantum Dynamics of Superconductors and Tunneling Between Superconductors," NATO ASI on Superconductivity, Percolation and Localization, Plenum, New York (1984).
4. A.O. Caldeira and A.J. Leggett, Physica 121A:587 (1983).
5. A.O. Caldeira and A.J. Leggett, Ann. Phys. 149:374 (1984).
6. R.P. Feynman and F.L. Vernon, Ann. Phys. 24:118 (1963).
7. G.W. Ford, M. Kac, and P. Mazur, J. Math. Phys. 6:504 (1965).
8. V. Hakim and V. Ambegaokar, to be published.
9. E. Wigner, Phys. Rev. 40:749 (1932).
10. F. Guinea, to be published.
11. A. Garg, private communication.

THE SCALING RELATION BETWEEN THE MICROSCOPIC (KINETIC) AND MACROSCOPIC (HYDRODYNAMIC) DESCRIPTIONS

L. Bânyai

Institut für Theoretische Physik
der Universität Heidelberg
D-6900 Heidelberg, W. Germany

INTRODUCTION

The ultimate goal of statistical mechanics is to deduce the properties of macroscopical bodies from the underlying microscopical dynamics of their constituents. However, this programme encounters a formidable obstacle, related to the long debated contradiction between microscopical reversibility and macroscopical irreversibility. To facilitate the task, an intermediate level of description is often introduced (kinetic level) that operates with microscopical entities, but within an irreversible description. To this category belong the Boltzmann equation, different Master equations, Langevin equations etc. Their justification was often based on intuition, but nevertheless they proved to be very useful in calculating different macroscopical properties of matter. After the big progresses of the equilibrium statistical mechanics, due to the proper understanding of the thermodynamic limit, there is a related progress[1] also in the foundation of the kinetic description.

The purpose of this lecture is, however, to discuss only the second step, i.e. the passage from the kinetic to the macroscopic (hydrodynamic) level. As we shall illustrate, this is also an extremely difficult task. Its importance lies not only in fundamental understanding, but in providing clues to the correct calculation of phenomenological constants.

In the frame of the Boltzmannian theory of neutral gases, already Hilbert was preoccupied by this problem, while Chapmann and Enskog gave an expansion method that proved to be very fruitful. Their basic idea was that after a certain time, the state approaches local equilibrium, in which the distribution function depends on time only through the densities of the conserved entities.

While working out the connection between his microscopical electrodynamics and Maxwell's macroscopical one, Lorentz felt the necessity of some smoothening in space of the microscopical entities. While the microscopical field, for example, varies rapidly on the interatomic distance, the corresponding macroscopical field varies only on a macroscopical scale.

These two basic ideas of "smoothening" in time and space are actually implemented in the new scaling theories[2-9] of the connection between the kinetic and hydrodynamic descriptions. "Looking" at the microscopical solutions on the macroscopical scale of length and time seems to be the key for

the fundamental understanding of the problem. The mathematical formulation was suggested first by the scale invariance of the solutions of the diffusion equation and then by the more complicated scale invariance of the solution of the diffusion-conduction equation describing the charge flow in a semiconductor. These ideas have been rigorously implemented sofar only in the discussion of linear hydrodynamics (small departure from equilibrium) of some exactly solvable kinetic models (neutral Lorentz gas[2-4], charged Boltzmann gas with instant thermalization[7] and charged hopping lattice gas[5-7]). Nevertheless, these exact results on essentially different models give already a hint to the basic mechanisms and the generality of the approach.

To get an idea of the problem we are facing, let us consider the diffusion of neutral particles in a medium. Macroscopically it is described by the diffusion equation

$$\frac{\partial n}{\partial t} = D \nabla^2 n$$

for the particle density $n(\vec{x},t)$, while microscopically we may have a Boltzmannian model for the distribution function $\varphi(\vec{x},\vec{v};t)$ of coordinates and velocities

$$\frac{\partial \varphi}{\partial t} + \frac{\partial \varphi}{\partial \vec{x}} \vec{v} = - B\varphi$$

with the collision term $B\varphi$ describing scattering on thermal scatterers. The particle density in this case is given by integrating over the velocities:

$$n^{micro}(\vec{x},t) = n_0 \int d\vec{v}\, \varphi(\vec{x},\vec{v};t)$$

(n_0 is the equilibrium density).

As one may see immediately, however, in the microscopic (kinetic) model the particle density at a given instant is determined by giving the full coordinate-velocity distribution function at some initial instant $t = 0$. In the macroscopical (hydrodynamical) description we renounce to know everything about the system, but the entity we retain ($n(\vec{x},t)$) depends only on its initial value.

Therefore we want less information, under the condition that the same amount of information can be discarded at $t = 0$!

To implement this idea it is clear that some operation has to be performed on the microscopical density.

As we shall see in what follows, the scale projection performs this task

$$\lim_{\lambda\to\infty} \lambda^3 n^{micro}(\lambda\vec{x}, \lambda^2 t) = n(\vec{x},t)$$

or

$$\lim_{\lambda\to\infty} \lambda^3 \left[n^{micro}(\lambda\vec{x}, \lambda^2 t) - n(\lambda\vec{x}, \lambda^2 t)\right] = 0$$

We shall see that, generally speaking, one has to scale all the macroscopical lengths including the Debye length, which is electromagnetical.

Such a scaling will accomplish also the transition to a continuum description for the latticial models we shall consider.

The plan of the presentation is the following: In the first section we shall discuss the macroscopical electrodynamical description of the charge flow in a semiconductor. In section 2 we introduce the scale projection conjecture and the general mathematical features of its implementation,

anticipating some common features of the microscopical (kinetic) models we analyse later in some detail. Section 3 is devoted to the detailed analysis of a neutral hopping model on a periodical lattice. Section 4 contains a condensed presentation of the results on the charged hopping model, while section 5 contains a simplified presentation of the proofs in the case of a charged Boltzmann gas with "instantly thermalizing" cross section.

MACROSCOPIC DESCRIPTION OF CHARGE EVOLUTION IN A SEMICONDUCTOR

We consider an infinite homogeneous semiconducting medium, characterized by the phenomenological constants: σ (conductivity), D (diffusivity), and ε (dielectric permitivity). We do not require isotropy, and therefore all these entities are actually 3 x 3 tensors. No external magnetic field is introduced and the internal magnetic field will be ignored (small internal currents). The relevant Maxwell equations are then: the continuity equation

$$\frac{\partial \rho_T}{\partial t} + \nabla \vec{j}_T = 0 \tag{1.1}$$

for the total charge and current densities, and the Poisson equation

$$\varepsilon_0 \nabla \vec{E} = \rho_T \tag{1.2}$$

relating the electric field $\vec{E}$ to the total charge density ρ_T. (Here ε_0 is the permitivity of the vacuum.) The total charge and current densities are made up of the external and internal ones

$$\begin{aligned} \rho_T &= \rho_{int} + \rho_{ext} \\ \vec{j}_T &= \vec{j}_{int} + \vec{j}_{ext} \end{aligned} \tag{1.3}$$

We chose to have no external magnetic fields, therefore

$$\vec{j}_{ext} = 0 \tag{1.4}$$

and

$$\frac{\partial \rho_{ext}}{\partial t} = 0 \tag{1.5}$$

Ignoring the internal magnetic field also implies

$$\vec{E} = -\nabla V \tag{1.6}$$

Now, the phenomenological relations specific for a semiconductor may be written as

$$\begin{aligned} \rho_{int} &= \rho - (\varepsilon - \varepsilon_0)\nabla \vec{E} \\ \vec{j}_{int} &= \sigma \vec{E} - D\nabla\rho + (\varepsilon - \varepsilon_0)\frac{\partial \vec{E}}{\partial t} \end{aligned} \tag{1.7}$$

Here ρ represents the so-called "free charge" density. We will see that actually it needs no definition. These phenomenological equations (1.7) together with the Maxwell equations (1.1), (1.2) completely determine all the physical entities everywhere at any instant, once the initial "free charge" density ρ is given.

Before writing down the combined equations for the two basic entities ρ and V, it is useful to remind that there is a universal relation (Einstein relation) between the conductivity and diffusivity

$$\sigma = \varepsilon_0 \ell^{-2} D \tag{1.8}$$

The electromagnetical length ℓ is given by

$$\ell = \left(\frac{e^2}{\varepsilon_0}\frac{\partial n_0}{\partial \mu}\right)^{-\frac{1}{2}}$$

where e, n_0 are the charge and the equilibrium density of "free" carriers, while μ is the chemical potential. Therefore, instead of two non-equilibrium phenomenological constants we may use a single non-equilibrium constant D and an equilibrium constant ℓ.

Then, introducing eq. (1.7) into eqs. (1.1), (1.2) we get

$$\frac{\partial \rho}{\partial t} = \nabla D \nabla \left(\rho + \frac{\varepsilon_0}{\ell^2} V\right)$$
$$\nabla \varepsilon \nabla V = -\rho - \rho_{ext} \tag{1.9}$$

which confirm our statement that the only initial condition to be given is $\rho|_{t=0}$. On the other hand, giving $\rho|_{t=0}$ is equivalent to giving $\rho_{int}|_{t=0}$. Therefore, indeed there is no need for any special definition of the "free charge" density. In the phenomenological theory it gives just a convenient way to describe the properties of the material. However, as it was anticipated by Lorentz, and as we will show in some examples, in a microscopical theory it has a well-defined physical meaning, corresponding to the charge that can be transported over macroscopical distances (in opposition to the polarization charge density).

As one may see from the inspection of eqs. (1.9), the so-called Debye length

$$\ell_D = \ell\left(\frac{\varepsilon}{\varepsilon_0}\right)^{\frac{1}{2}}$$

characterizes the space variation of $\rho(\vec{x},t)$. It can be seen that it coincides with the penetration depth of an external static electric field into the semiconductor. It is well-known that this penetration depth is of the order of centimeters for a typical semiconductor, i.e., it is macroscopical. This is due to the small concentration of "free" carriers.

On the contrary, in a metal, due to the big concentration of "free" carriers, ℓ_D is microscopical (angströms). Therefore phenomena occur on the surface and not in the bulk. The correct description of these phenomena, however, is not given by our equations. (Macroscopical variations on a microscopical scale just show the breakdown of our picture.)

In Fourier transforms the solution of eqs. (1.9) is immediate. We give here the expression of the total charge density:

$$\hat{\rho}_T(\vec{k},t) = \frac{\varepsilon_0 \vec{k}^2}{\vec{k}\varepsilon\vec{k}}\left[\hat{\rho}(\vec{k},0) + \frac{1}{1+\ell^2\frac{\vec{k}\varepsilon\vec{k}}{\varepsilon_0}}\hat{\rho}_{ext}(\vec{k})\right] e^{-\vec{k}D\vec{k}\left(1+\frac{\varepsilon_0}{\ell^2\vec{k}\varepsilon\vec{k}}\right)t} + \frac{\ell^2\vec{k}^2}{1+\ell^2\frac{\vec{k}\varepsilon\vec{k}}{\varepsilon_0}} \tag{1.10}$$

One may remark two interesting properties:

(i) $\hat{\rho}_T(\vec{k},0) \neq \hat{\rho}(\vec{k},0) + \hat{\rho}_{ext}(\vec{k})$
(The polarization charge is instantaneous.)

(ii) $\hat{\rho}_T(0,\infty) = 0$. (The external charge is screened in equilibrium. Non-conservation of charge occurs at the expense of surface charges at the infinity.)

The fundamental solution corresponds to choosing a point initial "free" charge e at the origin and a point external charge q at a distance $\vec{x}_0$.

$$\rho(\vec{x},0) = e\,\delta(\vec{x}) \qquad \text{or} \qquad \hat{\rho}(\vec{k},0) = e$$
$$\rho_{ext}(\vec{x}) = q\,\delta(\vec{x}-\vec{x}_0) \qquad\qquad \hat{\rho}_{ext}(\vec{k}) = q\,e^{i\vec{k}\vec{x}_0} \tag{1.11}$$

Then it is easy to see the scale invariance of the fundamental solution

$$\hat{\rho}_T(\tfrac{1}{\lambda}\vec{k}, \lambda^2 t; \lambda \ell, \lambda \vec{x}_0) = \hat{\rho}_T(\vec{k}, t; \ell, \vec{x}_0) \tag{1.12}$$

Similarly one may show that

$$\lambda^{-2}\hat{V}(\tfrac{1}{\lambda}\vec{k}, \lambda^2 t; \lambda \ell, \lambda \vec{x}_0) = \hat{V}(\vec{k}, t; \ell, \vec{x}_0) \tag{1.13}$$

In the coordinate space these correspond to the scale invariances:

$$\lambda^3 \rho_T(\lambda\vec{x}, \lambda^2 t; \lambda \ell, \lambda \vec{x}_0) = \rho_T(\vec{x}, t; \ell, \vec{x}_0) \tag{1.14}$$

$$\lambda V(\lambda\vec{x}, \lambda^2 t; \lambda \ell, \lambda \vec{x}_0) = V(\vec{x}, t; \ell, \vec{x}_0) \tag{1.15}$$

Therefore, all the macroscopical physical entities are invariant with respect to the simultaneous scaling of all lengths (including ℓ !) with λ, and of the time with λ^2 (up to an overall power of λ). Formulated in this way, the statement is valid for an arbitrary solution of our equations.

The reader may remark that the way we treated the macroscopical problem was such as to be as close as possible to the corresponding microscopical formulation. A peculiar aspect is that one may reconstruct all the physical entities (ρ_T, $\vec{j}_T$, $\vec{E}$) from the knowledge of the "free" charge density. This is particularly important because, generally speaking, in a microscopical theory the current density may not be uniquely defined. Our discussion has shown that we may restrict our attention to the charge densities.

THE SCALE PROJECTION CONJECTURE

Since the solution of the microscopical (kinetic) problem is not scale-invariant, we may only expect that its scale-invariant part may coincide with the solution of the macroscopical problem. The scale-invariant part is obtained by the scale projection

$$\lim_{\lambda\to\infty} \lambda^3 \rho_T^{micro}(\lambda\vec{x}, \lambda^2 t; \lambda \ell, \lambda \vec{x}_0)$$

If this limit exists, the resulting function is invariant with respect to further scaling with any finite λ.

The necessity of scale projection is obvious, but it is not at all obvious that it is sufficient to get the macroscopical result. Our conjecture [5,9] is that it is sufficient, i.e.

$$\lim_{\lambda\to\infty} \lambda^3 \rho_T^{micro}(\lambda\vec{x}, \lambda^2 t; \lambda \ell, \lambda \vec{x}_0) = \rho_T^{macro}(\vec{x}, t; \ell, \vec{x}_0) \tag{2.1}$$

(Of course, we then need to identify also the "free" carriers in the microscopical theory to be able to identify ℓ, but the important aspect is that once this choice has been made, the calculation of ℓ is immediate.)

The conjecture may also be formulated as

$$\lim_{\lambda\to\infty} \lambda^3 \left[\rho_T^{micro}(\lambda\vec{x}, \lambda^2 t; \lambda \ell, \lambda \vec{x}_0) - \rho_T^{macro}(\lambda\vec{x}, \lambda^2 t; \lambda \ell, \lambda \vec{x}_0)\right] = 0 \tag{2.2}$$

Under this form one may better see the physical meaning of the scaling. Equation (2.2) tells us that on the big (macroscopical) length (as λ) and

time (as λ^2) scale, the two solutions converge (faster than λ^{-3}). On this scale one does not observe the microscopical variations, and the unnecessary information is lost. Our conjecture is in perfect agreement with the ideas of both Lorentz and Chapmann-Enskog.

This conjecture has been successfully tested on two different kinds of microscopical (kinetic) exactly solvable models: a) Boltzmann gas of charged particles with "instantaneous thermalization" scattering cross section[8], and b) Charged hopping particles on a periodical (non-Bravais) lattice.[5-7,9] In both cases Coulomb interactions are treated in the self-consistent potential approximation.

These solvable models, which we shall analyse in further sections, have some common features which may be formalized as follows:
The state of the system is statistically characterized by some $\mathcal{P}(t)$ satisfying (not far from equilibrium) a linear non-homogeneous equation

$$\frac{d\mathcal{P}(t)}{dt} = -\mathcal{A}\,\mathcal{P}(t) + \mathcal{B}\,V^{ext} \tag{2.3}$$

the non-homogeneity being linear in the external potential. The internal charge density results through a linear operation on $\mathcal{P}(t)$:

$$\rho_{int}^{micro}(t) = \mathcal{H}\,\mathcal{P}(t) \tag{2.4}$$

The solution of eq. (2.3) may be written as

$$\mathcal{P}(t) = e^{-\mathcal{A}t}\left(\mathcal{P}(0) - \mathcal{P}(\infty)\right) + \mathcal{P}(\infty) \tag{2.5}$$

The existence of a unique $\mathcal{P}(\infty)$ is a result of the irreversibility property of eq. (2.3).

It is more advantageous for comparison with eq. (1.10) to formulate the problem in Fourier transforms with respect to the coordinates. Then we have

$$\hat{\rho}_T^{micro}(\lambda^2 t)_\lambda = \hat{\mathcal{H}}_\lambda\left[e^{-\hat{\mathcal{A}}_\lambda \lambda^2 t}\left(\hat{\mathcal{P}}(0)_\lambda - \hat{\mathcal{P}}(\infty)_\lambda\right) + \hat{\mathcal{P}}(\infty)_\lambda\right] + \hat{\rho}_\lambda^{ext} \tag{2.6}$$

(The lower index λ here means taking $\frac{1}{\lambda}\vec{k}, \lambda\ell, \lambda\vec{x}_0$.)

If we compare this with eq. (1.10), one may remark that there are formal similarities, but in the microscopical solution the spectrum of $\hat{\mathcal{A}}_\lambda$ is too rich and therefore it does not give a single exponential dependence.

What happens when we let λ tend to infinity? Generally speaking, the spectrum of $\hat{\mathcal{A}}_\lambda$ lies in the right half-plane with the following properties:

(i) There is one eigenvalue that goes to zero on the real axis as λ^{-2}.

(ii) There might be a non-degenerate eigenvalue that is vanishing for all λ.

(iii) The remaining part of the spectrum is bounded from below.

Therefore, for $\lambda\to\infty$ from $e^{-\hat{\mathcal{A}}_\lambda \lambda^2 t}$ only the two exponentials corresponding to the two vanishing eigenvalues will survive. (One of them corresponding to (i) is a true exponential, while that corresponding to (ii) is just a constant.)

This is the basic mechanism, but of course, one also shows that the surviving projectors and all the other terms in the $\lambda\to\infty$ limit give rise exactly to eq. (1.10) with definite expressions for D, ε, ℓ in terms of the microscopical characteristics.

Let us consider a system of neutral fermions on a lattice of sites $\vec{x}_i$. The average occupation number of the site i at the instant t is $\bar{n}_i(t)$. The energy of a fermion on the site i is ε_i. The state of the system is given by the ensemble of average occupation numbers, determined from the hopping rate equation

$$\frac{d\bar{n}_i}{dt} = -\sum_j \left[W_{ij}\, \bar{n}_i(1-\bar{n}_j) - W_{ji}\, \bar{n}_j(1-\bar{n}_i) \right] \tag{3.1}$$

where the site-to-site transition rates W_{ij} obey the "detailed balance" relation:

$$W_{ij} = W_{ji}\, e^{\frac{\varepsilon_i - \varepsilon_j}{k_B T}} \tag{3.2}$$

The rate equation (3.1) conserves the average total number of fermions $\sum_i \bar{n}_i(t) = N$ and keeps $0 \leqslant \bar{n}_i(t) \leqslant 1$ if they were in this interval at some initial moment.

The only microscopical local observable we may define is the particle density

$$n(\vec{x},t) = \sum_i \bar{n}_i(t)\, \delta(\vec{x} - \vec{x}_i) \tag{3.3}$$

There is no physically meaningful definition for the microscopical current density in this model, although many choices are possible.

Not far from equilibrium, described by the Fermi function $f(\varepsilon_i)$, we have

$$\bar{n}_i = f(\varepsilon_i) + \eta_i(t) \tag{3.4}$$

with small $\eta_i(t)$. The linearized rate equation is

$$\frac{d}{dt}\eta = -\Gamma \eta \tag{3.5}$$

with the solution $\eta(t) = e^{-\Gamma t}\eta(0)$
where

$$\Gamma_{ij} = \left(\delta_{ij} \sum_l \mathcal{W}_{il} - \mathcal{W}_{ij} \right) \frac{1}{f_j(1-f_j)} \tag{3.6}$$

and

$$\mathcal{W}_{ij} = f_i(1-f_j)\, W_{ij} = \mathcal{W}_{ji} \quad . \tag{3.7}$$

The operator Γ is real and hermitian in the scalar product

$$(\psi, \varphi) \equiv \sum_i \psi_i^* \varphi_i \frac{1}{f_i(1-f_i)} \quad .$$

Since

$$(\psi, \Gamma\psi) = \frac{1}{2} \sum_{i,j} \left| \frac{\psi_i}{f_i(1-f_i)} - \frac{\psi_j}{f_j(1-f_j)} \right|^2 \mathcal{W}_{ij} \geq 0 \tag{3.8}$$

this operator is positive ($\Gamma \geq 0$).

If any pair of sites may be connected by a sequence of non-vanishing transition rates ("connectivity"), then it results from eq. (3.8) that there is only one null-eigenvector:

$$\psi_i^{(0)} = \frac{f_i(1-f_i)}{\sqrt{\sum_j f_j(1-f_j)}} \tag{3.9}$$

and we have

$$\eta_i(t) \xrightarrow[t\to\infty]{} 0$$

since the initial condition was orthogonal to the null-eigenvector (3.9). ($\sum_i \eta_i(t)$ = 0 is equivalent to $(\psi^{(0)}, \eta(t)) = 0$.)

Equation (3.5) can be exactly solved on periodical lattices (not necessarily Bravais-type).

Let us consider that the position of the sites are given by

$$\vec{x}_{\vec{r},s} = \vec{r} + \vec{\xi}_s \qquad (s = 1, 2, \ldots S) \tag{3.10}$$

where $\vec{r}$ runs through a Bravais lattice, while $\vec{\xi}_s$ gives the position of the site within the elementary cell. Translational symmetry would require

$$\begin{aligned} \varepsilon_{\vec{r},s} &\equiv \varepsilon_s \\ W_{\vec{r}s;\vec{r}'s'} &\equiv W_{ss'}(\vec{r}-\vec{r}') \end{aligned} \tag{3.11}$$

It is then natural to use discrete Fourier transforms defined as

$$\tilde{f}(\vec{k}) = \sum_{\vec{r}} e^{i\vec{k}\vec{r}} f(\vec{r}) \quad ; \quad f(\vec{r}) = \frac{v}{(2\pi)^3} \int_{BZ} d\vec{k}\, e^{-i\vec{k}\vec{r}} \tilde{f}(\vec{k})$$

where v is the volume of the elementary cell, while the integration is carried out over the Brillouin zone.

Then, for the Fourier transform of η we have:

$$\tilde{\eta}_s(\vec{k},t) = \sum_{s'} \left(e^{-\tilde{\Gamma}(\vec{k})t} \right)_{ss'} \tilde{\eta}_{s'}(\vec{k},0) \tag{3.12}$$

For each $\vec{k}$ we have a finite-size matrix problem. The matrix $\tilde{\Gamma}(\vec{k})$ has the following important properties:

(i) $\tilde{\Gamma}(\vec{k}) = \tilde{\Gamma}(\vec{k})^+$ in the scalar product $(x,y) \equiv \sum_{s=1}^{S} x_s^* y_s \cdot p_s^{-1}$ $\left(p_s \equiv f_s(1-f_s) / \sum_{s'} f_{s'}(1-f_{s'}) \right)$

(ii) $\tilde{\Gamma}(\vec{k}) = \tilde{\Gamma}(-\vec{k})^*$ (from reality in $\vec{r}$-space),

(iii) $\tilde{\Gamma}(\epsilon\vec{k})$ is analytic in ϵ around $\epsilon = 0$ (due to the assumed finite range of $W_{ss'}(\vec{r})$),

(iv) $\tilde{\Gamma}(\vec{k}) > 0$ for $\vec{k} \neq 0$ and $\tilde{\Gamma}(0)$ has a single null-eigenvector (by the same connectivity argument we used before).

Then it results that all the eigenvalues $\gamma_\alpha(\vec{k})$ ($\alpha = 1, \ldots S$) are positive. As $\vec{k}$ goes to zero, all the eigenvalues remain strict positive with the exception of the lowest one that has to vanish quadratically (due to (iv) and (ii)):

$$\begin{aligned} &\gamma_\alpha(0) > 0 \qquad \text{for} \qquad \alpha \neq 1 \\ &\gamma_1(\vec{k}) \underset{\vec{k}\to 0}{\sim} \vec{k}\mathcal{D}\vec{k} \end{aligned} \tag{3.13}$$

The eigenprojector corresponding to the null-eigenvalue at $\vec{k} = 0$ is

$$P^{(1)}(0)_{ss'} = p_s \tag{3.14}$$

Now we are ready to discuss the scaling limit. We may write eq. (3.12) in its spectral decomposition form. With scaled parameters it looks as

$$\tilde{\eta}_s\left(\tfrac{1}{\lambda}\vec{k},\lambda^2 t\right)=\sum_{\alpha=1}^{S} e^{-\gamma_\alpha\left(\frac{1}{\lambda}\vec{k}\right)\lambda^2 t}\, P_\alpha\left(\tfrac{1}{\lambda}\vec{k}\right)_{ss'}\,\tilde{\eta}_{s'}\left(\tfrac{1}{\lambda}\vec{k},0\right)$$

When λ tends to infinity, only the term $\alpha = 1$ survives

$$\lim_{\lambda\to\infty}\tilde{\eta}_s\left(\tfrac{1}{\lambda}\vec{k},\lambda^2 t\right)=e^{-\vec{k}\mathcal{D}\vec{k}t}\,p_s\sum_{s'}\tilde{\eta}_{s'}(0,0)$$

To simulate the fundamental solution of the diffusion equation we must consider an initial condition that can be normalized to unity

$$\sum_{s,\vec{r}}\eta_s(\vec{r},0)=1$$

or

$$\sum_s\tilde{\eta}_s(0,0)=1$$

(A supplementary particle is put at a microscopical distance from the origin of coordinates.) Then,

$$\lim_{\lambda\to\infty}\tilde{\eta}_s\left(\tfrac{1}{\lambda}\vec{k},\lambda^2 t\right)=e^{-\vec{k}\mathcal{D}\vec{k}t}\,p_s \tag{3.15}$$

On the other hand, the Fourier transform of the density, defined according to eq. (3.3), (more exactly, its deviation from the constant equilibrium value) is given by

$$\hat{n}(\vec{k},t)=\sum_s\tilde{\eta}_s(\vec{k},t)\,e^{-i\vec{k}\vec{\xi}} \tag{3.16}$$

Therefore

$$\lim_{\lambda\to\infty}\hat{n}\left(\tfrac{1}{\lambda}\vec{k},\lambda^2 t\right)=e^{-\vec{k}\mathcal{D}\vec{k}t} \tag{3.17}$$

or

$$\lim_{\lambda\to\infty}\lambda^3 n(\lambda\vec{x},\lambda^2 t)=\frac{1}{(4\pi t)^{3/2}[\det\mathcal{D}]^{1/2}}\,e^{-\frac{\vec{x}\mathcal{D}^{-1}\vec{x}}{4t}} \tag{3.18}$$

which is exactly the fundamental solution of the diffusion equation with the diffusivity tensor $\mathcal{D}_{\mu\nu}$.

To obtain the solution corresponding to an arbitrary initial condition one has to scale also its geometrical characteristics (to become macroscopical).

One may see that eq. (3.15) expresses the Chapmann-Enskog limiting situation, where the state is depending on time only through the density of particles.

As a side remark, it is useful to retain that the small $\vec{k}$ development of $\hat{\Gamma}(\vec{k})$ (Moyal expansion) itself would give a wrong result, except on a Bravais lattice. (The equation itself does not approach the diffusion equation, but its solution approaches the solution of the diffusion equation!)

We could discuss here this simple neutral example to full understanding (although we have omitted some steps necessary for full rigour). Unfortunately, the case of charged hopping is much too complicated to fit within the alloted space. Therefore, in the next section we shall only give the basic ideas and main results, without any proof.

The rate equation for charged hopping fermions in the self-consistent potential approximation is written similarly to equation (3.1),

$$\frac{d\bar{n}_i}{dt} = -\sum_j \left\{ W_{ij}^t \, \bar{n}_i (1-\bar{n}_j) - W_{ji}^t \, \bar{n}_j (1-\bar{n}_i) \right\} \tag{4.1}$$

but with time-dependent transition rates, obeying an instantaneous "detailed balance" relation

$$W_{ij}^t = W_{ji}^t \, e^{\frac{E_i(t) - E_j(t)}{k_B T}} \tag{4.2}$$

where $E_i(t)$ is the self-consistent energy of the electron on site i at the instant t (determined by the average Coulomb configuration of the system at this instant). It can be decomposed into an unperturbed equilibrium and a non-equilibrium part:

$$E_i(t) = \epsilon_i + e\mathcal{V}_i(t) \tag{4.3}$$

where

$$\epsilon_i = \mathcal{E}_i + \frac{e^2}{4\pi\varepsilon_0} \sum_{j(\neq i)} \frac{f(\epsilon_j) - q_j}{|\vec{x}_i - \vec{x}_j|} \tag{4.4}$$

$$\mathcal{V}_i(t) = \frac{e}{4\pi\varepsilon_0} \sum_{j(\neq i)} \frac{\bar{n}_j(t) - f(\epsilon_j)}{|\vec{x}_i - \vec{x}_j|} + e\mathcal{V}_i^{ext} \tag{4.5}$$

After a double linearization (in $\eta_i(t) \equiv \bar{n}_i(t) - f(\epsilon_i)$ and $\mathcal{V}_i(t)$) we end up with the following equation:

$$\frac{d}{dt}\eta_i(t) = -\sum_j \Gamma_{ij} \left[\eta_j(t) + \frac{e f_j(1-f_j)}{k_B T} \mathcal{V}_j(t) \right] \tag{4.6}$$

where Γ_{ij} is given again by eq. (3.6) with the equilibrium transition rates.

Combining eqs. (4.6) and (4.5) we have

$$\frac{d}{dt}\eta_i(t) = -\sum_j A_{ij}\,\eta_j(t) + \sum_j B_{ij}\,\mathcal{V}_j^{ext} \tag{4.7}$$

with

$$A_{ij} = \sum_\ell \Gamma_{i\ell} \left(\delta_{\ell j} + \frac{e^2 f_\ell(1-f_\ell)}{4\pi\varepsilon_0 k_B T} \, \frac{1-\delta_{\ell j}}{|\vec{x}_\ell - \vec{x}_j|} \right) \tag{4.8}$$

$$B_{ij} = -\frac{e}{k_B T} \Gamma_{ij} \, f_j(1-f_j) \tag{4.9}$$

(compare with eq. (2.3)!).

An artefact of the self-consistent potential approximation is that the evolution operator A is not always positive. Nevertheless we shall see that under reasonable conditions the stability is assured.

Let us now consider again hopping on a periodical lattice. However, now we shall consider a more complicated connectivity structure than in the previous section, to provide our model with both "free" and "bound" electrons.

The sites within a given cell have a label corresponding to "free" or "bound". The "free" sites form a connected lattice, while the "bound" sites

are connected only between themselves and only within the same cell. Only the Coulomb interaction connects all the sites.

We shall then define the corresponding electromagnetic lengths

$$\frac{1}{\ell^2} = \frac{e^2}{\varepsilon_0}\frac{\partial n^0_{free}}{\partial\mu} \quad ; \quad \frac{1}{L^2} = \frac{e^2}{\varepsilon_0}\frac{\partial n^0_{bound}}{\partial\mu} \tag{4.10}$$

Equation (4.6) in discrete Fourier transforms looks as

$$\frac{d}{dt}|\tilde{\eta}(\vec{k},t)\rangle = -\tilde{A}(\vec{k})\,|\tilde{\eta}(\vec{k},t)\rangle - \frac{\varepsilon_0 v}{e}\tilde{\Gamma}(\vec{k})\,\nu\,|\tilde{v}^{ext}(\vec{k})\rangle \tag{4.11}$$

where an operator and ket notation was used for convenience and

$$\nu_{ss'} = \frac{e^2}{\varepsilon_0 v k_B T} f_s(1-f_s)\,\delta_{ss'} \tag{4.12}$$

One may also write

$$\tilde{A}(\vec{k}) = \tilde{\Gamma}(\vec{k})\left(1 + \nu\,\tilde{C}(\vec{k})\right) \tag{4.13}$$

where $\tilde{C}(\vec{k})$ is the discrete Fourier transform of the Coulomb matrix. It is completely determined by

$$C_{ss'}(\vec{\tau}) = \frac{v(1-\delta_{ss'}\,\delta_{\vec{\tau},\vec{\tau}'})}{4\pi\,|\vec{\tau}+\vec{\xi}_s-\vec{\xi}_{s'}|} = \frac{v}{(2\pi)^3}\int_{BZ} d\vec{k}\; e^{-i\vec{k}\vec{\tau}}\,\tilde{C}_{ss'}(\vec{k}) \tag{4.14}$$

although the inverse transform does not exist in the ordinary sense. It can be shown that $\tilde{C}_{ss'}(\vec{k})$ has the following structure:

$$\tilde{C}_{ss'}(\vec{k}) = \frac{e^{-i\vec{k}(\vec{\xi}_s-\vec{\xi}_{s'})}}{k^2} + \tilde{R}_{ss'}(\vec{k}) \tag{4.15}$$

where $\tilde{R}_{ss'}(\epsilon\vec{k})$ is analytical in ϵ around $\epsilon = 0$.

Using an old result of Onsager that

$$C \geq -\frac{v}{2\pi d}$$

where d is the smallest distance in the lattice, and choosing

$$\ell, L > \sqrt{\frac{v}{2\pi d}}$$

we may assure the positivity of the matrix

$$\nu^{-\frac{1}{2}}\tilde{X}\,\nu^{\frac{1}{2}} \equiv 1 + \nu^{\frac{1}{2}}\tilde{C}\,\nu^{\frac{1}{2}} \tag{4.16}$$

Then it makes sense to write the solution of eq. (4.11) as

$$|\tilde{\eta}(\vec{k},t)\rangle = e^{-\tilde{A}(\vec{k})t}\left[|\tilde{\eta}(\vec{k},0)\rangle + \frac{v\varepsilon_0}{e}\tilde{X}(\vec{k})^{-1}\nu\,|\tilde{v}^{ext}(\vec{k})\rangle\right] - \frac{v\varepsilon_0}{e}\tilde{X}(\vec{k})^{-1}\nu\,|\tilde{v}^{ext}(\vec{k})\rangle \tag{4.17}$$

The following properties of the evolution operator $\tilde{A}$ can be shown[6,9]:

a) $\tilde{A}(\vec{k})$ is diagonalizable, with real, positive eigenvalues bounded as functions of $\vec{k} \in BZ$. It has a single null-eigenvector related to the existence of "bound" electrons. (These properties are non-trivial, since $\tilde{A}$ is neither hermitian nor bounded.),

b) The behaviour of the spectrum and eigenproperties of $\tilde{A}_\lambda \equiv \tilde{A}(\underset{\sim}{\vec{k}}, \lambda\ell)$

for $\lambda \to \infty$ is an analytic perturbation problem. The eigenvalues a_λ are analytic in λ^{-1} around the origin, while the eigenprojectors may have poles. There is only one eigenvalue that goes to zero, as λ^{-2} .

The scaling of ℓ is an essential ingredient! Except for the zero eigenvalue due to "bound electrons" all the other eigenvalues of $\tilde{A}$ are strict positive even for $\vec{k} = 0$.

Now if we take neutral islands of "bound electrons"

$$\sum_s^{"b"} \eta_s(\vec{r},0) = 0 \quad ,$$

an initial state with a supplementary "free" electron in the cell $\vec{r} = 0$

$$\sum_s^{"f"} \eta_s(\vec{r},0) = \delta_{\vec{r},0}$$

and a point external charge q (in a point $\vec{x}_0$ not belonging to the lattice)

$$U_s^{ext}(\vec{r}) = \frac{q}{4\pi\varepsilon_0} \frac{1}{|\vec{r}+\vec{\xi}_s-\vec{x}_0|}$$

one can rigorously show[9] that

$$\lim_{\lambda\to\infty} \hat{\rho}_T^{micro}\left(\tfrac{1}{\lambda}\vec{k}, \lambda^2 t\,; \lambda\ell, \lambda\vec{x}_0\right) = \hat{\rho}_T^{macro}(\vec{k}, t\,; \ell, \vec{x}_0) \tag{4.18}$$

with $D_{\mu\nu}$ determined completely by entities pertinent to the "free electrons" (whose density also determines the e.m. length ℓ) and ε determined completely by entities pertinent to the "bound electrons".

While the diffusivity $D_{\mu\nu}$ is given again by the curvature at $\vec{k} = 0$ of the lowest eigenvalue of the Γ matrix (restricted to the "free electrons") as in section 3, the dielectric permitivity is given by

$$\left(\frac{\varepsilon-\varepsilon_0}{\varepsilon_0}\right)_{\mu\nu} = \frac{1}{L^2} \left< \xi_\mu q^{\frac{1}{2}} \pi \,, \left[1 + \frac{1}{L^2} \pi q^{\frac{1}{2}} \tilde{R}(0) q^{\frac{1}{2}} \pi\right]^{-1} \pi q^{\frac{1}{2}} \xi_\nu \right> \tag{4.19}$$

where

$$\pi \equiv 1 - |q^{\frac{1}{2}}\rangle\langle q^{\frac{1}{2}}|$$

and

$$q_s = f_s(1-f_s) \Big/ \sum_{s'}^{"b"} f_{s'}(1-f_{s'})$$

for s belonging to the "bound electron" sites and vanishing otherwise.

The "free electrons" provide the possibility for macroscopic current flow, while the "bound electrons" give the local polarization properties.

It can be seen that while the polarization charge does not set on instantaneously, the respective time is microscopical (vanishing after scaling).

The resulting dielectric constant contains the local field effects and therefore differs from the predictions of the naive linear response theory. It can be shown (in the case of cubic symmetry) that the Clausius-Mosotti

formula

$$\frac{\varepsilon-\varepsilon_0}{\varepsilon_0} = \frac{\varkappa}{1-\frac{1}{3}\varkappa}$$

where $\varkappa$ is the polarizability of the cell, holds in the sense of equality of asymptotic series in the inverse of the lattice constant.

This last result was recently extended to the quantum-mechanical theory of the electronic dielectric constant of crystals.[10]

SCALING IN A CHARGED BOLTZMANN MODEL

In this section we shall consider the scaling procedure in the kinetic model of charged particles described by the Boltzmann equation for the distribution function of coordinates and velocities $\varphi(\vec{x},\vec{v},t)$:

$$\left(\frac{\partial}{\partial t} + \vec{v}\nabla_{\vec{x}} + \frac{e}{m}\vec{E}(\vec{x},t)\nabla_{\vec{v}}\right)\varphi(\vec{x},\vec{v},t) = -\int d\vec{v}'\left[W_{\vec{v}\vec{v}'}\,\varphi(\vec{x},\vec{v},t) - W_{\vec{v}'\vec{v}}\,\varphi(\vec{x},\vec{v}',t)\right] \tag{5.1}$$

where the electric field is determined self-consistently

$$\varepsilon_0 \nabla_{\vec{x}} \vec{E}(\vec{x},t) = e n_0 \int d\vec{v}\left(\varphi(\vec{x},\vec{v},t) - f_0(v)\right) + \rho_{ext}(\vec{x}) \tag{5.2}$$

from the external charge density and the instantaneous deviation of the average internal charge density from its constant equilibrium value. Here f_0 is the normalized Maxwell distribution

$$f_0(v) = \left(\frac{m}{2\pi kT}\right)^{3/2} e^{-\frac{mv^2}{2k_BT}} \tag{5.3}$$

The distribution function φ is normalized on the volume Ω as

$$\frac{1}{\Omega}\int_\Omega d\vec{x}\int d\vec{v}\,\varphi(\vec{x},\vec{v},t) = 1 \tag{5.4}$$

We consider again a static external charge and ignore the internal magnetic field, therefore

$$\vec{E} = -\nabla V$$

The transition rates between different velocity states obey the "detailed balance" relation

$$W_{\vec{v}\vec{v}'} = W_{\vec{v}'\vec{v}}\, e^{\frac{m}{2k_BT}(v^2-v'^2)} \tag{5.5}$$

In this model all "electrons" are "free", therefore

$$\frac{1}{\ell^2} = \frac{e^2 n_0}{\varepsilon_0 k_B T} \tag{5.6}$$

We shall consider the "instant thermalizing" cross section in what follows

$$W_{\vec{v}\vec{v}'} = \frac{1}{\tau} f_0(v') \tag{5.7}$$

for which the linearized version of the model is exactly solvable.

A peculiarity of this cross-section is that in the homogeneous case, it reduces to the trivial constant-relaxation-time model. Indeed, for a homogeneous flow we have

$$\varphi = f(\vec{v},t) \qquad \text{and} \qquad \vec{E}(t) = \vec{E}_{ext}(t)$$

and using the pertinent normalization condition of $f(\vec{v},t)$, together with eq. (5.7), we get

$$\left(\frac{\partial}{\partial t} + \frac{e}{m}\vec{E}^{ext}(t)\nabla_{\vec{v}}\right) f(\vec{v},t) = -\frac{f(\vec{v},t) - f_0(v)}{\tau} \tag{5.8}$$

The standard text-book analysis of this case gives the frequency-dependent complex conductivity

$$\sigma(\omega) = \frac{n_0 e^2}{m}\frac{1}{i\omega + \tau^{-1}}$$

and from its low-frequency, real and imaginary parts one gets

$$\sigma = \frac{n_0 e^2}{m}\tau \quad , \qquad \varepsilon - \varepsilon_0 = \frac{n_0 e^2}{m}\tau^2 \tag{5.9}$$

Using the Einstein relation, the diffusion constant D results as

$$D = \frac{k_B T}{m}\tau \tag{5.10}$$

An apparent paradox is the non-vanishing of $\varepsilon - \varepsilon_0$ in a model containing only "free" electrons. We shall comment on it later, after the full analysis of the model.

For small deviations from equilibrium and small electric fields ($\eta \equiv \varphi - f_0$ and $\frac{e\ell|\vec{E}|}{k_B T}$ small) one may write down the linearized equations for the non-homogeneous case:

$$\left(\frac{\partial}{\partial t} + \vec{v}\nabla_{\vec{x}}\right)\eta(\vec{x},\vec{v},t) + \frac{e}{m}\vec{E}(\vec{x},t)\frac{\partial f_0}{\partial \vec{v}} = -\frac{1}{\tau}\left[\eta(\vec{x},\vec{v},t) - f_0(\vec{v})\int d\vec{v}'\,\eta(\vec{x},\vec{v}',t)\right]$$

$$\varepsilon_0 \nabla_{\vec{x}}\vec{E}(\vec{x},t) = e n_0 \int d\vec{v}\,\eta(\vec{x},\vec{v},t) + \rho_{ext}(\vec{x})$$

(5.11)

After Fourier transformation in coordinate space, they can be combined into a single equation for the Fourier transform of η:

$$\frac{\partial}{\partial t}\hat{\eta}(\vec{k},\vec{v},t) = -(i\vec{k}\vec{v} + \tau^{-1})\hat{\eta}(\vec{k},\vec{v},t) + \left(\tau^{-1} - \frac{i\vec{k}\vec{v}}{k^2\ell^2}\right)f_0(v)\int d\vec{v}'\,\hat{\eta}(\vec{k},\vec{v}',t) -$$
$$- \frac{i\vec{k}\vec{v}}{k^2\ell^2}f_0(v)\frac{\hat{\rho}_{ext}(\vec{k})}{e n_0}$$

(5.12)

or in operator notations

$$\frac{\partial}{\partial t}\hat{\eta} = -\hat{A}\hat{\eta} + \hat{b} \tag{5.13}$$

with

$$\hat{A}(\vec{k})_{\vec{v}\vec{v}'} = (i\vec{k}\vec{v} + \tau^{-1})\,\delta(\vec{v}-\vec{v}') - \left(\tau^{-1} - \frac{i\vec{k}\vec{v}}{k^2\ell^2}\right) f_0(v) \tag{5.14}$$

$$\hat{b}(\vec{k})_{\vec{v}} = -\frac{i\vec{k}\vec{v}}{k^2\ell^2} f_0(v) \frac{q}{en_0} e^{i\vec{q}\vec{r}_0} \tag{5.15}$$

In eq. (5.15) we explicitly introduced a point external charge

$$\rho_{ext}(\vec{x}) = q\,\delta(\vec{x}-\vec{x}_0) \tag{5.16}$$

Of course, we shall consider also initial conditions that would correspond to the macroscopical fundamental solution

$$\eta(\vec{x},\vec{v},0) = \frac{1}{n_0}\,\delta(\vec{x})\,g(\vec{v}) \tag{5.17}$$

with an arbitrary normalized function of velocity $g(\vec{v})$:

$$\int d\vec{v}\, g(\vec{v}) = 1 \tag{5.18}$$

It is convenient to work out this problem in Laplace transforms with respect to the time variable:

$$\hat{\hat{\eta}}(s) = \int_0^\infty dt\, e^{-st}\,\hat{\eta}(t) \tag{5.19}$$

From eq. (5.13) we get immediately

$$\hat{\hat{\eta}}(s) = \frac{1}{s+\hat{A}} \left(\hat{\eta}(0) + \frac{1}{s}\hat{b}\right) \tag{5.20}$$

Now, the evolution operator $\hat{A}$ can be split in two pieces

$$\hat{A} = T(\vec{k}) - J(\vec{k},\ell) \tag{5.21}$$

a diagonal (unperturbed) part

$$T(\vec{k})_{\vec{v}\vec{v}'} = (i\vec{k}\vec{v} + \tau^{-1})\,\delta(\vec{v}-\vec{v}') \tag{5.22}$$

and a non-diagonal (perturbation) part

$$J(\vec{k},\ell)_{\vec{v}\vec{v}'} = \Psi(\vec{k},\vec{v};\ell) \tag{5.23}$$

with

$$\Psi(\vec{k},\vec{v};\ell) \equiv \left(\tau^{-1} - \frac{i\vec{k}\vec{v}}{k^2\ell^2}\right) f_0(v) \tag{5.24}$$

As it is easy to see,

$$J^2 = \frac{1}{\tau} J \tag{5.25}$$

i.e., τJ is a one-dimensional projector (bounded in L_1).

Let us now consider the resolvent operator

$$R(z) \equiv \frac{1}{z+\hat{A}} \tag{5.26}$$

that appears in eq. (5.20). It can be written as

$$R(z) = R_0(z) \frac{1}{1 - J R_0(z)} \tag{5.27}$$

where

$$R_0(z) = \frac{1}{z + T} \tag{5.28}$$

is the "unperturbed" resolvent.

On the other hand, we have

$$J R_0(z) J = \mathcal{F}(z) J \tag{5.29}$$

with

$$\mathcal{F}(z) \equiv \int d\vec{v}\, \psi(\vec{k}, \vec{v}, \ell) \frac{1}{i\vec{k}\vec{v} + \tau^{-1} + z} \tag{5.30}$$

and therefore

$$R(z) = R_0(z) + \frac{1}{F(z)} R_0(z) J R_0(z) \tag{5.31}$$

We may thus conclude that the perturbed spectrum consists of the unperturbed one (a cut defined by $\mathrm{Re}\, z = -\tau^{-1}$) and poles determined by the zeroes of the function $F(z)$.

About these zeroes, the following properties may be shown:

(i) There are no zeroes outside the strip $-\tau^{-1} \leq \mathrm{Re}\, z < 0$.

(ii) For $\mathrm{Re}\, z > -\frac{1}{2\tau}$ there is at most one zero, which is real, and there exist k_0, ℓ_0 such that for $k < k_0,\ \ell > \ell_0$ such a zero $z(k,\ell)$ exists.

(iii) $\lim_{\lambda \to \infty} \lambda^2 z(\frac{1}{\lambda}\vec{k}, \lambda\ell) = -k^2 D (1 + \frac{1}{k^2\ell^2})$ with D given by eq. (5.10).

We will not give the proof[8] for these statements here, nor their implementation in the time evolution scaling, but instead we will directly examine the scaling limit of the solution in Laplace transforms. It is easy to show that

$$\lim_{\lambda \to \infty} \lambda^2 F(\tfrac{1}{\lambda^2} s, \tfrac{1}{\lambda}\vec{k}, \lambda\ell) = \left[(1 + \tfrac{1}{k^2\ell^2}) D k^2 + s\right] \tau \tag{5.32}$$

the rest of the limits involved being trivial. After a simple algebra one finds

$$\lim_{\lambda \to \infty} \lambda^{-2} \hat{\hat{\eta}}(\tfrac{1}{\lambda^2} s, \tfrac{1}{\lambda}\vec{k}, \lambda\ell, \lambda\vec{x}_0; \vec{v}) =$$

$$= \frac{f_0(v)}{(1 + \frac{1}{k^2\ell^2}) D k^2 + s} \left[\frac{1}{n_0} - \frac{1}{s} \frac{D}{\ell^2} \frac{q}{e n_0} e^{i\vec{k}\vec{x}_0}\right] \tag{5.33}$$

or for the scaled Laplace and Fourier transformed internal charge density

$$\lim_{\lambda \to \infty} \lambda^{-2} e n_0 \int d\vec{v}\, \hat{\hat{\eta}}(\tfrac{1}{\lambda^2} s, \tfrac{1}{\lambda}\vec{k}; \lambda\ell, \lambda\vec{x}_0; \vec{v}) =$$

$$= \frac{1}{(1 + \frac{1}{k^2\ell^2}) k^2 D + s} \left[e - \frac{D}{s\ell^2} q\, e^{i\vec{k}\vec{x}_0}\right] \tag{5.34}$$

It can be easily seen that this is indeed the Laplace transform in time of the macroscopic charge density of eq. (1.10) (minus the external charge) for $\varepsilon = \varepsilon_o$.

Therefore, according to our scaling analysis that actually covers the general solution of the problem, there is no polarization charge in this model and the permitivity is that of the vacuum. The contradiction with the second of eqs. (5.9) is only apparent. Actually, in the naive theory

$$" \frac{\varepsilon - \varepsilon_o}{\varepsilon_o} " = \frac{1}{3} \left(\frac{L}{\ell} \right)^2$$

where

$$L = \tau \sqrt{\langle v^2 \rangle_o}$$

is the mean free path, and therefore in the scaling limit the term is vanishing. (This is also physically correct, since $\ell \gg L$.) Only within this precision there is a unique ε in all possible experiments. Otherwise, there would be different ε's in different experiments, differing by terms that vanish when $\ell \to \infty$.

Comparing eqs. (5.33) and (5.34) one may see again the exact validity of the Chapmann-Enskog conjecture in the scaling limit.

REFERENCES

1. E.B. Davies, "Quantum Theory of Open Systems", Academic Press, London (1967).
   V. Gorini, A. Frigerio, M. Verry, A. Kossakowski, and E.C.G. Sudarshan, Properties of Quantum Markovian Master Equations, Rep. Math. Phys. (GB) 13:149 (1978).
   H. Spohn, Kinetic equations from Hamiltonian dynamics: Markovian limits, Rev. Mod. Phys. 52:569 (1980).
2. H.P. Kean, The Central Limit Theorem for Carleman's Equation, Israel J. Math. 21:54 (1975).
3. G. Papanicolau, Asymptotic analysis of transport processes. Bull. AMS 81:330 (1975).
4. H. Spohn, Lecture Notes, University of Leuwen preprint (1979).
5. L. Bányai, P. Gartner, Macroscopical evolution from microscopical equations in an exactly soluble hopping model, in: "Recent Advances in Statistical Mechanics", Proc. Brasov Int. School, Central Inst. of Physics, Bucharest (1979).
6. L. Bányai, P. Gartner, On the connection between the macroscopical and microscopical evolution in an exactly soluble hopping model I. Neutral particles, Physica 102A:357 (1980).
7. L. Bányai, P. Gartner, On the connection between the macroscopical and microscopical evolution in an exactly soluble hopping model II. Charged particles, Physica 103A:119 (1980).
8. L. Bányai, P. Gartner, and V. Protopopescu, Macroscopic behaviour of a charged Boltzmann gas, Physica 107A:166 (1981).
9. L. Bányai, P. Gartner, The macroscopic electrodynamic behaviour of a soluble hopping model, Physica 115A:169 (1982).
10. L. Bányai, P. Gartner, Clausius-Mosotti limit of the quantum theory of the electronic dielectric constant, Phys. Rev. B 29:728 (1984).

RETARDED SYSTEMS: THE ROAD TO CHAOS

J.M. Aguirregabiria and L. Bel

Laboratoire de Physique Théorique. Institut H. Poincaré

11 Rue Pierre et Marie Curie. 75231 Paris. France

SUMMARY. We show that one expects a cascade of dominant reductions (asymptotic approximation of a retarded equation by an ordinary one) of increasing orders in the process of evolution from the "laminar" regime to the "turbulent" one. We report also on some preliminary numerical evidence confirming the theoretical expectation for the first two steps of the cascade, in particular for optical hybrids with a time delay akin with non linear optical cavities.

INTRODUCTION

Predictive (ordinary) systems of differential equations associated with hereditary (pure retarded) ones have been used since 1965[1-12] to deal with the equations of motion of point particles in the framework of classical field theories (mainly electromagnetism). Evidence of spontaneous transition to a predictive system was brought to attention by one of us in 1982[13-15].

We point out in this paper that both concepts, together with the concept of cascading dominant reductions can be useful in understanding some physical systems considered in recent years[16-18] which involve a time delay and which have been proved to exhibit an organized evolution towards some type of chaos.

We summarize below the main theoretical results in connection with the concepts just mentioned. Let us consider a retarded differential equation of the following type:

$$dx/dt = W(x,\hat{x};t;G) \quad , \quad \hat{x} = x(t-r) \tag{1}$$

where r is the fixed retardation and G a tuning constant. We shall assume the function W smooth enough with respect to all its arguments and we shall assume also that for G=0 the function W becomes a function $W^{(0)}(x;t)$ independent of x. We shall refer to the differential equation:

$$dx/dt = W^{(0)}(x;t) \tag{2}$$

as the zero order predictive equation associated with the retarded Eq.(1).

Let us consider an s-th order ordinary differential equation:

$$d^s x/dt^s = \xi_{(s)}(x, dx/dt, \dots, d^{s-1}x/dt^{s-1}; t; G) \qquad (3)$$

and let:

$$x = \varphi_{(s)}(x_0, \dots, (d^{s-1}x/dt^{s-1})_0; t, t_0; G) \qquad (4)$$

be its general solution corresponding to initial conditions (x_0, ..., $(d^{s-1}x/dt^{s-1})_0$) for $t=t_0$. By definition Eq.(3) is an s-th order reduction of Eq.(1) if all solutions of Eq.(3) are solutions of Eq.(1). Or in other words, if for each set of initial conditions one has:

$$d\varphi_{(s)}/dt = W(\varphi_{(s)}, \varphi_{(s)}(x_0, \dots, (d^{s-1}x/t^{s-1})_0; t-r; t_0; G); t; G) \qquad (5)$$

This is a functional equation for the unknown $\varphi_{(s)}$ and therefore indirectly for the function $\xi_{(s)}$ defining an s-th order reduction of Eq.(1). Any given hereditary equation will have in general an infinity of reductions for each value of s. Some of them are remarkable in the sense that they are good asymptotic approximations of the hereditary equation, i.e., they are Dominant Reductions.

To define what we mean here by a Dominant Reduction of a hereditary equation, let us choose a protocol of agreement P between two functions of t as t tends to ∞ (we shall use the notation $f_1 \sim f_2$ (P)). For instance we can say that P_1) f_1 and f_2 are asymptotically equivalent if:

$$\lim_{t\to\infty} (f_1 - f_2) = 0 , \qquad (6)$$

or P_2) if:

$$\lim_{t\to\infty} (f_1 - f_2)/(|f_1| + |f_2|) \quad \text{whenever} \quad f_1 \cdot f_2 > 0 . \qquad (7)$$

Let us now consider one of the reductions $\xi_{(s)}$ of Eq.(1). We say that this reduction is a dominant reduction if:

$$\xi_{(s)}(\Psi, \dots, d^{s-1}\Psi/dt^{s-1}; t; G) \sim d^s\Psi/dt^s \ (P) \qquad (8)$$

Ψ beeing any solution constructed using the method of steps[19-21] and generic initial conditions (arbitrary functions of t in an interval of length r). To clarify what this means let us consider the linear equation:

$$dx/dt = K\,x + G\,\hat{x} \quad , \quad (K \text{ and } G \text{ constants}) \qquad (9)$$

Since the general solution of this equation is a linear superposition of solutions of the form exp(zt), z beeing any of the roots of the characteristic equation:

$$z = K + G \exp(-z \cdot r) , \qquad (10)$$

we can say for example that the reduction:

$$dx/dt = a\,x \qquad (11)$$

where a is the root (supposed to be real) with the largest real part, is a dominant reduction according to the protocol P_2 and also with respect to P_1 if a is the only root with positive real part. This example makes intuitive the concept of dominant reduction of a retarded equation and shows why it is necessary to refer to an explicit protocol. By the way, it is obvious that the concepts of reduction and dominant reduction make sense also for ordinary differential equations.

The first remarkable reduction is the so-called predictive equation associated with Eq.(1). As we have already mentioned this reduction was first considered in connection with the equations of motion in classical field theories and more generally by one of us[13-15]. Let us consider a first order differential equation like:

$$dx/dt = \xi(x;t;G) . \tag{12}$$

By definition this equation is the predictive equation associated with Eq.(1) if it is a reduction of the latter and the function ξ is a function of G smooth enough in the neighborhood of G=0. In general of course it will be difficult or impossible to find a closed form for Eq.(12), but it is easy to set up a perturbative algorithm. The main steps are the following: i) write Eq.(5) as (we drop the sub-index s=1):

$$\xi(x;t;G) = W(x, \varphi(x;t-r;t;G);t;G) , \tag{13}$$

ii) write the identity:

$$\varphi(x;t-r,t) = x + \int_t^{t-r} du\ \xi(\varphi(x;u,t;G);u;G) \tag{14}$$

and iii) consider the Taylor developpements:

$$W(x,\hat{x};t;G) = W^{(0)}(x;t) + G\ W^{(1)}(x,\hat{x};t)+... \tag{15}$$

$$\xi(x;t;G) = \xi^{(0)}(x;t) + G\ \xi^{(1)}(x;t)+... \tag{16}$$

$$\varphi(x;u,t;G) = \varphi^{(0)}(x;u,t) + G\ \varphi^{(1)}(x;u,t)+... \tag{17}$$

where $\xi^{(0)} = W^{(0)}$ and where $\varphi^{(0)}$ is the general solution of the zero order predictive equation (2). One can easily see that Eqs.(13) and (14) provide a recurrent algorithm to construct the series (16) and (17).

It has been proved, first up to second order of perturbation theory[13-15] and later to any order[22] that for G small enough the predictive equation is a dominant reduction of Eq.(1).

DOMINANT REDUCTIONS OF HIGHER ORDER

We consider here a second perturbative algorithm which does not rely on the smallness of a parameter but that instead is useful only in a neighborhood of a stationnary solution, if any, of Eq.(1). The stationnary solutions x=y are the solutions corresponding to the solutions y of the functional equation (we assume from now on that W does not depend explicitly on t):

$$W(y,y;G) = 0 . \tag{18}$$

Let us assume that y is a solution. Changing the origin of the variable x we can assume that y=0. Let:

$$W = \sum_{a+b=1}^{\infty} W_{ab}\ x^a\ \hat{x}^b \tag{19}$$

be the Taylor expansion of the function W with respect to x and $\hat{x}$. Let z_i (i=1,2,...,∞) be the roots of the characteristic equation of the linear

approximation of the r-h-t of Eq.(19), i.e., the roots of Eq.(10) with $K=W_{10}$ and $G=W_{01}$. Assuming that there are no multiple roots we know that: i) there exists one root which has the largest real part (or a pair if the root is complex). We shall note z_1 this root (or z_1 and z_2 if they are complex conjugate). ii) different roots have different real parts (except pairs of complex conjugates); if z_i are ordered according to decreasing values of the real part then this real part tends to $-\infty$ when i tends to ∞. This order corresponds also essentially to increasing values of the positive imaginary part which increases without limit.

Let us consider a countable and ordered set of variables y_i. We shall define a formal function $F(y_i)$ as the limit:

$$F(y_i) = \lim_{s\to\infty} \sum_{n_1+\ldots+n_s=1}^{\infty} c_{n_1\ldots n_s}\, y_1^{n_1} \ldots y_s^{n_s} \tag{20}$$

where n_i are positive integers and the c's are constants and in particular:

$$c_{n_1\ldots n_s} = 1 \text{ for } N = n_1 + \ldots + n_s = 1 . \tag{21}$$

We say that F is the generating function of the general solution of Eq.(1) if:

$$x = F(k_i \exp(z_i t)) \tag{22}$$

satisfies, order by order, Eq.(1) for any choice of the constants k_i.

For any value of s and for any values of n_μ $(\mu=1,..,s)$ the corresponding coefficients of Eq.(20) are given by:

$$c_{n_1\ldots n_s} = N_{n_1\ldots n_s}/D_{n_1\ldots n_s} \tag{23}$$

where:

$$D_{n_1\ldots n_s} = \sum n_\mu z_\mu - W_{10} - W_{01}\exp(-\sum n_\mu z_\mu r) \tag{24}$$

and where the N's can be obtained by the recurrent algorithm:

$$N_{n_1\ldots n_s} = \sum_{a+b=2}^{n_1+\ldots+n_s} W_{ab} \sum_{p_\mu+q_\mu=n_\mu} c_{a,p_1\ldots p_s}\, \hat{c}_{b,q_1\ldots q_s} \tag{25}$$

where:

$$c_{a,n_1\ldots n_s} = \sum_{p_\mu+q_\mu=n_\mu} c_{a-1,p_1\ldots p_s}\, c_{q_1\ldots q_s} ,$$

$$c_{1,n_1\ldots n_s} = c_{n_1\ldots n_s} \tag{26}$$

$$\hat{c}_{1,n_1\ldots n_s} = c_{n_1\ldots n_s} \exp(-\sum n_\mu z_\mu r) \tag{27}$$

$$\hat{c}_{a,n_1\ldots n_s} = \sum_{p_\mu+q_\mu=n_\mu} \hat{c}_{a-1,p_1\ldots p_s}\, \hat{c}_{q_1\ldots q_s} . \tag{28}$$

We justify this definition on the grounds that for linear equations, like Eq.(9), any solution constructed by the method of steps out of some initial function defined, say, in the interval [-r,0] can be represented as a convergent series of the type:

$$x = \sum c_i \exp(z_i t) \tag{29}$$

which corresponds to the generating function $F = \sum y_i$.

Using the notation:

$$F_{[s]} = \sum_{n_1+\ldots+n_s=1}^{\infty} c_{n_1 \ldots n_s} \, y_1^{n_1} \ldots y_s^{n_s} . \tag{30}$$

it is obvious that:

$$\varphi_{[s]} = F_{[s]}(k_\mu \exp(z_\mu t)) \tag{31}$$

are particular solutions of Eq.(1). Therefore, for each value of s, the differential equation:

$$d^s x/dt^s = \xi_{[s]}(x,\ldots,d^{s-1}x/dt^{s-1}) \tag{32}$$

having the function (31) as general solution is a reduction of Eq.(1). If:

$$\xi_{[s]} = \sum_{n_1+\ldots+n_s=1}^{\infty} a_{n_1\ldots n_s} \, x^{n_1} \ldots (d^{s-1}x/dt^{s-1})^{n_s} \tag{33}$$

is the Taylor developpement of $\xi_{[s]}$ the coefficients a can be calculated recurrently using:

$$\sum_{p_1+\ldots+p_s=N} E_{n_1\ldots n_s,p_1\ldots p_s} \, a_{p_1\ldots p_s} = c_{s,1,n_1\ldots n_s} - \sum_{1 \leq q_1+\ldots+q_s \leq N-1} E_{n_1\ldots n_s,q_1\ldots q_s} \, a_{q_1\ldots q_s} \tag{34}$$

where

$$c_{r,1,n_1\ldots n_s} = \left(\sum n_\mu z_\mu\right)^{r-1} c_{1,n_1\ldots n_s} \tag{35}$$

$$c_{r,a,n_1\ldots n_s} = \sum_{p_\mu+q_\mu=n_\mu} c_{r,a-1,p_1\ldots p_s} \, c_{r,1,q_1\ldots q_s} \tag{36}$$

$$E_{n_1\ldots n_s,p_1\ldots p_s} = \sum_{q_{1\mu}+\ldots+q_{s\mu}=n_\mu} c_{1,p_1,q_{11}\ldots q_{1s}} \ldots c_{s,p_s,q_{s1}\ldots q_{ss}} \tag{37}$$

We expect any of the series that we have introduced to converge in a neighborhood of a stationnary solution, and we expect also in particular that the solutions constructed by the method of steps will be representable as $F(k_i \exp(z_i t))$.

If all roots have negative real parts then the solution goes to zero at infinity. This is a plausible conclusion drawn from the fact that each term in the general solution goes to zero exponentially but it is a well known rigorous result also.

A similar plausibility argument leads to a more general conclusion. Let us assume that $S \geq 1$ roots have positive real parts. Then the usefulness of the expansion of the general solution we have considered is doubtful when $t \to \infty$. Let

$$d^s x/dt^s = \zeta_{[s]}(x,\ldots,d^{s-1}x/dt^{s-1},t,k_{s+1},k_{s+2},\ldots) \tag{38}$$

with s=S be the differential equation that one obtains by eliminating the constants of integration $k_1,\ldots,k_s$. It can be easily seen that the

quantities $\zeta_{[s]}$ are free of asymptotic divergences and tend to $\xi_{[s]}$ when $t \to \infty$. The differential equation (32) is therefore the dominant reduction according to the simplest protocol P_1.

Let us see now how the values of W_{10} and W_{01} affect the set of roots (spectrum) of the characteristic equation. To make this examination to depend on one parameter only we notice that making the substitutions

$$z' = r \cdot (z - W_{10}) \quad , \quad G' = r\, W_{01} \exp(-W_{10}\, r) \tag{39}$$

the characteristic equation becomes

$$z' \exp(z') = G', \tag{40}$$

where G' can be considered as an effective tuning constant. The actual dependence of G' on G and the particular stationnary solution depends on the retarded equation that one considers. We have to keep in mind also that a given stationnary solution may exist only for some interval of G. In what follows we assume that $|G'|$ increases monotonously with $|G|$.For G' positive there exists a single real root and this is the dominant one ,i.e., the root with the largest real part. If G' is negative and greater than $-1/e$ then there are two real roots and the greatest is the dominant root of the spectrum. In both cases the number of complex roots with positive real parts increases with the absolute value of G'. This general pattern makes plausible the following scenario to describe the evolution of the main characteristics of the solutions of Eq.(1) in a neighborhood of a stationnary solution when the effective tuning constant G' increases (or decreases) steadily starting from 0:

i) if the absolute value of G' is small enough we expect that the reduction (32) for s=1 will be a dominant reduction of Eq.(1). This in fact follows also more generally from the known result, mentioned before, which says that for $|G'|$ small enough any retarded equation spontaneously approximates its predictive associate equation.

ii) if G' increases (decreases) steadily beyond a critical value the value S will increase from 3 (2) to ∞ (two units by step) and therefore we expect to observe a cascade of dominant reductions.

What about turbulent solutions?. In our opinion systems governed by an equation of type (1) can develop three types of turbulence depending on either one of this situations: a) they have a dominant reduction with a low order (> 2) and this ordinary equation develops turbulence in the usual sense. b) The lowest order dominant reduction they have corresponds to a high order. Here turbulence would be synonymous of high but, so to speak, ordinary complication. And c) any of the main technical assumptions that we have made does not hold and the system can not be approximated by any ordinary equation of any order.

NUMERICAL RESULTS

We have considered, up to the second step of the cascade, several retarded equations obtaining similar results. These exhibit satisfactory numerical agreement with the theoretically expected reductions.

In particular we have considered the equation

$$dx/dt = -x + G^2(1 + 2\, B \cos \hat{x}) \ , \tag{41}$$

which according to Refs. 16-18 is an exact equation to describe an hybrid optical device with a time delay and an approximate one to describe a particular case of non-linear optical cavity, for B=0.3, r=3.5, and different values of G.

For $0 < G \lesssim 0.24$ the root of the characteristic equation with the largest real part is a negative real one and in consequence we expect that there will exist a first order dominant reduction (in fact, the predictive associated equation)

$$dx/dt = \xi_{[1]}(x;G) = \sum_{n=1}^{\infty} \xi_{[1],n}\, x^n \tag{42}$$

that describes the asymptotic behaviour of an arbitrary solution. After computing the coefficients $\xi_{[1],n}$, the agreement between the dx/dt given by the approximated predictive equation

$$dx/dt = \xi^{[N]}_{[1]}(x;G) = \sum_{n=1}^{N} \xi_{[1],n}\, x^n \tag{43}$$

and that obtained by numerical integration of the retarded equation can be tested. We have done this test for different values of G and N and for several initial conditions obtaining a very good agreement between these quantities, except for the very few initial steps of the integration.

For example, when G=0.1 the stationary solution is $x_0=1.599923\ 10^{-2}$. The dominant root of the characteristic equation is $z_1 = \xi_{[1],1}=-1.003215$. Other coefficients in the expansion in Eq.(43) are $\xi_{[1],2}=-3.755322$ and $\xi_{[1],3}=1.087809.10^3$ (for the initial conditions we have considered, it was sufficient to take N=3 but the coefficients of higher order can be easily computed). For instance, we have integrated numerically the solution of the translated equation

$$dx/dt = -(x+x_0)+G^2(1+2B\cos(\hat{x}+x_0)) \tag{44}$$

satisfaying the initial conditions $\Psi(t)=1$, $-r \leq t \leq 0$, and we have seen that for $t \geq 2r$ the test quantity

$$(\xi^{[3]}_{[1]}(\Psi(t);G) - \dot{\Psi}(t))/(\left|\xi^{[3]}_{[1]}(\Psi(t);G)\right| + |\dot{\Psi}(t)|) \tag{45}$$

is less than 10^{-3} and tends to zero (up to the precision of the numerical calculations) very quickly, being less than 10^{-7} for $t > 3r$.

For $G \gtrsim 0.24$ the dominant root z_1 of the characteristic equation becomes complex and thus according to the theoretical framework above mentioned we expect that the retarded equation in consideration has, for a certain interval of values of G, a second order dominant reduction of the type

$$d^2x/dt^2 = \xi^{[N]}_{[2]}(x,\dot{x};G) = -z_1\bar{z}_1x+(z_1+\bar{z}_1)\dot{x}-\sum_{i+j=2}^{N}\eta_{ij}(\bar{z}_1x-\dot{x})^i(z_1x-\dot{x})^j \tag{46}$$

The coefficients $\eta_{ij}=\bar{\eta}_{ji}$, $i+j\geq 2$, can be recurrently computed for each value of G. We have integrated numerically the retarded equation for several values of G and N, and different initial conditions. The results exhibit good agreement between the actual acceleration obtained by integration of the retarded equation and that given by Eq.(42). In particular, for G=1.5 the stationary solution is $x_0=1.862178$ and the dominant root $z_1=0.01151539+0.7207192\ i$. The coefficients η_{ij} are listed in Table 1. If $\Psi(t)$ is the solution of the retarded equation which satisfies

TABLE 1. Values of the coefficients $\eta_{ij} = \bar{\eta}_{ji}$ in the expansion of the second order reduction (46) of the retarded equation (44) for G=1.5, B=0.3, r=3.5 and x_0=1.862178.

i,j	$Re(\eta_{ij})$	$Im(\eta_{ij})$
2,0	$-4.526547\ 10^{-2}$	$-6.785052\ 10^{-3}$
1,1	$-4.037474\ 10^{-2}$	0
3,0	$-1.597720\ 10^{-1}$	$7.030389\ 10^{-2}$
2,1	$-6.162613\ 10^{-2}$	$-4.886361\ 10^{-3}$
4,0	$2.560899\ 10^{-2}$	$3.876095\ 10^{-2}$
3,1	$-3.323412\ 10^{-2}$	$-6.458984\ 10^{-3}$
2,2	$-3.093816\ 10^{-2}$	0
5,0	$5.216666\ 10^{-2}$	$8.122322\ 10^{-3}$
4,1	$-1.115723\ 10^{-1}$	$5.545775\ 10^{-2}$
3,2	$-3.129171\ 10^{-2}$	$2.468189\ 10^{-2}$
6,0	$4.138975\ 10^{-3}$	$-2.506283\ 10^{-2}$
5,1	$4.451497\ 10^{-2}$	$7.131775\ 10^{-2}$
4,2	$-2.915466\ 10^{-2}$	$-5.809669\ 10^{-3}$
3,3	$-5.039039\ 10^{-2}$	0
7,0	$-1.202039\ 10^{-2}$	$-1.212505\ 10^{-2}$
6,1	$1.026617\ 10^{-1}$	$1.636661\ 10^{-2}$
5,2	$-1.324542\ 10^{-1}$	$6.157138\ 10^{-2}$
4,3	$-3.697791\ 10^{-2}$	$4.740727\ 10^{-2}$

TABLE 2. Some values of $q(t,N) = 10^{-3} \cdot Q(t,N)$ for the solution $\Psi(t)$ of the retarded equation (44) with G=1.5, B=0.3, x_0=1.862178 and $\Psi(t)$=1 for $-r \leq t \leq 0$.

t	$\ddot{\Psi}(t)$	q(t,1)	q(t,3)	q(t,5)	q(t,7)
0.0	1.9098	-1000.00	-228.99	275.94	733.78
0.7	0.9484	-1000.00	-362.27	-221.48	-144.77
1.4	0.4710	-368.80	-278.21	-232.48	-208.57
2.1	0.2329	193.12	108.47	74.11	57.93
2.8	0.1161	558.28	489.63	454.30	434.44
3.5	0.7687	-270.47	-346.94	-379.89	-395.40
4.2	0.6114	-270.19	-137.05	-96.26	-80.47
4.9	-0.1090	1000.00	1000.00	953.39	714.84
5.6	-0.2857	308.92	171.30	153.10	146.32
6.3	-0.2485	-140.48	-82.59	-69.84	-61.87
...					
18.2	0.0854	-1000.00	-91.51	-18.98	-5.79
18.9	0.2249	-215.27	-54.44	-13.36	-0.08
19.6	0.2481	60.02	11.63	1.53	-0.62
20.3	0.2857	102.87	34.51	10.94	1.99
21.0	0.3795	-41.30	-29.12	-15.01	-6.98
21.7	0.3119	-99.73	-16.41	-0.42	3.12
22.4	0.0630	138.03	47.82	13.91	0.72
23.1	-0.1426	128.63	7.49	0.08	-2.04
23.8	-0.2456	-50.71	-14.76	-8.14	-4.29
24.5	-0.3371	-50.17	0.99	4.23	3.90

$\psi(t) = 1$ for $-r \leq t \leq 0$, we have seen for all tested t values if $t > 5r$ the quantity

$$Q(t,N) = (\ddot{\varphi}(t,N)-\ddot{\Psi}(t))/(|\ddot{\varphi}(t,N)|+|\ddot{\Psi}(t)|) ,$$

$$\ddot{\varphi}(t,N) = \xi^{[N]}{}_{[2]}(\Psi(t),\dot{\Psi}(t);G) \qquad (47)$$

is less than 10^{-2} for $N = 7$. Some typical values of this test quantity can be seen in Table 2. This table suggests that better values of the test quantity $Q(t,N)$ can be expected if higher orders are computed for Eq.(46).

For larger values of G we have not yet computed the higher order reductions to see the other steps of the cascade that we expect to appear as G increases.

REFERENCES

1. E. Kerner, Can the Position Variable be a Canonical Coordinate in Relativistic Many-Particle Theory ?, J. Math. Phys., 6:1218 (1965).
2. L. Bel, A. Salas and J. M. Sanchez-Ron, Approximate Solutions of Predictive Relativistic Mechanics for the Electromagnetic Interaction, Phys. Rev. D, 7:1099 (1973).
3. L. Bel and J. Martin, Approximate Solutions of Predictive Relativistic Mechanics for the Electromagnetic Interaction. II, Phys. Rev. D, 8:4347 (1973).
4. A. Salas and J. M. Sanchez-Ron, Predictive Solutions of Classical Electrodynamics, Nuovo Cimento, 20 B:209 (1974).
5. D. Hirondel, Free-particle-like formulation of Newtonian instantaneous action-at-a-distance electrodynamics, J. Math. Phys., 15:1689 (1974).
6. L. Bel and J. Martin, Approximate solutions of predictive relativistic mechanics for short-range scalar interactions, Phys. Rev. D, 9:2760 (1974).
7. L. Bel and X. Fustero, Mécanique relativiste prédictive des systèmes de N particules, Ann. Inst. H. Poincaré, XXV:411 (1976).
8. R. Lapiedra and A. Molina, Classical predictive electrodynamics of two charges with radiation: General framework. I, J. Math. Phys., 20:1308 (1979).
9. J. L. Sanz, Two charges in an external electromagnetic field: A generalized covariant Hamiltonian Formulation, Ann. Inst. H. Poincaré, XXXI:115 (1979).
10. M. Portilla, Momentum and angular momentum of two gravitating particles, J. Phys. A, 12:1075 (1979).
11. L. Bel, Th. Damour, N. Deruelle, J. Ibanez and J. Martin, Poincaré-Invariant Gravitational Field and Equations of Motion of two Point like Objects: The Postlinear Approximation of General Relativity, Gen. Gel. Grav., 13:963 (1981).
12. X. Fustero, L. Mas and R. Lapiedra, Predictive mechanics of magnetic monopoles and electric charges, Phys. Rev. D, 12:3474 (1977).
13. L. Bel, Spontaneous Predictivisation, in "Relativistic Action at a Distance: Classical and Quantum Aspects" (Lectures Notes in Physics No. 162), J. Llosa, ed., Springer-Verlag, Berlin (1982).
14. L. Bel, Prédictivisation spontanée des systèmes dynamiques hereditaires, C. R. Acad. Sci., 294:463 (1982).
15. L. Bel, Predictivizacion espontanea de los sistemas dinamicos hereditarios, in "Encuentros Relativistas 82," Universidad del Pais Vasco, Bilbao (1982).
16. K. Ikeda, Multiple-Valued Stationary State and its Instability of the Transmitted Light by a Ring Cavity System, Opt. Commun., 30:257 (1979).

17. K. Ikeda, H. Daido and O. Akimoto, Optical Turbulence: Chaotic Behaviour of Transmitted light from a Ring Cavity, Phys. Rev. Lett., 45:709 (1980).
18. H. M. Gibbs, F. A. Hopf, D. L. Kaplan and R. L. Shoemaker, Observation of Chaos in Optical Bistability, Phys. Rev. Lett., 46:474 (1981).
19. R. Bellman and K. L. Cooke, "Differential-Difference Equations," Academic Press, New York (1963).
20. L. E. El'sgol'ts and S. B. Norkin, "Introduction to the Theory and Application of Differential Equations with Deviating Arguments," Academic Press, New York (1973).
21. R. D. Driver, "Ordinary and Delay Differential Equations," Springer-Verlag, New York (1973).
22. J. M. Aguirregabiria and L. Bel, to be published.

QUANTUM OPTICS

Masers and lasers have provided an important tool for the study of statistical phenomena in physics. Quantum optical experiments have demonstrated properties such as phase transitions, symmetry breaking, bistability and multistability, and chaotic regimes of operation. Much study has been made of the statistical properties of radiation, and has led to notions such as photon antibunching and squeezed states. Production of such states has been suggested as a means to enhance the signal-to-noise ratio in sensitive optical measurements, such as interferometers to detect gravitational waves or ring-laser gyroscopes. The papers in this section are representative of work going on in this active and productive area.

THE ONE-ATOM MASER -

A TEST SYSTEM FOR SIMPLE QUANTUM ELECTRO-DYNAMIC EFFECTS

Herbert Walther

Sektion Physik, Universität München and
Max-Planck-Institut für Quantumoptik,
D-8046 Garching, Fed. Rep. of Germany

INTRODUCTION

The experimental demonstration of the maser has generated a large amount of interest in theoretical models describing the interaction of two-level atoms with a single mode of an electromagnetic field in a cavity.[1,2,3] The first models treated purely academic problems, but, modified versions were stimulated which then led to an understanding of a major part of the experimentally observed phenomena, including also the even larger variety of effects observed after the laser was invented. It is a characteristic feature of maser and laser experiments that large numbers of atoms and photons are present. One reason for this is the small size of the matrix elements describing the atom-radiation interaction. Therefore, when only a small amount of photons are involved in an experiment the atom-field evolution time usually becomes much longer than other characteristic times of the system, such as the atomic relaxation, the atom-field interaction time, and the cavity mode damping time. The fundamental theories of radiation-matter interaction involving single electromagnetic modes and small photon occupation numbers therefore could not be tested experimentally so far. They predict, however, some interesting and basic effects; these include:

(1) Modification of the spontaneous emission rate of a single atom in a resonant cavity
(2) Oscillatory energy exchange between a single atom and the cavity mode
(3) Disappearance and quantum revival of optical nutation induced on a single atom by a resonant field.

The situation concerning the experimental testing of these basic effects has changed drastically in the last few years since frequency tunable lasers allow study of highly excited Rydberg states of atoms. They are very suitable for observing these effects for three reasons. First, these states are very strongly coupled to the radiation field. Second, the transitions to neighboring levels are in the millimeter wave region, which allows one to build cavities with low-order modes that are sufficiently large to ensure rather long interaction times. Finally, the Rydberg atoms have relatively long spontaneous emission lifetimes, which means that in spite of their very strong coupling to microwave radiation, they can live for a very long time in excited states allowing precise spectroscopic investigations.

In the following recent experiments on the problem of single atom - single mode interaction will be discussed. In addition, a brief review of the properties of Rydberg atoms will be given; for details see, for example, Refs. 4 and 22)

PROPERTIES OF RYDBERG ATOMS

When a valence electron of an atom is excited into an orbit with high principal quantum number and therefore far from the ionic core, the energy levels of the atoms can simply be described by the Rydberg formula (Table 1). This is the reason why atoms in these highly excited states are often called Rydberg atoms.

The energy of the levels depends on the phenomenological quantum defect δ_ℓ of the states of angular momentum ℓ. For states of low ℓ, where the orbits of the classical Bohr-Sommerfeld theory are ellipses of high eccentricity, the penetration and polarization of the electron core by the valence electron lead to large quantum defects and strong departures from the hydrogenic behaviour. As ℓ increases, the orbits become more circular and the atom becomes more hydrogenic: δ_ℓ changes with ℓ^{-5}. In Table 1 the scaling laws for further properties of Rydberg atoms are compiled: The radius of the charge distribution of the valence electron scales as n^{*2}, and for $n^* = 50$ the linear dimension of the atom is already comparable with the wavelength of light in the visible region and competes with the size of larger biomolecules.

Table 1. Scaling laws for properties of Rydberg atoms

Energy:	$E_{n\ell} = R/(n-\delta_\ell)^2 = R/n^{*2}$ n^* effective quantum number, δ_ℓ quantum defect
Radius:	$\langle r\rangle \sim n^{*2}$
Lifetimes:	$\tau \sim n^{*3}$ (low angular momentum states) $\tau \sim n^{*5}$ (high angular momentum states)
Fine-structure interval:	$\Delta E \sim n^{*-3}$

The electric polarizability for the quadratic Stark effect increases as n^{*7}, and the diamagnetic interaction as n^{*4}. This allows one to perform experiments at field strengths high enough to make the interaction energy in the external electric or magnetic field comparable with or larger than the Coulomb energy of the atom. For practical reasons the corresponding field strengths for ground state atoms cannot be reached in the laboratory. The study of highly excited atoms in external electric and magnetic fields is therefore interesting in itself (For reviews see e.g. References 5 and 6). The sensitivity of Rydberg atoms to external electric fields also means that the atoms readily ionize in rather weak fields. This opens the possibility of very effective detection, as will be discussed later.

The large Rydberg atom orbitals are characterized by natural lifetimes much longer than those of less excited atoms. In the case of hydrogen Rydberg states, the dependence of the lifetime on n can be obtained by fully quantum mechanical radiation rate calculations involving hydrogenic coulombic wave-functions. For Rydberg states of other species the lifetimes (and the other radiative parameters) do not scale exactly as a power of n but rather as a power of n*. The n* scaling law can be determined by using calculations of the Bates and Damgaard type.[7] The lifetimes scale either as n^{*3} (when ℓ is small) or as n^{*5} (when $\ell \cong n$).

In the following we will give a simple explanation of this scaling law. (In this discussion we do not discriminate between n and n*.) The rate of spontaneous emission of radiation for a transition from a state n to n' is given by the Einstein A coefficient:

$$A_{n\to n'} = 16\pi^3\upsilon^3\, e^2 \langle r_{nn'}\rangle^2 \;/\; 3\varepsilon_o hc^3,$$

where υ is the transition frequency and $\langle r_{nn'}\rangle$ the matrix element of the electric dipole operator between the initial n and the final state n'. For the case n' << n one has a small $\langle r_{nn'}\rangle$ owing to the small radial overlap of the wave functions for n and n' and, as will be shown below, $A_{n\to n'} \sim n^{-3}$. If n' is close to n, the energy difference $E_n - E_{n'} \sim n^{-3}$ and $\langle r_{nn'}\rangle^2 \sim n^4$, and so $A_{n\to n'}$ becomes proportional to n^{-5}. The magnitude of the Einstein coefficient $A_{n\to n'}$ still depends on the angular momentum ℓ. This can be understood by simple classical arguments: For low angular momentum states (core penetration) the lifetime τ can be deduced from the third Kepler law. Accordingly, the electron orbiting period T is given by $T \cong (n^2 a_o)^{3/2} \cong n^3$ (in the classical picture T must be proportional to τ since transition to a lower orbit is always more probable when the electron approaches the core and undergoes maximum acceleration). For the case of high angular momentum orbitals, in the classical picure, the electron radiates continuously and lowers its radius. The acceleration of the electron is inversely proportional to the square of the radius of the orbit, and so the power of the emitted radiation scales as n^{-8}. The distance between neighbouring Rydberg levels changes as n^{-3}, giving a characteristic time requirement of $n^{-3}/n^{-8} = n^5$ for each step, which corresponds to the lifetime for states with large ℓ. The square of the matrix element $\langle r_{nn'}\rangle$ ($n \cong n'$) scales as n^4, showing a rather high transition probability for induced transitions. Rydberg atoms therefore strongly absorb microwave or far-infrared radiation. As a consequence, black-body radiation may cause strong mixing of the states. This is especially the case for states with high angular momenta since the spontaneous lifetimes for these are much longer and the induced transitions can therefore be saturated much more easily than for the lower ℓ states.

We now wish to discuss also the scaling laws relating to black-body-induced effects. The induced transition rate due to black-body radiation is proportional to $\langle r_{nn'}\rangle^2 S_\upsilon$, where S_υ is the energy flux of the black-body radiation per unit band width and unit surface area. At low frequencies (Rayleigh-Jeans limit) S_υ increases as υ^2. As the distance between the Rydberg states scales as n^{-3} (here again we conduct the discussion with n instead of n*), it is therefore found that S_υ is proportional to n^{-6}. Since $\langle r_{nn'}\rangle^2 \sim n^4$, it follows that the induced transition rate behaves as n^{-2}. Important in experiments is the ratio between the induced transition rate and the spontaneous rate, which changes as n^{-3} for low ℓ and as n^{-5} for high ℓ. This means that for a given atom and a given temperature there exists an n above which the blackbody-induced rate overcomes the spontaneous rate.

The sensitivity of Rydberg atoms to blackbody radiation can also be explained in the following terms. The blackbody radiation energy density can be expressed in terms of the number of photons per mode $\bar{n}$. For the Rayleigh-Jeans limit this gives $\bar{n} = kT/h\upsilon$. At 300 K it follows that $kT/h \cong 6 \times 10^{12}$ Hz; this means that for frequencies larger than kT/h, where $\bar{n} << 1$, no significant blackbody influence can be observed. However, for a Rydberg state with a transition frequency to a neighbouring state at 10^{11} Hz, where $\bar{n} = 60$, the blackbody-induced transition rates can be orders of magnitude larger than the spontaneous rates.

The effect of blackbody radiation on Rydberg atoms is mainly to induce transistions to nearby states. As a result the population will evolve as a function of time after pulsed laser excitation. Changes in the populations typically appear on a µs time scale.

In the first experimental verification of blackbody radiation effects Gallagher and Cooke[8] measured the lifetime of high-lying p-states of Na and found values three times smaller than theoretically predicted. When, however, room-temperature blackbody-induced population transfer to nearby states is included, good agreement with the experimental data was achieved.

Haroche et al.[9] observed that the decay of the 25s sodium state occurs with an important population transfer to nearby states, such as the 25p level, which cannot be accounted for by ordinary spontaneous emission (since the 25p level lies above the 25s one). This transfer of population can be quite consistently explained as a 20% blackbody radiation-induced mixing between the 25s and 25p states.

In a slightly different approach Beiting et al.[10] used field ionization of Rydberg atoms with a linearly rising electric field pulse. The various different Rydberg states ionize at different electric field strengths. With such a setup, the population distribution among the Rydberg states can be measured in a single shot. Beiting et al. measured this population distribution for various delays between the pulsed laser excitation and the ionizing electric field pulse. It was observed that the population initially prepared in a Rydberg state is transferred to higher-lying states, the more the longer the delay. Using a similar setup, Rempe[11] measured the population transfer from the initially laser-excited 23 $^1F_3$ state of strontium to nearby d and g states (Fig. 1). In these experiments the temperature of the environment was not varied.

The first direct observation of the temperature dependence of the population transfer was performed by Koch et al.[12] by heating the environment, and in an experiment by Figger et al.[13] where sodium atoms of a thermal beam in a cooled environment were excited to the 22d state using two continuous wave dye lasers. In the latter case the interaction region was inside a copper box cooled to 14 K (Fig. 2). The sodium Rydberg atoms could be exposed to the radiation of a blackbody source by opening a flap at the side of the box. After interacting with the blackbody radiation for about 50 µs the atoms entered a dc electric field which acted like an optical edge filter: all atoms in Rydberg states higher than the initially laser-excited 22d state are ionized and detected. The signal rises linearly with temperature, in good agreement with the Rayleigh-Jeans limit of Planck's radiation formula. The effect of the blackbody source mounted outside the box is reduced since the atoms only see the source at a small solid angle.

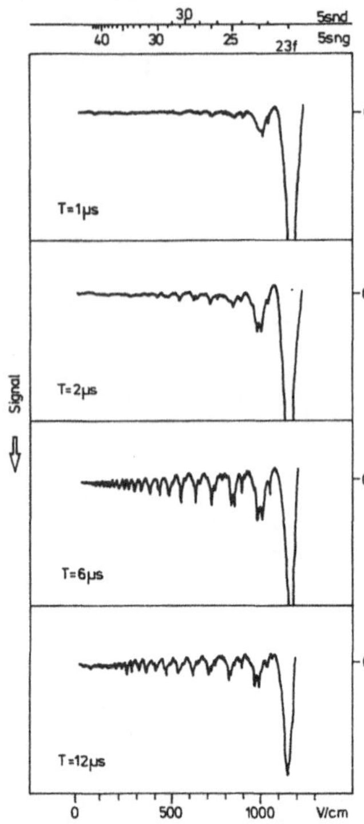

Fig. 1. Blackbody-induced transitions between Rydberg states. The 23 $^1F_3$ level of strontium is excited by the laser radiation. The detection of the Rydberg states is performed by an electric field increasing linearly in time. The field ramp starts 1, 2, 6, and 12 μsec after the pulsed laser excitation. The field ionization signal at smaller field strengths results from Rydberg levels populated by blackbody radiation of the apparatus at 300 K.

Aside from population transfer, blackbody radiation also affects Rydberg atoms in a more subtle way. The spectral energy density distribution of the blackbody radiation at 300 K has its maximum at about 2×10^{13} Hz. A typical electric dipole transition starting from the ground state of an atom has 10^{14} - 10^{15} Hz and a transition between two Rydberg states has about 10^{11} Hz. It is thus apparent that for a ground-state atom the blackbody radiation appears as a slowly varying field, whereas for a Rydberg atom it appears to be rapidly varying leading to an dynamic Stark shift of the Rydberg levels. The blackbody-induced ac Stark shift ΔW_n for a Rydberg atom in the state n can be expressed as [14]

$$\Delta W_n = \frac{e^2}{2h} \sum_{n'} \int_0^\infty \frac{\langle r_{nn'} \rangle^2 E^2_{\omega_b} \omega_{nn'}}{(\omega_{nn'} - \omega_b)(\omega_{nn'} + \omega_b)} d\omega_b ,$$

where E^2_ω is the squared electric field of the blackbody radiation in a band width $d\omega_b$ at a frequency ω_b. Accurate evaluation of the shift was

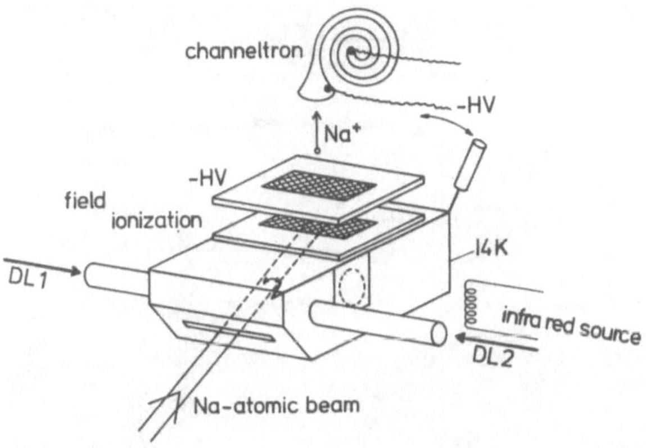

Fig. 2. Experimental setup for demonstrating the interaction of blackbody radiation with Rydberg atoms. The flap at the right side of the cooled box (14 K) allows the thermal radiation of the infrared source to enter the box. The 22d state was excited by cw laser radiation. The Rydberg atoms in the 22p level were detected by field ionization.

performed by Farley and Wing.[15] All Rydberg states experience roughly the same energy shift of about 2.4 kHz at 300 K.

Since the shift of all Rydberg states by blackbody radiation is the same, it can be detected only as a change of the optical transition frequency which connects to the ground state. Consequently, the line width and the stability of the dye laser used for excitation has to be 10^{-12} and the spectral resolution in the experiment has to be correspondingly high. This challenge was accepted by Hollberg and Hall.[16] They used Doppler-free two-photon absorption to excite the 5s-36s transition in rubidium atoms. With the Ramsey method of separate fields, the spectral width of the signal was decreased to 40 kHz. The line centre could be determined with an accuracy of 150 Hz. The atoms were exposed to the radiation of a blackbody source, and a chopper periodically blocked this radiation. Thus, the experiment was insensitive to long-term drifts. When the temperature of the blackbody source was raised to about 500 K, a shift in the line position of 1.4 KHz was observed. This is ten times larger than the uncertainty. A first study of the temperature dependence also shows agreement with the predicted T^2 dependence.

Gallagher et al.[17] measured the blackbody radiation-induced level shift in two-electron atoms where a state of a doubly excited configuration has an energy so close to a singly excited Rydberg state that the transition frequency is in the microwave region. Since the doubly-excited state is not shifted by the blackbody radiation, the shift of the Rydberg state is measurable at the comparatively low microwave frequency. As pointed out by Gallagher et al.[17], the measurement of blackbody-induced level shifts may in turn also be used to determine the absolute temperature of the environment.

SINGLE ATOM IN RESONANT CAVITY - MODIFICATION OF SPONTANEOUS EMISSION RATES

The energy levels of the combined two-level atom and field system can be described in the dressed atom picture (Haroche[18]; Haroche et al.[19]). The lowest energy of the system is represented by $|g,0\rangle$ describing the atom in its ground state $|g\rangle$ with no photon in the cavity. The higher energy levels are separated by the energy of a photon. The states $\pm n\rangle$ are a superposition of the states $|e,n\rangle$ (e stands for excited atomic state and n for the photon number) and $|g,n+1\rangle$ of the system without interaction between the cavity field and the atom:

$$|\pm n\rangle = [\ |e,n\rangle \pm |g,n+1\rangle]/\sqrt{2}.$$

The energy separation between the levels $|+n\rangle$ and $|-n\rangle$ is $2h\ \Omega\sqrt{n+1}$, where Ω is the coupling strength between the field and the atoms. There is a small change proportional to $\sqrt{n}$ when the field strength is increased.

In a realistic description of the interaction the dissipative processes also have to be considered. Since Rydberg atoms have lifetimes longer than the atom-field interaction time, their relaxation can generally be neglected. However, the relaxation of the cavity field is important: the harmonic oscillator representing the field is coupled to a thermal reservoir at temperature T representing, for example, the cavity walls. The scheme shown in Fig. 3 gives the corresponding "coupling constants". The thermal equilibrium of the field mode is obtained in the characteristic time Q/ω, where Q is the quality factor of the cavity and ω the frequency of the resonant mode (For a detailed discussion see Reference 22).

The behaviour of an atom entering an empty cavity (i.e. at T = 0 K) in the excited state $|e\rangle$ depends on the relative size of Ω and ω/Q. If $\Omega > \omega/Q$ (small damping of the cavity), the probability of finding the atom

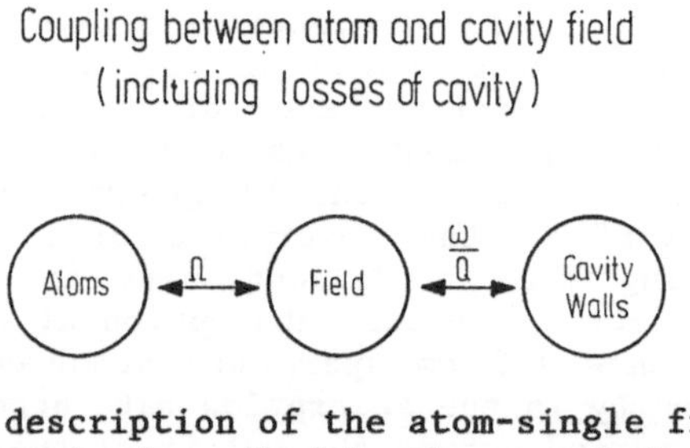

Fig. 3. Schematic description of the atom-single field mode system. The coupling is described by the one photon Rabi frequency Ω and the characteristic damping time Q/ω.

in the state $|e\rangle$ undergoes a damped oscillation. This regime can be considered as a self-induced Rabi nutation in the field of the single photon emitted and reabsorbed by the atom. If $\Omega < \omega/Q$, the probability decreases exponentially at a rate $\Gamma_{cavity} = 4\ \Omega^2 Q/\omega$. There is a cavity-enhanced decay rate which is related to the spontaneous rate in free space Γ_{spont} in the following way:

$$\Gamma_{cavity} = (3/4\pi^2) \cdot Q\lambda^3\ \Gamma_{spont}\ /\ V,$$

where V is the volume of the cavity and λ the wavelength of the radiation.

This relation was predicted long ago by Purcell[20]. Physically, the cavity enhances the strength of the vacuum fluctuations at the resonance frequency; as a consequence the transition rate is increased. ($\Gamma_{cavity}/\Gamma_{spont}$ is obtained when the number of oscillator modes per unit frequency interval in a resonant cavity is divided by the corresponding value in free space.)

The opposite effect, the decrease of the decay rate, is obtained when the cavity is detuned. If the transition frequency of the atom lies below the fundamental frequency of the cavity, spontaneous emission is significantly inhibited. In an ideal case no mode is available for the photon and therefore spontaneous emission cannot occur[21].

To change the decay rate of an atom, in principle no resonator has to be present; any conducting surface near the radiator affects the mode density and, therefore, the radiation rate. Parallel-conducting planes can somewhat alter the emission rate but can only reduce the rate by a factor of 2 because of the existence of TEM modes, which are independent of the separation. The effect of conducting surfaces on the radiation rate has been studied theoretically in a number of investigations (for details see Ref. 22)

To demonstrate experimentally the modification of the spontaneous decay rate, it is not necessary to go to single-atom densities in both cases. The experiments where the spontaneous emission is inhibited can also be performed with higher densities. However, in the opposite case, when the increase of the spontaneous rate is observed, a large number of excited atoms increases the field strength in the cavity and the induced transitions disturb the experiment.

The first experimental work on the inhibited spontaneous emission was done by Drexhage[23]. The fluorescence of a thin dye film near a mirror was investigated. Drexhage observed an alteration in the fluorescence lifetime arising from the interference of the molecular radiation with its surface image. An experiment with Rydberg atoms was recently performed by Vaidyanathan, Spencer and Kleppner[24]. They observed a wavelength-dependent cutoff in the absorption of black-body radiation by Rydberg atoms arising from a discontinuity in the density of modes between parallel-conducting plates. Absorption at a wavelength of 2/3 cm by atoms between planes 1/3 cm apart was measured at a temperature of 180 K. The discontinuity in the absorption rate occurred when the absorption wavelength was varied across the cutoff of the parallel-plate modes. The experiment was performed with Na atoms and the transition employed was 29d → 30p. For the tuning of the atomic resonance across the cutoff frequency a small electric field was applied to the parallel plates.

Inhibited spontaneous emission was observed clearly for the first time by Gabrielse and Dehmelt[25]. In these nice experiments on a single

electron stored in a Penning trap they observed that the cyclotron excitation shows a lifetime which is up to ten times larger than that calculated for a cyclotron orbit in free space. The electrodes of the trap form a cavity which decouples the cyclotron motion from the vacuum radiation field leading to a longer lifetime.

The first observation of enhanced atomic spontaneous emission in a resonant cavity was published by Goy, Raimond, Gross and Haroche [26]. Their experiment was performed with Rydberg atoms of Na excited in the 23s state in a niobium superconducting cavity resonant at 340 GHz. Cavity tuning-dependent shortening of the lifetime was observed taking advantage of the very strong electric dipole of these atoms and of the high Q value of the superconducting resonator. This cooling, necessary for superconducting operation, also had the advantage of totally suppressing the blackbody field effects (n = 0) required to test purely spontaneous emission effects in the cavity.

It was shown that the partial spontaneous emission probability on the 23S → 22P transition in Na is increased from its free space value $\Gamma_{spont} = 150\ s^{-1}$ up to $\Gamma_{cavity} = 8 \times 10^4\ s^{-1}$. This enhanced rate is still 35 times smaller than the damping rate $\omega/Q = 2.8 \times 10^6 s^{-1}$ of the field in the cavity. This means that the photon emitted in the mode is absorbed in the mirrors much faster than the atoms decay.

With a tenfold increase in Q, the values of Γ_{cavity} and ω/Q would be of the same size, so that the emitted photon would be stored in the cavity long enough for the atom to reabsorb it. This would approach the regime of quantum mechanical oscillations between a two-level atom and a single electromagnetic field mode. An experiment in this regime (maser operation) will be discussed in the following section.

SINGLE ATOM IN RESONANT CAVITY - OSCILLATORY REGIME

The Rydberg maser experiment[27] employs an atomic beam to ensure collision free conditions for the highly excited Rydberg states. A diagram of the vacuum chamber with atomic beam arrangement and microwave cavity is shown in Fig. 4. These parts are mounted inside a helium bath cryostat.

For the experiment rubidium atoms were used. The atomic beam oven is carefully shielded from the cavity by copper plates cooled by water, liquid nitrogen and liquid helium. The beam passes through small apertures into the liquid helium cooled part of the apparatus. There the atoms are pumped by the laser radiation to the upper maser level and enter the cavity. Behind the cavity the atoms are monitored by field ionization and a subsequent detection of the ejected electrons in a channeltron electron multiplier.

The Rydberg states were excited using the frequency doubled light of a continuous wave ring dye laser. The second harmonic was generated by means of a temperature stabilized ADA crystal. In a single pass, about 60 µW of ultraviolet radiation at 297 nm were obtained. This laser power was sufficient to excite up to 30 000 atoms/s in the upper maser level $63p_{3/2}$ of the more abundant (70%) ^{85}Rb isotope.

Fine structure splitting between $63p_{3/2}$ and $63p_{1/2}$ of ^{85}Rb amounts to 396 MHz. Therefore, it is no problem to excite a single fine structure level of ^{85}Rb with the narrowband ultraviolet radiation ($\Delta\upsilon \cong 2$ MHz). The

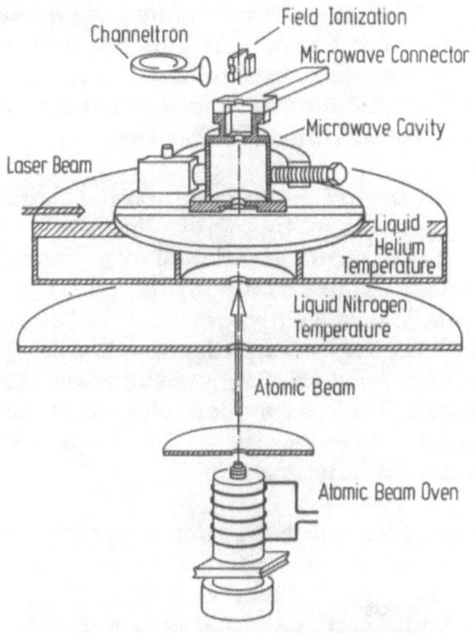

Fig. 4. Experimental setup for the observation of single-atom - single-photon interaction.

atomic beam passes through the cylindrical cavity along its axis, where only the TE_{1np} and TM_{1np} modes possess a nonvanishing transversal electric field. For the experiment the TE_{121} mode was used. This mode has a plane field distribution and is twofold degenerate in an ideal cylindrical cavity. The degeneracy is removed by slightly deforming of the circular cross section into an oval shape, which then determines the direction of polarization of the field mode. The deformation is achieved by squeezing the cylinder both with a screw and also by a piezo-electric (PZT) crystal ($\Delta\ell \cong 4\mu m/1500V$ at 2 K) for fine tuning. This causes one of the degenerate resonances to be shifted towards higher frequencies, whereas that for orthogonal polarization shifts to lower values. The field polarization important for our experiment is coupled to an external waveguide so that the cavity performance could be tested. The upper frequency branch used in our experiment could be mechanically tuned by about 15 MHz; the piezo drive can sweep the frequency by 0.5 MHz/1500 V.

The cylindrical cavity had a diameter of 24.7 mm and a length of 24 mm. It was manufactured from pure niobium rods in two parts: a cylindrical jar and a cover with the waveguide coupling. Constructing the cavity in this fashion has a negligible effect on the cavity Q. Both parts were recrystallized several hours in an ultra-high vacuum oven and then chemically polished ($\cong$ 100 μm). After assembling and electron-beam welding, the cavity was again chemically polished and then annealed for 8 hours in an ultra-high vacuum oven at 2000 °C.

The cavity was enclosed in a mu-metal shield to reduce the influence of ambient magnetic fields on the Q value due to frozen in flux. The properties of high frequency superconductivity were thoroughly investigated in a bath cryostat, where the cavity was cooled by direct contact with liquid helium. For these measurements a second waveguide coupling was attached to the jar and the cavity was evacuated. The surface resistance was obtained from the Q value (1089 Ω/Q), which was determined from the decay time of the stored energy (decrement method).

The temperature of the cavity in the atomic beam arrangement was measured using a germanium resistor. It could be varied from 4.3 to 2.0 K, corresponding to Q factors of 1.7×10^7 and 8×10^8, respectively.

To shield the 300 K thermal radiation of the test equipment a cooled flap with a temperature less than 10 K could be inserted into the waveguide at the upper end of the cavity. Therefore, the thermal background field inside the niobium cavity is essentially determined by the temperature of the walls. The average number of photons of the blackbody radiation per mode is given by $\bar{n} = [\exp(h\nu/kT)-1]^{-1}$ being about $\bar{n} = 3.8$ at 4.3 K and 1.5 at 2 K.

Field ionization is the standard technique to detect Rydberg atoms. The field strength necessary for ionization scales as n^{*-4}. Therefore it is possible to distinguish between Rydberg states belonging to different main quantum numbers.

In our case, the continuous wave excitation requires a continuous detection of the Rydberg atoms. Having left the maser cavity, the Rb atoms move into the inhomogeneous dc electric field of a plate capacitor. Propagating along the atomic beam, the Rydberg atoms reach the point of maximum field strength in front of a hole in the anode. If the maximum field strength is chosen properly (≅ 20V/cm), atoms in the 63p state are ionized and detected via counting the electrons with a channeltron electron multiplier. At the same field strength, atoms in the neighbouring 61d level are ionized with a smaller probability (≅ 15% of 63p state).

Transitions from the initially prepared $63p_{3/2}$ state to the $61d_{3/2}$ level are thus detected by a reduction of the electron count rate. The detector is placed 63 mm downstream from the point where the atoms are excited. Considering the lifetime of the $63p_{3/2}$ level, one finds that ≅ 70% of the excited atoms reach the detector.

To demonstrate maser operation, the cavity was tuned over the $63p_{3/2}$ - $61d_{3/2}$ transition by changing the voltage of the PZT; the field ionization signal was recorded simultaneously. The signal transients could be averaged by means of a PDP 11/10 computer. For most of the measurements about 10 - 20 transients were averaged. The signal obtained with a cavity temperature of 4.3 K is shown in Fig. 5. The reduction of the count rate on the maser resonance was about 28%. Up to particle fluxes of 22×10^3 atoms/s no change of the shape and the position of the resonance could be observed.

In the case of the measurements at a cavity temperature of 2 K (Fig. 6), a reduction of the $63p_{3/2}$ signal could be clearly seen for fluxes as small as 800 atoms/s. An increase of the flux causes power broadening and finally an asymmetry and a small shift. The shift has to be attributed to the ac Stark effect, caused predominantly by virtual transitions to the $61d_{5/2}$ level which is only 50 MHz away from the maser transition. The fact that the field ionization signal at resonance is independent of the particle flux (between 800 and 22×10^3 atoms/s) indicates that the

transition is saturated. This fact and the observed power broadening shows that there is a multiple exchange of photons between Rydberg atoms and the cavity field.

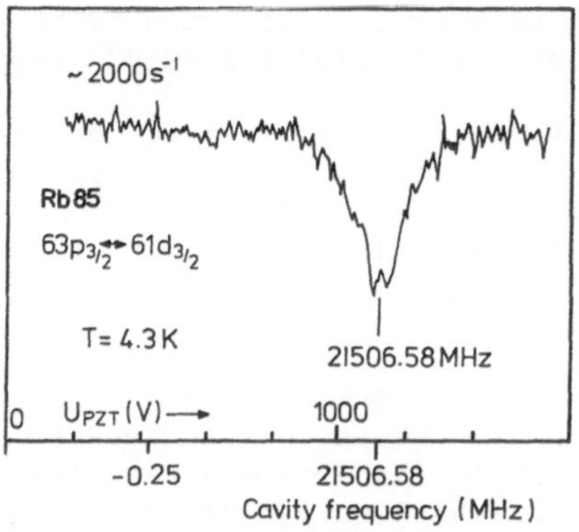

Fig. 5. Maser resonance at a cavity temperature of 4.3 K. Flux of the Rydberg atoms in the upper level N = 2000/s.

With an average transit time of the Rydberg atoms through the cavity of 80 μs one calculates for a flux of 800 atoms/s a probability of 0.06 of finding a Rydberg atom in the cavity. According to Poissonian statistics this implies that more than 99% of the events are single atom. This clearly demonstrates that single atoms are able to maintain continuous oscillation of the cavity.

Since the transition is saturated, half of the atoms excited initially in the $63p_{3/2}$ state leave the cavity in the lower $61d_{3/2}$ maser level. The decay to other levels can be neglected for the average transit time of 80 μs. The energy radiated by those atoms is stored in the cavity field for the characteristic cavity decay time, increasing the average field strength.

The average number of photons left in the cavity by the Rydberg atoms is given by

$$n = \tau_d \cdot N/2$$

where τ_d is the characteristic decay time of the cavity and N the number of Rydberg atoms in the upper maser level entering the cavity per unit time. For the highest particle flux used in our experiment $N = 22 \times 10^3$

atoms/s, one calculates n ≅ 55 photons at 2K ($\tau_d \approx$ 5 ms). At 4.3 K ($\tau_d \approx$ 0.13 ms) there is an average number of n ≅ 1.4 photons. This number is smaller than the average number of blackbody photons being about 4 at

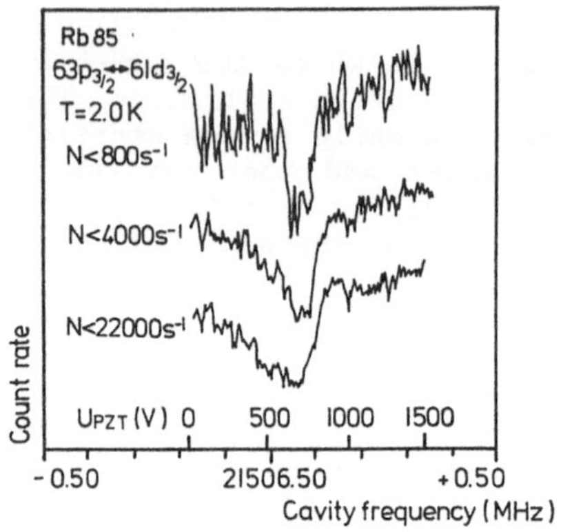

Fig. 6. Maser resonance at a cavity temperature of 2 K.

4.3 K. At a temperature of 2K the average number of blackbody photons is n ≅ 1.5. In the case of N ≅ 800 atoms/s one obtains n ≅ 2, that means that the energy of the radiation generated by the Rydberg atoms in the cavity is about of the same size as the energy of the blackbody radiation.

The coupling constant between atom and radiation is big enough so that also under those condition a multiple exchange of photons between the cavity mode and a single Rydberg atom must occur. It follows that under the conditions shown in Fig. 6 the atom performs on the average about 5 - 20 Rabi periods during it passes through the cavity.

Due to the velocity distribution of the atoms it is not possible to observe the Rabi nutation directly. Presently a Fizeau-type velocity selector is inserted between the atomic beam oven and the cavity. This will enable us to observe the Rabi nutation of the atoms directly, since then we will have a fixed interaction time of the atom with the cavity field. Changing the selected velocity leads to a different interaction time and leaves the atom in another phase of the Rabi cycle when it arrives at the detector. In this way more detailed studies of the atom-field interaction than in the present experiment will be possible. Especially there is a good chance that also the experimental observation of the predicted disappearance and revival of the Rabi nutation gets possible. These latter effects will be briefly discussed in the next Section.

SINGLE ATOM IN RESONANT CAVITY - DISAPPEARANCE AND REVIVAL OF OPTICAL NUTATION

At temperatures T > 0 K the cavity also contains thermal photons. The effects described above therefore become more complicated since the atom evolves through an oscillatory or irreversibly damped transient regime towards a final state distribution corresponding to the thermal equilibrium. The transient behaviour is again dependent on whether there is weak ($\Omega >> \omega/Q$) or strong ($\Omega << \omega/Q$) damping of the cavity. In the first case the transient regime can be described by a sum of elementary Rabi oscillations in a field in which the number of photons is a random quantity following the Bose-Einstein statistics. The distribution of Rabi frequencies results in an apparently random oscillation which for large n values very quickly collapses and then revives again [28,29]. This behaviour is typical of a chaotic quantum field; a semi-classical description of a random Gaussian field does not give this result. It has always been thought that this interesting phenomenon could not be observed experimentally. However, the possibilities now opened up by Rydberg atoms bring us close to its realization. Superconducting cavities with Q values in the range between 10^9 and 10^{10} can be realized and it should therefore be possible to keep the damping small enough so that the oscillations are not washed out before their revival occurs.

So far these phenomena have not yet been observed experimentally, but, with the experimental setup described in the previous section, this should be possible. Since the disappearance and revival of optical nutation depend on the probability distribution of the photon states in the cavity, the experimental setup will allow photon statistics to be performed at small photon numbers in the cavity. Furthermore, direct observation of the vacuum field will be possible.

References

1. E. T. Jaynes, F.W. Cummings, Proc. IEEE 51, 89 (1963)
2. L. Allen, J.H. Eberly, Optical Resonance and Two-Level Atoms, Wiley, New York (1975).
3. P.L. Knight, P.W. Milonni, Phys. Rev. C66, 21 (1980)
4. J.A.C. Gallas, G. Leuchs, H. Walther, H. Figger, in Advances in Atomic and Molecular Physics, Vol. 20, pp. 412-466, Academic Press, New York 1984.
5. D. Kleppner, M.G. Littman, M.L. Zimmerman, in Rydberg States of Atoms and Molecules, ed. by R.F. Stebbings and F.B. Dunning, pp. 73-116, Cambridge Univ. Press. London and New York (1983).
6. D. Delande, J.C. Gay, Comments At. Mol. Phys. 13, 275 (1983).
7. D.R. Bates, A. Damgaard, Philos. Trans. R. Soc. London 242, 101 (1949).
8. T.F. Gallagher, W.E. Cooke, Phys. Rev. Lett. 42, 835 (1979).
9. S. Haroche, C. Fabre, M. Goy, M. Gross, J.M. Raimond, in Laser Spectroscopy IV, ed by H. Walther and K.W. Rothe, Springer Series in Optical Sciences Vol. 21, Springer Verlag, Berlin and New York (1979).
10. E.J. Beiting, G.F. Hildebrandt, F.G. Kellert, G.W. Foltz, K.A. Smith, F.B. Dunning, R.F. Stebbings, J. Chem. Phys. 70, 3551 (1979).

11. G. Rempe, Diploma Thesis, University of Munich (1981).
12. P.R. Koch, H. Hieronymus, A.F.J. Van Raan, W. Raith, Physics Letters, 75A, 273 (1980).
13. H. Figger, G. Leuchs, R. Straubinger, H. Walther, Opt. Comm. 33, 37 (1980).
14. C.H. Townes, A.G. Schawlow, Microwave Spectroscopy, McGraw Hill, New York (1955).
15. J.W. Farley, W.H. Wing, Phys. Rev. A23, 2397 (1981).
16. L. Hollberg, J.L. Hall, in Laser Spectroscopy VI, ed. by H.P. Weber and W. Lüthy, Springer Series in Optical Sciences, Vol. 40, Springer Verlag, Berlin and New York (1983).
17. T.F. Gallagher, W. Sandner, K.A. Safinya, W.E. Cook, Phys. Rev. A23, 2065 (1981).
18. S. Haroche, Ann. Phys. (Paris) 6, 189 (1971).
19. S. Haroche, C. Fabre, J.M. Raimond, D. Goy, M. Gross, L. Moi, Journ. de Physique 43, C2-265 (1982).
20. E.M. Purcell, Phys. Rev. 69, 681 (1946).
21. D. Kleppner, Phys. Rev. Lett. 47, 233.
22. S. Haroche, J.M. Raimond, in Advances in Atomic and Molecular Physics Vol. 20, Academic Press, New York 1984.
23. K.H. Drexhage, in Progress in Optics Vol. 12, ed. by E. Wolf, p. 165, North-Holland, Amsterdam (1974).
24. A. Vaidyanthan, W. Spencer, D. Kleppner, Phys. Rev. Lett. 47, 1592 (1981).
25. G. Gabrielse, H. Dehmelt, to be published and G. Gabrielse,, R. Van Dyck Jr., J. Schwinberg, H. Dehmelt, Bull. Am. Phys. Soc. 29, 926 (1984).
26. P. Goy, J.M. Raimond, M. Gross, S. Haroche, Phys. Rev. Lett 50, 1903 (1983).
27. D. Meschede, H. Walther, G. Müller, Phys. Rev. Lett., in print.
28. P. Meystre, Thesis, Ecole Polytechnique Federale, Lausanne (1974).
29. J.H. Eberly, N.B. Narozhny, J.J. Sanchez-Mondragon, Phys. Rev. Lett. 44, 1323 (1980).

11. [illegible] (1981).
12. W.E. Kock, [illegible]
13. [illegible] H. Walther, Opt. Comm. [illegible]
14. [illegible] McGraw Hill, New York [illegible]
15. [illegible] Phys. Rev. A23, 2397 (1981).
16. [illegible] in Laser Spectroscopy V, ed. by A.R.W. McKellar [illegible] and [illegible] Springer Series in Optical Sciences, Vol. [illegible], Springer-Verlag, Berlin and New York (1981).
17. [illegible] Phys. Rev. A[illegible] 2063 (1984).
18. [illegible]
19. [illegible]
20. E.M. Purcell, Phys. Rev. 69, 681 (1946).
21. [illegible] Phys. Rev. Lett. [illegible]
22. [illegible] in Advances in [illegible] Academic Press, New York [illegible]
23. K.H. Drexhage, in Progress in Optics Vol. 12, ed. by E. Wolf, p. [illegible] North-Holland, Amsterdam [illegible]
24. [illegible] Phys. Rev. Lett. [illegible] (1985).
25. [illegible]
26. [illegible] Phys. Rev. Lett. [illegible] (1987).
27. D. Meschede, H. Walther, G. Müller, Phys. Rev. Lett. [illegible]
28. [illegible]
29. [illegible] (1983).

WHAT ARE SQUEEZED STATES REALLY LIKE?

Michael Martin Nieto*

Theoretical Division
Los Alamos National Laboratory
University of California
Los Alamos, New Mexico 87545

INTRODUCTION

This is work, the concept of which I have been mulling by myself and with Vern Sandberg over the last few years. Vern is now in the Medium Energy Physics Division at Los Alamos. He used to be in Kip Thorn's group at Caltech and before that he was with John Wheeler's group at Texas. Also, the stuff I'll be talking about tomorrow partially involves work which we jointly did with Bob Fisher, who is an experimental laser man.

In this lecture* I'm going to use what I call the "Bethe philosophy." I was a graduate student at Cornell during Hans Bethe's school of physics there. If there is one thing that the great man taught me, it is that if you really understand it, then you've got to be able to explain it simple. I remember Bethe could bring a seminar to a standstill when he would ask, with that slow teutonic voice of his, "But does this description give you the free particle?" Sometimes it didn't.

If you're doing mathematics, you can have fun with the superscripts to the subscripts. I enjoy that just as any other theoretician. Tomorrow, in the second lecture, we'll be playing that game.

However, today, after first giving the answer, I'm going to begin with the simple harmonic oscillator and some quantum mechanics. Then I'll review the special case of the squeezed states that you all know and love, the coherent states. I will start by discussing the coherent states from my simple picture, the Schrödinger point of view. This, I think, allows you to really understand things simply. Next, I'll describe the coherent states from the operator formalism. The operator formalism is more useful in the sense that once you understand what you're doing, you can calculate easier with it. Lastly today, I will describe the squeezed states from the Schrödinger point of view, and discuss their properties.

---

*This written version is based upon a recording of the two lectures actually delivered. The text has been edited to include topics raised by the questions and comments of the participants, both at the lectures and during breaks.

In my second lecture I'll begin by discussing harmonic motion, and coherent and "squeezed" states for general potentials. Then I'll move on to discuss the (harmonic oscillator) squeezed states from the operator point of view and some of their mathematical properties. I'll end by discussing the physics that they are useful for. In other lectures at this Institute, you'll get more detailed and perceptive discussions of the physics involved. My aim is to try to show you the path from this simple description to where you're really down to the nitty-gritty; detecting photon anti-bunching or gravitational waves. So with that and the memory that we're operating from a simple point of view, let me begin by giving the answer.

Here we go, folks. A coherent state, very simply, is a state in a harmonic oscillator potential which, at t=0, i) is a Gaussian that is displaced from the origin by x_0, ii) has a phase which is proportional to the position x, and iii) the width of the Gaussian is that of the ground state, σ_0. In particular, it is

$$\psi_{cs} = [2\pi\sigma_0^2]^{-1/4} \exp\left\{ - \left[\frac{x - x_0}{2\sigma_0}\right]^2 + \frac{i}{\hbar} p_0 x\right\} , \tag{1.1}$$

$$\sigma_0^2 = \frac{\hbar}{2m\omega} = \frac{1}{2a_0^2} . \tag{1.2}$$

A squeezed state is the same Gaussian but the width is different than that of the ground state,

$$\sigma_0 \to \sigma = s\sigma_0. \tag{1.3}$$

That's all it is, honest to God. You can go out to eat lunch in Santa Fe.

So why all the fuss? Well, part of the fuss, is that people like to make things complicated. (We can too, and we will a little later. As I indicated, it's fun to play with operators.) Furthermore, people are continuously rediscovering these states in different guises.

From the operator formalism, the operator transformation which creates a squeezed state,

$$a \to b = \mu a + \nu a^+ , \qquad |\mu|^2 - |\nu|^2 = 1 , \tag{1.4}$$

is basically an updated version of the Holstein-Primakoff transformation[1] which was used to discuss magnetic solids back in 1940, or the Bogolubov transformation[2] which was used to discuss superfluidity and superconductivity back in 1947. [A little cultural side-point. Note the journal Bogolubov published in, "Journal of Physics, U.S.S.R." I'm now dated. At the end of World War II, the Soviet Union still had journals in foreign languages--a tradition that went back to Czarist times. But apparently someone (Stalin?) decided that shouldn't be, and it stopped.]

In 1962, before he became the President of the University of California, David Saxon published a book on quantum mechanics.[3] On p. 162, in very concise Schrödinger wave function form, he describes the time evolution of squeezed states as we're going to be discussing it. The ideas involved trace back to Kennard's article.[4] Its simply that fans of squeezed states often do not realize this early work was done.

The first (widely quoted) "modern" (re-)discovery was by Takahasi[5] in 1965. He was interested in communication theory, discussing parametric amplifiers using wave-function theory. Next was David Stoler,[6] who was interested in general minimum-uncertainty packets from an operator point of view. Horace Yuan[7] was interested in two-photon coherent states. Hollenhorst[8] invented the term "squeezed states." (That's the term which I prefer and which I use. You'll see why later.)

But since then, there have been pulsating states,[9] correlated coherent states,[10] new coherent states,[11] twisted states,[12] etc., etc., etc. These articles actually illuminate and emphasize things from new viewpoints, which is why new names are sometimes given for the states. However, in those cases where the connection to squeezed states was not realized, papers then came out saying the states were already known. This is confusing, and so, as I say, that's part of the reason for all the fuss.

SOME QUANTUM MECHANICS

Before we describe Schrödinger's states in our terminology, let's review some quantum mechanics, starting with the simple harmonic oscillator. This is something which you all got in your first course in quantum mechanics. You have the Hamiltonian eigenvalue equation

$$H\psi_n = E_n\psi_n \quad , \tag{2.1}$$

$$H = -\frac{\hbar^2}{2m}\frac{d^2}{dx^2} + \frac{1}{2}m\omega^2x^2 \quad . \tag{2.2}$$

You all know that the eigenstates are the Hermite polynomials times a Gaussian, and that the energy levels are equally spaced:

$$\psi_n = \left(\frac{a_0}{\pi^{\frac{1}{2}}2^n n!}\right)^{\frac{1}{2}} \exp\,[-\tfrac{1}{2}\,a_0^2\,x^2]\,H_n(a_0x), \tag{2.3}$$

$$E_n = \hbar\omega(n + \tfrac{1}{2}) \quad . \tag{2.4}$$

Now is a good time to quickly introduce the operator formalism that we will use extensively, later. This formalism is "natural" in the harmonic oscillator system. It is natural because, for this equally-spaced energy-level system, the raising and lowering operators a^+ and a, are independent of n. That is, the same operator a^+ that takes you from the nth to the nth-plus-first state also takes you from the nth-plus-first to the nth-plus-second state. The operators are related to x and p as

$$x = \left(\frac{\hbar}{2m\omega}\right)^{\frac{1}{2}}(a + a^+) \quad , \tag{2.5}$$

$$p = \left(\frac{\hbar}{2m\omega}\right)^{\frac{1}{2}}\frac{1}{i}(a - a^+) \quad . \tag{2.6}$$

Then, since

$$[x, p] = i\hbar \quad , \tag{2.7}$$

one has that

$$[a, a^{+}] = 1 \quad . \tag{2.8}$$

Note, in particular, that what will be the standard notation for coherent states in the operator formalism is given by

$$\begin{aligned} \langle a\rangle_{cs} &= \alpha = \alpha_1 + i\alpha_2 \\ &= \langle x\rangle_{cs}\, a_0/2^{\frac{1}{2}} + i\, \langle p\rangle_{cs}/(\hbar 2^{\frac{1}{2}}) \quad . \end{aligned} \tag{2.9}$$

Commutators bespeak of uncertainty relations, so let me remind you about them. If we have two Hermitian operators X and P (not necessarily little x and little p), whose commutator is

$$[X, P] = iG \quad , \tag{2.10}$$

then we have that

$$(\Delta X)^2 (\Delta P)^2 \geq \tfrac{1}{4} \langle G\rangle^2 \quad . \tag{2.11}$$

Equality is obtained by those states which satisfy the eigenvalue equation

$$\left[X + \frac{i\langle G\rangle}{2(\Delta P)^2} P\right] \psi_{mus} = \left[\langle X\rangle + \frac{i\langle G\rangle}{2(\Delta P)^2} \langle P\rangle\right] \psi_{mus} \quad . \tag{2.12}$$

Because $\langle X\rangle$, $\langle X^2\rangle$, $\langle P\rangle$, and $\langle P^2\rangle$ are four parameters, equality in Eq. (2.11) reduces ψ to a function of three parameters.

That's going to be very important for us later. It will turn out that squeezed states satisfy (2.12) with x and p the operators, and coherent states set $(\Delta X)/(\Delta P)$ to a restricted value.

COHERENT STATES

Now let's go back to the beginning of coherent states. The beginning is Schrödinger's fourth paper in quantum mechanics.[13] Schrödinger discovered what we now call the coherent states for the simple harmonic oscillator in 1926 in an article in Naturwissenschaften. They became known as the "coherent states" especially due to the work of Glauber[14] and Klauder and Sudarshan[15] in quantum optics.

What Schrödinger was interested in was to find wave functions which follow the classical motion of a particle. He wanted the shapes to remain constant in time. But note; this was 1926. So this was before Hiesenberg discovered the uncertainty principle and this was before Bohr had explained it. This was also before Born had realized that the density is $\psi^*\psi$. Schrödinger found a solution which did what he wanted, and which amounts to satisfying the uncertainty relation with a packet of ground state width. But, how did he get it?

Schrödinger thought in analogy to electromagnetic waves. What he did was to take the real part of ψ. So, instead of getting the Gaussian wave-packet, he got oscillations inside a Gaussian envelope. But, he had it. He didn't have the coherent-state wave packets, but he had the coherent states. What the hell, that's pretty good for 1926.

Schrödinger's point of view has a wave function which follows the classical motion and keeps its shape with time. That is closest to the uncertainty point of view. The commutator (2.7), $[x,p] = i\hbar$, implies that

$$(\Delta x)(\Delta p) \geq \hbar/2 \quad . \tag{3.1}$$

The solution to the eigenvalue equation (2.12) is the Gaussian (1.3). If you further have that

$$(\Delta x)/(\Delta p) = 1/m\omega \quad , \tag{3.2}$$

this amounts to specifying that the width of the Gaussian is that of the ground state for the harmonic oscillator; i.e., Eq. (1.1). This is the Gaussian which Schrödinger discovered, which follows the classical motion, and which does not change its shape with time.

You also know that the coherent states can be defined as eigenstates of the annihilation operator. If you just try to solve this equation,

$$a|\alpha\rangle = \alpha|\alpha\rangle \quad , \tag{3.3}$$

you get that these coherent states $|\alpha\rangle$ can be written in terms of the number states as the particular combination

$$|\alpha\rangle = e^{-\frac{1}{2}|\alpha|^2} \sum_{n=0}^{\infty} \frac{\alpha^n}{\sqrt{n!}} |n\rangle \quad . \tag{3.4}$$

By using the Hermite polynomial generating function, one can verify that (3.4) is Eq. (1.1), up to a phase.

Notice in particular that α, which is defined by (2.9), is a two-parameter complex number, not the four parameters from $\langle x\rangle$, $\langle x^2\rangle$, $\langle p\rangle$, and $\langle p^2\rangle$. Satisfying the equality in Eq. (3.1) reduces the four by one, and giving the particular value of the width in Eq. (3.2) reduces the parameters to two. Eqs. (1.1) and (3.4) are equivalent definitions.

Finally, pay attention to a third method for defining coherent states. This method is going to be generalized to squeezed states. Consider the displacement operator,

$$D(\alpha) = \exp[\alpha a^{+} - \alpha^{*} a] \quad . \tag{3.5}$$

It takes the ground state $|0\rangle$ and it displaces it to $|\alpha\rangle$:

$$D(\alpha)|0\rangle = \alpha|\alpha\rangle \quad . \tag{3.6}$$

From the Schrödinger viewpoint, the ground-state Gaussian is displaced in phase space to a beginning value x_0 and a beginning value p_0. The Gaussian then oscillates back and forth without changing its shape, partially because the harmonic oscillator has equally-spaced levels. It turns out that you indeed get the classical motion, and the coherent states $|\alpha\rangle$ which you get from $D(\alpha)$ are the same ones we had before in (3.4).

You can sit down and have fun calculating that $\langle x(t)\rangle$ is equal to that of the classical motion, with the ground-state energy subtracted:

$$\langle\alpha|x(t)|\alpha\rangle = \left[\frac{2\hbar|\alpha|^2}{m\omega}\right]^{\frac{1}{2}} \sin(\omega t + \phi). \tag{3.7}$$

Furthermore, if you calculate $\Delta x(t)$ and $\Delta p(t)$, they are constant:

$$[\Delta x(t)]^2\ [\Delta p(t)]^2 = [\hbar/(2m\omega)][m\omega\hbar/2] = \hbar^2/4 \quad . \tag{3.8}$$

The coherent-state wave packet keeps its shape and goes back and forth forever.

SQUEEZED STATES IN THE SCHRÖDINGER PICTURE

Now we go on to squeezed states. The squeezed states at t=0 are minimum-uncertainty states which contain, as a special class, the minimum-uncertainty coherent states. That is, Eqs. (1.1)-(1.3). Recall that when we got the coherent states we started with a Gaussian with 3 parameters. (Satisfying the equality in the uncertainty relation had reduced the possible 4 parameters to 3.) Having the width be that of the ground-state reduced the Gaussian to two parameters.

Well, now we're not going to reduce out that last parameter. The state no longer has the width of the ground state. We have the three parameter system defined by (1.3), which I repeat:

$$\psi_{ss} = [2\pi\sigma^2]^{-\frac{1}{4}} \exp\left\{-\left[\frac{x - x_0}{2\sigma}\right]^2 + \frac{i}{h} p_0\, x\right\} \quad , \tag{4.1}$$

$$\sigma = s\sigma_0 \quad , \tag{4.2}$$

$$\sigma_0 = \frac{\hbar}{2m\omega} = \frac{1}{2a_0^{\ 2}} \quad . \tag{4.3}$$

σ_0 is the width of the ground state or the coherent state. "s" is what I call the "squeeze factor."

This shows you why I like the name "squeeze." What you're doing by putting that "s" in there is that you're either squeezing in the Gaussian (s<1) or you're squeezing it out (s>1), the momentum-space wave function being squeezed oppositely. You're taking the Gaussians and you're squeezing them. When "s" is equal to 1 you have a coherent state. When "s" is not equal to 1 you still have a squeezed state.

At t=0 both coherent states and squeezed states are minimum-uncertainty states. The difference is that for t≠0 a squeezed state will not remain a minimum-uncertainty state. What it does when it evolves away from t=0 is the basis of all the interest for the people in quantum optics and in gravitational wave detection. The fact that it is going to do something different is where the fun comes in.

You can sit down and, for t=0, you can calculate $\langle x\rangle$, $\langle x^2\rangle$, $\langle p\rangle$, and $\langle p^2\rangle$. (An exercise for the graduate student; he can crank it out.) You get that $[(\Delta x)^2(\Delta p)^2]$ is the same as for the coherent states.

But what happens in time? Well, there are two ways to do it. We'll do it both ways. The first way is to calculate things with the operators being time dependent. That means I want to calculate the expectation values of

$$\mathcal{O}(t) = \exp[iHt/\hbar]\ \mathcal{O}\ \exp[-iHt/\hbar] \quad , \tag{4.4}$$

where ϑ is x, x^2, p, and p^2. These are

$$\langle x(t)\rangle = x_0 \cos \omega t + \frac{p_0}{m\omega} \sin \omega t \quad , \tag{4.5}$$

$$\langle p(t)\rangle = p_0 \cos \omega t - m\omega x_0 \sin \omega t \quad , \tag{4.6}$$

$$\langle x^2(t)\rangle = (x_0^2 + \sigma^2) \cos^2 \omega t + \frac{\left(p_0^2 + \frac{\hbar^2}{4\sigma^2}\right)}{m^2\omega^2} \sin^2 \omega t + \frac{2x_0 p_0}{m\omega} \cos \omega t \sin \omega t \quad , \tag{4.7}$$

$$\langle p^2(t)\rangle = \left(p_0^2 + \frac{\hbar^2}{4\sigma^2}\right) \cos^2 \omega t + m^2\omega^2(x_0^2 + \sigma^2) \sin^2 \omega t - 2m\omega x_0 p_0 \cos \omega t \sin \omega t \quad . \tag{4.8}$$

Because of the $s = \sigma/\sigma_0$ in $\langle x^2\rangle$ and $\langle p^2\rangle$, this is already telling you that if I squeeze in x-space I squeeze in the opposite direction in p-space. When you calculate the uncertainty, you get

$$[\Delta x(t)]^2 = \frac{1}{2a_0^2}\left[s^2\cos^2\omega t + \frac{1}{s^2} \sin^2 \omega t\right] \quad , \tag{4.9}$$

$$[\Delta p(t)]^2 = \frac{\hbar^2 a_0^2}{2}\left[\frac{1}{s^2} \cos^2 \omega t + s^2 \sin^2 \omega t\right] \quad . \tag{4.10}$$

Now, because cosine rotates into sin after $\pi/2$, you can already see what's going to happen. The two uncertainties are going to oscillate in width, out of phase.

When you take the uncertainty product you get

$$[\Delta x(t)]^2[\Delta p(t)]^2 = \frac{\hbar^2}{4}\left[1 + \frac{1}{4}\left(s^2 - \frac{1}{s^2}\right)^2 \sin^2[2\omega t]\right] \quad . \tag{4.11}$$

Notice two things. When $s=1$, we have a coherent state and the uncertainty product is a constant in time, as advertised. The second thing, which also is as advertised, is that this uncertainty product oscillates in and out with an angular velocity which is twice the angular velocity of the particle going back and forth in the well.

To reemphasize, a squeezed state <u>does</u> <u>not</u> remain a minimum-uncertainty state as a function of time. At various times, whenever $\omega t = \pi/2$, it will be a minimum-uncertainty state. But it oscillates from being a minimum-uncertainty state. In a little bit, I'll give you a figure which describes how this comes about.

This squeezed-state Gaussian property of regaining its original shape turns out to be more general. It is the property of any wave packet in a simple harmonic oscillator well, because the energy levels

are equally spaced. Take any wave function, decompose it into eigenstates and consider it as a function of time:

$$\psi(x,t) = \sum_{n=0}^{\infty} c_n \psi_n(x) \exp[-i\omega t(n + \tfrac{1}{2})] \quad . \tag{4.12}$$

Because the decomposition is commensurate, every full period [$t = 2\pi j$], $|\psi,(x,t)|$ will return to its original shape and position, just because the eigenstates are evenly spaced. Every half period [$t = (2j + 1)\pi/\omega$], ψ will be mirrored in the form $|\psi(-x,0)|$, because it's the harmonic oscillator. What happens is that a packet will start on the left, be mirrored on the right one half period later, and return to the left as it started. Even a delta function (considered as a limit of a Gaussian) will do this.

Now! What is $\psi_{ss}(x,t)$? To find out, first do what I just said, decompose it into eigenvectors. Combining Eqs. (2.3), (2.9), (4.1), and (4.12),

$$\begin{aligned} c_n &= \langle n|\psi_{ss}\rangle \\ &= \left[\frac{2s}{(1 + s^2)2^n n!}\right]^{\frac{1}{2}} \exp\left[-\frac{\alpha_1{}^2 + \alpha_2{}^2 s^2}{1 + s^2}\right] \\ &\quad \times \left[\frac{1 - s^2}{1 + s^2}\right]^{n/2} H_n\left(\frac{2^{1/2}(\alpha_1 + i\alpha_2 s^2)}{[(1 - s^2)(1 + s^2)]^{\frac{1}{2}}}\right) \quad . \end{aligned} \tag{4.13}$$

Here I have changed to the notation $\alpha = \alpha_1 + i\alpha_2$ instead of x_0 and p_0 because that allows me to get rid of some of the constants. When s=1, c_n just becomes the normal Gaussian coherent state factor.

$$\lim_{s\to 1} c_n = \exp[-\tfrac{1}{2}|\alpha|^2] \frac{\alpha^n}{(n!)^{\frac{1}{2}}} \quad . \tag{4.14}$$

Now we can write down $\psi_{ss}(x,t)$ explicitly.

$$\psi_{ss}(x,t) = \sum_{n=0}^{\infty} c_n \psi_n(x) \exp[-i\omega t(n+\tfrac{1}{2})] \quad . \tag{4.15}$$

What does it look like? Well, again folks, this is an exercise for the graduate student (or in this case me, when I made God knows how many mistakes--you know how that goes). This is it:

$$\begin{aligned} \psi_{ss}(t) &= \left[\frac{1}{2^{\frac{1}{2}}\pi^{\frac{1}{2}}} \frac{1}{\Delta x(t)}\right]^{\frac{1}{2}} \exp\left[-\frac{[x - x_0(t)]^2}{4[\Delta x(t)]^2}\right] \\ &\quad \times \exp\left\{ix\left[\frac{p_0 \cos\omega t}{h} - \frac{a_0{}^2 x_0 \sin\omega t}{s^4}\right]\frac{\sigma^2}{[\Delta x(t)]^2}\right\} \\ &\quad \times \exp\left\{-i\,\frac{x^2(s^4 - 1)\cos\omega t \sin\omega t}{4s^2[\Delta x(t)]^2}\right\} \exp\{i\phi(t)\} \quad , \end{aligned} \tag{4.16}$$

$$\phi(t) = \left[\left(\frac{x_0^2 a_0^2}{s^2} - \frac{p_0^2 s^2}{\hbar^2}\right)(\sin \omega t \cos \omega t) \right. \tag{4.17}$$

$$\left. -2 \frac{x_0 p_0 a_0^2}{\hbar(1 + s^2)} (s^2 \cos^2 wt - \sin^2 wt)\right] [\Delta x(t)]^{-2} \quad .$$

Very interesting and ungodly. That's the reason why you eventually want to change to an operator formalism.

Let me note a few things. We've got the Gaussian shape and the width oscillates in time, just like we said it would. The momentum phase, in the second exponential of (4.16), follows the classical motion. The third exponential is technically why the packet does not remain a minimum-uncertainty state in time. The last piece is, for our purposes, an irrevelant phase factor, but to show all the people I actually did it, there it is.

Using $\psi_{ss}(x,t)$ we can now calculate $\langle x(t)\rangle$ in the opposite way. We can calculate $\langle\psi_{ss}(x,t)\ |x|\ \psi_{ss}(x,t)\rangle$. Believe it or not, it is the same. $\langle x(t)\rangle$ comes out to be what we had before. Also, $\langle x^2(t)\rangle$, $\langle p(t)\rangle$, $\langle p^2(t)\rangle$, $\Delta x(t)$, and $\Delta p(t)$ are all as before. So, indeed everything works.

SOME PROPERTIES OF SQUEEZED STATES

What we have now is a physical understanding of squeezed states. We start at t=0 with a displaced Gaussian with a squeezed shape. It will follow the classical motion with time, oscillating back and forth. But, what does it's shape explicitly do?

Let's start it at x=0. In Fig. 1, the little ellipses represent uncertainties in x and p space. They are called uncertainty ellipses. What happens is, the wave packet follows the large circle with angular frequency ω, but the shape oscillates with angular frequency 2ω. In x and p variables, the ellipse changes shape with a quadrapole motion. At 45 degrees it is a circle. The area, A(t), of this circle at $\omega t = \pi/4$ is bigger than the area of the ellipse at $\omega t = 0$, since

$$A(t) = \pi\Delta x(t)\Delta p(t) \quad . \tag{5.1}$$

Then at $\omega t = \pi/2$ it becomes the same shaped ellipse, only it's rotated. Note that when it's gone one-half period, it's got the same shape and orientation as it had before. That means that the uncertainty ellipse is oscillating with twice the angular frequency. In particular, a wave packet at t=0 will have the same uncertainties one-half period later. (The reader should not confuse this ellipse with the uncertainty ellipse of the $X_{1,2}$ variables of Sec. 9. That ellipse stands still in x,p coordinates and rotates like a frankfurter in its own coordinates.)

Since t=0 is really a phase, any Gaussian with the types of phases proportional to x and x^2 in Eq. (4.16) is a squeezed state away from minimum uncertainty; but it will rotate into minimum uncertainty in time. [Those phases prevent the Fourier transform of $\psi(x,t)$ yielding a minimum-uncertainty $\psi(p,t)$ for all t.]

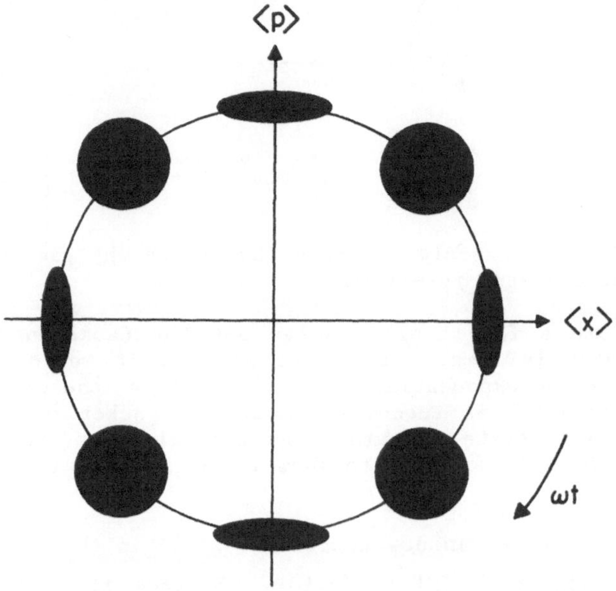

Figure 1. The motion and shape with time of a squeezed-state uncertainty ellipse.

So, the general squeezed state in time is not a mininum-uncertainty state, but a Gaussian state which will evolve into a minimum-uncertainty state. People often like to say that a coherent state is a Gaussian. Stop them! A coherent state is a Gaussian in a harmonic oscillator which at t=0 has the width of the ground state and a phase proportional to x. But a coherent state is <u>not</u> a Gaussian, <u>per se</u>!

A couple of final points, since I am reaching the end of my alloted time.

If I define the Hamiltonian $\mathcal{H}$ as being the normal Hamiltonian H divided by $\hbar\omega$, then for squeezed states

$$\langle\mathcal{H}\rangle_{ss} = |\alpha|^2 + \frac{1 + s^4}{4s^2} \quad , \tag{5.2}$$

$$\langle\mathcal{H}^2\rangle_{ss} = \langle N\rangle_{ss}{}^2 + \alpha_1{}^2 s^2 + \frac{\alpha_2{}^2}{s^2} + \frac{(1 - s)^2}{8s^4} \quad , \tag{5.3}$$

$$[\Delta\,\mathcal{H}]^2 = \alpha_1{}^2 s^2 + \frac{\alpha_2{}^2}{s^2} + \frac{(1 - s^4)^2}{8s^4} \quad . \tag{5.4}$$

This reduces to the harmonic oscillator coherent state results when s=1.

One last thing. Last night as I was putting my head on my pillow, it dawned on my that the free Gaussian should fit into this. It occurred to me that a free Gaussian must be a squeezed state, in the limit of $\omega \to 0$ harmonic oscillator coupling. So, I quickly got out my copy of Schiff.[16] If you look at Schiff, one has for $\langle x\rangle = \langle p\rangle = 0$ that the time evolution of the free Gaussian has the form

$$\psi^*(t)\psi(t) = [2\pi Q^2(t)]^{-\frac{1}{2}} \exp\left[-\frac{x^2}{2Q^2(t)}\right], \tag{5.5}$$

$$Q^2(t) = Q^2(0) + \frac{\hbar^2 t^2}{4m^2 Q^2(0)} \quad . \tag{5.6}$$

But this is the functional form of $\psi^*_{ss}(t)\psi_{ss}(t)$, with $\Delta x(t)$ replacing $Q(t)$. Then, since from (4.9)

$$\left(\begin{matrix}\lim \\ \omega,s \to 0 \\ (\omega/s^2) \to \Omega\end{matrix}\right) \quad [\Delta x(t)]^2 = \frac{\hbar}{2m\Omega}\,[1 + \Omega^2\, t^2 + 0\ \ldots] \quad , \tag{5.7}$$

taking

$$\Omega = \hbar/[2mQ^2(0)], \tag{5.8}$$

gives us that you can think of free Gaussian as being a special case of a squeezed state. It spreads out, but it never gets anywhere close to the limit where it would turn around and start contracting again. It's always falling apart.

So, with that being said, let me summarize. A coherent state at t=0 is a Gaussian in a harmonic oscillator, with the width of the ground state. It has a phase proportional to x and the expectation values of x and p correspond to its starting position and momentum in a harmonic oscillator. It oscillates back and forth following the classical motion in time and it keeps its proper shape forever.

A squeezed state is a slight relaxation. The relaxation is simply that you allow the width of the Gaussian to be different from that of the ground state. It, too, can be displaced and the wave packet will follow the classical motion in time.

The important difference is that the squeezed states' uncertainty will oscillate back and forth. The uncertainty will oscillate between an ellipse with the small width in the x direction, to one with the small width in the p direction. That aspect of this Schrödinger formalism is, when we translate it into the operator formalism, a very useful tool. It is what makes squeezed states important in the quantum theory of measurement for gravitational wave detection and for quantum optics.

OK. That's all I'm going to talk about in the Schrödinger representation. I will soon translate this to the operator formalism and show the connection. When we discuss the beautiful, if obtuse, operators, think back to the wave function if you are getting lost. The key will always be the parameter s, which tells you how squeezed or how unsqueezed the Gaussian is.

HARMONIC MOTION, COHERENT STATES AND SQUEEZED STATES FOR GENERAL POTENTIALS

What I am going to start with today will be modified from what I had originally intended. That's the result of the questions which I received yesterday, both here and at lunch. As we have observed, a lot of the interest in coherent states and squeezed states comes from the fact that these states follow the classical motion in a harmonic oscillator. Now it's known by everybody that the harmonic oscillator has very special properties. One is that the classical period of oscillation is independent of energy. Another is that the quantum eigenvalues are equally spaced. So, the question arises if all the special properties characteristic of coherent and squeezed states are due to these properties of the harmonic oscillator. The answer is, "It helps, but it's not entirely due to this." This answer can be understood from a discussion of what I call "classical and quantum harmonic potentials."

The classical period of oscillation in a potential V(x) is given by the integral

$$\tau_c(E) = 2\pi/\omega_c(E) = (2m)^{\frac{1}{2}} \int_{x_L}^{x_R} \frac{dx}{[E - V(x)]^{\frac{1}{2}}} . \tag{6.1}$$

If $\omega_c(E)$ is a constant independent of energy, I call V "a classical harmonic potential." For a given ω_c, there are an <u>uncountable</u> number of potentials which satisfy that condition; as a matter of principle. The simple harmonic oscillator with $\omega = \omega_c$ is one of them, so the simple harmonic oscillator is a classical harmonic potential.

Supposing, I now look at the quantum eigenvalue problem for a potential, V. If the energy eigenvalues are equally spaced, proportional to a given constant, $\omega_q = (E_{n+1}-E_n)/\hbar$, I will call V a "quantum harmonic potential." Again the simple harmonic oscillator is one. The reason I call such a potential "harmonic" is because after every period $\tau = 2\pi/\omega_q$ the wave packet will always reconstruct itself as we discussed around Eq. (4.12). Again there are an <u>uncountable</u> number of potentials which satisfy the condition for a given ω_q.

For a given $\omega_c = \omega_q = \omega_0$ are all of the classical harmonic potentials also quantum harmonic potentials, and the reverse? The answer is,[17] "No." For a given ω_0 there are three sets. There are classical harmonic potentials that are not quantum harmonic potentials, quantum harmonic potentials which are not classical harmonic potentials, and potentials which are both; and all three sets are uncountable.[17]

Let me give you an example of a quantum harmonic potential which is not a classical harmonic potential. I came across this particular potential by luck at the right time, when I first became interested in the question. (You've got to be good or you've got to be lucky, one of the two.) We had just finished the work on coherent states for general potentials where we had concentrated on classical motion, the harmonic oscillator being the most classical of all.

The potential was sent to me in a paper by Abraham and Moses;[18] and as I always say, "How can you not believe a paper by Abraham and Moses?" They were looking at the inverse method of Gel'fand and Levitan. This method allows you to do the following: If you have a potential which, as

a matter of principle, is solvable with a complete set of eigenvectors and associated eigenvalues, you can generate a new potential which has the same spectrum, but with any particular eigenvalue and eigenvector removed. In particular, if you give me the harmonic oscillator, I can generate a potential with the ground state removed. The eigenvalues are still equally spaced. So, it's still a quantum harmonic potential, but you can show that the new potential does not have a period of oscillation which is independent of energy.

How you actually construct this is another talk. I just give you the (dimensionless) results that the potential

$$v = \frac{1}{2} y^2 + \Phi [\Phi - 2y] , \tag{6.2}$$

$$\Phi = \frac{\exp[-y^2]}{\int_y^{\infty} dt \exp[-t^2]} , \tag{6.3}$$

has the orthonormal eigenvectors

$$\chi_n(y) = \psi_n(y) - (\tfrac{2}{n})^{\frac{1}{2}} \Phi(y) \psi_{n-1}(y) , \tag{6.4}$$

with eigenvalues $\varepsilon_n = (n + \frac{1}{2})$, n = 1, 2, . . ., where $\psi_n(y)$ are the harmonic oscillator eigenfunctions of Eq. (2.3).

If you now put Eq. (6.2) into our integral Eq. (6.1), you find that the classical period of oscillation is a slightly varying function of E.[17] [$\omega_c(E)$ never differs from ω_q by more than 7%.] Thus, V is not a classical harmonic potential but it still is a quantum harmonic potential. And you can think of an uncountable number of these things.

The reversed case is even more easily seen,[19] as I mentioned at lunch yesterday. Take two half-harmonic oscillators of different width placed together. Since the harmonic oscillator classical period of oscillation is independent of energy, a particle in one side of the half-harmonic oscillator has a period independent of energy, and also when it is on the other side. Thus, it's independent of energy over both sides. However, if you think of the quantum problem, you have a parabolic cylinder function for an eigenvector on one side, which is trying to match with a differenct parabolic cylinder function on the other side. Matching their slopes and values at the origin, gives a solution that is <u>not</u> a parabolic cylinder function equal to a Hermite polynomial. This means that the eigenvalues are not equally spaced.

To summarize, the special features of the simple harmonic oscillator are partially due to its "quantum" and "classical" harmonic nature, but not entirely. It is not only the simple harmonic oscillator which can have ω_c and $(E_{n+1} - E_n)/h = \omega_q$ equal to a constant.

Another question was raised by Marlan, about if there are "squeezed states for general potentials." The answer is in the papers we did on the coherent states for general potentials.[20] Let me first tell you what that program was. Then I can say, "Yes, there are squeezed states for general potentials, and they're already written down in our papers; one step from the end." Our program was to find what we called "minimum-uncertainty coherent states" (MUCS).

Give me a potential. Consider the classical problem. I can find a transformation from x and p to what I call X_c and P_c (which are functions of x and p) such that they vary as $\sin[\omega_c(E)t]$ and $\cos[\omega_c(E)t]$. What that amounts to is changing variables so that the classical Hamiltonian looks like the harmonic oscillator Hamiltonian, $H \sim \frac{1}{2}X_c^2 + \frac{1}{2}P_c^2$. It's switching variables so that the classical phase-space path is a circle in the new variables, X_c and P_c.

Now I change these classical variables into quantum operators, X and P. $[X,P] = iG$ defines the uncertainty relation $(\Delta X)^2(\Delta P)^2 \geq \frac{1}{4}\langle G\rangle^2$ of Eq. (2.11). Find those states, which we call "minimum-uncertainty states," (MUS) which satisfy the equality in the uncertainty relation. This is a three parameter set, complex C and real B. Our last step was to set B such that the ground state is a member of that set. Those are our "minimum-uncertainty coherent states" for general potential.

The squeezed states are one step back. Those are what I called "minimum-uncertainty states." They are an arbitrary B-shape, such that the ground-state "B" is a special case.

That's exactly the same thing as for the harmonic oscillator. There, for squeezed states you've got a general Gaussian that is a "minimum-uncertainty state." Then you set the parameter so that it has the width of the ground state to make a coherent state. So, to answer Marlan's question again, "Yes, a procedure can be carried out to obtain squeezed states, for general potentials. We've already done it, except that we were interested in the last step. The squeezed states are one step back." You can look in our papers[20] and see that it can all be done analytically for solvable potentials.

SQUEEZED STATES IN THE OPERATOR FORMALISM

Recall what I have emphasized, that the coherent states are simply Gaussians displaced from the origin with the width of the ground state and a complex phase. The squeezed state is the same Gaussian, but with a width different from the ground state, labeled by $s \neq 1$, s being what I call the "squeeze parameter." That's all there was to it.

Remember that folks, because now we're going to go into the operator formalism. As I said, its a lot easier to calculate with, but much less transparent in a physical sense than wave functions are.

In Sec. III we recalled that in the operator formalism the coherent state $|\alpha\rangle$ can be generated by displacing the ground state:

$$|\alpha\rangle = D(\alpha)|0\rangle \quad , \tag{7.1}$$

$$D(\alpha) = \exp[\alpha a^{+} - \alpha^{*} a] \tag{7.2}$$

$$= \exp[-\tfrac{1}{2}|\alpha|^2] \exp[\alpha a^{+}] \exp[-\alpha^{*} a] \quad . \tag{7.3}$$

What I've done is to take the ground-state Gaussian and displace it. But how do I "squeeze" the ground-state Gaussian? In other words, what is the "squeeze operator?" It is this beastie:

$$S(z) = \exp[\tfrac{1}{2}z a^{+}a^{+} - \tfrac{1}{2}z^{*} aa] \quad , \tag{7.4}$$

with z the complex parameter

$$z = re^{i\theta} \quad , \qquad \theta = 2\omega t_0 \quad , \tag{7.5}$$

$$r \equiv \ln s \quad . \tag{7.6}$$

θ amounts to an initial time, t_0. If θ is not 0, it means you've started at some $t \neq 0$. r is equal to the logarithm of s, the squeeze factor which is in the Gaussian. That's the connection between the operator formalism and the Schrödinger formalism, and it makes sense. If s=1, then r=0 so z=0, so there's no squeeze.

A squeezed state can be produced one of two ways. The first way, which is analogous to Eq. (4.1), is to squeeze and then displace:

$$\psi_{ss} \rightarrow |\alpha,z\rangle = D(\alpha)S(z)|0\rangle. \tag{7.7}$$

That's the way I tend to do it because I tend to think more in terms of gravitational waves than in terms of quantum optics and also because this gives you the Schrödinger wave function (4.1) in terms of the parameters α and s. But you can equally as well do it the opposite way. The quantum optics people usually like to do it that way, a point I'll come to later.

The next question is, what is S(z) in normal-ordered form? That's an important question. Well, there are two ways you can do it. One is to use something which is called McCoy's Theorem, which allows you to write an exponential of a general second-order polynomial in a and a^+, as a normal-ordered object. (You put S(z) in the formula and you get out the right answer.) McCoy was a very smart fella. But the method is obscure.

It is much more fun to do it by matrices,[21] and that's what I'm going to do right now. First note the following identification with the algebra SU(1,1):

$$\tfrac{1}{2}a^+a^+ \equiv L_+ = \begin{bmatrix} 0 & 1 \\ 0 & 0 \end{bmatrix} , \tag{7.8}$$

$$-\tfrac{1}{2}aa \equiv L_- = \begin{bmatrix} 0 & 0 \\ 1 & 0 \end{bmatrix} , \tag{7.9}$$

$$\tfrac{1}{2}a^+a+\tfrac{1}{4} \equiv L_3 = \begin{bmatrix} \tfrac{1}{2} & 0 \\ 0 & -\tfrac{1}{2} \end{bmatrix} . \tag{7.10}$$

These matrices and these operators satisfy the SU(1,1) commutation relations

$$[L_+,L_-] = 2L_3, \ [L_3L_\pm] = \pm L_\pm \quad . \tag{7.11}$$

Therefore, the squeeze operator can be written as

$$S(z) = \exp \begin{bmatrix} 0 & z \\ z^* & 0 \end{bmatrix} = I \sum_{n=0}^{\infty} \frac{|z|^{2n}}{(2n)!} + \begin{bmatrix} 0 & z \\ z^* & 0 \end{bmatrix} \sum_{n=0}^{\infty} \frac{|z|^{2n}}{(2n+1)!}$$

$$= \begin{bmatrix} \cosh r & e^{i\theta}\sinh r \\ e^{-i\theta}\sinh r & \cosh r \end{bmatrix} \tag{7.12}$$

$$= \begin{bmatrix} 1 & e^{-i\theta}\tanh r \\ 0 & 1 \end{bmatrix} \begin{bmatrix} 1/\cosh r & 0 \\ 0 & \cosh r \end{bmatrix} \begin{bmatrix} 1 & 1 \\ e^{-i\theta}\tanh r & 1 \end{bmatrix} .$$

Now you work backwards and write these matrices as exponents of matrices giving

$$S(z) = \exp[e^{i\theta}(\tanh r)\ L_+]\ \exp[-2\ \ln(\cosh r)\ L_3]\ \exp[e^{-i\theta}(\tanh r)L_-] . \tag{7.13}$$

Finally, by using the Baker-Campbell-Hausdorff-type formula

$$\exp(xa^+a) = \sum_{n=0}^{\infty} \frac{(e^x - 1)^n}{n!} (a^+)^n (a)^n \tag{7.14}$$

on the middle exponential in Eq. (7.13), we get

$$S(z) = \exp[\tfrac{1}{2}e^{i\theta}(\tanh r)\ a^+a^+]\ (\cosh r)^{-\frac{1}{2}} \sum_{n=0}^{\infty} \frac{(\mathrm{sech}\ r - 1)^n}{n!} (a^+)^n (a)^n$$

$$\times \exp[-\tfrac{1}{2}e^{i\theta}(\tanh r)\ aa] \quad . \tag{7.15}$$

Note that in Eq. (7.15), even though you have cross terms, effectively you're dealing with pairs of a's and (a^+)'s. That's where the concept of "two-photon coherent states" comes in, especially when you take expecation values. The quantum optics people like to think of squeezed states as two-photon coherent states; two excitations.

MATHEMATICAL PROPERTIES OF SQUEEZED STATES

Another property of squeezed states comes from this well known theorem:

$$e^{\zeta A}\ Be^{-\zeta A} = B + \zeta[A,B] + \frac{\zeta^2}{2!}\ [A,[A,B]] + \ldots \tag{8.1}$$

Using this theorem and summing the infinite series you can obtain

$$b^+ = S^+\ a^+\ S = (\cosh r)\ a^+ + e^{-i\theta}(\sinh r)\ a \quad , \tag{8.2}$$

$$b = S^+\ a\ \ S = (\cosh r)\ a\ + e^{i\theta}\ (\sinh r)\ a^+ \quad , \tag{8.3}$$

$$[b,\ b^+] = 1 \quad . \tag{8.4}$$

This is just the Bogolubov transformation[2] discussed many years ago for superconductivity and superfluidity.

But this is also one of the many new discoveries of the squeezed states, called twisted states.[11] They are the same thing. Why? You look for the eigenvector of

$$b = \mu a + \nu a^{+} \quad , \quad |\mu^2| - |\nu^2| = 1 \quad . \tag{8.5}$$

Well, $[\cosh^2 r - \sinh^2 r]$ is equal to one. Basically, this result was realized in Stoler's papers back around 1970.[2] (A lot of this stuff that I'm talking about right now, Vern Sandberg and I are going to be writing up.)

Finally, from the same theorem, we have two further results. Any particular function of a and a^+, when you surround it by squeeze operators, is equal to the same operator, only as a function of b and b^+:

$$S^{-1}(z)\, f(a,a^{+})\, S(z) = f(b,b^{+}) \quad . \tag{8.6}$$

This means that D does not commute with S, but, rather that

$$D(\alpha)\, S(z) = S(z)\, D(\gamma) \quad , \tag{8.7}$$

$$\gamma = \alpha \cosh r - \alpha^{*} e^{i\theta} \sinh r \quad . \tag{8.8}$$

So, it is not fundamentally important whether you prefer the gravitational wave point of view, where you apply first S then D, or whether you like to do it the quantum optics way where you first displace and then you squeeze. The two view are mathematically related by a complex scale transformation. You have to be conscious and careful about it, and which you apply first. But if you are, there is no fundamental problems.

Going on, we've really got three parameters α_1, α_2, and $|z| = r$, with θ corresponding only to an initial time. Now we've learned the game. Take an exponent of a and a^+, and that displaces. Take an exponent of (aa) and (a^+a^+), and that squeezes. So what about exponents of $(a)^3$ and $(a^+)^3$, or $(a)^n$ and $(a^+)^n$, for that matter? Well, let me define a k-photon unitary operator,

$$\begin{aligned} U_k(z_k) &\equiv \exp[z_k(a^{+})^k - z_k^{*}(a)^k] \\ &= \exp[iA_k] \quad . \end{aligned} \tag{8.9}$$

U_1 is D, U_2 is S. What about U_3, U_4, U_5, etc.? It turns out that they don't exist.[21]

You can show that very simply. Take the ground-state expectation value of U_k. Expand U_k and collect all the terms. It's very straightforward and looks something like this:

$$\begin{aligned} \langle 0|U_k|0\rangle = 1 &- |z_k|^2 \frac{k!}{2!} + |z_k|^4 \frac{1}{4!} [(k!)^2 + (2k)!] \\ &- |z_k|^6 \frac{1}{6!} \left\{ (k!)^3 + 2k!(2k)! + \frac{1}{k!} [(2k)!]^2 + (3k)! \right\} \\ &+ \cdots + (-1)^n |z_k|^{2n} \frac{1}{(2n)!} C_n + \cdots \quad . \end{aligned} \tag{8.10}$$

C_n has many terms, all of which are greater than 0. The largest is of order (kn)! as a matter of fact, all of them are of this order. That means that

$$\lim_{n\to\infty} |z_k|^{2n} \frac{1}{(2n)!} C_n \neq 0 \tag{8.11}$$

for all k>2 and $z_k \neq 0$. Therefore, the terms of this series do not decrease monotonically. The terms are bounded below by (kn)!/(2n)! and so, by the ratio test, there is no convergent limit as $n \to \infty$. The series is divergent and so $\langle 0|U_k(z_k)|0\rangle$ is undefined by the series for k>2.

If you want to put in some fancy languange, and I thank John Klauder for emphasizing to me that you really should put it in fancy language, consult footnotes 4 and 20 of Ref. 21. It means the following: Even though A_k is Hermitian, so U_k is unitary, one is assuming that the ground state is an analytic vector of the generator A_k. It is not! In other words, $A_{k>2}$ is not self-adjoint in the harmonic oscillator basis.

This is interesting. This is one place where the difference between self-adjoint and Hermicity makes a difference. Here we have a potential which is proportional to x^2. So for large $|x|$ the wave function dies out as $\exp[-x^2]$. That means that D, which is an exponent of x, just moves the center of the wave function. S, which is an exponent of x^2, changes the shape. But $U_{k>2}$, which is an exponent of x^k, can make the wave function blow up. This is an operator statement. It is saying that the expectation value of the operator expansion (of say $\exp[-x^4]$) does not converge. In general, if you have a potential which rises as $|x|^n$, then the wave function dies out as $\exp[-|x|^{(n/2)+1}]$, so you could go up to $U_{k\geq[n/2]+1}$.

Note that it was rather critical that when I talked about the ratio test, it gave you no answer for k=2. Remember the series summed as $(\tanh r)<1$ for $r<\infty$. From this point of view, the displacement operator converges well, the squeeze operator just barely makes it, and higher orders just blow up. The odd-ball thing is that if you now go to the harmonic oscillator with para-bose statistics, even the squeeze operator diverges.[22] That's a cute little result.

THE PHYSICS OF SQUEEZED STATES

Let's begin with a little bit of physics as to why there is a usefulness in the differing viewpoints of quantum optics and gravitational waves. First, let's look at $SD|0\rangle$. Applying the displacement operator can be thought of as filling an optical cavity. Then applying the squeeze operator can be considered as processing the light trying to make anti-bunched light, which is a low-noise object. Anti-bunching is basically a quantum-optics phenomena. Bunching can be "classical" in concept. [Remember the Hanbury-Brown Twiss experiment? That was a "classical" experiment. They had these two dishes looking at a faint object in the middle of the Australian desert. If they got a photon in one dish, they tended to get a photon in the other dish.] At this Institute, Leuchs[23] will describe anti-bunching experiments, and why he feels objections to these experiments are not valid.

Something that I know hasn't been done is to detect gravitational waves. Gravitational wave people want to apply S first and then D. They

want to take a Weber bar and perform a quantum non-demolition experiment, starting with a squeeze of the ground state. They want to read out the bar, measuring what amounts to a combination of position and momentum. Then a highly classical gravity wave comes along and displaces the bar, which is D. Without the squeeze, the noise would make the measurement impossible.

When the gravitational people at Caltech were thinking of this process, they wanted to convince their great man, Feynman, that this really is the correct way of looking at things. Of course Feynman always likes analogies to classical physical. So they thought of this one: Consider a pendulum. Since it is a harmonic oscillator, it will always leave and come back to the center in half a classical period, no matter what it's energy (or momentum). Let's say at t=0 the thing is standing still and I take a hammer, and I go whack! One half of a period later I go whack again. Now, I don't care how hard I hit it, all I care is that I went whack half a period later and the pendulum was there. But one time I go whack and the pendulum is not there. That means something's happened to it while it was flying around.

As a matter of principle, that's how they want to detect gravitational radiation. They want to monitor a time-dependent operator of position and momentum (which I'll come to) in very precise time steps. If one time the measurement does not give what it's supposed to, that means a gravitational wave has passed. They've got to do this at such a low noise level because the signal is so weak, as we discuss below.

Let me close by making some final comments on gravitational wave detection. (For a good understanding you should really see the article by Caves, Thorn, Drever, Sandberg, and Zimmerman in RMP.[24]) Gravitational wave people consider a couple of explicity time-dependent operators:

$$X_1 \equiv x \cos \omega t - \frac{p}{m\omega} \sin wt \quad , \tag{9.1}$$

$$X_2 \equiv x \sin \omega t + \frac{p}{m\omega} \sin \omega t \quad , \tag{9.2}$$

$$[X_1, X_2] = \frac{i\hbar}{m\omega} \quad . \tag{9.3}$$

Practically, they propose that one should measure x, say with a transducer modulated by the bar's classical ω, and simultaneously measure p, say with another transducer modulated by the classical ω but with the phase 90 degrees ahead of the first. They want to couple, mix, and amplify the two signals. What they want to do is to measure X_1. In principle, these measurements can be made arbitrarily quick. (People have also discussed Gedanken experiments.)

As a matter of principle, people believe they can do it, but they have not done it yet. The real nitty-gritty details are very involved, and I refer you to Ref. 24. When I read it, I say I understand it. Two weeks later I ask myself, "What-in-the-hell did I read?" But at least I think I understand the principle.

Now notice I said what they are trying to do is to measure X_1. But remember,

$$x(t) = e^{iHt/\hbar} x e^{-iHt/\hbar}$$

$$= x \cos \omega t + \frac{p}{m\omega} \sin \omega t \tag{9.4}$$

That means that X_1, which they're trying to measure, is $x(-t)$. So

$$X_1(t) = e^{iHt/\hbar} X_1 e^{-iHt/\hbar}$$

$$= x \quad . \tag{9.5}$$

They have rotated to a useful coordinate system.

Remember I was talking about the rotating uncertainty ellipse? (See Fig. 1.) $\langle x(t) \rangle$ is rotating clockwise. Of course, you'll make the "starting point" very narrow. In $\langle X_1 \rangle = \langle x(-t) \rangle$ the time is going counterclockwise. But once you put the implicit time dependence in, $\langle X_1(t) \rangle$, you add a clockwise rotation, so you "stand still" in time. So, you're always measuring something with the narrow width on the right. But you've got to do that by coupling these transducers. That's the trick.

If you can do this in practice, then you'll have a real chance of measuring gravitational waves. They haven't done it yet, not by a long shot. A sensitivity to a relative change in length of the bar of $\Delta L/L \sim 10^{-15}$ is claimed. They've got a ways to go yet, but that's the idea.

Thank you very much.

REFERENCES

1. T. Holstein and H. Primakoff, Field Dependence of the Intrinsic Domain Magnetization of a Ferromagnet, Phys. Rev. 58:1098 (1940).
2. N. Bogolubov, On the Theory of Superfluidity, J. Phys. USSR 11:292 (1947).
3. D. S. Saxon, "Elementary Quantum Mechanics, Preliminary Edition," Holden-Day, San Francisco (1964), p. 162.
4. E. H. Kennard, Zur Quantenmechanik einfacher Bewegungstypen, Zeit. Phys. 44:326 (1927).
5. H. Takahasi, Information Theory of Quantum-Mechanical Channels, Ad. Communication Systems 1:227 (1965).
6. D. Stoler, Equivalence Classes of Minimum Uncertainty Packets, Phys. Rev. D 1:3217 (1970); 4:1925 (1974).
7. H. P. Yuen, Two-photon coherent states of the radiation field, Phys. Rev. A 13:2226 (1976).
8. J. N. Hollenhorst, Quantum limits on resonant-mass gravitational-radiation detectors, Phys. Rev. D 19:1669 (1979).
9. I. Fugiwara and K. Miyoshi, Pulsating States for Quantual Harmonic Oscillator, Prog. Theor. Phys. 64:715 (1980).
10. V. V. Dodonov, E. V. Kurmyshev, and V. I. Man'ko, Generalized Uncertainty Relation and Correlated Coherent States, Phys. Lett. 79A:150 (1980).
11. P. A. K. Rajogopal and J. T. Marshall, New coherent states with applications to time-dependent systems, Phys. Rev. A 26:2977 (1982).
12. H. P. Yuen, Contractive States and the Standard Quantum Limit for Monitoring Free-Mass Positions, Phys. Rev. Lett. 51:719 (1983).

13. E. Schrödinger, Der stetige Übergang von der Mikro- zur Makromechanik, Naturwiss. 28:664 (1926).
14. R. J. Glauber, Coherent and Incoherent States of the Radiation Field, Phys. Rev. 131:2766 (1963).
15. J. R. Klauder and E. C. G. Sudarshan, "Fundamentals of Quantum Optics," Benjamin, New York (1968).
16. L. I. Schiff, "Quantum Mechanics, 2nd. Ed.," McGraw Hill, New York (1955), p. 58.
17. M. M. Nieto and V. P. Gutschick, Inequivalence of the classes of classical and quantum harmonic potentials: Proof by example, Phys. Rev. D 23:922 (1981); M. M. Nieto, Uncountability of the sets of harmonic potentials, Phys. Rev. D 24:1030 (1981).
18. P. B. Abraham and H. E. Moses, Changes in potentials due to changes in the point spectrum: Anharmonic oscillators with exact solutions, Phys. Rev. A 22:1333 (1980).
19. G. Ghosh and R. W. Hasse, Inequivalence of the classes of quantum and classical harmonic potentials: Proof by example, Phys. Rev. D 24:1027 (1981).
20. M. M. Nieto and L. M. Simmons, Jr., Coherent States for general potentials. I. Formalism, Phys. Rev. D 20:1321 (1979), and five other articles in the series leading up to M. M. Nieto, L. M. Simmons, Jr., and V. P. Gutschick, Coherent states for general potentials. VI. Conclusions about the classical motion and the WKB approximation, Phys. Rev. D 23:927 (1981).
21. R. A. Fisher, M. M. Nieto, and V. D. Sandberg, Impossibility of naively generalizing squeezed coherent states, Phys. Rev. D 29:1107 (1984).
22. T. S. Santhanam and V. V. Satyanarayana, Impossibility of squeezed coherent states for a para-Bose oscillator, Phys. Rev. D (in press).
23. G. Leuchs, Status and Future Prospects of Squeezed State and Antibunching Experiments, these Proceedings.
24. C. M. Caves, K. S. Thorne, R. W. P. Drever, V. D. Sandberg, and M. Zimmerman, On the measurement of a weak classical force coupled to a quantum-mechanical oscillator. I. Issues of principle, Rev. Mod. Phys. 52:341 (1980).

QUANTUM STATISTICS OF NONLINEAR OPTICS

D.F. Walls*

Institute for Theoretical Physics
University of California
Santa Barbara, California 93106

1. INTRODUCTION

In these lectures we shall discuss a number of intrinsically quantum mechanical features of light. We begin with a discussion of photon statistics, including photon antibunching, and sub-Poissonian photon statistics.[1] We then describe phase sensitive quantum fluctuations and the property of squeezing.[2] Squeezed states of light have fewer fluctuations in one quadrature than a coherent state. This offers the possibility of an enhanced signal to noise ratio in optical communication systems compared to the quantum limit imposed using coherent light sources. Squeezed states also have potential applications in the detection of very weak forces such as gravitational radiation, where the quantum noise in the detector is comparable to the signal strength.

Squeezing and photon antibunching are intrinsically quantum mechanical features of light. In Sec. 3 we shall consider photon correlation experiments involving more than one mode of the radiation field which provide a test of the Cauchy Schwartz and Bell's inequalities.[3] Violation of these classical inequalities is another manifestation of the quantum nature of the electromagnetic field. We show that photon antibunching, squeezing, and the violation of the Cauchy Schwartz and Bell's inequalities are associated with the non-existence of a positive Glauber-Sudarshan P distribution.

A number of present proposals to generate squeezed states of light involve the use of optical cavities. A full treatment of this problem requires an input-output formulation with the input and output fields as well as the intracavity field quantized. Previous treatments, for example, of the photon statistics of the laser assumed the same statistics for the output light as that for the cavity mode. This is correct for photon number statistics; however, the phase-sensitive correlation functions important for squeezing may be different inside and outside the cavity. In Sec. 4 we present a general formulation of this problem and in Sec. 5 apply this formalism to obtain the squeezing spectrum for the output of a degenerate parametric oscillator. We conclude these lectures in Sec. 6 with a discussion of the effect of dissipation in quantum coherence. There is considerable interest in achieving an experimental demonstration of a macroscopic superposition of quantum states.[4] When one includes the interaction with

*On leave from: Department of Physics, University of Waikato, Hamilton, New Zealand

the environment, it is seen that rapid dephasing effects may make observation of such macroscopic superpositions extremely difficult.

2. STATISTICAL PROPERTIES OF THE ELECTROMAGNETIC FIELD

In the quantum theory of the electromagnetic field the field operator may be written as[5]

$$E(\underset{\sim}{r},t) = E^{(+)}(\underset{\sim}{r},t) + E^{(-)}(\underset{\sim}{r},t) \tag{2.1}$$

where $E^{(+)}(\underset{\sim}{r},t)$ is given by

$$E^{(+)}(\underset{\sim}{r},t) = i \sum (\tfrac{1}{2}\hbar\omega_k)^{\frac{1}{2}} a_k u_k(\underset{\sim}{r}) e^{-i\omega_k t}$$
$$E^{(-)} = \{E^{(+)}\}^\dagger \tag{2.2}$$

where $u_k(\underset{\sim}{r})$ is a vector mode function and $a_k(a_k^\dagger)$ are the annihilation (creation) operators for the kth mode of the electromagnetic field. These obey the usual boson commutation relations

$$[a_k, a_{k'}^\dagger] = \delta_{kk'} \ . \tag{2.3}$$

Experiments involving the detection of the electromagnetic field with a photomultiplier and subsequent correlation of the output of the photomultiplier correspond to photon correlation or intensity correlation measurements. Typical of such measurements are the intensity of the electromagnetic field $\langle E^{(-)}(\underset{\sim}{r},t)\, E^{(+)}(\underset{\sim}{r},t)\rangle$ and the second order correlation function $\langle E^{(-)}(\underset{\sim}{r},t)\, E^{(-)}(\underset{\sim}{r},t+\tau)\, E^{(+)}(\underset{\sim}{r},t+\tau)\, E^{(+)}(\underset{\sim}{r},t)\rangle$. Measurements of photon number correlations serve to distinguish between fields with different photon number distributions. Such measurements, however, do not give any information on phase fluctuations. To investigate such phase-dependent properties will require measurements of correlation functions such as $\langle E^{(+)2}\rangle$ and $\langle E^{(-)2}\rangle$. We may illustrate many of the quantum effects we wish to consider by reference to a single mode field. The results may be readily generalized to multimode fields.

An arbitrary density operator of the electromagnetic field may be expressed in the following diagonal coherent state representation[5,6]

$$\rho = \int P(\alpha)|\alpha\rangle\langle\alpha| d^2\alpha \ . \tag{2.4}$$

$P(\alpha)$ is a quasiprobability distribution and, for certain states of the electromagnetic field, may be highly singular. When $P(\alpha)$ exists as a non-singular positive function, the statistical properties of the electromagnetic field may be simulated by a classical stochastic process. Intrinsically quantum mechanical behavior will occur in a regime in which no non-singular $P(\alpha)$ exists. For a coherent state $P(\alpha)$ is a delta function. Thus, there is a sense in which the coherent state represents the border between quantum and classical behavior. When the P representation is singular, other quasiprobability distributions may be used, for example, the Q and Wigner functions. Another representation which is often useful is a generalization of the P representation given by Drummond and Gardiner[7]

$$\rho = \int P(\alpha,\alpha^\dagger) \frac{|\alpha\rangle\langle(\alpha^\dagger)^*|}{\langle(\alpha^\dagger)^*|\alpha\rangle} d\,\mu\{\alpha,\alpha^\dagger\} \tag{2.5}$$

where α and $\alpha^\dagger$ are independent complex variables and $d\mu(\alpha,\alpha^\dagger)$ is the integration measure. We shall now discuss some quantum effects which may be exhibited by light fields.

2.1 Photon Antibunching

Photon antibunching is a property associated with the photon counting statistics of a light beam. To describe the photon statistics of light, it is sufficient to use a diagonal representation in the number states

$$\rho = \sum P_n |n> \tag{2.6}$$

where P_n is the probability that there are n photons in the mode.

We may define a normalized second-order correlation function

$$g^{(2)}(0) = \frac{<a^\dagger a^\dagger aa>}{<a^\dagger a>^2} \quad . \tag{2.7}$$

This is related to the variance V(n) in the photon number by

$$g^{(2)}(0) - 1 = \frac{V(n)-\bar{n}}{\bar{n}^2} \tag{2.8}$$

where $\bar{n} = <a^\dagger a>$ is the mean photon number. Clearly, for light with super Poissonian statistics ($V(n) > \bar{n}$), $g^{(2)}(0)$ is greater than unity, whereas for light with sub-Poissonian statistics ($V(n) < \bar{n}$) $g^{(2)}(0)$ is less than unity. Light with $g^{(2)}(0)$ less than unity is said to exhibit photon antibunching in contrast with photon bunching $g^{(2)}(0) > 1$ observed in the Hanbury-Brown-Twiss effect.[8] Photon antibunching and sub-Poissonian statistics have been observed in resonance fluorescence from a two-level atom.[9,10] We may express $g^{(2)}(0)$ in terms of the P representation as follows

$$g^{(2)}(0) = 1 + \frac{\int P(\alpha)(|\alpha|^2-<|\alpha|^2>)^2 d^2\alpha}{[\int P(\alpha)|\alpha|^2 d^2\alpha]^2} \quad . \tag{2.9}$$

It is clear that photon antibunching ($g^{(2)}(0) \leq 1$) may only occur if $P(\alpha)$ assumes negative values. Thus, photon antibunching represents an intrinsically quantum mechanical effect since it cannot be described by a classical stochastic process with a positive distribution function.

2.2 Squeezing

Whereas photon antibunching is a property associated with photon number statistics, squeezing is a property associated with phase-sensitive fluctuations. We may write the electric field for a single mode as

$$E(t) = \frac{\lambda}{2} (X_1 \cos\omega t + X_2 \sin\omega t) \tag{2.10}$$

where λ is a constant including the spatial wave functions. X_1 and X_2 are hermitian operators for the amplitudes of the two quadrature phases of the field.

$$X_1 = \frac{a+a^\dagger}{2i}$$
$$X_2 = \frac{a-a^\dagger}{2i} \quad . \tag{2.11}$$

They obey the commutation relation

$$[X_1, X_2] = \frac{i}{2} . \tag{2.12}$$

From this commutation relation we may deduce the following results for the uncertainties

$$\Delta X_i = [V(X_i)]^{\frac{1}{2}} \quad \text{in} \quad X_1 \quad \text{and} \quad X_2$$

$$\Delta X_1 \, \Delta X_2 \geq \frac{1}{4} . \tag{2.13}$$

A family of minimum uncertainty states is defined by taking the equal sign. One such class of minimum uncertainty states is the coherent states which have $V(X_1) = V(X_2) = 1/4$. A broader class of minimum uncertainty states may have unequal variances in each quadrature. These are the so-called squeezed states. The condition for squeezing is

$$V(X_i) < \frac{1}{4} \qquad i = 1 \text{ or } 2 . \tag{2.14}$$

It is sometimes convenient, especially for multimode fields, to write the condition in terms of the normally ordered variance

$$:V(X_i): < 0 \qquad i = 1 \text{ or } 2 . \tag{2.15}$$

We may express the variance in one quadrature in terms of the P representation

$$V(X_1) = \frac{1}{4} \left\{ 1 + \int P(\alpha)[(\alpha + \alpha^*) - (\langle\alpha\rangle + \langle\alpha^*\rangle)]^2 d^2\alpha \right\} . \tag{2.16}$$

The condition for squeezing $[V(X_1) < 1/4]$ requires that $P(\alpha)$ be a non-positive function. In this sense, squeezing like photon antibunching is a nonclassical property of the electromagnetic field.

We may formally define a squeezed state as follows.[11]

$$|\alpha, \zeta\rangle = D(\alpha) S(\zeta) |0\rangle \tag{2.17}$$

where $S(\zeta)$ is the squeeze operator

$$S(\zeta) = \exp(\tfrac{1}{2}\zeta^* a^2 - \tfrac{1}{2}\zeta a^{\dagger 2}) \tag{2.18}$$

and

$$\zeta = re^{i\theta}$$

and $D(\alpha)$ is the displacement operator

$$D(\alpha) = e^{-\frac{1}{2}|\alpha|^2} e^{\alpha a^\dagger} e^{-\alpha^* a} . \tag{2.19}$$

An alternative but equivalent characterization of squeezed states has been given by Yuen.[17] If one defines a mode b by

$$b = \mu a + \nu a^\dagger \tag{2.20}$$

and $\mu^2 - \nu^2 = 1$, then if mode a is in a coherent state, mode b will be in a squeezed state. The variances in the squeezed state $|\alpha, \zeta\rangle$ are given by

$$V(Y_1) = \frac{1}{4} e^{-2r}$$

$$V(Y_2) = \frac{1}{4} e^{2r} \tag{2.21}$$

where $Y_1 + iY_2 = (X_1 + iX_2)\ e^{-(i\theta)/2}$ is a rotated complex amplitude. The mean photon number in the squeezed state $|\alpha,\zeta\rangle$ is

$$\langle n\rangle = |\alpha|^2 + \sinh^2 r \ . \tag{2.22}$$

clearly, the variances $V(X_i)$ are independent of the field amplitude α. Thus, squeezing is a quantum mechanical effect which may occur in fields with arbitrary intensity.

In the limit where the coherent amplitude greatly exceeds the squeezing ($|\alpha|^2 \gg \sinh^2 r$) we find the following relation between the photon statistics and the squeezing

$$\begin{aligned} \frac{V(n)-\bar{n}}{\bar{n}} &= 4\ :V(X_1): \\ &= (e^{-2r} - 1) \end{aligned} \tag{2.23}$$

where we have chosen α real so that the amplitude is carried by X_1. For $r > 0$ we have reduced amplitude fluctuations $:V(X_1): < 0$ and sub-Poissonian statistics whereas for $r < 0$ we have an increase in amplitude fluctuations and super Poissonian statistics.

No such simple relation between photon statistics and squeezing exists for all values of α. For example, in the opposite limit of $\alpha \ll 1$, that is, a squeezed vacuum $|0,r\rangle$

$$g^{(2)}(0) = 1 + \frac{\cosh 2r}{\sinh^2 r} \ . \tag{2.24}$$

Thus photons in a squeezed vacuum are always bounded irrespective of the sign of the squeeze parameter.

A simple example of a system which may exhibit both photon antibunching and squeezing is the degenerate parametric amplifier which may be described by the following interaction Hamiltonian

$$H = i\hbar g\ (a^2 - a^{\dagger 2}) \ . \tag{2.25}$$

This clearly generates squeezed states [see Eq. (2.8)] and for an initially highly excited coherent state will also show photon antibunching.

3. VIOLATION OF CLASSICAL INEQUALITIES

3.1. Two Mode Fields

We shall now consider some classical inequalities which may be violated in two mode quantum fields. Consider two modes characterized by the operators a and b.

A weak inequality is imposed by quantum mechanics. Since $a^\dagger a$ is a hermitian operator, we must have $\langle(a^\dagger a + \lambda b^\dagger b)^2\rangle \geqslant 0$ for all real λ. The left-hand side of this inequality is a quadratic in λ and the inequality means this quadratic has no real roots. Thus, its discriminant must be less than or equal to zero, which gives the following inequality

$$\langle a^\dagger a b^\dagger b\rangle^2 \leqslant \langle(a^\dagger a)^2\rangle\langle(b^\dagger b)^2\rangle \ . \tag{3.1}$$

A stronger inequality may be derived which holds for fields with a positive P representation. We may write the normally ordered expectation values as follows

$$
\begin{aligned}
\langle a^{\dagger}a^{\dagger}aa\rangle &= \int |\alpha|^4 \, P(\alpha,\beta)d^2\alpha \; d^2\beta \\
\langle b^{\dagger}b^{\dagger}bb\rangle &= \int |\beta|^4 \, P(\alpha,\beta)d^2\alpha \; d^2\beta \qquad (3.2)\\
\langle a^{\dagger}b^{\dagger}ab\rangle &= \int |\alpha\beta|^2 P(\alpha,\beta)d^2\alpha \; d^2\beta .
\end{aligned}
$$

Then using standard methods we may prove the Cauchy-Schwarz inequality

$$\langle|\alpha\beta|^2\rangle^2 \leqslant \langle|\alpha|^4\rangle\langle|\beta|^4\rangle \qquad (3.3)$$

from which the following operator inequality follows

$$\langle a^{\dagger}ab^{\dagger}b\rangle^2 = \langle a^{\dagger}a^{\dagger}aa\rangle\langle b^{\dagger}b^{\dagger}bb\rangle \quad . \qquad (3.4)$$

For situations with symmetry between the modes a and b we may write the weak inequality (3.1) as

$$\langle a^{\dagger}ab^{\dagger}b\rangle \leqslant \langle a^{\dagger}a^{\dagger}aa\rangle + \langle a^{\dagger}a\rangle \qquad (3.5)$$

whereas the strong inequality (3.4) may be written as

$$\langle a^{\dagger}ab^{\dagger}b\rangle \leqslant \langle a^{\dagger}a^{\dagger}aa\rangle \quad . \qquad (3.6)$$

Zubairy has discussed the violation of this strong inequality in the two-photon laser.[13]

A simple example of a system which violates the strong inequality is a non-degenerate parametric amplifier

$$H = i\hbar g \, (a^{\dagger}b^{\dagger} - ab) \; . \qquad (3.7)$$

In fact, for this system

$$\langle a^{\dagger}ab^{\dagger}b\rangle = \langle a^{\dagger}a^{\dagger}aa\rangle + \langle a^{\dagger}a\rangle \qquad (3.8)$$

which is the maximum violation of Eq. (3.6) allowed by Eq. (3.5).

Analyses of the non-degenerate parametric oscillator including pumping and loss terms have been given by Graham[14] who used the Wigner function and by McNeil and Gardiner[15] who used the generalized P function.

Recently Graham[16] has proved the following equality satisfied by fields which have an interaction of the form of Eq. (3.7) plus pumping and loss terms. In the case of equal losses for both modes he finds the equality

$$\langle a^{\dagger}ab^{\dagger}b\rangle = \langle a^{\dagger}a^{\dagger}aa\rangle + \frac{1}{2}\langle a^{\dagger}a\rangle \quad . \qquad (3.9)$$

This equality is seen to be obeyed in the analysis of McNeil and Gardiner[15] which included losses.

3.2 Four Mode Fields

Clearly further quantum features may be uncovered as we consider fields with a larger number of modes. The four mode case is worthy of study since the Bell's inequalities[3b,17] may be formulated in this case. We shall follow a discussion by Reid and Walls[18] and generalize the Hamiltonian (3.7) to two pairs of coupled modes

$$H = i\hbar g\,(a_1^\dagger b_1^\dagger + a_2^\dagger b_2^\dagger - a_1 b_1 - a_2 b_2) \ . \tag{3.10}$$

Here a_1 and b_1 represent two photons of the same polarization traveling in opposite directions. The modes a_2 and b_2 travel in opposite directions and have orthogonal polarization to modes a_1 and b_1.

We may define the following correlation coefficient

$$C^{(1)} = \frac{\langle a_1^\dagger a_1 b_1^\dagger b_1 - a_1^\dagger a_1 b_2^\dagger b_2 \rangle}{\langle a_1^\dagger a_1 b_1^\dagger b_1 + a_1^\dagger a_1 b_2^\dagger b_2 \rangle} \ . \tag{3.11}$$

Thus the state created by a single photon pair emission

$$H|0\rangle = \frac{1}{\sqrt{2}}\,(|1100\rangle + |0011\rangle) \tag{3.12}$$

has $C^{(1)} = 1$.

The basis of Bell's inequalities is to create a correlated state of photons such as that described by Eq. (3.12). One then places polarizers at angles θ and ϕ in the paths of photons traveling in the +z and -z directions, respectively. After the polarizers, the fields are described by the related modes

$$\begin{aligned} c_+ &= a_1\cos\theta + a_2\sin\theta \\ c_- &= -a_1\sin\theta + a_2\cos\theta \\ d_+ &= b_1\cos\phi + b_2\sin\phi \\ d_- &= -b_1\sin\phi + b_2\cos\phi \ . \end{aligned} \tag{3.13}$$

Placing photocounters after the polarizers one is able to measure the photon number correlation functions and calculate the correlation function

$$E(\theta,\phi) = \frac{1}{\chi}\ \langle (c_+^\dagger c_+ - c_-^\dagger c_-)(d_+^\dagger d_+ - d_-^\dagger d_-)\rangle \tag{3.14}$$

and

$$\begin{aligned} \chi &= \langle (c_+^\dagger c_+ + c_-^\dagger c_-)(d_+^\dagger d_+ + d_-^\dagger d_-)\rangle \\ &= \langle (a_+^\dagger a_+ + a_-^\dagger a_-)(b_+^\dagger b_+ + b_-^\dagger b_-)\rangle \ . \end{aligned} \tag{3.15}$$

This correlation function may be written, for the field under discussion, as

$$E(\theta,\phi) = C^{(1)}\cos 2(\theta - \phi) \ . \tag{3.15}$$

The Bell's inequalities are based on a realistic local hidden variable theory. A form of the Bell's inequalities may be written as

$$2 \leqslant E(\theta,\phi) + E(\theta',\phi) - E(\theta,\phi') + E(\theta',\phi') \leqslant 2 \tag{3.16}$$

with the choice of angles

$$\Phi = \theta-\phi = \theta'-\phi = \theta'-\phi' = \frac{1}{3}(\theta-\phi'). \tag{3.17}$$

This may be written as

$$|3E(\Phi) - E(3\Phi)| \leqslant 2 . \tag{3.18}$$

Using the quantum mechanical result [Eq. (3.15)] for $\Phi = \pi/8$ we find

$$|3E(\Phi) - E(3\Phi)| = 2\sqrt{2}\, C^{(1)} . \tag{3.19}$$

Hence we see that Bell's inequalities are violated by the predictions of quantum mechanics for states with $C^{(1)} > 1/\sqrt{2}$.

The origin of the different predictions of quantum and classical theory may usefully be discussed using the P representation. In terms of the P representation the correlation function $E(\theta,\phi)$ may be written in the form[18]

$$E(\theta,\phi) = \frac{1}{\chi} \int P(\underset{\sim}{\alpha})(|\xi_+|^2 - |\xi_-|^2)(|\zeta_+|^2 - |\zeta_-|^2)d^2\alpha_1 d^2\alpha_2 d^2\beta_1 d^2\beta_2 \tag{3.20}$$

where

$$\xi_+ = \alpha_1\cos\theta + \alpha_2\sin\theta$$

$$\xi_- = -\alpha_1\sin\theta + \alpha_2\cos\theta$$

$$\zeta_+ = \beta_1\cos\phi + \beta_2\sin\phi$$

$$\zeta_- = -\beta_1\sin\phi + \beta_2\cos\phi .$$

This is the same form as in hidden variable theory with the variable α representing the state of the system assuming the role of a hidden variable. The representation is local since $\xi_\pm$ is a function of θ and $\zeta_\pm$ is a function of ϕ. Then provided $P(\alpha)$ is a positive distribution function, Bell's proof (3.6) leads to the inequalities (3.20). We see that a breakdown of the proof and the analogy with a hidden variable theory occurs when $P(\alpha)$ becomes negative. Thus, the nonclassical behavior indicated by a violation of Bell's inequalities may be associated with a negative $P(\alpha)$.

We note that experiments demonstrating a violation of Bell's inequalities have been realized.[19,20]

4. INPUT-OUTPUT FORMULATION OF OPTICAL INTERACTIONS IN CAVITIES

We wish to study the quantum statistics of light produced from nonlinear interactions inside optical cavities. The laser itself is the most celebrated example of such a system. The quantum theory of the laser which was developed employing a number of different techniques, e.g. master equations, Fokker-Planck equations, or Langevin equations enabled the photon statistics of the light inside the cavity to be computed. Measurements of the photon statistics are, of course, made on the light which

emerges from the laser cavity. The photon statistics of the light external to the laser cavity has been assumed to be the same as that calculated inside the cavity and this has been borne out by photon counting experiments. It is true that where measurements of photon number statistics are concerned the result is unchanged whether the light is internal or external to the cavity.

The situation is, however, markedly different when one is considering phase-sensitive measurements such as the variance in the quadrature phases of the field. The amount of squeezing present will depend strongly on whether one considers internal or external cavity modes. The reason for this is that squeezing is very sensitive to the presence of vacuum fluctuations. Thus, a calculation of squeezing in a cavity configuration must include all modes entering and leaving the cavity including the vacuum modes. The vacuum modes which make no contribution to photon number statistics can drastically affect the squeezing.

The treatment we shall present of this problem is a general formulation developed by Collet and Gardiner.[21,22] We shall discuss their formulation, derive some general results, and present an application in Sec. 5.

4.1 General Formulation

We consider the internal cavity mode a as the system which is coupled to the external cavity modes which may be treated as reservoir. The Hamiltonian describing this system reservoir interaction is

$$\begin{aligned} H &= H_{SYS} + H_B + H_{INT} \\ H_B &= \hbar \int d\omega\, b^\dagger(\omega)\, b(\omega) \\ H_{INT} &= i\hbar \int d\omega\, \kappa(\omega)(b^\dagger(\omega)a - a^\dagger b(\omega)) \, . \end{aligned} \tag{4.1}$$

a is the annihilation operator for the cavity mode, $b(\omega)$ are the bath operators, and H_{SYS} is left unspecified. Solving first for the bath operators

$$\dot{b}(\omega) = - i\omega b(\omega) + \kappa(\omega)a \tag{4.2}$$

$$b(\omega) = e^{-i\omega(t-t_0)} b_0(\omega) + \kappa(\omega) \int_{t_0}^{t} e^{-i\omega(t-t')} a(t')dt' \, . \tag{4.3}$$

The cavity mode operator then obeys the equation

$$\begin{aligned} \dot{a} &= - \frac{i}{\hbar} [a,H_{SYS}] - \int d\omega\, \kappa(\omega)b(\omega) \\ &= - \frac{i}{\hbar} [a,H_{SYS}] - \int d\omega\, \kappa(\omega)e^{-i\omega(t-t_0)} b_0(\omega) \\ &\quad - \int d\omega\, \kappa^2(\omega) \int_{t_0}^{t} a(t')e^{-i\omega(t-t')} dt' \, . \end{aligned} \tag{4.4}$$

Choosing

$$\kappa^2(\omega) = \frac{\gamma}{2\pi} \, , \tag{4.5}$$

using

$$\int_{-\infty}^{\infty} d\omega\, e^{-i\omega(t-t')} = 2\pi\, \delta(t-t') \tag{4.6}$$

with the convention

$$\int_{t_0}^{t} c(t')\, \delta(t-t')dt' = \frac{1}{2}\, c(t) \tag{4.7}$$

and defining

$$b_{IN} = -\frac{1}{\sqrt{2\pi}} \int d\omega\, e^{-i\omega(t-t_0)}\, b_0(\omega) \tag{4.8}$$

we find

$$\dot{a} = -\frac{i}{\hbar}[a,H_{SYS}] - \frac{\gamma}{2}\, a + \sqrt{\gamma}\, b_{IN} \quad . \tag{4.9}$$

This has the form of a quantum Langevin equation where b_{IN} would be a noise operator. In this discussion, however, we consider b_{IN} as the operator describing the input field. Equation (4.3) expressed the bath operator in terms of its value at the initial time t_0. It could, however, equally be expressed in terms of its value at some final time t_1.

$$b(\omega) = e^{i\omega(t-t_1)}\, b_1(\omega) - \kappa(\omega) \int_t^{t_1} e^{-i\omega(t-t')}\, a(t')dt' \quad . \tag{4.10}$$

Defining

$$b_{OUT} = \frac{1}{\sqrt{2\pi}} \int d\omega\, e^{-i\omega(t-t_1)}\, b_1(\omega) \tag{4.11}$$

this gives

$$b_{IN} + b_{OUT} = \frac{i}{\sqrt{2\pi}} \int d\omega\, \kappa(\omega) \int_{t_0}^{t_1} e^{-i\omega(t-t')}\, a(t')dt'$$

$$= \sqrt{\gamma}\, a \quad . \tag{4.12}$$

The time reversed equation is

$$\dot{a} = -\frac{i}{\hbar}[a,H_{SYS}] + \frac{\gamma}{2}\, a - \sqrt{\gamma}\, b_{OUT} \quad . \tag{4.13}$$

Equations (4.9) and (4.13) are two equivalent ways of describing the time evolution of the cavity mode. Equation (4.9) involves b_{IN} and the time reversed equation (4.13) involves b_{OUT}.

4.2. Two Time Correlation Functions

We shall now derive explicit relations connecting the two time correlation functions of the input and output fields. First, using

$$
\begin{aligned}
b_{IN}(t) &= -\frac{1}{\sqrt{2\pi}} \int d\omega\, e^{-i\omega(t-t_0)}\, b_0(\omega) \\
&= -\frac{1}{\sqrt{2\pi}} \int d\omega \left[b(\omega) - \kappa(\omega) \int_{t_0}^{t} e^{-i\omega(t-t')}\, a(t')dt' \right] \\
&= -\frac{1}{\sqrt{2\pi}} \int d\omega\, b(\omega) + \frac{\sqrt{\gamma}}{2}\, a(t) \quad .
\end{aligned}
\tag{4.14}
$$

Then for an arbitrary internal operator c

$$
[c(t), \sqrt{\gamma}\, b_{IN}(t)] = \frac{\gamma}{2} [c(t), a(t)] \tag{4.15}
$$

since c must commute with $b(\omega)$. Now since $c(t)$ can only be a function of $b_{IN}(t')$ for earlier times $t' < t$, and the input field operators must commute at different times

$$
[c(t), \sqrt{\gamma}\, b_{IN}(t')] = 0 \qquad t' > t \quad . \tag{4.16}
$$

Similarly, one may show

$$
[c(t), \sqrt{\gamma}\, b_{OUT}(t')] = 0 \qquad t' < t \tag{4.17}
$$

so that, using $b_{IN} + b_{OUT} = \sqrt{\gamma}\, a$, we have

$$
\begin{aligned}
&[c(t), \gamma\, a(t') - \sqrt{\gamma}\, b_{IN}(t')] = 0 \qquad t' < t \\
&[c(t), \sqrt{\gamma}\, b_{IN}(t')] = [c(t), a(t')] \qquad t' < t \quad .
\end{aligned}
\tag{4.18}
$$

Combining Eqs. (4.15)-(4.18) gives the commutator of any system operator with the input field operator at any time as

$$
[c(t), \sqrt{\gamma}\, b_{IN}(t')] = \gamma\, u(t-t')[c(t), a(t')] \quad . \tag{4.19}
$$

The commutator of the output field operator may be shown to be equal to the commutator of the input field as required

$$
[b_{OUT}(t), b_{OUT}^{\dagger}(t')] = [b_{IN}(t), b_{IN}^{\dagger}(t')] \quad . \tag{4.20}
$$

For the case of a coherent or vacuum input, it is now possible to express the variances of the output field entirely in terms of those of the internal system. For an input field of this type, the moments $\langle b_{IN}^{\dagger}(t) b_{IN}(t')\rangle$ will factorize. The moments $\langle b_{IN}^{\dagger}(t) a(t')\rangle$ also factorize, so that

$$
\begin{aligned}
\langle b_{OUT}^{\dagger}(t), b_{OUT}(t')\rangle &= \langle (\sqrt{\gamma}\, a\, (t) - b_{IN}^{\dagger}(t)), (\sqrt{\gamma}\, a(t') - b_{IN}(t'))\rangle \\
&= \gamma\, \langle a^{\dagger}(t), a(t')\rangle
\end{aligned}
\tag{4.21}
$$

where we use the notation

$$
\langle U,V\rangle = \langle UV\rangle - \langle U\rangle\langle V\rangle \quad .
$$

Also,

$$<b_{OUT}(t), b_{OUT}(t')> = <(\sqrt{\gamma}\, a(t) - b_{IN}(t))(\sqrt{\gamma}\, a(t') - b_{IN}(t'))>$$

$$= \gamma <a(t), a(t')> - \sqrt{\gamma}\, [b_{IN}(t), a(t')]$$

$$= \gamma <a[\max(t,t')], a[\min(t,t')]> . \tag{4.22}$$

Thus the two time correlation functions of the output field are related to the time ordered two time correlation functions of the cavity fields.

4.3. Spectrum of Squeezing

To calculate the squeezing spectrum we take the Fourier transform of the normally ordered two time correlation function

$$<:X_i^{OUT}(t), X_i^{OUT}(0):>$$

where X_i are the amplitudes of the quadrature phases

$$X_1 = \frac{a+a^\dagger}{2} , X_2 = \frac{a-a^\dagger}{2i} . \tag{4.23}$$

The spectrum of the output field is given by

$$:S_{X_i}^{OUT}(\omega): = \gamma \int T <:X_1(\tau), X_1(0):> e^{i\omega\tau} d\tau \tag{4.24}$$

where T denotes time ordering and γ is the loss rate at the output mirror.

$$T <:X_1(\tau) X_1(0):> = \frac{1}{4} \{<a(\tau) a(0)> + <a^\dagger(\tau) a(0)>$$

$$+ <a^\dagger(0) a^\dagger(\tau)> + <a^\dagger(0) a(\tau)>\} . \tag{4.25}$$

Now given that the $P(\alpha)$ for the cavity mode obeys the Fokker-Planck equation

$$\frac{\partial P(\alpha)}{\partial t} = \left\{ \frac{\partial}{\partial\alpha_i} A_{ij}\alpha_j + \frac{1}{2}\frac{\partial^2}{\partial\alpha_i\partial\alpha_j} D_{ij} \right\} P(\alpha) \tag{4.26}$$

and defining the correlation matrix

$$C = \begin{bmatrix} <a(\tau) a(0)> & <a^\dagger(\tau) a(0)> \\ <a^\dagger(0) a(\tau)> & <a^\dagger(0) a^\dagger(\tau)> \end{bmatrix} . \tag{4.27}$$

The spectrum of fluctuations is given by[23]

$$S(\omega) = (A+i\omega)^{-1} D(A^T-i\omega)^{-1} \tag{4.28}$$

where

$$S_{ij}(\omega) = \int e^{i\omega\tau} C_{ij}(\tau) d\tau .$$

The correlation functions obtained from the Fokker-Planck equation for the Glauber P function correspond to time and normal ordering. The spectrum of the squeezing in the output field is given by

$$:S_i^{OUT}(\omega): = \frac{\gamma}{4}\{(S_{12} + S_{21}) \pm (S_{11} + S_{22})\} . \tag{4.29}$$

Note for classical fields $:S_i(\omega): > 0$. Squeezing at frequency ω occurs when $:S_i(\omega): < 0$. In the following section we shall consider an example of the above formalism.

5. SQUEEZING SPECTRUM OF THE DEGENERATE PARAMETRIC OSCILLATOR

We consider a parametric amplifier placed inside an optical cavity. The pump field is treated classically. The Hamiltonian describing this system is

$$H = h\omega a^{\dagger} a + \frac{i\hbar}{2}(\varepsilon a^{\dagger 2} - \varepsilon^{*} a^{2}) + a\Gamma^{\dagger} + a^{\dagger}\Gamma \tag{5.1}$$

where $\varepsilon = \varepsilon_p \chi$ and ε_p is the amplitude of the pump and χ is the nonlinear susceptibility of the medium. Γ is the reservoir operator representing cavity losses. Using standard techniques[31] the following Fokker-Planck equation for $P(\alpha,\alpha^{\dagger})$ may be derived

$$\frac{\partial P(\alpha\alpha^{\dagger})}{\partial t} = \left\{ \varepsilon^{*}\left(\frac{\partial}{\partial\alpha^{\dagger}}\alpha - \frac{1}{2}\frac{\partial^2}{\partial\alpha^{\dagger 2}}\right) + \varepsilon\left(\frac{\partial}{\partial\alpha}\alpha^{\dagger} - \frac{1}{2}\frac{\partial^2}{\partial\alpha^2}\right) + \frac{\gamma}{2}\left(\frac{\partial}{\partial\alpha^{\dagger}} + \frac{\partial}{\partial\alpha}\alpha\right)\right\} P(\alpha,\alpha^{\dagger}) \tag{5.2}$$

where the reservoir has been assumed to be at zero temperature. This equation has a stationary solution given by[24]

$$P(\alpha,\alpha^{\dagger}) = N \exp\left(2\alpha\alpha^{\dagger} - \frac{\gamma}{2\varepsilon}\alpha^2 - \frac{\gamma}{2\varepsilon^{*}}\alpha^{\dagger 2}\right) . \tag{5.3}$$

This function would not be normalizable if $\alpha^{\dagger}_{*}$ was taken to be α^{*}. However, if the pump phase is chosen to be real $\varepsilon = \varepsilon^{*} = |\varepsilon|$, this solution is normalizable for α and $\alpha^{\dagger}$ both along the real axis

$$P(\alpha,\alpha^{\dagger}) = N \exp\left(-\frac{1}{2}\underset{\sim}{\alpha}^{T}\underset{\sim}{B}^{-1}\underset{\sim}{\alpha}\right) \tag{5.4}$$

where

$$\underset{\sim}{\alpha} = \begin{pmatrix}\alpha \\ \alpha^{\dagger}\end{pmatrix}$$

and

$$\frac{1}{2}B^{-1} = \begin{pmatrix} \frac{\gamma}{2|\varepsilon|} & -1 \\ -1 & \frac{\gamma}{2|\varepsilon|} \end{pmatrix} .$$

The covariance matrix

$$C(\alpha,\alpha^\dagger) = \begin{pmatrix} \langle\alpha^2\rangle - \langle\alpha\rangle^2 & \langle\alpha^\dagger\alpha\rangle - \langle\alpha^\dagger\rangle\langle\alpha\rangle \\ \langle\alpha^\dagger\alpha\rangle - \langle\alpha^\dagger\rangle\langle\alpha\rangle & \langle\alpha^{\dagger 2}\rangle - \langle\alpha^\dagger\rangle^2 \end{pmatrix} \tag{5.5}$$

is

$$C(\alpha,\alpha^\dagger) = \frac{1}{2}\,\frac{1}{(\gamma/2)^2+|\varepsilon|^2} \begin{pmatrix} \frac{\gamma}{2}|\varepsilon| & |\varepsilon|^2 \\ |\varepsilon|^2 & \frac{\gamma}{2}|\varepsilon| \end{pmatrix} \tag{5.6}$$

from which the results for the variances in the quadrature phases may be deduced.

$$\begin{aligned} :V(X_1): &= \frac{1}{4}\,\frac{|\varepsilon|}{\frac{\gamma}{2}-|\varepsilon|} \\ :V(X_2): &= -\frac{1}{4}\,\frac{|\varepsilon|}{\frac{\gamma}{2}+|\varepsilon|} \end{aligned} \tag{5.7}$$

At threshold for oscillation $|\varepsilon| = \frac{\gamma}{2}$

$$\begin{aligned} :V(X_1): &\to \infty \\ :V(X_2): &\to -\frac{1}{8} \quad . \end{aligned} \tag{5.8}$$

This result is confirmed by a calculation which quantizes the pump field.[25,26] This analysis shows that the variance of fluctuations in X_2 inside the cavity may be reduced by at most a factor of two.

Now we shall consider the variance of fluctuations outside the cavity. For a double-ended cavity with decay rates γ_1 and γ_2 at each mirror, the Fokker-Planck equation for the internal mode is given by Eq. (5.2) with $\gamma = \gamma_1 + \gamma_2$. The drift and diffusion matrices are

$$A = \begin{pmatrix} \frac{\gamma_1+\gamma_2}{2} & \varepsilon^* \\ \varepsilon & \frac{\gamma_1+\gamma_2}{2} \end{pmatrix}, \quad D = \begin{pmatrix} -\varepsilon & 0 \\ 0 & \varepsilon^* \end{pmatrix} \quad . \tag{5.9}$$

Using the formula (4.29), the spectrum of the squeezing in the two quadratures may be obtained directly.

$$:S_1^{OUT}(\omega): = \frac{\varepsilon\,\frac{\gamma_1}{2}}{\left(\frac{\gamma_1+\gamma_2}{2}+\varepsilon\right)^2+\omega^2} \tag{5.10}$$

$$:S_2^{OUT}(\omega): = \frac{-\varepsilon \frac{\gamma_1}{2}}{\left(\frac{\gamma_1+\gamma_2}{2} - \varepsilon\right)^2 + \omega^2} \tag{5.11}$$

Maximum squeezing occurs at threshold $\varepsilon = \frac{\gamma_1+\gamma_2}{2}$.

$$:S_1^{OUT}(\omega): = \frac{\gamma_1}{2} \frac{\frac{\gamma_1+\gamma_2}{2}}{\omega^2} \tag{5.12}$$

$$:S_2^{OUT}(\omega): = -\frac{1}{4} \frac{\frac{\gamma_1}{2} \frac{\gamma_1+\gamma_2}{2}}{\frac{\gamma_1+\gamma_2}{2} + \omega^2} \tag{5.13}$$

A graph of $:S_2^{OUT}(\omega):$ at threshold is plotted in Fig. 1 for both single ($\gamma_2 = 0$) and double-ended ($\gamma_1 = \gamma_2$) cavities.

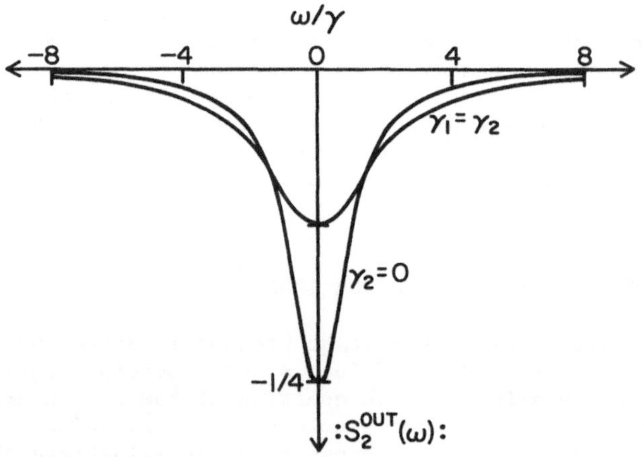

Fig. 1. Squeezing spectrum $:S_2^{OUT}(\omega):$ at threshold for $\gamma_2 = 0$ and $\gamma_1 = \gamma_2$.

For $\omega = 0$

$$:S_2^{OUT}(0): = -\frac{1}{8} \quad \gamma_1 = \gamma_2$$

$$= -\frac{1}{4} \quad \gamma_2 = 0 \tag{5.14}$$

Thus for a single-ended cavity maximum squeezing may be achieved in the output frequency band around $\omega = 0$.[21,27,28] If there are losses at the other mirror or indeed losses in the medium the squeezing will be reduced from its optimal value. We note that the spectrum of fluctuations in the other quadrature $:S_1^{OUT}(\omega):$ diverge at $\omega = 0$.

The squeezing in the total field is found by integrating over ω

$$:S_2^{TOT}: = \int d\omega - \frac{1}{4}\frac{\gamma_1}{2}\frac{\frac{\gamma_1+\gamma_2}{2}}{\left(\frac{\gamma_1+\gamma_2}{2}\right)^2+\omega^2}$$

$$= -\frac{\gamma_1}{8} \tag{5.15}$$

at threshold.

The squeezing in the total field inside and outside the cavity are connected by the relation

$$:V(X_1):_{OUT} = \gamma_1 :V(X_1):_{INT} \tag{5.16}$$

since the variance in the total field is given by the equal time commutators. Thus the

$$:V(X_1):_{INT} = -\frac{1}{8} \tag{5.17}$$

in agreement with the result of the master technique. Thus, by frequency analyzing the output light from the cavity ideal squeezing may be found in the frequency components around $\omega = 0$. The frequency analysis may be accomplished by a heterodyne detection scheme. In the presence of losses, the squeezing is reduced from the ideal values obtained above. Quantum fluctuations in the medium will further restrict the optimum squeezing attainable as has been shown by Reid and Walls[29] in an analysis of four wave mixing.

6. EFFECT OF DISSIPATION ON QUANTUM COHERENCE

There is considerable interest at present in detecting a macroscopic superposition of quantum states.[4] Such a superposition state would exhibit quantum interference effects. The question of how the interaction with the environment will affect such a coherent superposition has been considered recently by Caldeira and Leggett.[30] To illustrate this problem we consider a simple system, namely a harmonic oscillator prepared in a linear superposition of two coherent states.

$$|\psi\rangle = |\alpha\rangle + |\beta\rangle \quad . \tag{6.1}$$

After a time t the density operator is

$$\rho(t) = |\alpha(t)\rangle\langle\alpha(t)| + |\beta(t)\rangle\langle\beta(t)| + |\alpha(t)\rangle\langle\beta(t)|$$
$$+ |\beta(t)\rangle\langle\alpha(t)| \tag{6.2}$$

where

$$|\alpha(t)\rangle = |\alpha\, e^{-i\omega t}\rangle$$

$$|\beta(t) = |\beta\, e^{-i\omega t}\rangle \ .$$

To look for spatial interference fringes we consider the matrix elements of ρ in coordinate space $\langle x|\rho(t)|x\rangle$. We consider the situation where $\alpha = -\beta = \sqrt{\omega/2\hbar}\, x_0$, where x_0 is real and represents the initial position of the two wave packets. Using the relation[31]

$$\langle x|\alpha\rangle = \left(\frac{\omega}{\pi\hbar}\right)^{1/4} \exp\left[-\frac{\omega}{2\hbar}x^2 + \sqrt{\frac{2\omega}{\hbar}}\,\alpha x - \frac{1}{2}|\alpha|^2 - \frac{1}{2}\alpha^2\right] \tag{6.3}$$

we find

$$\langle x|\rho(t)|x\rangle = I_+ + I_- + 2\sqrt{I_+I_-}\cos\phi \tag{6.4}$$

where

$$I_\pm = \left(\frac{\omega}{\pi\hbar}\right)^{1/2} \exp\left\{-\frac{\omega}{h}(x \pm x_0 \cos\omega t)^2\right\}$$

$$\phi = \frac{2\omega}{\hbar}\sin\omega t\; x_0\; x \quad .$$

For times $t = \frac{n\pi}{\omega}$, the interference term is proportional to $e^{-\frac{\omega}{\hbar}x^2}\, e^{-\frac{\omega}{\hbar}x_0^2}$ which for initially well-separated wave packets is very small. At times $t = (n + \frac{1}{2})\pi/\omega$ the interference term is proportional to

$$e^{-\frac{\omega}{h}x^2} \cos\frac{2\omega x_0 x}{h} \quad .$$

At these times the two wave packets overlap and we obtain interference fringes due to the cosine term.

We shall now investigate how the interference terms survive when the oscillator is damped via an interaction with the environment. We essentially follow the treatment of Milburn and Walls.[32]

The master equation in the interaction picture for the damped harmonic oscillator at zero temperature is

$$\frac{\partial\rho}{\partial t} = \lambda(2a\rho a^\dagger - a^\dagger a\rho - \rho a^\dagger a) \tag{6.5}$$

where λ is the damping coefficient. An exact solution for the density operator $\rho(t)$ has been given by Srinivas and Davies.[33] The solution is

$$\rho(t) = \sum_{n=0}^{\infty} N_t(m)\, \rho(0) \tag{6.6}$$

where

$$N_t(m) = \int_0^t dt_m \int_0^{t_m} dt_{m-1} \cdots \int_0^{t_2} S_{t-t_m}\, J\, S_{t_m - t_{m-1}} \cdots J\, S_{t_1}\, \rho(0)$$

and

$$J\rho = \lambda a \rho a^{\dagger}$$

$$S_t \rho = \exp\left(-\frac{\lambda t}{2} a^{\dagger} a\right) \rho \exp\left(-\frac{\lambda t}{2} a^{\dagger} a\right) .$$

We consider an arbitrary element of the initial density operator

$$\rho_{\alpha\beta}(0) = N \, |\alpha\rangle\langle\beta| \tag{6.7}$$

where $N = \langle\beta|\alpha\rangle^{-1}$. Using the relation

$$e^{\frac{-\lambda a^{\dagger} a t}{2}} |\alpha\rangle = \exp\left\{-\frac{|\alpha|^2}{2}(1 - e^{-\lambda t})\right\} |\alpha e^{-\frac{\lambda t}{2}}\rangle \tag{6.8}$$

one can show that

$$S_t |\alpha\rangle\langle\beta| = \exp\left[-\frac{|\alpha|^2}{2}(1 - e^{-\lambda t}) - \frac{|\beta|^2}{2}(1 - e^{-\lambda t})\right]$$

$$|\alpha e^{-\frac{\lambda t}{2}}\rangle\langle\beta \, e^{-\frac{\lambda t}{2}}| \; . \tag{6.9}$$

The solution for $\rho_{\alpha\beta}$ at time t is then given by

$$\rho_{\alpha\beta}(t) = N \exp\left[-\frac{|\alpha|^2}{2}(1 - e^{-\lambda t}) - \frac{|\beta|^2}{2}(1 - e^{-\lambda t})\right]$$

$$|\alpha e^{-\frac{\lambda t}{2}}\rangle\langle\beta e^{-\frac{\lambda t}{2}}|$$

$$\left[\sum_{m=0}^{\infty} (\lambda\alpha\beta^*)^m \int_0^t e^{-\lambda t_m} dt_m \int_0^{t_m} dt_{m-1} \, e^{-\lambda t_{m-1}} \cdots \int_0^{t_2} dt_1 \, e^{-\lambda t_1}\right] \tag{6.10}$$

Using time ordering[31] this becomes

$$\rho_{\alpha\beta}(t) = N \exp\left[-\frac{|\alpha|^2}{2}(1 - e^{-\lambda t}) - \frac{|\beta|^2}{2}(1 - e^{-\lambda t})\right] |\alpha e^{-\frac{\lambda t}{2}}\rangle\langle\beta e^{-\frac{\lambda t}{2}}|$$

$$\sum_{m=0}^{\infty} \left(\frac{\lambda\alpha\beta^*}{m!}\right)^m \left(\int_0^t e^{-\lambda t'} dt'\right)^m$$

$$= N \exp\left[(e^{-\lambda t} - 1)\left(\frac{|\alpha|^2}{2} + \frac{|\beta|^2}{2} - \alpha\beta^*\right)\right] |\alpha e^{-\frac{\lambda t}{2}}\rangle\langle\beta e^{-\frac{\lambda t}{2}}| \tag{6.11}$$

Then for an initial state given by Eq. (6.1), the density operator at time t is

$$\rho(t) = N\left\{|\alpha e^{-\frac{\lambda t}{2}}\rangle\langle\alpha e^{-\frac{\lambda t}{2}}| + |\beta e^{-\frac{\lambda t}{2}}\rangle\langle\beta e^{-\frac{\lambda t}{2}}|\right.$$

$$+ \exp\left[(1 - e^{-\lambda t})\left(-\frac{|\alpha|^2}{2} - \frac{|\beta|^2}{2} + \alpha\beta^*\right)\right] |\alpha e^{-\frac{\lambda t}{2}}\rangle\langle\beta e^{-\frac{\lambda t}{2}}|$$

$$\left.+ \exp\left[(1 - e^{-\lambda t})\left(-\frac{|\alpha|^2}{2} - \frac{|\beta|^2}{2} + \alpha^*\beta\right)\right] |\beta e^{-\frac{\lambda t}{2}}\rangle\langle\alpha e^{-\frac{\lambda t}{2}}|\right\} \quad (6.12)$$

where

$$N = \left[1 + \left(\exp\left(-\frac{|\alpha|^2}{2} - \frac{|\beta|^2}{2} + \alpha\beta^*\right) + cc\right)\right]^{-1} .$$

In fact, for an arbitrary initial state

$$|\psi\rangle = \sum f_\alpha |\alpha\rangle \quad (6.13)$$

the density operator at time t may be expressed in the form

$$\rho(t) = \sum_{\gamma\gamma'} f_\gamma f_{\gamma'} \langle\gamma|\gamma'\rangle^{(1-e^{-\lambda t})} |\gamma e^{-\frac{\lambda t}{2}}\rangle\langle\gamma' e^{-\frac{\lambda t}{2}}| . \quad (6.14)$$

The off diagonal elements in the coherent state basis are reduced by the prefactor $\langle\gamma|\gamma'\rangle^{(1-e^{-\lambda t})}$.

Let us once again consider the interference experiment with $\alpha = -\beta = \sqrt{\omega/2\hbar}\ x_0$ in the presence of damping. We find

$$\langle x|\rho(t)|x\rangle = I_+ + I_- + 2\sqrt{I_+ I_-}\cos\phi\, e^{-\chi} \quad (6.15)$$

where

$$I_\pm = \left(\frac{\omega}{\pi\hbar}\right)^{1/2} \exp\left(-\frac{\omega}{\hbar}(x \pm x_0 e^{-\frac{\lambda t}{2}}\cos\omega t)^2\right)$$

$$\phi = \frac{2\omega}{\hbar} \sin\omega t\, x_0 e^{-\frac{\lambda t}{2}} x$$

$$\chi = \frac{\omega}{\hbar} x_0^2 (1 - e^{-\lambda t})$$

We see that there is now a rapid damping of the interference term by the factor $e^{-\chi}$. The large the initial excitation x_0, the more rapid the dephasing. This will make the observation of a macroscopic superposition of quantum states very difficult in practice. This example illustrates the role of the environment in rapidly reducing a pure superposition state to a mixture.

ACKNOWLEDGMENTS

The work reported in these lectures represents a survey of recent research performed in conjunction with my colleagues and students at the University of Waikato. In particular, I wish to acknowledge the important contributions made by M.J. Collett, C.W. Gardiner, G.J. Milburn, and M.D. Reid. I also wish to thank A. Caldeira and W. Zurek for enlightening discussions on the material presented in Sec. 6. This work was partially supported by the National Science Foundation under Grant No. PHY77-27084 (supplemented by funds from the National Aeronautics and Space Administration), the United States Army through its European Research Office, and the Claude McCarthy Foundation.

REFERENCES

1. For a review see H. Paul, Rev. Mod. Phys. 54:1061 (1982); D.F. Walls, Nature 280:451 (1979).
2. For a review see D.F. Walls, Nature 306:141 (1983).
3. For a review see (a) R. Loudon, Rep. Prog. Phys. 48:58 (1980); (b) J.F. Clauser and A. Shimony, Rep. Prog. Phys. 41:1881 (1978).
4. A.J. Leggett, "Proc. Int. Symp. Foundations of Quantum Mechanics," Tokyo (1983).
5. R.J. Glauber, Phys. Rev. 130:2529 (1963); Phys. Rev. 131:2766 (1963).
6. E.C.G. Sudarshan, Phys. Rev. Lett. 10:277 (1963).
7. P.D. Drummond and D.F. Walls, J. Phys. 13A:2353 (1980).
8. R. Hanbury Brown and R.Q. Twiss, Nature 177:27 (1956).
9. H.J. Kimble, M. Dagenais, and L. Mandel, Phys. Rev. A 18:201 (1978).
10. R. Short and L. Mandel, Phys. Rev. Lett. 51:384 (1983).
11. C.M. Caves, Phys. Rev. D 23:1693 (1981).
12. H.P. Yuen, Phys. Rev. A 13:2226 (1976).
13. M.S. Zubairy, Phys. Lett. 87A:162 (1982).
14. R. Graham, Phys. Lett. 32A:373 (1970).
15. K.J. McNeil and C.W. Gardiner, Phys. Rev. A 28:1560 (1983).
16. R. Graham, Phys. Rev. Lett. 52:117 (1984).
17. J.S. Bell, Physics 1:195 (1965); Rev. Mod. Phys. 38:447 (1983).
18. M.D. Reid and D.F. Walls, Phys. Rev. Lett. 53:955 (1984).
19. J.F. Clauser, Phys. Rev. D 9:853 (1974).
20. A. Aspect, P. Grangier, and G. Roger, Phys. Rev. Lett. 49:91 (1982).
21. M.J. Collett and C.W. Gardiner, Phys. Rev. A 30:1386 (1984).
22. M.J. Collett, M.Sc. Thesis, University of Waikato (1984).
23. S. Chaturvedi, C.W. Gardiner, I.S. Matheson, and D.F. Walls, J. Stat. Phys. 17:469 (1977).
24. G.J. Milburn, Ph.D. Thesis, University of Waikato (1983).
25. G.J. Milburn and D.F. Walls, Optics Comm. 39:401 (1981).
26. L.A. Lugiato and G. Strini, Optics Comm. 41:67 (1982).
27. B. Yurke, Phys. Rev. A 29:408 (1984).
28. C.M. Savage and C.W. Gardiner, Optics Comm. 50:173 (1984).
29. M.D. Reid and D.F. Walls, Optics Comm. 50:406 (1984).
30. A. Caldeira and A.J. Leggett, to be published.
31. W.H. Louisell, "Quantum Statistical Properties of Radiation," Wiley, New York (1973), p. 109.
32. G.J. Milburn and D.F. Walls, to be published.
33. M.D. Srinivas and E.B. Davies, Optica Acta 28:981 (1981).

PHOTON STATISTICS, ANTIBUNCHING AND SQUEEZED STATES

Gerd Leuchs*

Joint Institute for Laboratory Astrophysics, University of Colorado and National Bureau of Standards, Boulder, Colorado 80309 USA

1. INTRODUCTION

Since the pioneering experiment by Hanbury Brown and Twiss[1] and the developments in the quantum description of the electromagnetic radiation field initiated by Glauber[2] and Sudarshan[3] the field of photon statistics and correlation has gained increasing attention. Today some applied techniques based on light correlation measurements are well established, e.g., in astronomy[4] and diagnostics of particle motion.[5] One of the frontiers in present research is concerned with nonclassical properties of the radiation field referred to in the literature as photon antibunching and squeezed states. In the following discussions we attempt to describe the status and address future prospects of experiments on these phenomena.

Starting in Sec. 2, light correlation is presented in the framework of classical electrodynamics. It will become apparent that some aspects of the radiation field cannot be described in classical terms. Consequently, Sec. 3 deals with the extension to quantized radiation fields. Having thus introduced nonclassical states of the field the question of whether and how photon antibunching and squeezed states are related is addressed in Sec. 4. After this introduction to the basic principles it is important to learn about the effect of linear attenuation (beam splitters, neutral density filters, detector quantum efficiency) on the detected signal (Sec. 5). Experiments on the change of photon statistics by the nonlinear interaction of radiation fields with matter are described in Sec. 6. The next three sections then deal with experimental observations of antibunching and sub-Poissonian photon statistics in resonance fluorescence and with possible schemes for the generation and detection of squeezed states.

Presently, the interest in nonclassical properties of the radiation field is primarily that of experimental proof of existence, and that of comparison between theory and observations. However, some potential

---

*Heisenberg Fellow of the Deutsche Forschungsgemeinschaft.

future applications have been proposed already and are addressed briefly in the final section (Sec. 10).

Whenever possible, simple pictures are used in this paper; for rigorous derivations and calculations the reader is referred to the original literature. Various subjects discussed here have been treated in several reviews which are suggested for complementary reading.[6-11]

2. CLASSICAL RADIATION FIELD

In classical electrodynamics, a monochromatic wave is often described using the complex representation

$$E(r,t) = E^{-}(r,t) + E^{+}(r,t)$$

$$= \frac{1}{2}\left[\varepsilon(r,t)\, e^{-i\omega t} + \varepsilon^{*}(r,t)\, e^{i\omega t}\right] \quad . \qquad (2.1)$$

With this convention, the intensity of the wave is $I(r,t) \propto \langle E^2(r,t)\rangle = (1/2)\,\varepsilon\varepsilon^{*}$. The two terms in Eq. (2.1) are called positive and negative frequency parts. In the classical theory this complex description is merely a tool for simplifying some calculations. Physical significance is gained in the quantum description, where ε and ε^{*} are replaced by the lowering and raising operators of the field.

As an alternative to Eq. (2.1) the monochromatic wave can be described as a linear superposition of a sine and a cosine wave,

$$E(r,t) = \frac{\varepsilon + \varepsilon^{*}}{2} \cos \omega t + \frac{\varepsilon - \varepsilon^{*}}{2i} \sin \omega t \quad , \qquad (2.2)$$

with real coefficients $\varepsilon_1 \equiv (\varepsilon+\varepsilon^{*})/2$ and $\varepsilon_2 \doteq (\varepsilon-\varepsilon^{*})/(2i)$. In this description, the state of the field can be represented graphically by a two-dimensional phase diagram (Fig. 1). A monochromatic plane wave may be represented by one point. The length of the vector from the origin to this point is the amplitude of the field and its angle with respect to the ε_1-axis is the phase of the cosine wave. If ε_1 and ε_2 are slowly varying in time, the point in Fig. 1 will migrate accordingly. This migration characterizes the field.

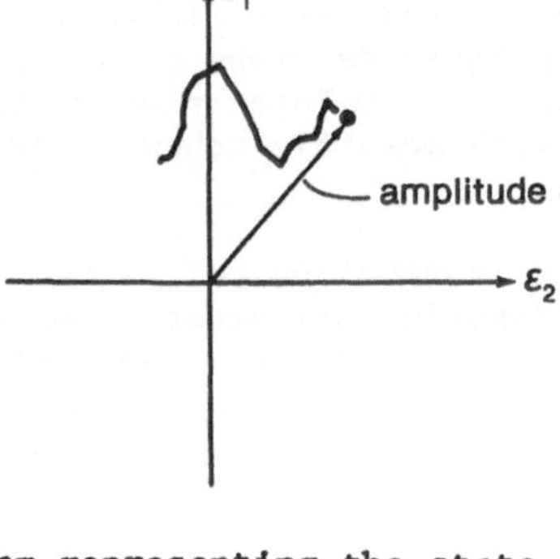

Fig. 1. Phase diagram representing the state of an electromagnetic wave having slowly varying amplitude and phase.

In practice, one often deals with fields that are essentially stationary, i.e., independent against translation in time. In that sense, the migration in Fig. 1 is only one particular realization of the field in some time interval. For stationary fields the time average of an observable is often replaced by an average over an ensemble of N radiation fields ($N \to \infty$) having all the same characteristics (ergodic principle[12]). A unique way to characterize a radiation field is by its correlation functions.[6,13]

2.1. First-order Correlation Function

A standard way to describe part of the characteristics of a radiation field is by its spectrum $S(\omega)$ which is the Fourier-transform of the first-order correlation function

$$G^{(1)}(r_1,t_1,r_2,t_2) = \langle \varepsilon^*(r_1,t_1)\, \varepsilon(r_2,t_2)\rangle \quad . \tag{2.3}$$

It is obtained by multiplying the electric field amplitudes at two space-time points and taking the ensemble average which is indicated by the brackets. For $r_1 = r_2$ and $t_1 = t_2$, $G^{(1)}$ is proportional to the intensity at this point. Therefore, it is called first order in intensity. Any detector that we use to measure the field relies on a quantum absorption process. Here, it is modeled by a two-level atom with a ground and excited state, the latter having a finite width $\hbar\Gamma$ (Fig. 2).

The rate R of absorption of radiation by the atom is given by time-dependent perturbation theory as

$$R \propto \int_{-\infty}^{\infty} d\tau \, \langle g|V(t')|e(t')\rangle \, \langle e(t)|V(t)|g\rangle \quad , \tag{2.4}$$

where $V(t) = dE(t)$ is the interaction operator in the electric dipole approximation, and $\tau = t-t'$. The time evolution of the excited state $|e(t)\rangle$ is given by $\exp\{-i\omega_o t\}$. Taking $E(t)$ from Eq. (2.1), leaving out terms oscillating at $\approx 2\omega$, and including the damping Γ, one obtains[6]

$$R \propto \int_{-\infty}^{\infty} d\tau \, \exp\{i(\omega_o-\omega)\tau - \Gamma\tau\}\, G^{(1)}(\tau) \quad . \tag{2.5}$$

There are two limiting cases to this expression. For Γ much smaller than the bandwidth, b, of the radiation field, the rate of absorption, R, is just the Fourier-transform $S(\omega_o)$ of the first-order correlation function

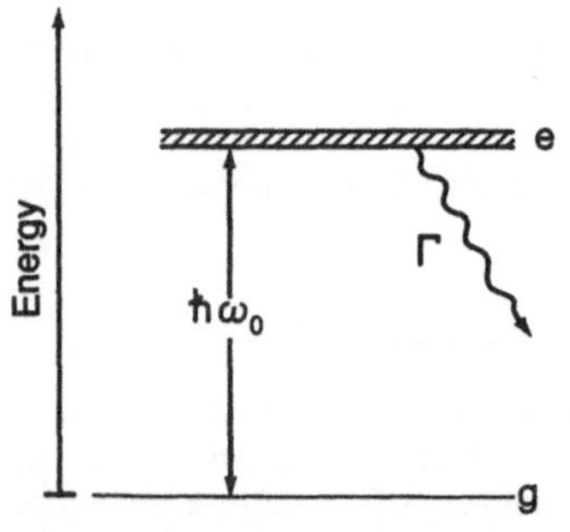

Fig. 2. Energy level diagram of a two-level atom. The width of the excited state is $\hbar\Gamma$, where Γ^{-1} is the lifetime of the excited state.

$G^{(1)}(r,t,r,t') \equiv G^{(1)}(\tau)$. In the other limiting case, $\Gamma >> b$, the exponential in Eq. (2.5) can be replaced by the δ function, yielding

$$R \propto G^{(1)}(0) \quad , \quad \text{for} \quad \Gamma >> b \quad . \tag{2.6}$$

2.2. Second-Order Correlation Function: Intensity Correlation versus Two-Photon Absorption

A radiation field cannot, in general, be unambiguously defined by the first-order correlation function, rather higher-order correlation functions are often needed.[6] The general form of the second-order correlation function is

$$G^{(2)}(r_1t_1,r_2t_2,r_3t_3,r_4t_4) = \langle \varepsilon^*(r_1t_1)\varepsilon^*(r_2t_2)\varepsilon(r_3t_3)\varepsilon(r_4t_4)\rangle \quad . \tag{2.7}$$

Two special cases will be considered here. In the first case, the intensity of the radiation field is measured at two space-time points using two different one-photon absorption detectors (r_1 is set equal to r_2 for simplicity). These two intensities are then multiplied and ensemble averaged

$$G_I^{(2)}(\tau) = 4\langle I(t_1)I(t_2)\rangle = \langle \varepsilon^*(t_1)\varepsilon(t_1)\varepsilon^*(t_2)\varepsilon(t_2)\rangle \quad . \tag{2.8}$$

The subscript I indicates that this is the intensity correlation function.[1] Here $\tau = t_2 - t_1$.

In the second case, we consider a two-photon absorber. Again this can be an atom which is excited only by the simultaneous absorption of two photons. The rate of two-photon absorption[14] can be derived analogously to the one-photon absorption rate [Eq. (2.5)]

$$R \propto \int_{-\infty}^{\infty} d\tau \, \exp\{i(\omega_o - 2\omega)\tau - \Gamma\tau\} \, G_{TP}^{(2)}(\tau) \quad . \tag{2.9}$$

Since the interaction operator for nonresonant two-photon absorption is $V(t) \propto E^2(t)$, the second-order correlation function describing the two-photon absorption process (TP) is

$$G_{TP}^{(2)}(\tau) = \langle \varepsilon^*(t_1)\varepsilon^*(t_1)\varepsilon(t_2)\varepsilon(t_2)\rangle \quad . \tag{2.10}$$

Note the difference between Eqs. (2.10) and (2.8). For narrow band detection $\Gamma << b$ the two-photon absorption rate is proportional to the Fourier-transform of the second-order correlation function

$$R \propto \int_{-\infty}^{\infty} d\tau \, \exp\{i(\omega_o - 2\omega)\tau\} \, G_{TP}^{(2)}(\tau) \quad . \tag{2.11}$$

For a broad band two-photon absorber, on the other hand, one finds $R \propto G_{TP}^{(2)}(0)$. For $\tau = 0$, however, $t_1 = t_2$ and $G_{TP}^{(2)}(0) = G_I^{(2)}(0)$. Consequently, a broad band two-photon absorber can be used to measure the $\tau = 0$ value of the intensity correlation function. This has been used in experiments described in Sec. 6.

2.3. Examples of Radiation Fields

We will now give the explicit forms of some correlation functions for the two common light fields. A chaotic or thermal field is defined by the emission of an ensemble of N identical oscillators ($N \rightarrow \infty$), the phases of which are randomly distributed.[7] For such a field the mean value of the electric field is zero. A typical phase diagram of such a field is shown in Fig. 3a. Thermal fields are emitted by all incandescent lamps and electrical discharges. Ideal laser radiation above

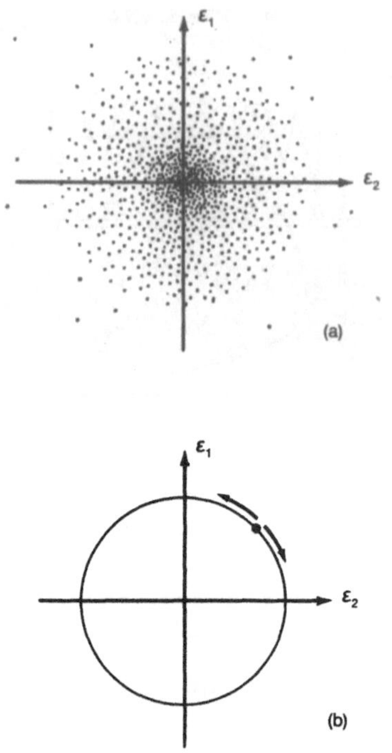

Fig. 3. Phase diagram for a chaotic or thermal field (a) and for a phase diffusion laser field (b). The dots in (a) are only one possible realization of the field. The state of the phase diffusion field (b) is represented by the dot which exhibits a Brownian motion along the circle.

threshold, on the other hand, has a mean electric field different from zero and corresponds to only one point in the phase diagram. The stable amplitude of the laser field is the result of an equilibrium between optical gains and losses. However, there is no restoring force for the phase[15]; phase changes are caused, e.g., by spontaneous emission. As a result, the momentary state of the laser field will correspond to a point in the phase diagram somewhere on a circle around the origin (Fig. 3b). Mathematically, this is described by the phase diffusion model which assumes that phase changes are delta function correlated[6,16,17]:

$$\left\langle \frac{d\phi(t)}{dt} \frac{d\phi(t+\tau)}{dt} \right\rangle = 2b\ \delta(\tau) \quad . \tag{2.12}$$

These characteristics lead to a Lorentzian power spectrum with a full width at half maximum (FWHM) of 2b. In most practical situations, the laser bandwidth is dominated by amplitude and frequency fluctuations. Extreme stabilization is required to reach the phase diffusion limit. Figure 4 shows the correlation functions (calculated according to formulae from Zoller[18]) characterizing the thermal and the phase diffusion laser field both having a coherence time $\tau_c = 1/b$. The correlation functions are normalized to the intensity

$$g^{(i)}(\tau) = G^{(i)}(\tau)/\langle\varepsilon^*(t)\varepsilon(t)\rangle^i \quad .$$

For the ideal laser (stable phase) all $g^{(i)}(\tau)$ are of course τ independent. We note that the two fields cannot be distinguished by looking only at $g^{(1)}(\tau)$. For the second-order correlation functions, however, this is quite different. It is also apparent that the phase diffusion field cannot be distinguished from the ideal laser by intensity correlation measurements only.

One remarkable feature in Fig. 4 is that for the phase diffusion field the second-order correlation function describing two-photon absorption falls off faster than for the thermal field. In practice, such

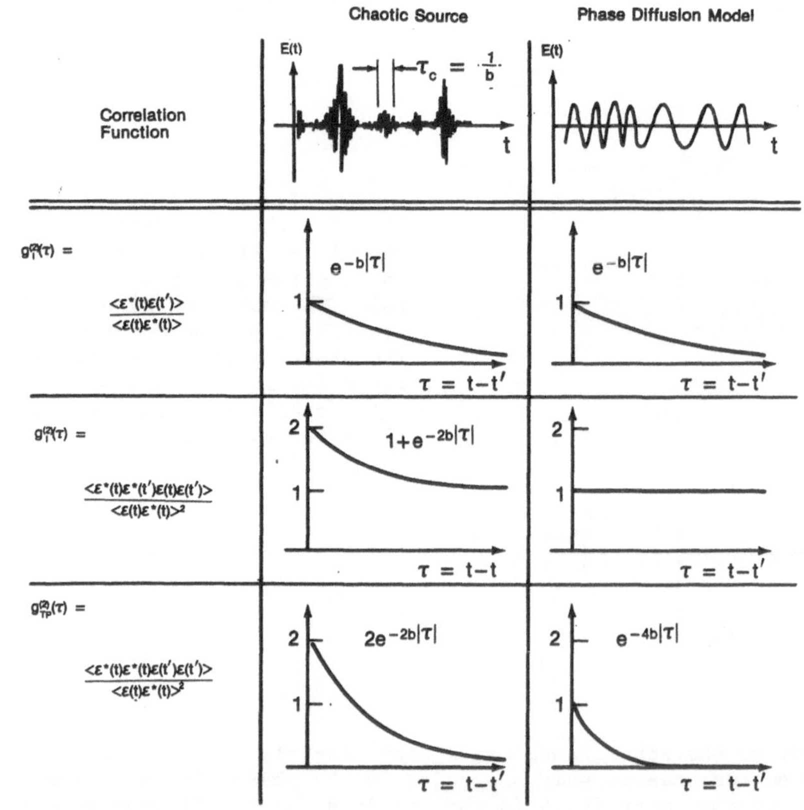

Fig. 4. Correlation functions for the thermal and the phase diffusion field. Possible momentary realizations of E(t) are given at the top of the figure. For the phase diffusion model the time scale is expanded to illustrate the changes in the optical phase.

delta-function correlated phase changes are not very realistic since there will be a cutoff at high frequencies. A more realistic model[19,20] is obtained by replacing $2\delta(\tau)$ in Eq. (2.12) with $\beta \exp\{-\beta|\tau|\}$. For $\beta \gg b$ the delta function is approached and $g^{(2)}_{TP}(\tau)$ is the same as for the pure phase diffusion field in Fig. 4. In the other limit, $\beta \ll b$, i.e., very slow phase changes, $g^{(2)}_{TP}(\tau)$ falls off only as $\exp\{-2b|\tau|\}$, like the one for the thermal source. The fast decay of $g^{(2)}_{TP}(\tau)$ for the delta function correlated phase diffusion field ($\beta \gg b$) leads to a two-photon absorption line [Eq. (2.11)] which is twice as broad as might be expected from the power spectrum. This result was theoretically predicted by Mollow[14] and has recently been experimentally verified by Elliott et al.[21]

2.4. Intensity Fluctuations

The classical equivalents to photon statistics are intensity fluctuations, which can be characterized by the probability distribution for

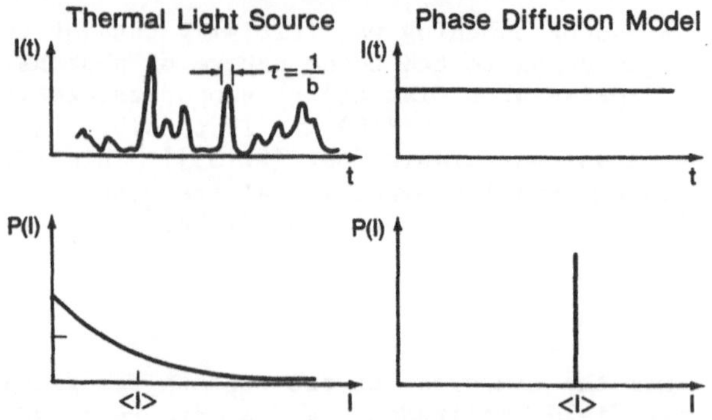

Fig. 5. Possible realizations of the intensity are shown along with the corresponding probability distributions. In the case of the thermal field, $P(I) \propto \exp(I/I_o)$.

the intensity P(I) (Fig. 5) or by the intensity correlation function [Eq. (2.8)]. The intensity can be written as the mean intensity $\langle I \rangle$ plus a fluctuation $\Delta I(t)$ for which the mean value is zero. The mean value of the squared intensity is then $\langle I^2 \rangle = \langle I \rangle^2 + \langle \Delta I^2 \rangle$. This has immediate consequences for possible values of the normalized intensity correlation function

$$g_I^{(2)}(\tau) = \langle I(t)I(t+\tau)\rangle / \langle I(t)\rangle^2 = \begin{cases} 1 + \langle \Delta I^2 \rangle / \langle I \rangle^2 & , \quad \tau = 0 \\ 1 & , \quad \tau = \infty \end{cases} . \quad (2.13)$$

For $\tau \to \infty$, the intensities are not correlated and $g_I^{(2)}(\tau)$ approaches 1. For the other limit, we can give an important result. Since $\langle \Delta I^2 \rangle / \langle I \rangle^2 \geq 0$, it follows that $g_I^{(2)}(0) \geq 1$, since P(I) is a classical probability distribution and, therefore, positive definite. A value of $g_I^{(2)}(0)$ smaller than one is classically not possible (representative cases are shown in Fig. 4). The maximum of $g_I^{(2)}(\tau)$ at $\tau = 0$ is called photon bunching, and was first observed by Hanbury Brown and Twiss[1]; Fig. 6 is a schematic of their experimental setup. The light of a low-pressure spectral lamp was measured by two photomultipliers sitting at roughly the same distance from the beam splitter. Apertures ensured that the spatial coherence condition was fulfilled. The two detector currents were multiplied and averaged. The time delay τ was introduced by moving one of the detectors.

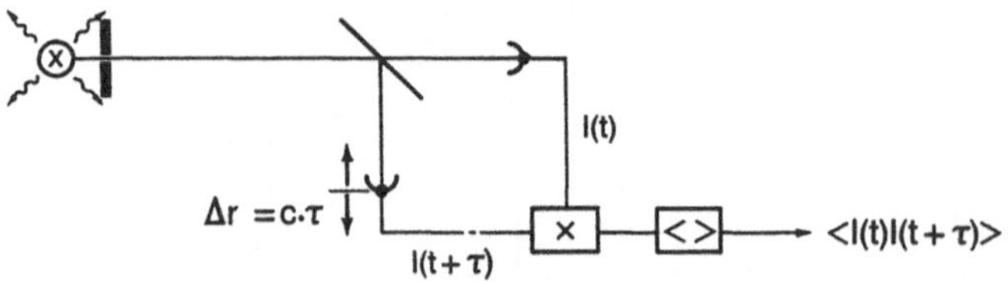

Fig. 6. Experimental setup for the original experiment by Hanbury Brown and Twiss (Ref. 1).

The effect of photon bunching was initially thought to be an inherent property of light owing to the boson nature of photons. Therefore, it came as a big surprise when Glauber[13,6] showed theoretically that for the quantized radiation field, $g_1^{(2)}(0) < 1$ is possible, in direct contradiction to the classical formula [Eq. (2.13)]! This effect, known as photon antibunching, and its experimental observation constitutes an unambiguous proof of the quantum nature of light.[8]

3. QUANTIZED RADIATION FIELD

When going from the classical to the quantum description of the field, the electric field amplitudes ε and ε^* are replaced by $\hat{a}$ and $\hat{a}^+$, the photon annihilation and creation operators.[22] The quantization of the field is a direct consequence of the commutation relation $[\hat{a},\hat{a}^+] = \hat{1}$. In analogy to Eq. (2.3) the electric field operator can be written as

$$\hat{E} \propto \frac{\hat{a} + \hat{a}^+}{2} \cos \omega t + \frac{\hat{a} - \hat{a}^+}{2i} \sin \omega t \quad . \tag{3.1}$$

Again we define new operators $\hat{a}_1 = (\hat{a}+\hat{a}^+)/2$ and $\hat{a}_2 = (\hat{a}-\hat{a}^+)/2i$ which have real eigenvalues describing the amplitude of the cosine and sine part of the monochromatic wave. Based on $[\hat{a},\hat{a}^+] = \hat{1}$ one can derive a commutation relation for $\hat{a}_1$ and $\hat{a}_2$, $[\hat{a}_1,\hat{a}_2] = (i/2)\ \hat{1}$. Whenever two operators do not commute, there is an uncertainty relation for the expectation values; in the present case, it is

$$\langle\Delta\hat{a}_1^2\rangle \langle\Delta\hat{a}_2^2\rangle \geq 1/16 \quad . \tag{3.2}$$

Here $\langle\Delta\hat{a}_i^2\rangle = \langle\hat{a}_i^2\rangle - \langle\hat{a}_i\rangle^2$. We now return to the phase diagram introduced in Fig. 1 and relabel the axes with the expectation values of $\hat{a}_1$ and $\hat{a}_2$. A monochromatic wave is now no longer represented by a point but by a region of uncertainty (Fig. 7). The uncertainty in the radial direction corresponds to amplitude fluctuations and the uncertainty in the peripheral direction corresponds to phase fluctuations. The state of the radiation field where both phase and amplitude fluctuations are symmetrically minimized, $\langle\Delta\hat{a}_1^2\rangle = \langle\Delta\hat{a}_2^2\rangle = 1/4$, best approximates the classical wave with well-defined phase and amplitude. Without loss of generality,

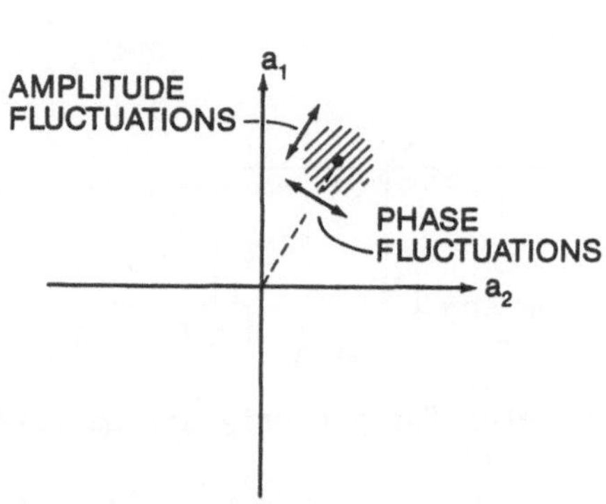

Fig. 7. Phase diagram for the quantum mechanical field amplitudes $\langle\hat{a}_1\rangle$ and $\langle\hat{a}_2\rangle$. Owing to Heisenberg's uncertainty relation for the noncommuting operators $\hat{a}_1$ and $\hat{a}_2$ the experimentally measured values will scatter (shaded region).

we will from here on discuss only states of the field having an uncertainty region centered on the a_1 axis. The state with symmetrical minimum uncertainty is the coherent state describing laser radiation and is an eigenstate of $\hat{a}$. Its photon number fluctuation can be calculated from the amplitude fluctuation by using $n \approx a_1^2$, $n = \langle n \rangle + \Delta n$, $a_1 = \langle a_1 \rangle + \Delta a_1$:

$$\langle \Delta n^2 \rangle \approx \langle (2\langle a_1 \rangle \Delta a_1)^2 \rangle \approx \langle a_1 \rangle^2 \approx \langle n \rangle \quad .$$

This fluctuation, $\langle \Delta n^2 \rangle = \langle n \rangle$, suggests that the coherent state has a Poissonian photon number distribution $P(n) = \langle n \rangle^n \, e^{-\langle n \rangle}/n!$. For the rigorous derivation, see Glauber[2] and Sudarshan.[3]

Other possible minimum uncertainty states have unsymmetrical uncertainty regions (Fig. 8) where the uncertainty is squeezed into mainly one component.[23,24] These states, which are called squeezed states, are defined by $\langle \Delta \hat{a}_1^2 \rangle < 0.25$ or $\langle \Delta \hat{a}_2^2 \rangle < 0.25$. They are also known as "two-photon coherent states."[23] Sometimes "squeezed state" also refers to a state which is not a minimum uncertainty state but where nevertheless the mean squared fluctuation in one of the components is less than 0.25.

The phase diagram on the left side of Fig. 8 shows a state where a_1 is squeezed, apparently resulting in reduced amplitude and increased phase fluctuations. Here, reduced amplitude fluctuations mean sub-Poissonian photon number fluctuations. The photons arrive more equally spaced,[25] i.e. less bunched, than for a coherent state which already has $g_I^{(2)}(0) = 1$. Therefore, sub-Poissonian number fluctuations correspond to photon antibunching. As a result, it is obvious that squeezed states and photon antibunching are related (see Sec. 4).

3.1. Ordering of Field Operators

As in the classical description, the photon statistics are tied to the correlation functions. Before we can be more quantitative, however, we have to address the question of how the field operators have to be ordered. For the classical intensity correlation function there was no problem

$$G_I^{(2)}(t'-t) = \langle I(t')I(t) \rangle = \langle \varepsilon^*(t')\varepsilon(t')\varepsilon^*(t)\varepsilon(t) \rangle \quad ;$$

the electric field amplitudes could have been arranged in any order without influencing the result. This is not true in the quantum description since the field operators do not commute. We find that the right ordering is determined by the experiment. The intensity correlation function

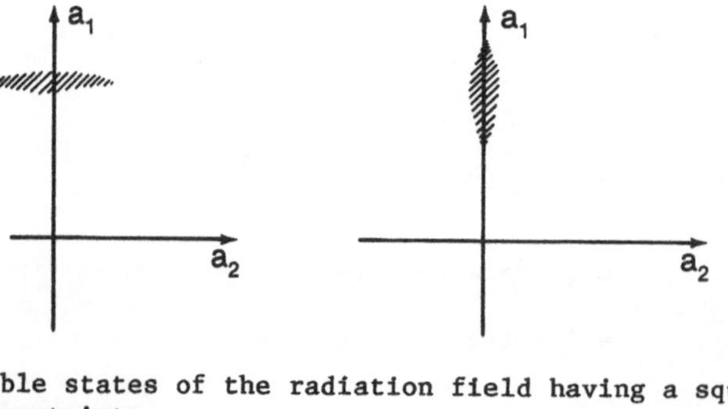

Fig. 8. Possible states of the radiation field having a squeezed region of uncertainty.

is measured by detecting two photons, one each at times t and t'. The probability amplitude for this process is given by the corresponding matrix element $\langle f|\hat{a}(t')\hat{a}(t)|i\rangle$. The final state $|f\rangle$ of the radiation field has two photons less than the initial state $|i\rangle$. The probability for this process, i.e., $G_I^{(2)}(t'-t)$, is obtained by taking the absolute squared and summing over all final states[6]

$$G_I^{(2)}(t'-t) \propto \sum_f \langle i|\hat{a}^+(t)\hat{a}^+(t')|f\rangle \langle f|\hat{a}(t')\hat{a}(t)|i\rangle \quad .$$

The sum extends over a full basis set, so that

$$\sum_f |f\rangle \langle f| = \hat{1} \quad .$$

States which cannot be reached do not contribute to the sum since the matrix element will be zero. As a result, all creation operators have to be to the left of all annihilation operators and we have

$$G_I^{(2)}(t'-t) = \langle i|\hat{a}^+(t)\hat{a}^+(t')\hat{a}(t')\hat{a}(t)|i\rangle \tag{3.3a}$$

and likewise

$$G_{TP}^{(2)}(t'-t) = \langle i|\hat{a}^+(t')\hat{a}^+(t')\hat{a}(t)\hat{a}(t)|i\rangle \quad . \tag{3.3b}$$

Knowing the right ordering of the operators, one may attempt some simple calculations.

3.2. Photon Statistics

Before evaluating the intensity correlation function, we recall some properties of the photon number state $|n\rangle$. These states have a well-defined photon number and, therefore, are the states with the smallest photon number fluctuation. They also form a complete basis set of eigenstates. The photon annihilation and creation operators act on the number state so as to lower or raise the photon number

$$\hat{a}|n\rangle = \sqrt{n}\ |n-1\rangle \quad \text{and} \quad \hat{a}^+|n\rangle = \sqrt{n+1}\ |n+1\rangle \quad . \tag{3.4}$$

In a phase diagram, a number state is represented by a circle, the phase being completely uncertain. Using Eq. (3.4) the intensity correlation function [Eq. (3.3)] can be evaluated for the case where the initial state is a pure number state (t = t')

$$\langle n|\hat{a}^+\hat{a}^+\hat{a}\hat{a}|n\rangle = n^2 - n \quad . \tag{3.5}$$

Generally speaking, the state of the radiation field will be a mixture or superposition of number states which can be described using the density matrix,[22]

$$\begin{aligned} g^{(2)}(0) &= \frac{Tr\{\hat{\rho}\ \hat{a}^+\hat{a}^+\hat{a}\hat{a}\}}{(Tr\{\hat{\rho}\ \hat{a}^+\hat{a}\})^2} \\ &= \sum_n \rho_{nn}(n^2-n)/(\sum_n \rho_{nn}n)^2 \\ &= (\langle n^2\rangle - \langle n\rangle)/\langle n\rangle^2 \\ &= 1 + (\langle \Delta n^2\rangle - \langle n\rangle)/\langle n\rangle^2 \quad . \end{aligned} \tag{3.6}$$

For a pure number state, $\langle \Delta n^2 \rangle = 0$ and $g^{(2)}(0) < 1$, which contradicts classical theory as remarked in Sec. 2.4. Therefore, the number state is also a nonclassical state of the field.[11] For a coherent state we have $\langle \Delta n^2 \rangle = \langle n \rangle$, yielding $g^{(2)}(0) = 1$, the same as the classical result for the ideal laser.

3.3. Classical vs. Quantum Description

The difference between the quantum description, which allows $g^{(2)}(0) \geq 0$, and the classical description, which restricts $g^{(2)}(0)$ to values larger than one, is at first sight somewhat surprising. The reason for this discrepancy is that in the classical measurement it is assumed that the intensity of the field can be probed without perturbing the field. This is, of course, not true if the detector absorbs one photon which is then missing in the field, an effect taken into account in the quantum description. As a result, $(n-1)(n)$ appears in Eq. (3.5) instead of n^2. For increasing $\langle n \rangle$, the quantum effect is less and less pronounced, since $g^{(2)}(0) \geq 1-1/\langle n \rangle$. In the limit of $\langle n \rangle \to \infty$, $g^{(2)}(0) \geq 1$ holds even in the quantum description. Therefore, $g^{(2)}(0) < 1$ or photon antibunching can be pronounced only for weak fields. The situation is different if the normalized variance

$$Q = \langle \Delta n^2 \rangle / \langle n \rangle \tag{3.7}$$

is observed directly, e.g. by measuring the photon number distribution. For classical fields it must have a width equal to or larger than the Poissonian distribution with the same mean photon number. Consequently, the values of Q not allowed classically are $0 \leq Q < 1$ whereas in the quantum description $Q = 0$ is possible (namely for a number state) even in the limit of $\langle n \rangle \to \infty$. In this limit, therefore, it seems appropriate to measure Q and not $g^{(2)}(0)$.

In view of the difference between the classical and quantum mechanical expression for $g^{(2)}(0)$, one may question whether Eqs. (2.13) and (3.6) give the same result for the zero value of the intensity correlation function in the classically allowed region. As already mentioned above both equations lead to the same value for the ideal laser radiation[22] (classically $\langle \Delta I^2 \rangle = 0$ and for the quantized field $\langle \Delta n^2 \rangle = \langle n \rangle$).

The same can be shown to hold for a thermal field for which the probability distribution of the intensity is $\exp\{-I/I_o\}$. For $\langle I^2 \rangle$ one finds

$$\begin{aligned} \langle I^2 \rangle &= \int I^2 e^{-I/I_o} dI \Big/ \int e^{-I/I_o} dI \\ &= 2\langle I \rangle^2 \quad . \end{aligned} \tag{3.8}$$

In a quantized field there is the same probability distribution for the photon number.[22] The averaged value of $\langle n^2 \rangle$, however, is obtained by evaluating a discrete sum instead of the integral:

$$\begin{aligned} \langle n^2 \rangle &= \sum n^2 e^{-n/n_o} \Big/ \sum e^{-n/n_o} \\ &= 2\langle n \rangle^2 + \langle n \rangle \quad . \end{aligned} \tag{3.9}$$

The differences between Eqs. (2.13) and (3.6) and those between Eqs. (3.8) and (3.9) cancel to give $g^{(2)}(0) = 2$ in both cases. In general, the results of the classical and quantum calculations are the same as long as $g^{(2)}(0)$ is in the classically allowed region.

4. RELATIONSHIP BETWEEN SQUEEZED STATES AND ANTIBUNCHING

In connection with the discussion of Fig. 8 it was suggested that squeezed states and antibunching are related. In particular, we noted that for a coherent state, squeezing of the variable carrying the coherent excitation results in reduced amplitude fluctuations, i.e., sub-Poissonian statistics and, therefore, in antibunching. The state with the largest possible antibunching, however, is the number state, represented by a circle in the phase diagram. It has large fluctuations in both components a_1 and a_2, and is therefore not a squeezed state.[11] Thus we see that antibunching does not imply squeezing. Weak squeezing of a coherent state, on the other hand, may result in antibunching (see introduction to Sec. 3), and if the state is squeezed harder, the a_1 component will be better defined (see Fig. 9). The amplitude fluctuations ΔA, however, will start increasing and there will be a transition from antibunching to bunching. The amount of squeezing needed to cause this transition has been calculated by Caves.[24] Here we use a simple argument to derive the threshold condition (Fig. 9).

The amplitude A corresponding to any point in the phase diagram is given by $(a_1^2 + a_2^2)^{1/2}$. The transition from squeezing to antibunching will occur when the mean squared amplitude fluctuation has reached the value for the coherent state $\langle \Delta A^2 \rangle = 1/4$. This condition is sensible since the coherent state is just at the borderline between sub- and super-Poissonian statistics. For strong squeezing a_1 is well defined and a_2 is fluctuating around zero. Therefore, we can write

$$(\langle A \rangle + \Delta A)^2 = a_1^2 + \Delta a_2^2 \quad .$$

Solving for $\langle \Delta a_2^2 \rangle$, and using $\langle A \rangle >> \langle \Delta A^2 \rangle^{1/2}$ and $\langle A \rangle \approx \langle a_1 \rangle$, we find the condition for the transition to bunching is

$$\langle \Delta a_2^2 \rangle \approx \langle a_1 \rangle \quad . \qquad (4.1)$$

In the limit of large a_1 this is the same result as obtained by Caves.[24]

It can be concluded that although antibunching and squeezing are related, one does not imply the other, which has been pointed out also by

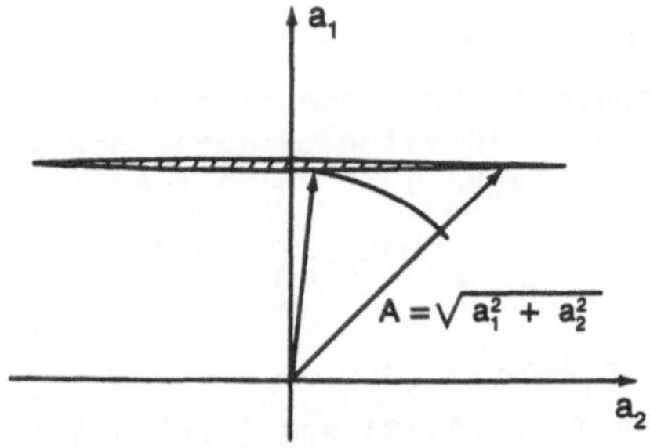

Fig. 9. Squeezing the region of uncertainty of a coherent state slightly may result in reduced amplitude fluctuations. Squeezing it more and more, however, will lead to increased fluctuations of the amplitude $A = \langle A \rangle + \Delta A$. Note that in the limit of large a_1 the fluctuations in a_2 are exaggerated by orders of magnitude.

Mandel.[26] Thus, the detection schemes necessary to determine the existence of one or the other are quite different. Antibunching or sub-Poissonian statistics can be detected via the intensity correlation function or the probability distribution of the photon number. Squeezing, on the other hand, is a property of either a_1 or a_2 and phase sensitive homo- or heterodyne detection is needed.[27] The latter will be discussed further in Sec. 8.

5. LINEAR ATTENUATION OF THE RADIATION FIELD AND COHERENCE EFFECTS

An important question for the experimentalist is whether, and how, the various states of the radiation field change under linear attenuation. Such attenuation is brought into every experiment by beam splitters, neutral density filters, Fresnel losses, etc. Also the quantum efficiency of the detector can be treated as a linear loss. We will use η to describe intensity losses; field amplitudes are then attenuated by $\eta^{1/2}$.

In Eq. (3.5), for example, there is a product of four field operators, consequently the expectation value scales down by η^2 under linear attenuation

$$\langle n^2_{eff}\rangle - \langle n_{eff}\rangle = \eta^2 \left(\langle n^2\rangle - \langle n\rangle\right) \quad , \tag{5.1}$$

where $\langle n_{eff}\rangle = \eta\langle n\rangle$, and n denotes the number of photons after attenuation. Note that simply replacing each n in Eq. (3.5) by n_{eff} would lead to the wrong result. This illustrates the importance of looking at the set of field operators defining the measurement. The normalized intensity correlation function has the product of four field operators both in the numerator and in the denominator of Eq. (3.6) and the η^2 cancel out. As a result, normalized intensity correlation functions are insensitive to linear attenuations.

It now remains to be seen whether mean squared fluctuations are affected by linear attenuation. From everyday laboratory experience we suspect that laser beams and thermal radiation sources do not change their characteristics when undergoing linear attenuation. This can be easily verified. For a coherent state we have $\langle\Delta n^2\rangle = \langle n\rangle$. Inserting this into Eq. (5.1) yields $\langle\Delta n^2_{eff}\rangle = \langle n_{eff}\rangle$. After linear attenuation, the radiation field still shows Poissonian statistics, but now at a photon number level reduced by the factor η. For a thermal radiation field, $\langle\Delta n^2\rangle = \langle n\rangle^2+\langle n\rangle$ [see Eq. (3.9)]. Using Eq. (5.1) again, we find $\langle\Delta n^2_{eff}\rangle = \langle n^2_{eff}\rangle+\langle n_{eff}\rangle$, demonstrating that a thermal field also stays thermal after linear attenuation.

The structure of Eq. (5.1), however, shows that coherent and thermal light fields are the only ones that do not change their photon statistics when they are attenuated. In general, the photon statistics may well be changed as can be demonstrated for the photon number state, having zero mean squared fluctuation $\langle\Delta n^2\rangle = 0$. From Eq. (5.1) one obtains

$$\langle\Delta n^2_{eff}\rangle = (1-\eta)\ \langle n_{eff}\rangle \tag{5.2}$$

for the attenuated field. Thus, linear attenuation introduces fluctuations into the initially non-fluctuating number state. For a more mathematical treatment of linear attenuation, see Bandilla,[28] Teich and Saleh,[29] Yuen and Chan,[30] or Schumaker.[31]

A simple general scaling law applying to all states of the radiation field can be obtained for the normally ordered normalized variance[32]

$$:Q: = \frac{\langle \Delta n^2 \rangle - \langle n \rangle}{\langle n \rangle}$$

$$= [g_I^{(2)}(0) - 1] \langle n \rangle \quad . \tag{5.3}$$

Nonclassical states of the radiation field are characterized by negative values of :Q:. Under linear attenuation, this normally ordered variance scales proportional to η, since $g_I^{(2)}(0)$ is invariant

$$(:Q:)_{eff} = [g_I^{(2)}(0) - 1] \, \eta \langle n \rangle \quad . \tag{5.4}$$

As will be seen in Sec. 8, squeezed states can be detected in hetero- or homodyne detection of the radiation field. Squeezing is observed by measuring either $g_I^{(2)}(0)$ or :Q: in the hetero- or homodyned field, and the effects of linear losses can then be calculated using the above formulae.

In the case of weak squeezing of a coherent state, $\langle \Delta a_2^2 \rangle << \langle a_1 \rangle$, the mean squared fluctuations in the photon number $\langle \Delta n^2 \rangle$ are related to $\langle \Delta a_1^2 \rangle$. Since the number of photons is $n(t) = (\langle a_1 \rangle + \Delta a_1(t))^2$, one finds $\langle \Delta n^2 \rangle \approx 4 \langle a_1 \rangle^2 \langle \Delta a_1^2 \rangle$. Noting that $\langle n \rangle = \langle a_1 \rangle^2$, we find that Eqs. (5.3) and (5.4) yield an expression for the mean squared field fluctuations in the squeezed component (Walls[11])

$$\langle \Delta a_1^2 \rangle_{eff} - \frac{1}{4} = \frac{\langle n \rangle}{4} \eta [g^{(2)}(0) - 1] \quad . \tag{5.5}$$

This shows that, at least in the limit of weak squeezing, the normally ordered mean squared fluctuation $\langle :\Delta a_1^2: \rangle \equiv \langle \Delta a_1^2 \rangle - 1/4$ also scales linearly with the linear attenuation coefficient η.

5.1. Effect of Coherence Properties

Equations (5.3) to (5.5) hold only for pure states of the radiation field showing no fluctuation, i.e. $g_I^{(2)}(\tau) = g_I^{(2)}(0)$ for all τ. Examples are the coherent state and the number state. For many practical radiation fields, however, $g_I^{(2)}(\tau)$ varies with τ and when this is taken into account,[6,33] the resulting variance will depend on the counting time interval T. Instead of $[g_I^{(2)}(0) - 1]$ the mean value of $[g_I^{(2)}(\tau) - 1]$ averaged over T has to be inserted in Eqs. (5.3) to (5.5). We can consider two limiting cases, T much larger or much smaller than the coherence time τ_c of the radiation field:

$$\frac{1}{T}\int_0^T d\tau \, [g_I^{(2)}(\tau) - 1] \approx [g_I^{(2)}(0) - 1] \begin{cases} 1 & , \text{ for } T < \tau_c \\ \tau_c/T & , \text{ for } T > \tau_c \end{cases} \quad . \tag{5.6}$$

This relation holds only approximately. The simplifying assumption is that $g_I^{(2)}(\tau) = g_I^{(2)}(0)$ for $\tau < \tau_c$ and $g_I^{(2)}(\tau) = 1$ for $\tau > \tau_c$. The average number of photons $\langle n \rangle$ now has to be replaced by the photon counting rate R times the counting time interval T. Inserting this into Eq. (5.3) one finds

$$:Q: \approx [g_I^{(2)}(0) - 1] \, R \begin{cases} T & , \text{ for } T < \tau_c \\ \tau_c & , \text{ for } T > \tau_c \end{cases} \quad . \tag{5.7}$$

This shows that :Q: will be rather small, i.e. close to the value for Poissonian statistics, if the counting time interval is too small. The variance :Q: reaches its extreme value only if T is chosen to be larger than τ_c (Mandel[33]). The spatial coherence properties can be treated similarly[34-36] and by also including the effect of linear losses, η, one finally obtains

$$(:Q:)_{eff} \approx \eta\,[g_I^{(2)}(0) - 1]\; R_\Omega \begin{cases} T\Omega & , \text{ for } T,\Omega < \tau_c,\Omega_c \\ \tau_c\Omega_c & , \text{ for } T,\Omega > \tau_c,\Omega_c \end{cases} \quad . \qquad (5.8)$$

Here Ω is the solid angle of emitted radiation collected on the detector and Ω_c is the coherence solid angle for which light emitted from different transverse positions of the source has a path difference of less than $\lambda/2$. The quantity R_Ω is thus the number of photons emitted per unit solid angle and time. Equation (5.5) has to be modified accordingly to take into account the coherence property of the radiation field.

6. PHOTON STATISTICS IN NONLINEAR MATTER FIELD INTERACTION

Let us consider a thermal radiation field of intensity I_o (see Fig. 5) entering a nonlinear medium and ask how the characteristics of the field are changed in the transmitted beam I_T. Basically, there are two different nonlinearities. $I_T(I_o)$ may have positive or negative curvature (Fig. 10). An example of the first case is a saturable absorber; the latter describes a property of, e.g., multiphoton absorption and second harmonic generation. In comparison to a case where I_T depends linearly on I_o, a positive curvature results in increased amplitude fluctuations in the transmitted beam whereas a negative curvature leads to reduced amplitude fluctuations. Figure 10 gives a classical argument for fields showing large amplitude fluctuations around zero, which we expect to predict the right behavior at least qualitatively.

In an experiment on the saturable absorber Krasinski _et al._[37] used a broadband pulsed Rhodamine 6G dye laser to represent a thermal light

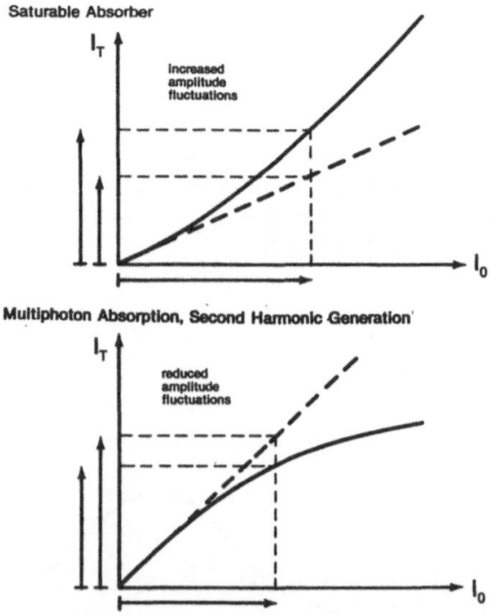

Fig. 10. The intensity I_T transmitted through a nonlinear medium is plotted versus the incident intensity I_o for two different nonlinear phenomena (solid curve). The way fluctuations are transferred from I_o to I_T is compared with the case of a linear interaction with the same low intensity absorption coefficient (dashed curve).

source (Fig. 11). A jet stream of malachite green dissolved in a mixture of ethylene glycol and glycerol served as a saturable absorber. The laser intensity was varied by moving the jet around the focus of lens L1. The transmitted beam was split and photomultiplier PM1 measured the average intensity <I> per pulse. The beam transmitted by the beam splitter was focused into a solution of dimethyl popop dissolved in dioxane which served as a strong broadband two-photon absorber. The two-photon absorption rate was determined by measuring the filtered fluorescence with photomultiplier PM2.

As discussed at the end of Sec. 2, the rate R_{TP} for broadband two-photon absorption is proportional to $G^{(2)}(0)$, the $\tau = 0$ value of the intensity correlation function. Therefore, $R_{TP}/\langle I\rangle^2$ is proportional to $g^{(2)}(0)$ in the present case. Figure 12 shows the experimental results for $g^{(2)}(0)$ as a function of laser intensity I_o. At $I_o = 0$ the curve was normalized to two, the $g^{(2)}(0)$ value for a thermal source (see end of Sec. 3). The rise of $g^{(2)}(0)$ with intensity reflects the increased fluctuations in the photon number predicted by Fig. 10, the largest increase occurring at the saturation intensity where the nonlinearity between I_T and I_o is largest.

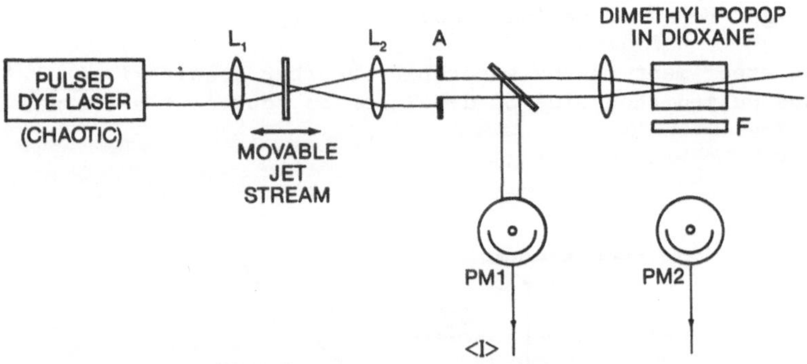

Fig. 11. Experimental setup used by Krasinski et al. (Ref. 37) for investigating the effect of a saturable absorber on the photon statistics. The pulsed dye laser was broadband, thus modeling a chaotic (thermal) field.

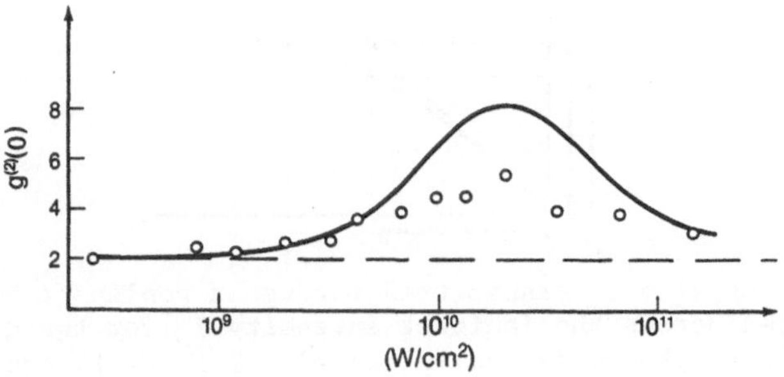

Fig. 12. Change of the $\tau = 0$ value of the second-order correlation function as a function of the incident intensity for the case of a saturable absorber (Krasinski et al., Ref. 37).

In an earlier experiment by Krasinski and Dinev[38] the change of the photon statistics has been studied for two-photon absorption. The experimental setup was similar to the one shown in Fig. 11, the difference being that the jet stream was replaced by a two-photon absorption cell. Here the curvature of $I_T(I_o)$ is negative and a decrease in the photon number fluctuation is expected as indicated in Fig. 10. Indeed, this behavior was found in the experiment (Fig. 13), however, the reduction of $g^{(2)}(0)$, was smaller than predicted theoretically.[39] A possible explanation offered by Krasinski and Dinev[38] is that other nonlinear processes of slower response time could become important at the high laser intensity of 5×10^9 W/cm^2, contributing to the absorption but not changing the photon statistics. Chaturvedi et al.[40] showed that the discrepancy between theory and experiment could also be due to the partially coherent nature of the input laser field.

So far, there exist no experimental results demonstrating the change of photon statistics in second harmonic generation. However, a study of an experimental model has been performed by Wagner et al.[41] They used electronic feed forward to a Pockels cell to simulate the nonlinearity. Owing to the comparatively low speed of the electronics, a change of photon statistics could be observed only for fluctuations at correspondingly low frequencies. To achieve this a He-Ne laser beam was sent through a rotating frosted glass plate. An aperture could be adjusted to pass only one or many coherence areas to the nonlinear interaction, thus modeling a thermal or a coherent light source respectively. Again, as in the case of two-photon absorption, the fluctuations of the fundamental laser beam were reduced. In this model case it even resulted in sub-Poissonian statistics.

Another interesting feature emerged from the Wagner et al. experiment. The laser beam modeling the second harmonic beam showed increased fluctuations for the thermal field but reduced (sub-Poissonian) fluctuations for the coherent fundamental field. In the latter case, the reduced fluctuations are in contradiction to our expectation based on Fig. 10, since the second harmonic intensity $I_{2\omega}(I_o)$ has a positive curvature. Note, however, the simple arguments based on Fig. 10 that have been made above relied on a fluctuating thermal input field having an electric field amplitude fluctuating around zero.

The surprising difference in the second harmonic photon statistics for thermal and coherent fundamental fields has been predicted by Kozierowski and Tanaś,[42] and explained qualitatively by Wagner et al.[41]

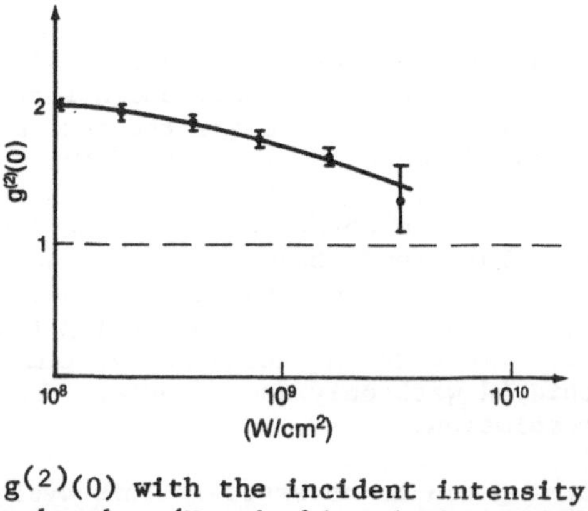

Fig. 13. Change of $g^{(2)}(0)$ with the incident intensity for the case of a two-photon absorber (Krasinski and Dinev, Ref. 38).

The probability distributions of photon numbers is quite different for these two fields. For the thermal field P(I) falls off exponentially with increasing intensity I [see Eq. (3.8)]. Second harmonic generation is more effective at larger I reducing the exponential tail of P(I) for the fundamental and increasing it for the second harmonic field. Since in both cases the maximum of P(I) is at $I = 0$, photon number fluctuations in the fundamental and second harmonic beam are decreased and increased respectively, which is in accordance with the above arguments based on Fig. 10. In the case of an incident coherent field, however, P(I) is a narrow Poisson distribution centered at a large mean value. Second harmonic generation and transmission of the fundamental will occur preferentially on the high and low intensity side of the initial distribution, respectively. Consequently, both distributions will be narrowed.

This result suggests that in an experiment it may be advantageous to look for photon antibunching in the second harmonic and not in the fundamental beam (Lugiato et al.[43]). The transmitted fundamental beam will always have a larger number of photons per mode $\langle n \rangle$ than the second harmonic beam, reducing $1 - g^{(2)}(0) \leq 1/\langle n \rangle$ (see Sec. 3.3). Apart from the hybrid system for modeling of second harmonic generation, all experiments discussed in this section dealt with photon statistics in the classically allowed region ($g^{(2)}(0) \geq 1$). We now come to the only system for which nonclassical states of the radiation field have been observed.

7. PHOTON ANTIBUNCHING AND SUB-POISSONIAN STATISTICS IN RESONANCE FLUORESCENCE

The experiments discussed in this section deal with the resonance fluorescence of a monochromatically driven two-level atom.[44] The two-level system is realized by first optically pumping sodium atoms into the 3s $^2S_{1/2}$, $F = 2$, $M_F = 2$ ground state and then exciting with circularly polarized light from there to the 3p $^2P_{3/2}$, $F = 3$, $M_F = 3$ state, which can only decay back into the 3s $^2S_{1/2}$, $F = 2$, $M_F = 2$ state. Under the influence of the monochromatic laser field of intensity I the atom undergoes Rabi oscillations between the lower and the upper state, the Rabi frequency being $\Omega \propto I^{1/2}$.

7.1. Photon Antibunching

After it had been independently proposed by Carmichael and Walls,[45] and by Cohen-Tannoudji[46] that photon antibunching could be observed in resonance fluorescence, Kimble et al.[47] performed the first experiment. In the following we will concentrate on the more recent experiments by Rateike et al.[48]

Figure 14 shows a schematic of the experimental setup in which the thermal sodium atomic beam, the laser beam and direction of observation are mutually perpendicular. A lens images the interaction region through a beam splitter onto two detectors. Apertures, not shown in the schematic, ensure that only the fluorescence of atoms near the center of the Gaussian laser beam profile is observed, where the laser intensity is essentially constant. The atomic beam density is so low that, on the average, only one atom is in the observation region. The intensity correlation function $g_I^{(2)}(\tau)$ is obtained by measuring the probability of finding one photon at time t and a second one at time $t + \tau$. In principle, this can be achieved with only one detector; two are used here to increase the time resolution.

Photon antibunching can be observed in this system because of a built-in dead time.[8] Immediately after the one atom in the interaction

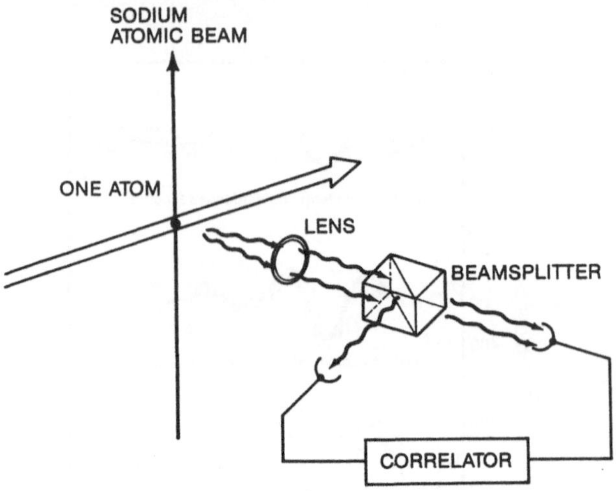

Fig. 14. Experimental setup used for the observation of photon antibunching in resonance fluorescence.

region has emitted a photon it must be in the ground state and cannot possibly emit a second photon right away. It takes some time, on the order of half a period of the Rabi oscillation, until the atom has an appreciable probability to be in the excited state again and thus be able to emit a second photon. Therefore, the curve we expect for $g_I^{(2)}(\tau)$ has a minimum at $\tau = 0$ and then rises to exhibit a damped periodic variation around the value of $g_I^{(2)}(\infty) = 1$ due to the Rabi frequency $\Omega \propto I^{1/2}$. Experimental results by Rateike et al.[48] are shown as histograms in Fig. 15 for different average atom number densities, $\langle N \rangle$, in the interaction volume. The solid line is the result of a least-squares fit of the theoretical formula given by Carmichael and Walls[45] to the data. The minimum of $g_I^{(2)}(\tau)$ at $\tau = 0$ is clearly below the long time average, showing that $g^{(2)}(0) < 1$! The level of accidental coincidences is due to laser light scattered by the apparatus. At larger values of $\langle N \rangle$ the minimum at $\tau = 0$ is less pronounced since fluorescence light from different atoms is uncorrelated.

One problem with this experiment is that for values of τ larger than the transit time of the thermal atoms through the observation region ($\approx$500 ns), the atom that emitted the photon defining $\tau = 0$ will surely have left the interaction region. For such large values of τ the number of coincidences [$\propto g_I^{(2)}(\tau)$] falls off to the uncorrelated background[49,50] [$=g^{(2)}(0)$]. In the original experiment by Kimble et al.[47] this effect was even more pronounced, owing to a much smaller observation region. The transit time effect could be avoided by repeating the same experiment on trapped atoms or ions.[51] In this case, a correlated signal can be expected for τ as large as the trapping time, which can be days and longer.

7.2. Direct Observation of sub-Poissonian Photon Statistics

If a light source emits antibunched radiation we also expect to observe sub-Poissonian photon statistics [Eq. (5.3)]. For resonance fluorescence, this has been investigated by Short and Mandel.[52] The experimental setup was slightly modified (Fig. 16) and the atomic beam density was kept very low to avoid two atoms being simultaneously in the observation region. One photomultiplier was used to detect the presence

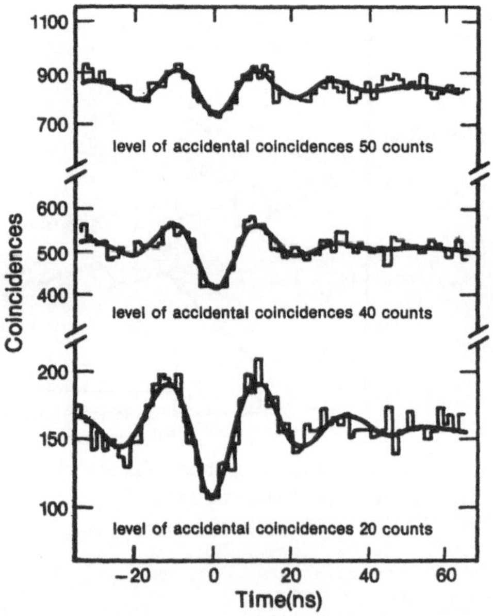

Fig. 15. Histograms of the measured intensity correlation function $g^{(2)}(\tau)$ for different average number density of atoms $\langle N \rangle$ in the observation region; $\langle N \rangle$ = 6.1, 3.4 and 1.6, from top to bottom). The $\tau = 0$ value is clearly below the long time average (Ref. 48).

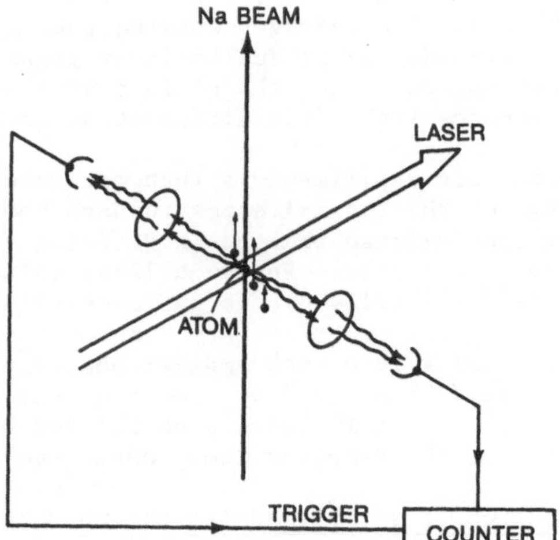

Fig. 16. Experimental setup used for the direct observation of photon statistics in resonance fluorescence by Short and Mandel (Ref. 52).

of an atom and to trigger gated counting for T = 200 ns. The photon number distribution P(n) was observed and compared with the Poisson distribution having the same mean photon number. The observed number of times, P(2), two photons were counted in the count interval of 200 ns was significantly below the respective number for the Poisson distribution. Based on Eq. (5.3), using $g^{(2)}(0) \approx 0$ and $\langle n \rangle \approx 1$, we expect the normally ordered variance to be $:Q: \approx -1$. The value calculated from the experimentally observed photon number distribution, however, is $:Q: = -0.0025$. The fact that both numbers are negative indicates that the radiation field is in a nonclassical state. The two v lues differ by a large amount because we forgot to take into account the effects of linear losses and coherence [Eq. (5.8)].

In the present experiment the linear losses were the transmission of the optics (0.4), and the quantum efficiency of the photomultiplier (0.15). These are multiplied to give a total loss of $\eta = 0.06$. For a laser intensity close to saturation of the two-level atom the rate of photon emission is one half the spontaneous decay rate Γ. Since the light source is a single atom, $\Omega_c = 4\pi$ and $R_\Omega \Omega_c = \Gamma/2$. In the experiment, however, only a fraction of the emitted light could be collected, $\Omega/\Omega_c = 0.067$. The counting time interval, T, on the other hand, was larger than the coherence time $\tau_c \approx \Gamma^{-1}$. Therefore the variance expected from Eq. (5.8) is $(:Q:)_{eff} \approx -0.0020$, which is rather close to the experimental value. The more accurate calculation[52] leads to an even better agreement with the experiment.

This experiment best demonstrates how drastically linear losses and coherence effects may alter the variance of the photon number distribution, an effect also important when considering possible schemes for observing squeezed states. However, before addressing the generation of squeezed states the detection problem will first be discussed.

8. DETECTION OF SQUEEZED STATES

As we have already mentioned in the preceding sections, phase sensitive detection such as homo- or heterodyning is needed for the observation of squeezed states.[27] In this scheme the signal beam, E_s, propagating the squeezed state is superimposed on a radiation field $E_{\ell o}$ of well-defined phase (local oscillator) and the beat is observed. The frequency of the signal field and the local oscillator can be the same or different resulting in a dc or ac beat signal (homo- or heterodyne detection), respectively.

A natural candidate for the local oscillator is a laser. Since lasers with absolutely stable phases are not available, it may be best to use only one laser beam, squeeze one part and use the other part as a local oscillator (Fig. 17). In this way, a fixed phase relation will be preserved, and the otherwise disastrous effects of phase fluctuations[53] are avoided. A glass wedge or an electro-optic crystal can then be used to change the relative phase between signal and local oscillator radiation. If the local oscillator field, $E_{\ell o}$, is treated as a classical wave, we may write $E_{\ell o} = \varepsilon_{\ell o} \cos \omega t$. For the signal field a_s the quantum description is used [Eq. (3.1)]. The number of photons in the region where both beams are superposed is

$$n_d(t) \propto \langle (E_{\ell o} + E_s(t))^2 \rangle = \frac{1}{2} \varepsilon_{\ell o}^2 + \varepsilon_{\ell o}(\langle a_{s,1} \rangle + \Delta a_{s,1}(t)) \tag{8.1}$$

and the fluctuating part is $\Delta n_d(t) = n_d(t) - \langle n_d \rangle = \varepsilon_{\ell o} \Delta a_{s,1}(t)$. The second subscript "1" denotes the coefficient of the cosine wave [Eq. (3.1)].

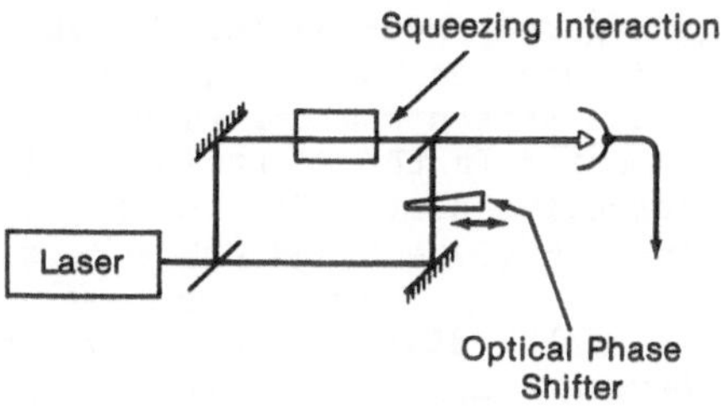

Fig. 17. Scheme for homodyne detection of a squeezed state. One laser is used as a primary light source to eliminate effects owing to fluctuation in the optical phase.

It is assumed that the local oscillator is much stronger than the signal field and terms quadratic in the signal field are neglected. It is clearly shown that the fluctuations in the intensity are due solely to the fluctuation in one quadrature of the signal field ($a_{s,1}$). As it turns out, the neglect of local oscillator quantization is quite misleading, since the amplitude fluctuations of $E_{\ell o}$ are not negligible compared to the ones in the squeezed signal field amplitude a_s. Assuming that the local oscillator is in a coherent state that is centered on the a_1-axis in the phase diagram, we have $\langle(\Delta a_{\ell o}^2)^{1/2}\rangle = 1/2$, and the fluctuation of photon number in the superposed beam is (Fig. 18)

$$\Delta n_d(t) \approx \langle a_{\ell o}\rangle \{\Delta a_{\ell o}(t) + \Delta a_{s,1}(t)\} \quad . \tag{8.2}$$

Instead of using a 50% beam splitter to superpose the beams, one could use a beam splitter transmitting the fraction $\eta < 0.5$ of the local oscillator intensity. This would reduce the contribution of local oscillator noise to the signal, but the local oscillator intensity would likewise be reduced.[31] In this case, the assumption that the signal amplitude is negligible in comparison to the local oscillator amplitude eventually becomes questionable.

A more elegant solution to this problem has been proposed by Yuen and Chan.[30] In their scheme, the signal is measured in both arms where the beam splitter superposes the input beams (Fig. 19). With the input ports assigned as in Fig. 19, the part of the local oscillator beam that is reflected off the denser medium undergoes a phase shift of 180° with respect to the transmitted part. The reflected and transmitted signal beams, however, will be in phase. Since $\cos(\omega t + 180°) = -\cos\omega t$ the detected photon number fluctuations Δn_d in the two arms are

$$\Delta n_{d,1}(t) \approx \langle a_{\ell o}\rangle (\Delta a_{\ell o}(t) - \Delta a_{s,1}(t))$$

$$\Delta n_{d,2}(t) \approx \langle a_{\ell o}\rangle (\Delta a_{\ell o}(t) + \Delta a_{s,1}(t)) \quad . \tag{8.3}$$

Here, the minus sign in the first equation does not lead to a reduction of the noise, since local oscillator and signal fluctuations are uncorrelated. By subtracting the intensities in both arms, however, the terms quadratic in the local oscillator amplitude cancel, and only the interference terms between signal and local oscillator survive. The fluctuation of this difference is

$$\Delta(n_{d,2}(t) - n_{d,1}(t)) \approx 2\langle a_{\ell o}\rangle\Delta a_{s,1}(t) \quad . \tag{8.4}$$

When both intensities are added, the interference terms cancel and the fluctuations will be

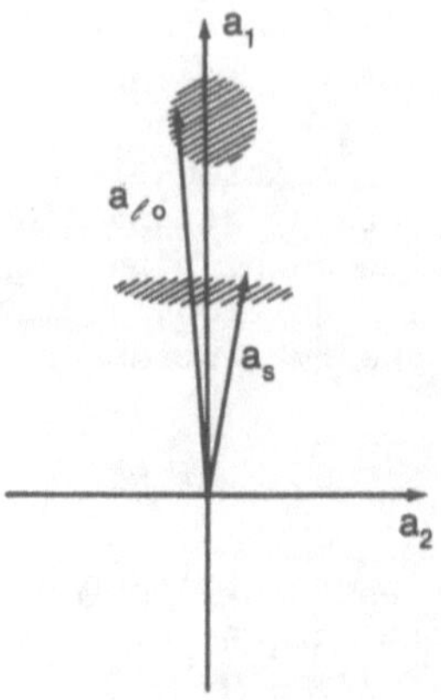

Fig. 18. Phase diagram of local oscillator and signal field. The ratio of amplitudes $a_{\ell o}/a_s$ shown is not in scale.

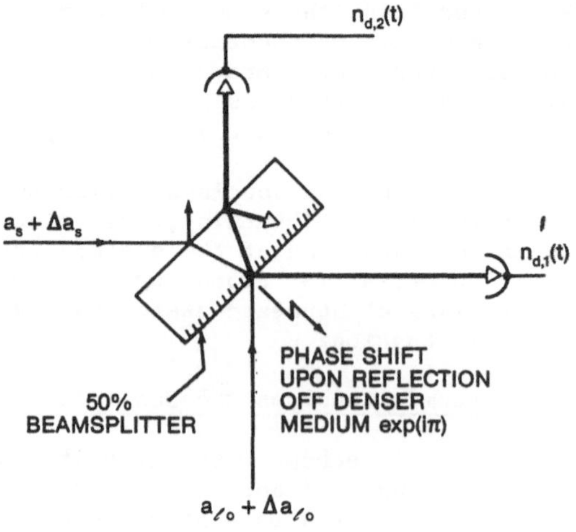

Fig. 19. Schematic explaining Yuen and Chan's (Ref. 30) proposal for two-port homodyning which allows one to suppress local oscillator noise.

$$\Delta(n_{d,2}(t) + n_{d,1}(t)) \approx 2\langle a_{\ell o}\rangle \Delta a_{\ell o}(t) \quad . \tag{8.5}$$

When both the signal and the local oscillator beams are in a coherent state, one has $\langle(\Delta a_{\ell o}^2)^{1/2}\rangle = \langle(\Delta a_s^2)^{1/2}\rangle = 1/2$, irrespective of the individual intensities, and the fluctuations in $(n_{d,2}+n_{d,1})$ and $(n_{d,2}-n_{d,1})$ have the same absolute magnitude. Furthermore, this fluctuation is also the same as the ones in Eqs. (8.3), since a_s and $a_{\ell o}$ are uncorrelated.

For the signal beam that is in a squeezed state, the fluctuation of the difference signal [Eq. (8.4)] will become smaller or larger depending on which component a_1 or a_2 is squeezed. An optical phase shift of 90° in either the local oscillator or the signal beam (see Fig. 17) can then change the difference signal from reduced to increased fluctuations or vice versa. This interesting and promising scheme for cancelling out the local oscillator noise is called "two-port homo- or heterodyning" and has also been discussed by Schumaker.[31]

It is worthwhile to note that the above arguments hold only for the local oscillator radiation being in a coherent state. If the local oscillator shows large additional fluctuations ($\Delta a_{\ell o} \approx a_{\ell o}$), then the interference term between the signal amplitude and the local oscillator noise ($a_s \Delta a_{\ell o}$) can no longer be neglected and it will contribute to the photon number fluctuation of the difference signal. Therefore, it is crucial to assume that $\Delta a_{\ell o}$ is of the order of unity.

9. PROSPECTS FOR SQUEEZING

On the one hand, the task of discussing possible future experiments on squeezed states is difficult, simply because so far there is no direct experimental evidence that such states of the radiation field exist. On

the other hand, there is much room for speculation since there are already a large number of theoretical publications suggesting various mechanisms for squeezed state generation.[11] No attempt is made in the following to discuss all of these suggestions. A few schemes will be discussed that we hope are representative of the present status of the field.

Since the two nonclassical properties of the radiation field, squeezing and antibunching, are related, it is probably straightforward to investigate some of the schemes promising antibunching with respect to their potential for producing squeezed states. In addition, there is the other class of squeezed states having reduced phase but increased amplitude fluctuation.

9.1. Atomic Vapor as the Nonlinear Medium

The only scheme with which photon antibunching has been observed so far is resonance fluorescence. Again we consider the two-level atoms driven by a narrow band single mode dye laser. Walls and Zoller[54] have shown theoretically that, for low values of laser intensity, the emitted fluorescence radiation will indeed be in a squeezed state. However, they did not find full squeezing ($\langle \Delta a_1^2 \rangle = 0$), but rather $\langle \Delta a_1^2 \rangle \gtrsim 3/16$. It has been pointed out by Collett et al.[55] that the amount of squeezing is different for the different spectral components of the fluorescent light. The above quoted minimum value is obtained only if the central frequency component is measured.

If one were to attempt an experimental verification, there would be problems in addition to the relatively small size of the effect. As the light source is only one atom, the coherence solid angle is $\Omega_c = 4\pi$; therefore any deviation from Poissonian photon statistics is reduced by Ω/Ω_c, Ω being the solid angle collected by the detector [Eq. (5.8) and Sec. 7.2]. In an atomic beam experiment there will be some fluctuation in the position of the atoms with respect to laser beam and detection system (Fig. 20). Phase changes imposed on the radiation field by variations in the optical path will destroy any squeezing.[56] Therefore, the position of the atoms should be fixed with an accuracy much smaller than one optical wavelength. The latter could be achieved by using ions held in a radio-frequency or Penning trap.[51,57] It would require, however, operation well inside the Dicke regime (i.e., ion position confined to much less than an optical wavelength). Even so, the other problems remain: small size of the squeezing, finite quantum efficiency of the detector and photon collection angle $\Omega < 4\pi$. This discussion suggests that it may be advantageous to employ a coherent interaction, where phase problems are reduced and where the squeezed radiation field is a collimated beam allowing for $\Omega \gtrsim \Omega_c$.

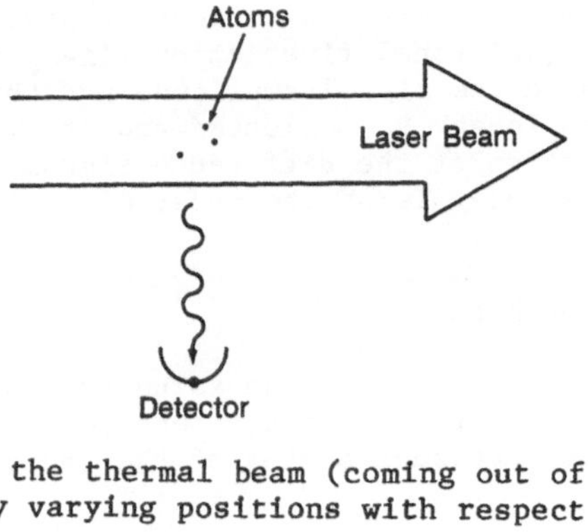

Fig. 20. The atoms in the thermal beam (coming out of the paper) have statistically varying positions with respect to laser beam and detector.

One of the nonlinear interactions proposed by Yuen and Shapiro[58] for generation of squeezed states is degenerate four-wave mixing. A state with squeezed phase fluctuations is produced if the transmitted probe and the phase conjugate beam are superposed on a beam splitter. Some experimental effort in this direction has been made using sodium vapor as the nonlinear medium.[59-61] A vapor usually shows high nonlinearities only in the vicinity of a resonance, owing to the limited atom number density that can be reached. Near resonance and close to saturation, however, spontaneous emission becomes important, which, in this case, tends to act against squeezing.[62] Therefore, the interaction should not be treated using a nonlinear susceptibility (effective Hamiltonian) but rather using a microscopic model[34,63] (density matrix formalism).

Reid and Walls[63] find squeezing if the laser intensity S, normalized to the saturation intensity, is much larger than the detuning δ, normalized to the resonance line width. To avoid having spontaneous emission become important, the relations $S \ll \delta^2$ and $10\ S^2 \ll \delta^3$ should be fulfilled in addition. Unfortunately, losses of the light beams when passing through the nonlinear medium tend to noticeably reduce the squeezing.[61,63] Recently, Kumar and Shapiro[64] pointed out that the effect of losses is substantially reduced when forward instead of backward degenerate four-wave mixing is used, where transmitted probe and phase conjugate beam are copropagating. Reid and Walls[62] could not verify this advantage when taking into account atomic fluctuations through their microscopic model.

Another interesting point has been raised by Reynaud and Heidmann[34-36] concerning the spatial coherence property of the squeezed state. If no cavity is used around the nonlinear atomic vapor, the radiated field will occupy many modes, but the coherence solid angle Ω_c will be small and of the order of the pump beam divergence, almost washing out the squeezing.

Another rather exotic scheme for squeezed state generation involves the Jaynes-Cummings model[65] for the coupling of a two-level atom with only one mode of the radiation field. It is remarkable that this most simple of all models in quantum optics reveals a rich variety of phenomena (see e.g., Eberly et al.[66]). Recently, Meystre and Zubairy[67] showed that squeezing may be obtained in the Jaynes-Cummings model. For a long time, the Jaynes-Cummings model was considered as a playground exclusively dedicated to theorists. Recent developments in the field of microwave-Rydberg-atom interaction, however, promise some experimental tests of the Jaynes-Cummings model in the near future.[68-70] Whether these experiments will ever result in a squeezed microwave field is questionable, especially since the expected squeezing[67] is only $\langle \Delta a_1^2 \rangle \approx 0.20$.

In summary, nonlinear interaction between light and free atoms does not appear promising for squeezed state generation. The main difficulty is spontaneous emission. The recipe seems to be off resonant interaction in order to avoid population in the excited state. The decrease in nonlinearity resulting from the detuning may be compensated for by an increase of laser power and atom density. This suggests that gas phase interactions may not be well suited to the purpose of squeezed state generation and that one should rather favor nonlinear crystals which offer interaction far off resonance and solid state density.

9.2. Interaction in Crystals

It has been shown that many nonlinear interactions lead to squeezing of the radiation field. As a typical example, we will discuss second harmonic generation. A coherent light wave traveling through a frequency doubling crystal has been calculated to yield sub-Poissonian statistics both in the fundamental and the second harmonic beam (Kozierowski and

Tanas[42]). Owing to the short interaction time, the decrease of amplitude fluctuations, especially in the second harmonic beam, is predicted to be very small. Mandel[32] has shown that this sub-Poissonian statistics is accompanied by squeezing which is also small in magnitude. For practical reasons, the crystal length -- and with it the single pass interaction time -- cannot be increased beyond certain limits. Therefore, it has been proposed by Lugiato et al.[43] that the crystal be placed inside an optical resonator, thus increasing the interaction time by one to two orders of magnitude. This resonator is required to be resonant for both the fundamental and the second harmonic beam. The experimental difficulty that the nodes of the two standing waves (ω and 2ω) do not coincide anywhere inside the crystal can be overcome by making use of the dispersion in air[71] or by using properly designed mirrors.[72]

Surprisingly, the calculation by Lugiato et al.[43] showed that the squeezing was at most about $\langle \Delta a_1^2 \rangle \gtrsim 0.1$. The minimum dispersion $\langle \Delta a_1^2 \rangle$ is reached at the threshold of oscillation.[73] The difference between this value and the one for Poissonian statistics, $\langle \Delta a_1^2 \rangle - 0.25 \gtrsim -0.15$, however, is only about one half of the corresponding number for ideally squeezed states: In view of potential applications, a rather discouraging fact.

A similar story could be told, e.g., for the degenerate parametric amplifier/oscillator,[74-78,53] optical bistability[79-81] and multiphoton absorption.[82-84] Again, with the nonlinear medium inside a symmetrical linear cavity ($R_1 = R_2$ in Fig. 21a), the mean squared fluctuations in either a_1 or a_2 were at best about a factor of two smaller than for Poissonian statistics. For a while, the future of squeezing did not look very promising. A noise reduction by at best a factor of two with respect to a coherent state would not find many applications. This situation changed when Yurke[85] showed that for an asymmetric cavity optimal squeezing is possible, i.e., $\langle \Delta a_i^2 \rangle = 0$ with $i = 1$ or $i = 2$. In multimode calculations by Collett and Gardiner[86] and Gardiner and Savage,[87] similar large squeezing in the output beam was found for the central frequency components. Consequently, some effort is being made to calculate this spectrum of squeezing for various nonlinear interactions (Collett and Walls[88]).

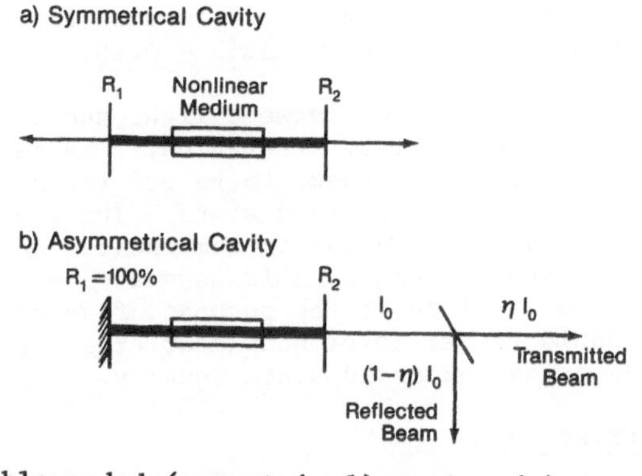

Fig. 21. A double-ended (symmetrical) cavity (a) is not capable of producing fully squeezed output fields. Part (b) shows the optical equivalent of the symmetrical cavity, built by using a single-ended cavity which may put out fully squeezed radiation.

The cumbersome factor of two seems to be a property of the double ended linear cavity, since this factor can be overcome by going to a single ended linear or ring cavity, independent of the nonlinear medium inside. We will now try to understand this difference by constructing an optical equivalent of the symmetrical cavity by using only the asymmetrical cavity. Figure 21a shows a nonlinear medium inside a linear symmetrical cavity (mirror reflectivity $R_1 = R_2 = R$). The nonlinear interaction produces squeezing of the radiation inside the cavity. The radiation leaks equally out of both mirrors. For symmetry reasons, both output beams must show the same squeezing, which will be different, however, from the amount of squeezing inside the resonator. The optical equivalent is sketched in Fig. 21b. The output of an asymmetrical cavity is sent through a beam splitter transmitting η and reflecting $1-\eta$ of the incident intensity. Let us now assume that the a_2 component of the cavity output is fully squeezed ($\langle\Delta a_2^2\rangle = 0$). According to Eq. (5.5) the effect of the beam splitter will be to change the a_1-fluctuation in the transmitted beam to $\langle\Delta a_2^2\rangle = (1-\eta)/4$ and that in the reflected beam to $\langle\Delta a_2^2\rangle = \eta/4$. For $\eta = 1$ the transmitted beam will be perfectly squeezed and the "reflected" beam will be the vacuum state with Poissonian statistics. For the optical equivalent of Fig. 21a with $R_1 = R_2$, η has to be one half. This is just the factor two found in the theoretical calculations![86]

In the general case of a linear cavity with two mirrors having different reflectivities R_1 and R_2, the ratio of the output intensities will be T_1/T_2 with $T_i = 1-R_i$. The amount of squeezing in these two beams is, of course, not expected to be the same. It can be estimated, according to Fig. 21b and Eq. (5.5), by setting $\eta = T_1/(T_1+T_2)$. Accordingly, the squeezing in the radiation transmitted by mirror R_1 is

$$\langle\Delta a_2^2\rangle = \frac{1}{4} - \frac{T_1}{4(T_1+T_2)} \quad . \tag{9.1}$$

Collett and Gardiner[86] found the same result for the central frequency component [see their Eq. (50)].

At this point, we can produce a recipe for squeezed state generation and detection: Place a nonlinear medium inside a resonator with only one input/output coupler and increase the input intensity until the critical point for the first instability is reached (self pulsing in the case of second harmonic generation), then use Yuen and Chan's scheme[30] of two-port homo- or heterodyning (Sec. 8) with highly efficient photodetectors (Sec. 5). Using homodyning or heterodyning with the right intermediate frequency, the frequency component having the largest squeezing can be selected (Gardiner and Savage[87]).

The experimental effort so far has been concentrated on four-wave mixing in optical fibers.[89] It remains to be seen which scheme will lead to the first experimental verification of squeezing and which one will produce a large enough effect that it will also be interesting for applications.

Another type of squeezed state generation should be briefly mentioned. The recent success at Orsay[90] in operating a storage ring free-electron laser in the visible has given new momentum to this widely tunable source of coherent radiation. In this context, the theoretical finding that, under certain conditions, the free-electron laser radiation can show photon antibunching and squeezing[91-93] is quite interesting.

10. PROPOSED APPLICATIONS FOR NONCLASSICAL STATES OF THE RADIATION FIELD

Possible applications for light beams showing sub-Poissonian statistics or squeezing may be found for some measurements which involve sensitive detection at the shot noise level. We will briefly discuss some examples. One area where a reduction of amplitude fluctuations may improve the detection sensitivity is optical communication (Yuen[23]). An interesting proposal in this direction has been made by Shapiro.[94] If an optical fiber is used to communicate messages to several places it would, of course, be advantageous if the fiber could be tapped without breaking it up. Such a tap can be realized by bringing a second fiber close to the first one, thus forming a directional coupler. Each directional coupler will attenuate the signal in the fiber. Shapiro[94] has calculated that this loss of signal intensity in the main fiber can be minimized if the tapping fiber carries radiation in a squeezed state. For the detection, the signal field in the tapping fiber is homodyned, the local oscillator being in phase with the component showing the reduced amplitude fluctuations.

This scheme can be shown to be back-action evading,[95] since fluctuations are imposed only to that quadrature of the main fiber field which is not probed. Thus the tapped fiber is an example for a quantum nondemolition measurement.[96,97] The claim is that this scheme would allow a large number of users to take information from the one fiber. A potential pitfall may be, however, that along with squeezing one quadrature in the tapping fiber field the other quadrature will show increased fluctuations which are, of course, coupled into the main fiber. It seems the lossless tapping of the fiber by many users will only work if all users detect the same quadrature component from the field in the main fiber. Thus, there would be the additional problem of also communicating this phase between the different users.

One type of experiment in physiology, where the photon statistics play a crucial role, deals with tests of the human visual system (Teich _et al._[98]). A light source emitting photons regularly, one by one, would be ideal for threshold measurements. As discussed in Sec. 7, such antibunched light has so far only been generated using a rather complicated experimental setup. Also, the time scale for which the antibunching has been observed — the transit time of the atom through the observational region — is at least six orders of magnitude too small for the physiological experiments.

Before such experiments become feasible, other simpler and more appropriate sources of antibunched light have to be found. In this context, Teich _et al._[99] have proposed to observe the atomic fluorescence in a Frank-Hertz experiment. When the tube is operated in the space charge limit,[100,101] the current shot noise is smaller than expected from Schottky's formula. The electrons drift antibunched toward the anode. Under favorable conditions, the background gas atoms which are excited by these electrons emit antibunched light.[99] Since the phase of the emitted photons is not controlled, this source does not, of course, produce squeezed states.

As a final example, we discuss gravitational wave detection.[96] Unlike electric fields, which can be rather easily detected through their different action on positive and negative charges, gravitational waves have so far never been observed. Gravitational waves accelerate different nearby masses in the same way, owing to the equality of the gravitational and the inertial mass.[102] The change in the distance between two test masses due to a gravitational wave is proportional to this distance. Therefore, one has to rely on a large spatial separation of the

test masses. One of the detection schemes uses a Michelson interferometer, where the beam splitter and the two end mirrors represent three distant masses, the separation of which can be sensitively monitored by observing the interference fringes (see Schleich and Scully[103] and references therein).

The gravitational wave produces a quadrupole-type strain pattern,[102] so that the length of one of the properly oriented interferometer arms is increased and the other is decreased. The sensitivity can be enhanced by using multiple pass geometry between the end mirrors and auxiliary mirrors mounted on the beam splitter mass. Theoretically, the sensitivity is limited by photon counting statistics at the low laser intensity and by fluctuating radiation pressure on the end mirrors at high laser intensity (Caves[104] and Loudon[105]). By choosing the right light intensity, the so-called standard quantum limit can be reached, $\Delta z = (h\tau/\pi m)^{1/2}$, where τ is the duration of the measurement, m the mass of the end mirrors and Δz the smallest detectable change in path difference. The fluctuations of the radiation pressure in the two interferometer arms are uncorrelated owing to the nature of a beam splitter that can also be viewed as an attenuator [Eq. (5.1)].

An alternative way to picture the uncorrelated fluctuations in each interferometer arm has been put forward by Caves.[24,104] The vacuum field -- for which the mean value of the field strength is zero but which has nonzero mean squared fluctuations [Eq. (3.2)] -- enters the interferometer through the otherwise unused entrance port. Unfortunately, the laser power for which the standard quantum limit is reached[104] is rather high (300 W). Caves[24] pointed out, however, that this power level can be lowered by sending a squeezed radiation field through the unused port of the interferometer.

The resolution Δz of today's best Michelson interferometers is limited essentially by photon counting statistics. In the experiment by Maischberger et al.[106] the demonstrated sensitivity was $\Delta z/z = 7 \times 10^{-17}$ for a 1 kHz detection bandwidth at a frequency of 7 kHz. The interferometer arm length was 3 m and the laser beam was reflected 138 times back and forth between the end and auxiliary mirrors. The laser power of 0.1 W used in the experiment was several orders of magnitude lower than the power for which the standard quantum limit is reached.

Gravitational wave intensities are estimated to induce a relative change in the interferometer arm length of $\Delta z/z \approx 10^{-21}$ for periodic waves at frequencies around 1 kHz (Thorne[102]). This value is about ten times larger than the standard quantum limit for τ = 1 ms, 10 kg end mirror masses, and an effective arm length of 1 km. The experimental number, so far 10^5 times too high, can be lowered by increasing the laser power and the interferometer arm length.[107] However, since the photon counting limit has already been reached, improvements could also be achieved by entering a squeezed radiation field through the otherwise unused entrance port of the interferometer.[24,35,36,96,108] For an increase of the interferometer arm length beyond the laboratory scale one faces exponentially growing engineering problems and costs. Thus, further efforts toward squeezed state generation seem economically viable and desirable.

ACKNOWLEDGMENTS

I am grateful to G. Alber and J. C. Bergquist for carefully reading this manuscript and improving it by making many helpful suggestions. I want to thank W. M. Itano, H. J. Kimble, L. Schnupp and D. F. Walls for

several discussions during the preparation of these lecture notes. I also would like to acknowledge the cooperation of J. L. Hall, D. Hils and J. J. Snyder while investigating the feasibility of producing squeezed states in second harmonic generation. It is a particular pleasure to thank the members of the JILA Scientific Reports Office for their extremely fast and efficient help.

REFERENCES

1. R. Hanbury Brown and R. Q. Twiss, Nature 177, 27 (1956); 178, 1447 (1956).
2. R. J. Glauber, Phys. Rev. Lett. 10, 84 (1963).
3. E. C. G. Sudarshan, Phys. Rev. Lett. 10, 277 (1963).
4. R. Hanbury Brown, The Intensity Interferometer (Taylor & Francis, London, 1974).
5. E. O. Schulz-DuBois, Photon Correlation Techniques in Fluid Dynamics, Springer Ser. Opt. Sci., Vol. 38 (1983).
6. R. J. Glauber, in Quantum Optics and Electronics, edited by C. deWitt, A. Blandin and C. Cohen-Tannoudji (Gordon and Breach, Edinburgh, 1964).
7. L. Mandel and E. Wolf, Rev. Mod. Phys. 37, 231 (1965).
8. D. F. Walls, Nature 280, 451 (1979).
9. R. Loudon, Rep. Prog. Phys. 43, 913 (1980).
10. H. Paul, Rev. Mod. Phys. 54, 1061 (1982).
11. D. F. Walls, Nature 306, 141 (1983).
12. J. R. Klauder and E. C. G. Sudarshan, Fundamentals of Quantum Optics (Benjamin, New York and Amsterdam, 1968).
13. R. J. Glauber, Phys. Rev. 130, 2529 (1963); 131, 2766 (1963).
14. B. R. Mollow, Phys. Rev. 175, 1555 (1968).
15. M. Sargent III, M. O. Scully and W. E. Lamb, Jr., Laser Physics (Addison Wesley, Reading, Mass., 1974).
16. W. H. Louisell, Quantum Statistical Properties of Radiation (John Wiley, New York, London, Sydney, Toronto, 1973).
17. H. Haken, Handbuch der Physik, Vol. XXV/2c (Springer, Berlin, 1970).
18. P. Zoller, in Laser Physics, edited by D. F. Walls and J. Harvey (Academic, Sydney, New York, 1980).
19. S. N. Dixit, P. Zoller and P. Lambropoulos, Phys. Rev. A 21, 1289 (1980).
20. J. J. Yeh and J. H. Eberly, Phys. Rev. A 24, 888 (1981).
21. D. S. Elliott, M. Hamilton, K. Arnett and S. J. Smith, Phys. Rev. Lett. 53, 439 (1984).
22. R. Loudon, Rep. Prog. Phys. 43, 913 (1980).
23. H. P. Yuen, Phys. Rev. A 13, 2226 (1976).
24. C. M. Caves, Phys. Rev. D 23, 1693 (1981).
25. D. F. Walls, in Laser Physics, edited by D. F. Walls and J. Harvey (Academic, Sydney, New York, 1980).
26. L. Mandel, Phys. Rev. Lett. 49, 136 (1982).
27. H. P. Yuen and J. H. Shapiro, Opt. Lett. 4, 334 (1979).
28. A. Bandilla, Opt. Commun. 23, 299 (1977).
29. M. C. Teich and B. E. A. Saleh, Opt. Lett. 7, 365 (1982).
30. H. P. Yuen and V. W. S. Chan, Opt. Lett. 8, 177 (1983).
31. B. L. Schumaker, Opt. Lett. 9, 189 (1984).
32. L. Mandel, Opt. Commun. 42, 437 (1982).
33. L. Mandel, Opt. Lett. 4, 205 (1979).
34. S. Reynaud and A. Heidmann, Opt. Commun. 50, 271 (1984).
35. A. Heidmann, Thèse de 3ème cycle (Paris, 1984, unpublished).
36. S. Reynaud and A. Heidmann, Ann. de Physique, to be published, 1984.
37. J. Krasinski, C. Radzewcz and J. Christowski, Phys. Lett. 70A, 287 (1979).
38. J. Krasinski and S. Dinev, Opt. Commun. 18, 424 (1976).

39. H. P. Weber, IEEE J. Quantum Electron. 7, 189 (1971).
40. S. Chaturvedi, P. Drummond and D. F. Walls, J. Phys. A 10, L187 (1977).
41. J. Wagner, P. Kurowski, W. Martienssen, Z. Physik B 33, 391 (1979).
42. M. Kozierowski and R. Tanaś, Opt. Commun. 21, 229 (1977).
43. L. A. Lugiato, G. Strini and F. DeMartini, Opt. Lett. 8, 256 (1983).
44. J. D. Cresser, J. Häger, G. Leuchs, M.-F. Rateike and H. Walther, in Dissipative Systems in Quantum Optics, edited by R. Bonifacio (Springer, Berlin, Heidelberg, 1982).
45. H. J. Carmichael and D. F. Walls, J. Phys. B 9, L43 (1976); 9, 1199 (1976).
46. C. Cohen-Tannoudji, in Frontiers in Laser Spectroscopy, Vol. 1, edited by R. Balian, S. Haroche and S. Liberman (North Holland, Amsterdam, 1977).
47. H. J. Kimble, M. Dagenais and L. Mandel, Phys. Rev. Lett. 39, 691 (1977); M. Dagenais and L. Mandel, Phys. Rev. A 18, 2217 (1978).
48. M.-F. Rateike, G. Leuchs and H. Walther, results reported in Refs. 8 and 44.
49. E. Jakeman, E. R. Pike, P. N. Pusey and J. M. Vaughan, J. Phys. A 20, L257 (1977).
50. H. J. Carmichael, P. Drummond, P. Meystre and D. F. Walls, J. Phys. A 11, L121 (1978).
51. D. J. Wineland, W. M. Itano and R. S. Van Dyck, Jr., Adv. Atom. Mol. Phys. 19, 135 (1983).
52. R. Short and L. Mandel, Phys. Rev. Lett. 51, 384 (1983).
53. K. Wódkiewicz and M. S. Zubairy, Phys. Rev. A 27, 2003 (1983).
54. D. F. Walls and P. Zoller, Phys. Rev. Lett. 47, 709 (1981).
55. M. J. Collett, D. F. Walls and P. Zoller, "Spectrum of Squeezing in Resonance Fluorescence," preprint (1984).
56. R. Loudon, Opt. Commun. 49, 24 (1984).
57. W. Neuhauser, M. Hohenstatt, P. Toschek and H. Dehmelt, Phys. Rev. Lett. 41, 233 (1978).
58. H. P. Yuen and J. H. Shapiro, Opt. Lett. 4, 334 (1979).
59. P. Kumar, J. H. Shapiro and R. S. Bondurant, Opt. Commun. 50, 183 (1984).
60. R. E. Slusher, B. Yurke and J. F. Valley, J. Opt. Soc. Am. B 1, 525 (1984).
61. R. S. Bondurant, P. Kumar, J. H. Shapiro and M. Maeda, Phys. Rev. A 30, 343 (1984).
62. M. D. Reid and D. F. Walls, Generation of Squeezed States via Degenerate Four Wave Mixing," preprint (1984).
63. M. D. Reid and D. F. Walls, Opt. Commun. 50, 406 (1984).
64. P. Kumar and J. H. Shapiro, Phys. Rev. A 30, 1568 (1984).
65. E. T. Jaynes and F. W. Cummings, Proc. IEEE 51, 89 (1963).
66. J. H. Eberly, N. B. Narozhny and J. J. Sanchez-Mondragon, Phys. Rev. Lett. 44, 1323 (1980).
67. P. Meystre and M. S. Zubairy, Phys. Lett. 89A, 390 (1982).
68. D. Meschede, Dissertation, Ludwig-Maximilians Universität, München, 1984.
69. J. A. C. Gallas, G. Leuchs, H. Walther and H. Figger, Adv. At. Mol. Phys. 20, 413 (1985); see also H. Walther, this volume.
70. S. Haroche and J. M. Raimond, Adv. At. Mol. Phys. 20, to be published, 1985.
71. J. M. Yarborough, J. Falk and C. B. Hitz, Appl. Phys. Lett. 18, 70 (1971).
72. K. M. Baird, K. M. Evenson, G. R. Hanes, D. A. Jennings and F. R. Petersen, Opt. Lett. 4, 263 (1979); and G. R. Hanes, "Doubly-Resonant Intracavity Generation of Second Harmonic of 1153 nm Radiation with Stabilization on h.f.s. Components of I_2^{127}," preprint.
73. P. D. Drummond, K. J. McNeil and D. F. Walls, Optica Acta 27, 321 (1980).

74. H. Takahashi, Adv. Commun. Syst. 1, 227 (1965).
75. D. Stoler, Phys. Rev. Lett. 33, 1397 (1974).
76. G. Milburn and D. F. Walls, Opt. Commun. 39, 401 (1981).
77. L. A. Lugiato and G. Strini, Opt. Commun. 41, 67 (1982).
78. A. Lane, P. Tombesi, H. J. Carmichael and D. F. Walls, Opt. Commun. 48, 155 (1983).
79. L. A. Lugiato and G. Strini, Opt. Commun. 41, 374 (1982).
80. L. A. Lugiato and G. Strini, Opt. Commun. 41, 447 (1982).
81. M. D. Reid and D. F. Walls, Phys. Rev. A 28, 332 (1983).
82. R. Loudon, Opt. Commun. 49, 67 (1984).
83. A. Bandilla and H.-H. Ritze, Opt. Commun. 19, 169 (1976).
84. M. S. Zubairy, M. S. K. Razmi, S. Iqbal and M. Idress, Phys. Lett. 98A, 168 (1983).
85. B. Yurke, Phys. Rev. A 29, 408 (1984).
86. M. J. Collett and C. W. Gardiner, Phys. Rev. A 30, 1386 (1984).
87. C. W. Gardiner and C. M. Savage, Opt. Commun. 50, 173 (1984).
88. M. J. Collett and D. F. Walls, "Squeezing Spectra for Nonlinear Optical Systems," preprint (1984).
89. M. D. Levenson, J. Opt. Soc. Am. B 1, 525 (1984).
90. M. Billardon, P. Elleaume, J. M. Ortéga, C. Bazin, M. Bergher, M. Velghe, Y. Petroff, D. A. G. Deacon, K. E. Robinson and J. M. J. Madey, Phys. Rev. Lett. 51, 1652 (1983).
91. W. Becker, M. O. Scully and M. S. Zubairy, Phys. Rev. Lett. 48, 475 (1982).
92. G. Dattoli and M. Richetta, Opt. Commun. 50, 165 (1984).
93. S. Rai and S. Chopra, Phys. Rev. A 30, 2104 (1984).
94. J. H. Shapiro, Opt. Lett. 5, 351 (1980).
95. B. Yurke and L. R. Corruccini, Phys. Rev. A 30, 895 (1984).
96. P. Meystre and M. O. Scully, Quantum Optics, Experimental Gravitation and Measurement Theory (Plenum, New York, 1983).
97. G. J. Milburn and D. F. Walls, Phys. Rev. A 30, 56 (1984).
98. M. C. Teich, P. R. Prucnal, G. Vannucci, M. E. Breton and W. J. McGill, Biol. Cybern. 44, 157 (1982).
99. M. C. Teich, B. E. A. Saleh and D. Stoler, Opt. Commun. 46, 244 (1983); and M. C. Teich, B. E. A. Saleh and J. Perina, J. Opt. Soc. Am. B 1, 366 (1984).
100. W. Schottky and E. Spenke, Wiss. Veroeff. Siemens-Werken 16, 1 (1937).
101. B. J. Thomson, D. O. North and W. A. Harris, RCA Review 4, 269 (1940).
102. K. S. Thorne, Rev. Mod. Phys. 52, 285 (1980).
103. W. Schleich and M. O. Scully, in Modern Trends in Atomic and Molecular Physics, edited by R. Stora and G. Grynberg (North-Holland, Amsterdam, in press); W. Schleich, Optische Tests der Allgemeinen Relativitätstheorie, Max Planck Institut für Quantenoptik, Report No. 89 (1984).
104. C. M. Caves, Phys. Rev. Lett. 45, 75 (1980).
105. R. Loudon, Phys. Rev. Lett. 47, 815 (1981).
106. K. Maischberger, A. Rüdiger, R. Schilling, L. Schnupp, W. Winkler and H. Billing, in Laser Spectroscopy V, edited by A. R. W. McKellar, T. Oka and B. P. Stoicheff (Springer-Verlag, Berlin, 1981), Springer Ser. Opt. Sci. Vol. 30.
107. A proposal for a space interferometer with a baseline of up to 10^6 km has been made by J. Faller, P. L. Bender, Y. M. Chan, J. L. Hall, D. Hils and J. Hough, in General Relativity and Gravitation, edited by B. Bertotti, F. de Felice and A. Pascolini (Consiglio Nazionale delle Richerce, Rome, 1983), pp. 960-962.
108. W.-T. Ni, "On quantum-mechanical noise and squeezed state technique in an interferometer," preprint (1984).

EMPTY NONLINEAR OPTICAL RESONATORS

A. Dorsel,[1] W. J. Firth,[2] A. Guzman de Garcia,[3] J. D. McCullen,[4] F. Marquis,[3] P. Meystre,[3] D. P. J. O'Brien,[3] G. Reiner,[3] M. Sargent III,[3,5] K. Ujihara,[3,6] E. Vignes,[7] H. Walther,[1,3] and E. M. Wright[3]

1. Sektion Physik, Universität München, D-8046 Garching, F. R. Germany
2. Department of Physics, Heriot-Watt University, Edinburgh, United Kingdom
3. Max-Plank-Institut für Quantenoptik, D-8046 Garching, F. R. Germany
4. Department of Physics, University of Arizona, Tucson, AZ 85721, USA
5. Optical Sciences Center, University of Arizona, Tucson, AZ 85721, USA
6. University of Electro-communications, Chofu, Tokyo 182, Japan
7. Department of Physics, UCSD, La Jolla, CA 92093 USA

1. Introduction

Ever since the invention of the laser, there has been considerable interest in the study of nonlinear optical resonators (NOR). A NOR is any optical resonator containing at least one nonlinear element. In the laser, this element is the active, inverted medium, and in conventional optical bistability, it is a passive absorptive or dispersive material. In these notes, we concentrate on two kinds of NOR's that have received considerable attention in recent years. They distinguish themselves from more conventional ones by the fact that they are empty, one of the mirrors itself being the source of the nonlinear behaviour. The simplest such NOR is a radiation-pressure driven optical resonator, which we showed recently[1] to exhibit optical bistability. Section 2-1 briefly reviews the physics of this system, and shows its analogy - and differences - with an optical resonator filled with a Kerr medium. We also discuss mirror confinement, i.e., the possibility of trapping a macroscopic mirror in the potential well due to the combined effects of radiation pressure and the restoring force of the mirror. An improved, three-mirror cavity version of the system[2] is then discussed in Sec. 2-2, and a noise analysis including both the effects of white and ground noise is presented in Sec. 2-3.

Our second empty nonlinear optical resonator uses a nonlinear optical phase conjugator[3] to provide a phase-conjugate mirror with reflection coefficient larger than unity.[4] Section 3-1 briefly reviews the physics of nonlinear optical phase conjugation via degenerate four-wave mixing, and presents an "exact" analysis of near-collinear degenerate four-wave mixing in a Kerr medium. Our theory includes pump depletion and yields simple closed form solutions for the nonlinear reflection and transmission coefficients of arbitrary diffraction order. Section 3-2 applies these results to the static analysis of an optical resonator consisting of a normal and a phase-conjugate mirror. The dynamic stability analysis is discussed in Sec. 3-3. Both optical bistability and period doubling to chaos are predicted. Section 4 presents a summary and conclusions.

2. Radiation Pressure Driven Optical Resonators

2-1. Fabry-Perot Resonator

Consider a plane Fabry-Perot interferometer with one fixed mirror and one very light mirror, suspended to swing as a pendulum. The light intensity in the cavity exerts pressure on the light mirror, driving it towards an equilibrium, with the gravitational and inertial forces balancing the pressure. At steady state, the mirror displacement x from its rest position in the absence of light is then proportional to the intracavity intensity. Since x determines the interferometer spacing, there is a one-to-one correspondance between this situation and usual bistability in the presence of a Kerr medium, except that we now have a change of the physical cavity length instead of its optical length. But contrary to standard dispersive bistability, this system cannot be described uniquely by a nonlinear susceptibility $\chi^{(3)}$ because the effect depends only upon the cavity dephasing from a transmission maximum rather than on the absolute length of a nonlinear medium. Reference 1 shows that this system can display a bistable response (radiation pressure OB). In addition, an effect expected from extensive computer studies has been confirmed by the experiments: when the incident light is sufficiently intense, the movable mirror becomes extremely stable, and mechanically resonant motion is suppressed. We call this behaviour 'mirror confinement.' Mirror confinement can be simply understood in terms of the potential describing the moving mirror motion. This potential can be derived from the dynamical equation for the moving mirror

$$\ddot{x} + \gamma\dot{x} + x = P(x), \tag{1}$$

where x measures the displacement of the mirror from its rest position in the absence of light, γ is the pendulum damping constant, and P(x) is the radiation pressure force per unit mass. For scaling convenience, x is measured in units of half wavelengths of the incident light (assumed to be monochromatic) and time is measured in units of the pendulum period Ω. For the two-mirror system, P(x) is given by

$$P_2(x) = \frac{K_2}{1 + F_2\sin^2(\phi/2)}, \tag{2}$$

where $\phi = 2\pi x - \phi_0$, ϕ_0 is the detuning of the resonator in the absence of light, and the constants K_2 and F_2 are given by

$$K_2 = \frac{4R'(1-R)W}{(1-\sqrt{RR'})^2 mc\lambda\,\Omega^2} \; ; \quad F_2 = \frac{4\sqrt{RR'}}{(1-\sqrt{RR'})^2}, \tag{3}$$

where R and R' are the intensity reflectivities of the fixed and movable mirrors, respectively, W is the input intensity, and m is the movable mirror mass.

The potential corresponding to the dynamical equation (1) with γ=0 is given by

$$V(x) = \frac{x^2}{2} - \int_0^x dy\, P(y) \tag{4}$$

and is plotted in Fig. 1a for parameter values of R = 0.99, R' = 0.95, K_2 = 109 (we use these values for examples throughout). This value of K_2 corresponds to an applied light power of 0.5W, and Ω=7 sec^{-1}, with a mirror mass of 60g. The potential minima correspond to possible stable states of the system. It is clear that if the mirror is captured into one of these wells, it thereafter responds significantly only to driving forces whose frequencies are near the oscillation frequency in the well. As the input power, and thus the internal power, increases, the depth and curvature of the well increase, and the effective resonant frequency of the system departs from the pendulum frequency. For sufficient input powers the system can therefore be made extremely stable against mechanically resonant oscillations.

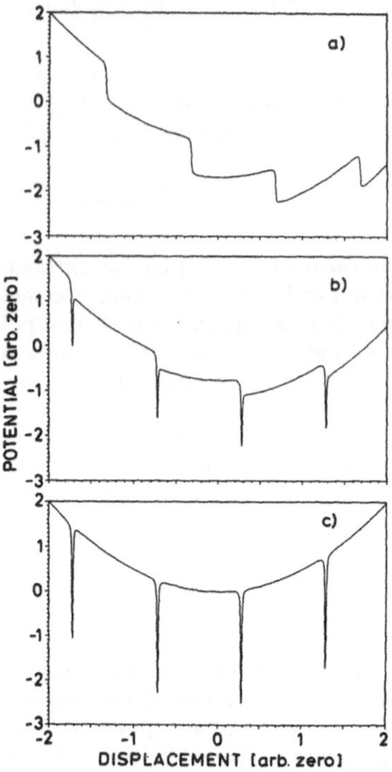

Fig. 1. Potential given by Eq. (4) versus mirror displacement in units of $\lambda/2$ for parameter values of $R = 0.99$, $R' = 0.95$, $K_2 = 109$ (we use these values for examples throughout). (a) Two-mirror case. (b) Three-mirror case with one-sided pumping. (c) Three-mirror case with incoherent symmetric pumping (after Ref. 2).

2-2. Three-Mirror System

The chief drawback presented by the two-mirror system in confining the mirror is that the radiation pressure pushes the mirror in only one direction. Thus on one side of the equilibrium point the force changes rapidly with position (on a scale determined by the cavity finesse), while on the other side the restoring force comes from the usual pendulum forces, and varies much more slowly with position. The situation can be significantly improved by suspending the movable mirror inside a fixed tuned Fabry-Perot cavity, thereby creating two coupled interferometers.[2] Depending upon which of the cavities is closer to resonance, the radiation pressure force can now be in either direction. By appropriately tuning the main cavity, reversal of the direction of the radiation pressure occurs when the overall transmission is large, providing maximal radiation forces to confine the pendulum.

Equation (1) remains valid for the three-mirror system, but the expression for the radiation pressure force changes. It may be derived using matrix methods[5] or simply from the boundary conditions. Its form is simplified by assuming the fixed mirrors have the same reflectivity R. Then putting $f = \sqrt{R'/R}$, we find

$$P_3 = \frac{K_3[f(1+R) - 2\cos(\phi-\phi_1)]}{1 + F_3[f\cos(\phi-\phi_1/2) - \cos(\phi_1/2)][fR\cos(\phi-\phi_1/2) - \cos(\phi_1/2)]}, \tag{5}$$

where we have taken the input field to be incident on one side of the fixed Fabry-Perot only. In this equation ϕ_1 is the round-trip phase detuning of the main cavity, and F_3 and K_3 are given by

$$F_3 = \frac{4R}{(1-R)^2} \; ; \; K_3 = \frac{1}{f}\left[\frac{1-fR}{1-R}\right]^2 K_2 \; . \qquad (6)$$

The denominator in Eq. (5) has a sharp minimum, similar to the denominator in Eq. (2). The value of ϕ at which it minimizes, however, is $\sin(\phi - \phi_1/2) = 0$, rather than $\sin \phi/2 = 0$. The value of ϕ_1 which makes this minimum the smallest is given by

$$\cos(\phi_1/2) = \frac{f(1+R)}{2} \; . \qquad (7)$$

With this choice of ϕ_1, the numerator of Eq. (4) passes through zero at the same value of ϕ for which the denominator is minimized. The force thus changes sign in a region of maximum transmission. Roughly speaking, this causes the equilibrium points of the mirror to occur at nodes of the standing wave of the light resonant in the main cavity. The cavity is, however, not quite resonant, and contributions from the pendulum restoring force alter this statement slightly.

Potential curves for the three-mirror system are plotted in Figs. 1b and 1c. Curve 1c is obtained with equal intensities incident on both sides of the fixed Fabry-Perot with the assumption that the light beams are mutually incoherent; the force then consists of the sum of two terms, each of the form of Eq. (5):

$$P_3 = 2K_3 \sin(\phi_1/2) \sin(\phi - \phi_1/2)/D \; , \qquad (8)$$

where D is the denominator in Eq. (5).

The curves for the three-mirror system show potential wells around equilibrium points of the movable mirror which are much narrower than their two-mirror counterparts. The examples in Fig. 1 are for the choice of cavity phase given by Eq. (7), which results in the narrowest wells. With values of ϕ_1 greater than this, the wells become deeper, and the spacing between the steep walls increases, making the potential more or less flat on the bottom. For a given choice of ϕ_1, the depth of the well depends upon the intensity of the incident light. The shape, however, depends on the cavity parameters. For the case shown in Fig. 1c, the width at half maximum is about 0.0117 of the incident wavelength, or 8.5 nm with 500 nm incident light.

The effective resonant frequencies for the three-mirror system can be estimated by a linearized analysis (see Sec. 2-3) of the dynamics represented in Eq. (1). This shows that for fixed ϕ, the frequency is nearly proportional to $\sqrt{W}$ so that the device may be tuned to higher frequencies by increasing the input power. For the examples shown in Fig. 1, the "captured" frequency is about 170 times the pendulum frequency for one-sided driving, and 240 times the pendulum frequency for light incident on both sides of the cavity. The damping is not affected in this approximation, so that the mirror acts as a resonator with greatly increased Q.

Alternatively the cavity tuning parameter ϕ_1 may be changed to change the frequency at which the mirror responds. Going away from the optimum confinement value of ϕ_1 given by Eq. (7) in general lowers the response frequency and makes the response broad-banded.

2-3. Noise Analysis

a) White Noise

This section investigates the effect of an external white noise source on the two- and three-mirror systems. A delta-correlated noise term is then added to the right hand side of Eq. (1), giving the Langevin equations

$$\dot{x} = u \qquad (9)$$

$$\dot{u} = -\gamma u - \left[\frac{\partial V}{\partial x} + \Omega^2 x\right] + F(t) \; , \qquad (10)$$

where u is the movable mirror velocity in units of $u_0 = \lambda\Omega/2$, and F(t) obeys

$$\langle F(t)\rangle = 0 \tag{11}$$

$$\langle F(t)F(t')\rangle = 2D\delta(t-t') . \tag{12}$$

Here the diffusion coefficient D can be written

$$D = \gamma(u_{th}/u_0)^2 , \tag{13}$$

where u_{th} is the thermal velocity associated with the noise.

The effect of white noise on Eqs. (9) and (19) can be investigated by solving the corresponding Fokker-Planck equation

$$\frac{\partial \Pi}{\partial t} = \left[-u\frac{\partial}{\partial x} + \frac{\partial}{\partial u}(\gamma u + \frac{\partial V}{\partial x}) + D\frac{\partial^2}{\partial u^2}\right] \Pi , \tag{14}$$

where $\Pi(x,u,t)$ is the probability density in position-velocity space.

In steady state $\partial(\Pi/\partial t = 0)$ the solution of this equation is given by[6]

$$\Pi(x,u) = N\, e^{-(u_0/u_{th})^2[u^2/2 + V(x)]} , \tag{15}$$

where N is a normalization constant. Integration over velocities yields the probability density Q(x) of finding the particle between x and x+dx. We find

$$Q(x) = \frac{e^{-(u_0/u_{th})^2 V(x)}}{\int_{-\infty}^{\infty} dy\, e^{-(u_0/u_{th})^2 V(y)}} , \tag{16}$$

which has the form of a partition function. We are interested in the confinement of the movable mirror when it is initially prepared in one of the potential minima. If this well corresponds to essentially the same potential value as one of its close neighbors, then, according to Eq. (16), the steady-state probabilities for the mirror to be in either well are equal. This arises since in the limit $t\to\infty$, tunneling produces the distribution (16). However, on more realistic time scales (say hours), tunneling is negligible if Q(x) for the initial well is narrow compared to the well width. We therefore consider individual wells and treat the remainder of the potential curve as that of the gravitational potential. In Fig. 2 we have plotted Q(x) as a function of x and u_{th}/u_0, along with a corresponding contour plot, for both the two-mirror (Fig. 2a and 2b) and three-mirror (Fig. 2c and 2d) cases. In each case, we choose the deepest potential well (see Fig. 1a and 1b). Figure 2a shows that the stability of the two-mirror system against noise is somewhat limited due to the slowly varying gravitational potential. This feature leads to fast tunneling from the initial well to the neighboring well on the slow side of the potential already for relatively low values of u_{th}/u_0. In contrast the three-mirror system shows far superior stability against noise. Figure 2c clearly shows that the probability density remains essentially constant in width, right up until a critical point at which the profile broadens abruptly and stability is lost. This feature is more clearly displayed in Fig. 2d, where the contours abruptly diverge at u_{th}/u_0. For the case of incoherent two way driving of the three-mirror system, the results are similar to those in Fig. 2c and 2d, except the curves are more symmetrical.

b) Ground Noise

A white noise analysis, though useful for comparing the relative stability and confinement properties of the various systems, is inappropriate to the description of ground noise, which is the main noise source as far as applications are concerned. To evaluate its effects, we use a linearized form of Eq. (1) in which P(x) is expanded to first order in x around an equilibrium point $x_0=0$. In the presence of the ground noise source G(t), this equation becomes (x is now the displacement from the chosen equilibrium point)

$$\ddot{x} + \gamma\dot{x} + \Omega_e^2 x = G(t), \tag{17}$$

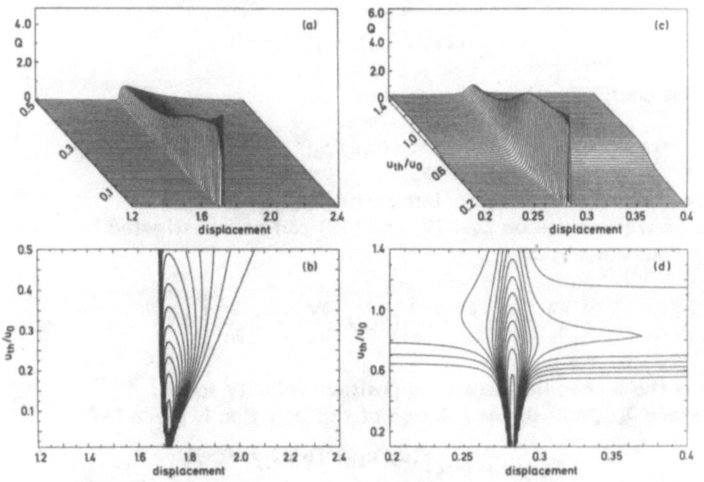

Fig. 2. Probability density $Q(x)$ versus x and u_{th}/u_0, along with a corresponding contour plot for the two-mirror ((a) and (b)) and three-mirror ((c) and (d)) cases. In each case, we choose the deepest potential well (see Fig. 1a and 1b) (after Ref. 2).

where the effective frequency Ω_e is found from

$$\Omega_e^2 = 1 + \left.\frac{\partial P}{\partial x}\right|_{x_0} , \tag{18}$$

which is around $\Omega_e^2 \simeq 1.147 \times 10^6$ W for the parameters considered. This harmonic approximation is clearly most appropriate for the three-mirror system (see Figs. 1b and 1c). In the case of two-way pumping, Ω_e^2 becomes

$$\Omega_e^2 = \frac{4\pi\sqrt{4 - f^2(1+R^2)}}{fT^2T'} \, W . \tag{19}$$

Assuming that the ground noise causes a translational motion z(t) of the whole system through stationary air, the noise term in Eq. (17) can be written

$$G(t) = -\gamma\dot{z} - \ddot{z} . \tag{20}$$

Fourier transforming (17) for a unit impulse displacement $z(t)=\delta(t)$ yields[7] the frequency transfer function $H(\omega)$

$$H(\omega) = \frac{\omega^2 - i\gamma\omega}{\Omega_e^2 - \omega^2 + i\gamma\omega} . \tag{21}$$

The spectral density $S_x(\omega)$ is then given by

$$S_x(\omega) = |H(\omega)|^2 S_z(\omega) , \tag{22}$$

where S_z is the spectral density of the ground noise, and the mean squared displacement of the moving mirror is given by

$$\langle x^2 \rangle = \int S_x(\omega)\, d\omega \,. \tag{23}$$

Ground noise is characterised by a spectral density of the general form $S_z(\omega) \approx C/\omega^4$, C being a constant dependent on location. For the Munich area $C \approx 6\times10^{-11}/\lambda^2\Omega^3$, the λ and Ω dependence arising from scaling.[8] For $\lambda = 500$ nm, we obtain

$$\langle x^2 \rangle = \frac{250}{\Omega^3} \int_{\omega_c}^{\omega_u} \frac{1}{\omega^4} \left[\frac{\omega^4 + \Omega^2\omega^2}{\Omega_e{}^2 - \omega^2 + \gamma^2\omega^2} \right] d\omega \,, \tag{24}$$

where ω_c is taken as a lower bound of the earth vibration frequencies, which has a value of about 1/min. In the scaled units we are using, this gives $\omega_c \approx 0.1/\Omega$. The upper value ω_u is taken as a few times Ω_e, say 10, then $\omega_u \approx 10^4\sqrt{C}$.

These values yield an X_{rms} displacement of the moving mirror of about 15 Å for $\gamma = 0.5$. With higher values of damping, X_{rms} slowly increases, because the effective resonator Q is decreased. This demonstrates that confinement to a small range of motion within the confines of the potential well is possible.

These results show, however, that control of the position of a movable mirror to within a range of a few Ångstroms by using radiation pressure should be readily achievable. This opens the way to a number of applications, since such a system represents a very sensitive, narrow-banded tunable transducer between mechanical and optical signals.

3. Phase-Conjugate Resonators

We now turn to a second example of an empty nonlinar optical resonator. Here instead of the nonlinear moving mirror of Sec. 2, we use a nonlinear optical phase conjugate mirror (PCM). There is currently much interest in such phase conjugate resonators (PCR) due to their ability to reduce or cancel phase distortions within the resonator.[9]

The mode characteristics of PCR's in the undepleted pump beam approximation have been discussed by several authors.[10,11] In particular, half-axial modes[9] are predicted to arise and have been observed experimentally.[12] These theories cannot predict the dynamic behaviour above the threshold for self-oscillations.[9] Large signal analyses of PCR's have been published,[13,14] but these studies consider only the static solutions, that is, the spontaneous signals that build up at the frequency of the pump beams.

In this section, we present a nonlinear analysis of the self-oscillations in a PCR, in which the phase-conjugate mirror (PCM) is a Kerr medium having instantaneous response. Section 3-1 gives a plane-wave analysis of near-collinear degenerate four-wave mixing in a Kerr medium, including pump depletion.[15] In Sec. 3-2 we extend these results to include the effects of feedback provided by a conventional mirror on the static behaviour of this system. We show that optical multistability occurs. Section 3-3 treats the dynamic behaviour of the system in the limit of a medium with length short compared to that of the external resonator. For certain parameter ranges, the static solution becomes unstable and oscillations follow. Initially, these oscillations correspond to the excitation of the half-axial modes predicted by the linear theories.[9] However, further change of parameters can lead to a period-doubling sequence which evolves to chaotic response.[16]

3-1. Near-Collinear Degenerate Four-Wave Mixing Plane-Wave Analysis

We consider the situation illustrated in Fig. 3. An isotropic, lossless Kerr medium, characterized by the intensity-dependent index of refraction $n(I) = n_0 + n_2 I$, is irradiated by two counterpropagating pump fields of input intensities I_F and I_B and a probe field of input intensity I_P. The paraxial wave equations for the slowly-varying envelopes F(z,x) and B(z,x) of the propagating fields are at steady state

$$\frac{\partial F}{\partial z} = \frac{i}{2k}\frac{\partial^2 F}{\partial x^2} + i[|F|^2 + (1+h)|B|^2]F, \tag{25}$$

$$-\frac{\partial B}{\partial z} = \frac{i}{2k}\frac{\partial^2 B}{\partial x^2} + i[(1+h)|F|^2 + |B|^2]B. \tag{26}$$

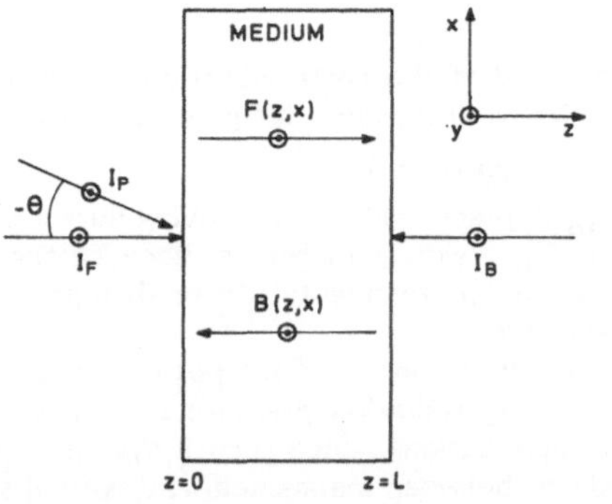

Fig. 3. Isotropic lossless Kerr medium pumped by two counterpropagating fields of intensities I_F and I_B and a probe field of intensity I_P incident at a small angle $-\theta$. All beams have the same frequency ω, are polarized in the y-direction, and are taken as plane waves. The medium surfaces are assumed to be anti-reflection coated (from Ref. 15).

Here the dimensionless intensities $|F|^2$ and $|B|^2$ are in units of $c/\omega n_2 L$ (as are I_F, I_B, and I_P), where $\omega = kc/n_0$ is the light frequency, c the speed of light, and L the length of the medium. We have taken $n_2>0$ without loss of generality. The coefficient $0 \leq h \leq 1$ describes nonlinear nonreciprocity[17] arising from the small-period ($\simeq 1/2k$) phase gratings formed by standing-wave effects in the medium. Equations (25) and (26) include all possible gratings. All fields are taken to be plane waves. For a small angle $-\theta$ between the probe field and the z-axis, second-order derivatives in the field equations can be neglected. However we keep the explicit x-dependence in the fields to be able to separate the various orders of scattering geometrically. This is the Raman-Nath approximation, which neglects the transverse structure of the fields. The limits of validity of this approximation are discussed in Ref. 18, where it is shown to be restricted to the case of thin samples. With this simplification, Eqs. (25) and (26) can be solved exactly[19,20] in terms of the known fields F(0,x) and B(L,x) at the boundaries of the medium. Noting the conservation laws $\partial|F|^2/\partial z = \partial|B|^2/\partial z = 0$, we find readily

$$F(L,x) = \exp[i(|F(0,x)|^2 + (1+h)|B(L,x)|^2)]F(0,x)\ , \tag{27}$$

$$B(0,x) = \exp[i((1+h)|F(0,x)|^2 + |B(L,x)|^2)]B(0,x)\ . \tag{28}$$

All propagation effects, including pump depletion, are contained within these equations.

For the geometry at hand, we have

$$F(0,x) = \sqrt{I_F} + Ee^{-ik\theta x}\ , \tag{29}$$

$$B(L,x) = \sqrt{I_B}\,, \tag{30}$$

where we have used the small angle approximation to write $\mathbf{k}_p - \mathbf{k}_F \equiv \Delta\mathbf{k} \simeq -k\theta\,\mathbf{x}$, for which $\mathbf{k}_F$ and $\mathbf{k}_p$ are the wave vectors of the forward pump and probe beam, respectively, $\mathbf{x}$ is the unit vector in the x direction, and $|E|^2 = I_P$. In writing Eqs. (29) and (30), we have assumed that the phases of the pump beams at their respective input planes are zero. With these boundary conditions Eqs. (27) and (28) become

$$F(L,x) = \sum_{n=-\infty}^{\infty} F_n\, e^{ink\theta x} \tag{31}$$

$$B(0,x) = \sum_{n=-\infty}^{\infty} B_n\, e^{ink\theta x}\,, \tag{32}$$

where

$$\begin{aligned} F_n &= Z\sqrt{I_F}\left[\frac{iE^*}{\sqrt{I_P}}\right]^n \left[J_n(2\sqrt{I_F I_P}) + i\sqrt{\frac{I_P}{I_F}}\,J_{n+1}(2\sqrt{I_F I_P})\right],\ n\geq 0 \\ &= Z\,E\left[\frac{iE}{\sqrt{I_P}}\right]^{-n-1} \left[J_{-n-1}(2\sqrt{I_F I_P}) + i\sqrt{\frac{I_P}{I_F}}\,J_{-n}(2\sqrt{I_P I_F})\right],\ n<0, \end{aligned} \tag{33}$$

and

$$\begin{aligned} B_n &= Z'\sqrt{I_B}\left[\frac{iE^*}{\sqrt{I_P}}\right]^n J_n\left[2(1+n)\sqrt{I_F I_P}\right],\ n\geq 0 \\ &= Z'\sqrt{I_B}\left[\frac{iE}{\sqrt{I_P}}\right]^{-n} J_{-n}\left[2(1+n)\sqrt{I_F I_P}\right],\ n<0. \end{aligned} \tag{34}$$

Here $Z = \exp[i(I_F+I_P+(1+h)I_B)]$ and $Z' = \exp[i((1+h)(I_F+I_P)+I_B)]$ describe phase shifts due to nonlinear refraction and $J_n(x)$ are Bessel functions of the first kind. The components F_n and B_n in Eqs. (31) and (32) give the field components of the forward and backward scat-

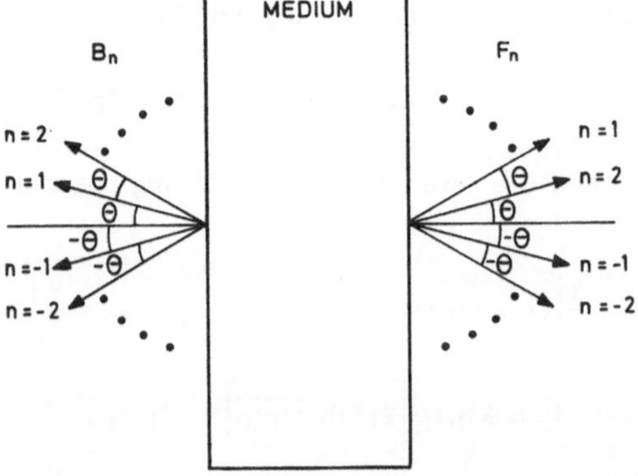

Fig. 4. Diagram of the F_n and B_n components of the forward and backward fields scattered at various angles $n\theta$ (after Ref. 15).

tered fields, respectively, at the angle $n\theta$, as illustrated in Fig. 4. We note that such Bessel function solutions are typical of scattering off periodic gratings. It is straightforward to verify the intensity conservation laws

$$\sum_{n=-\infty}^{\infty} |F_n|^2 = I_F + I_P ; \quad \sum_{n=-\infty}^{\infty} |B_n|^2 = I_B , \tag{35}$$

as well as to show that these results reduce to the small-signal theory in the weak probe case.[4]

3-2. The Phase-Conjugate Resonator: Static Solution

We now consider the PCR shown in Fig. 5. The backscattered field B_1 is fed back into the PCM as an input, all other possible backscattered fields (at angles $-\theta$, $\pm m\theta$, $m = 2,3,...$)

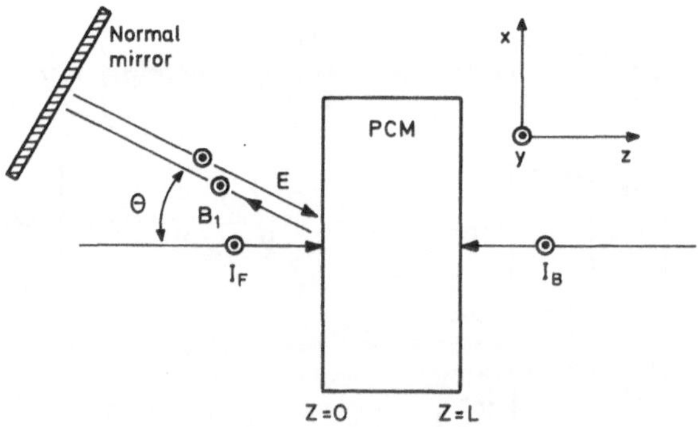

Fig. 5. Kerr medium PCM pumped by two counterpropagating beams of intensities I_F and I_B and frequency ω. The normal mirror allows for feedback of fields scattered at angle θ back into the PCM. All other possible backscattered fields (at angles $-\theta$, $\pm m\theta$, $m = 2,3,...$) are lost from the resonator (after Ref. 21).

are assumed to be lost from the resonator. At steady state, both E and B_1 have frequency ω and self-consistency of E requires[21]

$$E = \sqrt{R}\, e^{i\phi_0} B_1, \tag{36}$$

where R is the intensity reflectivity of the normal mirror and ϕ_0 is the phase shift for the field at frequency ω in the external resonator. By defining

$$S_F = \sqrt{I_F} e^{i\phi_F} ; \; S_B = \sqrt{I_B} e^{i\phi_B} ; \; E = \sqrt{I_P} e^{i\phi_P} \tag{37}$$

and substituting Eq. (34) into (36), we obtain the following phase and intensity self-consistency equations

$$\phi_P = m\pi + \frac{1}{2}\left[(1+h)(I_F+I_P) + I_B + \phi_F + \phi_B + \phi_0 + \frac{\pi}{2}\right] (\text{mod } 2\pi) \tag{38}$$

and

$$I_P = R I_B J_1{}^2\left[2(1+h)\sqrt{I_F I_P}\right] \equiv G(I_P) \, . \tag{39}$$

Equation (38) demonstrates the well known result that in steady state, the circulating fields are not only frequency synchronized, but also phase synchronized (in time) to the phases of the pump fields.[9] The transcendental equation (39) can be solved graphically for

I_P for given values of h, R, I_B, and I_F by plotting the two forms of $G(I_P)$ given by Eq. (39)

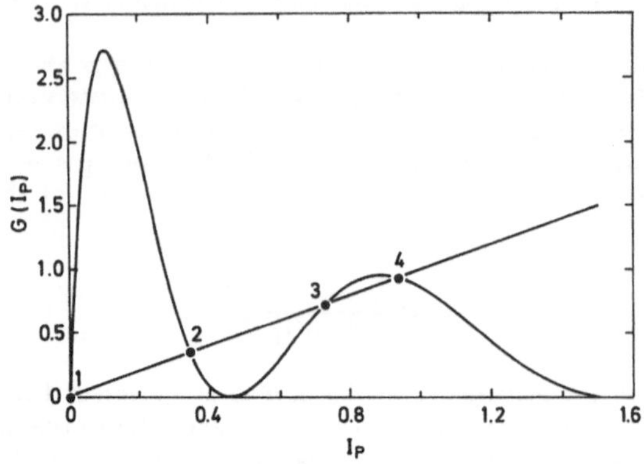

Fig. 6. Graphical solution of intensity self-consistency Eq. (39). Points 1-4 give possible solutions. Multiple solutions are evident (after Ref. 21).

and finding their intersection. This construction is illustrated in Fig. 6. Multiple solutions are clearly seen (note that $I_P = 0$ is always a solution of Eq. (39)). For illustrative purposes, we consider the case of h, R, and I_F fixed. Equation (39) can then be inverted to yield I_B explicitly as a function of I_P:

$$I_B = \frac{I_P}{R J_1^2\left[2(1+h)\sqrt{I_F I_P}\right]} \; . \tag{40}$$

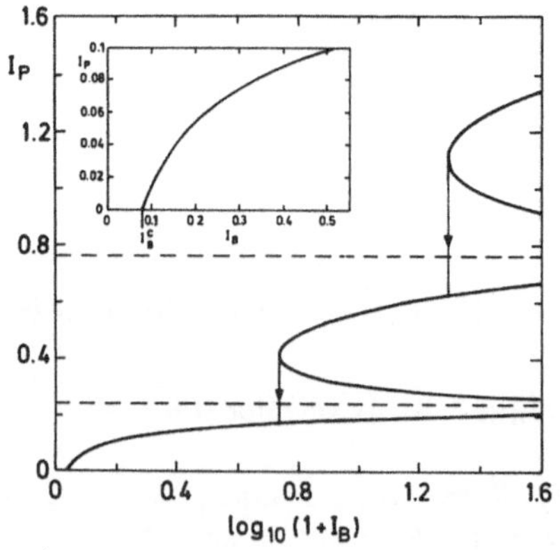

Fig. 7. Intensity I_P versus I_B given by Eq. (40) for h = 1, R = 0.5, and I_F = 2. Dashed line gives limiting intensities discussed in text. The inset shows the second-order-like phase transition for $I_B = I_B^c$ (after Ref 21).

This equation is plotted in Fig. 7 for h = 1, R = 0.5, and I_F = 2. We note that a nonzero solution for I_P appears only above a critical value of I_B (inset in Fig. 7), for which the solution displays a second-order-like phase transition. This critical value I_B^c is found by letting $I_P \to 0$ in Eq. (40), giving the critical condition

$$\left[I_B^c \, I_F (1+h)^2\right] R = 1 \; . \tag{41}$$

The term in square brackets is simply the small signal phase conjugate gain. Thus Eq. (41)

states that the critical condition occurs when the phase conjugate gain exceeds the resonator losses. For $I_B > I_B^c$, the null solution $I_P = 0$ becomes unstable against small perturbations as is shown in the next section. A macroscopic nonzero solution for the internal field then evolves from noise.

Further features are evident from Fig. 7. First, due to the zeros of the Bessel function in the denominator of Eq. (40), there exist limiting values of I_P (denoted I_j^ℓ for the jth value) which are shown as dashed lines in Fig. 7. These limiting intensities separate regions of the I_P versus I_B characteristic which are distinct except in the limit $I_B \to \infty$. Denoting the arguments of $J_1(x)$ that give $J_1(x) = 0$ as $\{O_j\}$, we find that the values of the limiting intensities are given by

$$I_j^\ell = \frac{O_j^2}{4(1+h)^2 I_F} \, . \tag{42}$$

Second, we observe optical multistability in the curves of I_P versus I_B. The negative slope regions are unstable as shown in the next section. However due to the presence of the limiting intensities, it is possible to connect the various branches only by decreasing I_B starting from a solution on an upper branch.

3-3. Dynamic Solutions

We consider dynamic solutions in the limit $L/L_C \to 0$, L_C being the resonator length, with n_2L remaining finite. This allows for two simplifications: (1) temporal evolution effects in the medium can be ignored, though we retain spatial effects, i.e., pump depletion, (2) the spectral response of the PCM will be very much greater than the axial mode spacing of the external resonator, so that to a good approximation all frequencies will experience the same reflection coefficient.[9] Equation (34) can then be treated as a time-dependent equation of the form

$$B_1(t) = i\bar{Z}(t)S_B \frac{S_F}{\sqrt{I_F}} \frac{E^*(t)}{|E(t)|} J_1\left[2(1+h)\sqrt{I_F}|E(t)|\right] , \tag{43}$$

where $\bar{Z} = \exp[i(1+h)(I_F + |E(t)|^2) + iI_B]$. We then obtain a difference equation relating the value of E at any time to that one resonator round-trip time earlier:

$$E((n+1)t_R) = \sqrt{R}\, e^{i\phi_0} B_1(nt_R), \tag{44}$$

with n = 0, 1, 2,.... Defining $I_n = |E(nt_R)|^2$, we obtain from Eq. (44) the return map

$$I_{n+1} = RI_B J_1^2(2(1+h)\sqrt{I_F I_n}) \, . \tag{45}$$

To investigate the stability of the static solution, we set $I_n = I_P + \delta_n$, where I_P is the solution of Eq. (39) for given pump fields I_F and I_B, and $|\delta_n| \ll I_P$. We find

$$\frac{\delta_{n+1}}{\delta_n} = g = 2RI_B(1+h)\sqrt{\frac{I_F}{I_P}}\, J_1(S)\left[J_0(S) - \frac{J_1(S)}{S}\right] , \tag{46}$$

where $S = 2(1+h)\sqrt{I_F I_P}$. The static solution becomes unstable for $|g| > 1$. It is straightforward to show that the case $g \geq 1$ corresponds to the negative slope regions in Fig. 7. Furthermore in the limit $I_P \to 0$, Eq. (46) reduces to

$$g = I_B I_F (1+h)^2 R \, . \tag{47}$$

Thus comparing this with Eq. (41) shows that the solution $I_P = 0$ is unstable for $I_B > I_B^c$.

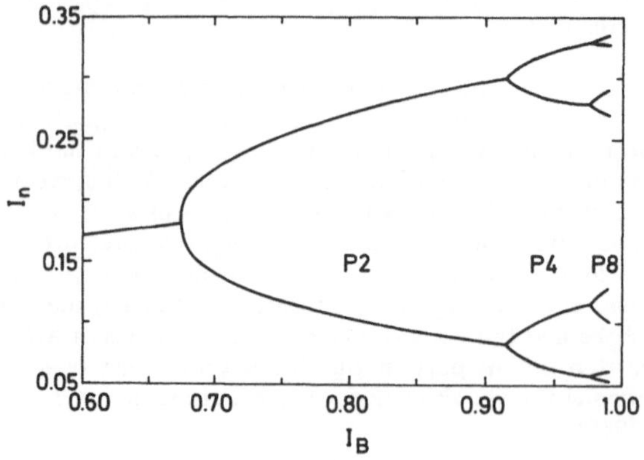

Fig. 8. Bifurcation diagram of the difference Eq. (45) in the case of a Feigenbaum route to chaos for the parameter values $I_F = 2$, $h = 1$, and $R = 1$. For clarity, only the first three period-doubling bifurcations are shown (after Ref. 21).

For $g \leq -1$, oscillatory instabilities occur. As the pump intensity I_B is increased, the system exhibits a series of period doubling bifurcations to chaos. This behaviour is illustrated in Fig. 8, which gives the bifurcation diagram of the difference Eq. (45) for the parameter values $I_F = 2$, $h = 1$, and $R = 1$. The symbols P2, P4, and P8 denote the period 2, 4, and 8 regimes, respectively. The P2 solution corresponds precisely to the excitation of half-axial modes previously predicted using a linear theory of phase-conjugate resonators.[3,4] Physically, this instability has the same origin as the Ikeda instabilities arising in optical bistable devices[22,23]: oscillation arises from beating between the signal at the pump beam frequency ω and two adjacent cavity modes parametrically generated through nondegenerate FWM. When the parametric gain exceeds the cavity losses, the system becomes unstable and oscillations follow. The excitation of half-axial modes is thus a form of Ikeda instability.

In the example presented here, the route to chaos follows the Feigenbaum scenario,[24,16] as can readily be seen from the following argument. Due to the formal identity between Eqs. (40) and (45), Fig. 6 gives the form of the return map (45), provided that the x- and y-axis be replaced by $\sqrt{I_n}$ and I_{n+1}, respectively. It is well-known[15] that a single-peaked return map with parabolic extremum leads to a period-doubling route to chaos. This is the case here, provided the parameters in Eq. (45) are such that the system is always confined (except maybe for a transient) to the first lobe of the Bessel function in Eq. (45). This leads to the condition

$$RI_B < \frac{O_1^2}{\mathrm{Max}(J_1^2)} \simeq 41.17 \,, \tag{48}$$

which is fulfilled in the example of Fig. 8. A numerical analysis of the difference Eq. (45) in this limit yields for the Feigenbaum number $\delta = 4.6692$, in excellent agreement with the theoretical prediction based on renormalization group theory.[24] When condition (48) is not fulfilled, the system can become quite complex, with the coexistence of two or more basins of attraction. A detailed study of this regime will be the subject of future work.

4. Summary and Conclusions

In this paper we have reviewed recent work performed in our group on the dynamics of optical resonators with nonlinear mirrors. These systems distinguish themselves from standard nonlinear resonators in that the nonlinearity now appears in the boundary conditions of the system, rather than in a medium within the resonator. Both systems presented here can lead to multistability and chaos,[25] and both also lend themselves to a number of potential applications, e.g., in photonic logic. In particular, radiation pressure-driven resonators can be used as a very sensitive transducer between mechanical and optical signals. Finally we note that the return map Eq. (45) offers a possibility of studying the transition between the Feigenbaum route to chaos and the coexistence of several bassins of attraction.

This work was carried out in part in the framework of an operation launched by the commission of the European Communities under the experimental phase of the European Stimulation Action (1983-1985).

References

1. A. Dorsel, J. D. McCullen, P. Meystre, E. Vignes, H. Walther, Phys. Rev. Lett. **51**, 1550 (1983)
2. J. D. McCullen, P. Meystre, and E. M. Wright, Opt. Lett. **9**, 193 (1984)
3. Optical Phase Conjugation, Ed. by R. A. Fisher, Academic, NY (1983)
4. A. Yariv and D. M. Pepper, Opt. Lett. **1**, 16 (1977)
5. R. Dändliken and T. Tschudi, Appl. Opt. **8**, 1119 (1969)
6. H. Risken, The Fokker-Planck Equation, Methods of Solutions and Applications, Springer-Verlag, Heidelberg (1984)
7. J. S. Bendat and A. G. Piersol, Measurement and Analysis of Random Data, John Wiley, NY (1966)
8. H. Billing, W. Winkler, R. Schilling, A. Rüdiger, K. Maischberger, and L. Schnupp in Quantum Optics, Experimental Gravitation, and Measurement Theory, Ed. by P. Meystre and M. O. Scully, Plenum Press (1983)
9. A. E. Siegman, P. A. Belanger, and A. Hardy, in Optical Phase Conjugation, Ed. by R. A. Fisher, Academic Press, NY (1983)
10. J. Au Yeung, D. Fekete, D. M. Pepper, and A. Yariv, IEEE J. Quant Electron. **QE-15**, 1180 (1979)
11. P. Belanger, A. Hardy, and A. E. Siegman, Appl. Opt. **19**, 602 (1980)
12. R. C. Lind and D. G. Steel, Opt. Lett. **6**, 554 (1984)
13. Y. Jian-quan, Z. Guosheng, and A. E. Siegman, Appl. Phys. **B30**, 11 (1983)
14. A. Hardy and Y. Silberberg, J. Opt. Soc. Am. **73**, 594 (1983)
15. E. M. Wright and P. Meystre, submitted to Opt. Lett.
16. E. Ott, Rev. Mod. Phys. **53**, 655 (1981)
17. R. Y. Chiao, P. L. Kelley, and E. Garmire, Phys. Rev. Lett. **17**, 1198 (1966); R. L. Carman, R. Y. Chiao, Phys. Rev. Lett. **17**, 1281 (1966)
18. A. M. Lazaruk, Opt. Spectrosc. **53**, 633 (1982)
19. E. M. Wright, PhD thesis, Heriot-Watt University, Edinburgh (1983)
20. Solutions similar to these have also been employed by W. J. Firth to investigate the stability of bistable devices, see e.g., W. J. Firth, Opt. Comm. **39**, 343 (1981)
21. E. M. Wright and P. Meystre, to be published
22. K. Ikeda, Opt. Comm. **30**, 257 (1979)
23. W. J. Firth, E. M. Wright, and E. J. D. Cummins, in Optical Bistability 2, C. M. Bowden, H. M. Gibbs, and S. L. McCall, Eds., Plenum, NY (1984)
24. M. J. Feigenbaum, J. Stat. Phys. **19**, 25 (1978) and **21**, 669 (1979)
25. Chaos in radiation pressure bistability has been studied by W. J. Firth and D. Hewitt, unpublished

AN EXAMPLE OF SYMMETRY BREAKING IN NONLINEAR OPTICS

János Bergou* and Marlan O. Scully**

Max-Planck-Institut für Quantenoptik
D-8046 Garching, Fed. Rep. of Germany

INTRODUCTION

Symmetry breaking is becoming one of the most studied phenomena common in many areas of physics. The notion extends from gauge field theories[1] to solid-state physics[2] and beyond. What is usually meant by symmetry breaking is the following. The equation governing the behaviour of a given system has a number of symmetry properties but the states of the system do not exhibit all of these symmetries. A well known example is provided by ferromagnetic materials[2]: above the Curie temperature the materials have no magnetization and the states exhibit all symmetry properties of the Hamiltonian (rotational symmetry in particular). Below the Curie temperature the materials acquire a finite magnetization vector which points in a well-defined spatial direction. The rotational symmetry is obviously broken in this case. In some sense, however, the system still remembers its symmetry since the direction of the magnetization is completely random. It is not predetermined by the system itself but by external perturbations which are always present in a realistic system. Furthermore, in most cases symmetry breaking occurs in some abstract space (Hilbert space of the states or phase space of the system, etc.) with only a few exceptions, of which ferromagnetism is again one of the most significant examples. Here the breaking of the symmetry takes place in real space. Very few other examples are known where symmetry breaking occurs in the three dimensional coordinate space.

Below we shall study a relatively simple nonlinear optical system having this property namely a waveguide with nonlinear boundaries. This example turns out to be very instructive since its linear counterpart (a linear waveguide) not only also exists (there is nothing like a linear ferromagnet) but is one of the best studied optical systems. Thus, we can immediately compare our results to the linear waveguide theory and see what new physical phenomena are brought about by introducing nonlinearity into the system. One more specific feature of our example is that already

---

* Alexander von Humboldt Fellow on leave from the Central Research Institute for Physics, H-1525 Budapest, P.O. Box 49, Hungary (permanent address)

** also Institute for Modern Optics, University of New Mexico, Albuquerque, NM 87131, USA

a weak nonlinearity is capable of causing dramatic changes in the behaviour of the system. This is so because nonlinearity influences the phase shift at reflection from the boundery and a small change in phase results in destructive interference in the guiding part instead of constructive, thus altering the features of the system completely.

For the sake of laser comparison we give a short overview of the linear waveguide theory[3].

SUMMARY OF LINEAR WAVEGUIDE THEORY

We consider the configuration shown in Figure 1. A guiding slab of width 2d and dielectric constant ε_2 is sandwiched between media of dielectric constant ε_1 such that $\varepsilon_1 < \varepsilon_2$ (symmetric slab waveguide). For simplicity we consider TE waves only, with electric field polarized in the y direction. The waves are propagating in the z direction, with propagation constant β. All the allowed values of the propagation constant are confined to the interval $\varepsilon_1 k^2 < \beta^2 < \varepsilon_2 k^2$ $(k=\omega/c)$. The allowed field patterns are functions of x (transverse coordinate) and they are either completely symmetric

$$E_s(x) = \begin{cases} A\cos Kx; |x|<d; \; K^2=\varepsilon_2 k^2-\beta^2 \\ A\cos Kd e^{-\gamma|x|}; |x|>d; \gamma^2=\beta^2-\varepsilon_1 k^2 \end{cases} \tag{1}$$

or completely antisymmetric

$$E_a(x) = \begin{cases} A\sin Kx; |x|<d \\ \mathrm{sgn}(x) A\sin Kd e^{-\gamma|x|}; |x|>d \end{cases} \tag{2}$$

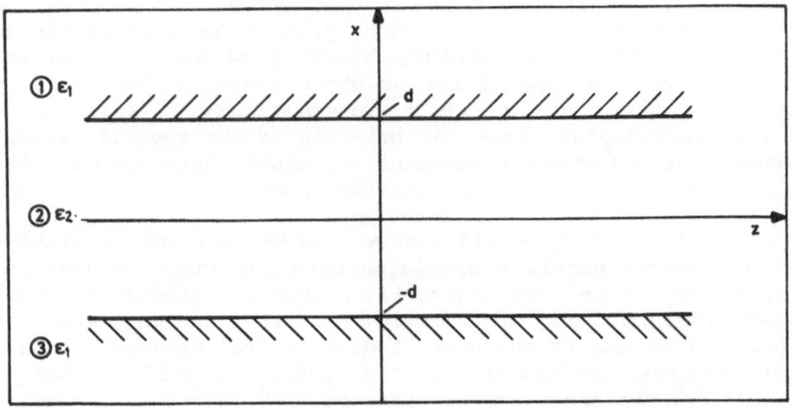

Fig. 1. Geometry of the symmetric slab waveguide. The guiding slab of dielectric constant ε_2 and width 2d occupies the space between x=d and x=-d. The slab is surrounded on both sides by identical media of dielectric constant ε_1. Waves are propagating along the z axis and are polarized along y (perpendicular to the plane of the Figure).

in the coordinate system of Fig. 1. From the requirement of continuity of the field derivatives at the slab boundaries at $x=\pm d$ we obtain the eigenvalue equation for the propagation constant

$$\tan Kd = \gamma/K \tag{3}$$

for the symmetric modes and

$$\cot an Kd = -\gamma/K \tag{4}$$

for the antisymmetric modes.

In the following we shall see that in the nonlinear case asymmetric field patterns are also allowed and the eigenvalue equations for the propagation constant will explicitly depend on the intensity ($\sim A^2$) of the corresponding field pattern. In connection with the asymmetric field patterns we note that asymmetric field patterns are also allowed in the linear case, e.g. $c_1E_s+c_2E_a$ ($c_1\neq 0$, $c_2\neq 0$) is clearly a solution of the wave equation, owing to linearity of the system. In the nonlinear case, however, neither the symmetric nor the antisymmetric part of a given asymmetric pattern is a solution of the wave equation, i.e. asymmetry becomes an intrinsic property of the field distriubtions, owing to nonlinearity. We shall see later that asymmetry occurs in the following way. The center of a symmetric or antisymmetric field distribution is simply shifted away from the origin, $x=0$, by a finite amount to $x=d_o$ ($\neq 0$). This shift can be equally likely positive or negative, i.e. the sign of the shift is completely random. In this sense the system still remembers its symmetry properties and the direction of the shift is determined by small perturbations which are always present in a physical system. This is again analogous to the ferromagnets where the direction of the magnetization is completely random, i.e. it can point into any spatial direction, thus remembering its own symmetry properties.

TE MODES OF THE SLAB WAVEGUIDE WITH NONLINEAR BOUNDARIES

We consider the same geometry again as for the linear waveguide. The only difference is that the dielectric function ε_1 of the surrounding media depends on the intensity. We assume a Kerr type nonlinearity in which case the dielectric function of our system can be given as

$$\varepsilon(x) = \begin{cases} \varepsilon_2;\ |x|<d \\ \varepsilon_1+\eta E(x)^2;\ |x|>d \end{cases} \tag{5}$$

In the following we assume that $\eta>0$.

As in the linear case we again look for a TE polarized wave propagating along z of the form $E=(0,E_y,0)$ where

$$E_y = E(x)\exp[i(\beta z-\omega t)].$$

Starting from Maxwell's equations one can derive the following scalar wave equation for $E(x)$

$$E(x)''+[k^2\varepsilon(x)-\beta^2]\, E(x) = 0 \tag{6}$$

Boundary conditions simply require continuity of E and E' at the slab boundaries, $x = \pm d$.

The solution of Eq. (6) in the linear regime, $|x|<d$, is trivial and can be written as

$$E(x) = \begin{cases} A^{\cos}_{\sin} \; [K_1(x-d_o)]; \; K_1^2=k^2\varepsilon_2-\beta^2>0 \\ A^{\cosh}_{\sinh} \; [K_2(x-d_o)]; \; K_2^2=\beta^2-k^2\varepsilon_2^2>0 \end{cases} \qquad (7)$$

Here A and d_o are constants of integration.

The method of solution of Eq. (6) in the nonlinear regime, $|x|>d$, is described in detail elsewhere[4], here we merely recall the result

$$E(x) = \pm(2/\eta)^{1/2}\gamma/k\cosh[\gamma(x-x_{1,3})] \qquad (8)$$

where $\gamma^2=\beta^2-k^2\varepsilon_1>0$ and x_1 and x_3 are integration constants in regions 1 and 3, respectively.

One can say at this point that all basic new physics of the nonlinear waveguide comes from Eq. (8). In a linear medium the evanescent waves are purely exponentially decaying ones. In a nonlinear medium, however, the evanescent waves (i.e. waves which decay exponentially at large distance from the boundary) can have two qualitatively different behaviours depending on the actual value of x_1 in medium 1 (or x_3 in medium 3). If $x_1<d$ ($x_3>-d$) the maximum at x_1 (x_3) is virtual and only the exponentially decaying tail of the field distribution (8) contributes to the wave in the nonlinear medium. The actual field pattern is similar to that of a linear waveguide. If $x_1>d$ ($x_3<-d$) the maximum at x_1 (x_3) becomes real and the field distribution in the nonlinear medium is very different from that of the linear waveguide.

The solutions (7) and (8) contain altogether 5 integration constants: A, β, x_1, x_3, and d_0. They appear in the solutions as parameters. The boundary conditions yield four equations to determine the parameter values. This does not cause a difficulty in the linear case where, due to the linearity of the system, the amplitude parameter A cancels from the continuity equations. Its actual value is determined from outside, usually by some input condition (e.g. the input power fixes the mode intensity and this last quantity is proportional to the square of the amplitude). The remaining four parameters are then uniquely determined by the four continuity equations. The structure of these equations is such that x_1, x_3, and d_o are completely fixed by them, namely $x_1=-x_3\to-\infty$, $d_o=0$. In other words, the evanescent waves in regions 1 and 3 are purely exponentially decaying ones in the linear case and the field patterns are either completely symmetric or completely antisymmetric in the transverse direction with respect to the origin, $x = 0$. The remaining parameter β is determined from the so-called eigenvalue equation, Eq. (3), for the symmetric modes and Eq. (4) for the antisymmetric modes. These equations also follow from the boundary equations but, unlike to the equations for x_1, x_3, and d_0, they may have more than one solution (eigenvalue). All these values lay in the $k^2\varepsilon_1<\beta^2<k^2\varepsilon_2$ interval, i.e. no hyperbolic solutions (lower two lines of Eq. (7)) exist in the slab region in the linear case.

In the nonlinear case, however, the situation is different. The amplitude A becomes an intrinsic parameter (power) then we can use the boundary equations to determine the dependence of the other parameters in the solution (x_1, x_3, d_o, and β) on the input parameter A. This yields the following qualitative behaviour. For small values of A the system is similar to its linear counterpart. The only possible value of d_o, for instance, is $d_o = 0$. The possible field distributions are either completely

symmetric or completely antisymmetric. Increasing the intensity of the mode ($\sim A^2$), at a certain threshold intensity suddenly $d_o \neq 0$ solutions appear. This means that, above a certain intensity, asymmetric field patterns may appear in a symmetric structure. The symmetry of the structure, reflection with respect to the yz plane, is broken.

We can discuss the dependence of x_1, x_3 on A in a similar way. For low input power $x_1=-x_3$, i.e. the field patterns are symmetric or antisymmetric in accordance with the results for d_o. Furthermore, for extremely low intensities $x_1=-x_3 \rightarrow -\infty$, i.e. the system reduces exactly to a linear waveguide. Increasing the intensity the relationship $x_1=-x_3$ still holds but both parameters will acquire finite values. Above the same threshold intensity where $d_o \neq 0$ solutions become possible, $x_1 \neq -x_3$ solutions become also possible and they again branch away continuously from the $x_1=-x_3$ solutions. It is also interesting to observe that $x_1>d$ ($x_3<-d$) solutions are also possible. This means that the actual field pattern has a maximum outside the slab region. In the case of a linear waveguide the guiding mechanism is total internal reflection from the slab boundary, exclusively. Consequently, the intensity of the guided waves is always higher in the slab region than outside. In the nonlinear waveguide, however, a second guiding mechanism becomes also possible. The increasing of the intensity makes the effective refraction index of the surrounding media equal to the refraction index of the slab owing to the intensity dependence of ε_1(c.f. Eq. (5)). The boundary becomes transparent and the intensity 'leakes out' of the slab region. In the nonlinear medium, however, due to the self-focussing property of the medium the intensity gets trapped. Indeed, x_1(x_3) never exceeds d (-d) by a large amount. The maxima of the corresponding field distribution thus always remains in the vicinity of the slab. These waves are therefore also guided waves, the guiding mechanism being self-focussing in the nonlinear material, instead of total internal reflection from the slab boundary.

This short discussion of the parameters x_1, x_3, and d_0 enables us to understand the resulting field distributions. Some typical field patterns are plotted in Fig. 2, clearly showing asymmetry and maxima outside the slab.

The discussion of the dependence of β on A reveals further interesting properties of the nonlinear waveguide system and is of fundamental importance in understanding the underlying physics. We therefore discuss these nonlinear eigenvalue equations separately in the next Section.

BISTABILITY AND HYSTERESIS

From the requirement of continuity of E and E' at x=±d one can easily derive equations relating the parameters x_1, x_3, d_o, A, and β. After eliminating the parameters x_1, x_3, and d_o from these equations one obtains one equation which relates β to A. We consider this equation as the generalized eigenvalue equation. It is the following (for $K_1^2>0$ in Eq. (7))

$$\tan K_1 d = (\gamma/K_1)\ (1-\eta A^2 k^2/2\gamma^2 \cos^2 K_1 d)^{1/2} \qquad (9)$$

for the symmetric modes,

$$\cot\mathrm{an} K_1 d = (\gamma/K_1)(1-\eta A^2 k^2/2\gamma^2 \sin^2 K_1 d)^{1/2} \qquad (10)$$

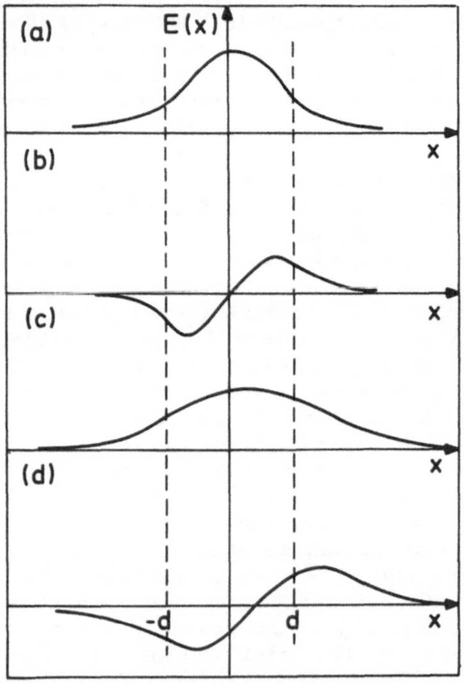

Fig. 2. Field patterns in a symmetric slab waveguide with nonlinear boundaries: a) symmetric distribution; b) antisymmetric distributions; c) asymmetric distribution branching away from the symmetric pattern; d) asymmetric distribution branching away from the antisymmetric pattern and exhibiting a maximum outside the slab.

for the antisymmetric modes, and

$$\text{contan}K_1(d-d_o) = (\gamma/K_1)(1-\eta A^2k^2/2\gamma^2 \sin^2K_1(d-d_o))^{1/2} \quad (11)$$

for the asymmetric ($d_o \neq 0$) solutions. To obtain the eigenvalue equations in the $K_2^2>0$ domain, we have to replace K_1 by K_2 and the trigonometric functions with the corresponding hyperbolic functions in Eqs. (9)-(11).

Discussing the properties of the above eigenvalue equations we first note the following. For vanishing nonlinearity (η=0) Eq. (9) reduces to Eq. (2) ($\tan K_1 d$ must be positive), Eq. (10) reduces to Eq. (3) (with the requirement that $\cot\text{an}K_1 d$ must be negative) and Eq. (11) (which has no

counterpart in the linear case) has no solution at all. Thus, in the limit of vanishing nonlinearity the system reproduces the properties of a linear waveguide, as it should. There are only completely symmetric and completely antisymmetric eigenmodes, the eigenvalue equations are identical with those of a linear waveguide. In order to understand what new features are brought about by nonlinearity we plotted the intensity dependence of β as resulting from Eq. (10) in Fig. 3. On the vertical axis we plotted the dimensionless intensity parameter $a^2 \equiv \eta A^2 k^2 d^2/2$ and on the horizontal axis the dimensionless parameter $b_1 = K_1 d$ which is uniquely related to the propagation constant β through the relation $b_1^2 = (\varepsilon_1 k^2 - \beta^2) d^2$.

The figure reflects the chracteristic new features of the nonlinear waveguide system. The waveguide is parametrized with the dimensionless parameter v such that $v^2 = (\varepsilon_2 - \varepsilon_1) k^2 d^2$. It measures the width in units of the wavelength in vacuum and is proportional to the differences of the (linear) dielectric constants, i.e. it is a measure of the effective optical thickness of the waveguide. The bigger is the value of v the more modes can be excited in the waveguide. The figure shows the dispersion of the antisymmetric modes for v=10, which corresponds to a very thick waveguide, which can support five antisymmetric modes in the linear case corresponding to the five distinct curves of the dispersion relation.

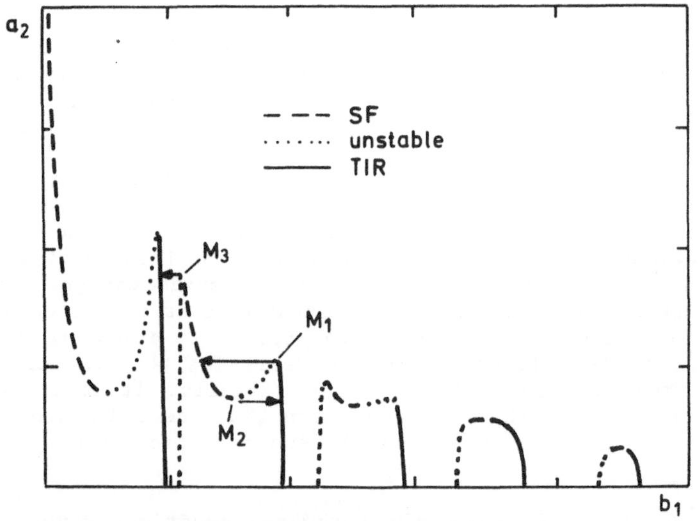

Fig. 3. Dispersion curve of the antisymmetric mode from Eq. (10). We have plotted the dimensionless intensity parameter $a^2 \equiv \eta A^2 k^2 d^2/2$ on the vertical axis and the dimensionless propagation parameter $b_1 = K_1 d$ on the horizontal axis for the waveguide parameter v=10. The solid line corresponds to the regimes of total internal reflection (TIR), the broken line to self-focussing (SF). The dotted part of each dispersion curve is unstable. Transitions between TIR and SF regimes exhibit hysteresis jumps as indicated by arrows at M_1 and M_2. Transitions to higher order modes correspond to multistability (also indicated by arrow at M_3).

Each curve originates on the b_1 axis and goes back there. Each curve has three characteristic parts. A part with negative slope (full line) the intersection of which with the b_1 axis corresponds to an eigenmode of the linear waveguide since the intersection point on the axis corresponds to vanishing nonlinearity ($a^2=0$). Increasing the intensity we can reach the first local maximum of the dispersion curve at point M_1. The dotted part between M_1 and the minimum at M_2 of the dispersion curve is obviously unstable. After the minimum at M_2 there is a second stable part of the dispersion curve (dashed line) until the second local maximum at M_3. After this maximum the curve falls again, this part being unstable (second dotted line). Now, we can discuss what happens if we begin to increase the intensity, i.e. we start to excite the nonlinear waveguide system. Initially we move along the first stable branch of a given dispersion curve. At the maximum in M_1 the system may jump to the second stable branch of the dispersion curve where we can increase the intensity further If we start to decrease the intensity along this second branch the system jumps back to the first stable branch at M_2. That is the system exhibits bistable behaviour with hysteresis jumps. The physical origin of the bistability is the following. In a nonlinear waveguide, in addition to total internal reflection (TIR) from the slab boundary, there is a second mechanism for guiding a wave. This second mechanism is self-focussing in the nonlinear medium, i.e. outside the slab. The slab serves in this case as giving directionality to the system. Along the first stable branch of the dispersion curve the field patterns are similar to those of a linear waveguide. All the maxima of the field distribution are inside the slab, the guiding mechanism is obviously TIR from the slab boundary. Along the second stable branch guiding is maintained via self-focussing in the nonlinear medium and the field distribution has additional extrema outside the slab. The transition from one guiding mechanism to the other is the physical origin of the bistable behaviour.

Another interesting property of the system, which is also related to the nonlinearity, is the following. In a linear waveguide the different modes can be excited simultaneously, a superposition of the eigenmodes is still a solution of the Maxwell's equations since the superposition principle is a consequence of the linearity of the system. In a nonlinear system the superposition principle does not hold any more, i.e. in a nonlinear waveguide only one mode can be excited at a time. Assume that we excite the mode described previously. Increasing the intensity we reach the maximum of the second stable branch. At this point the mode becomes overall cutoff. However, there still remains the possibility that the system makes a jump to the next allowed mode as indicated by the arrow in Fig. 3. In other words, besides the previously discussed bistability, the system exhibits multistability, as well. The physical origin of the later is different from that of the bistability. It is related to the multimode structure of the solution (more than one modes are allowed for the same intensity) and the lack of the superposition principle in a nonlinear system.

A closer look at Eq. (11) and the boundary conditions reveals, furthermore, the origin of the asymmetric solutions. They branch away from the symmetric resp. antisymmetric solutions in the point where the effective refraction indices of the slab and the surrounding media are equalized due to the intensity dependence of the index of refraction in the nonlinear media. It is also interesting that the nonlinear waveguide provides a self-classification of its own thickness. If $v<1$ the waveguide can maintain one symmetric and one asymmetric mode (which branches away from the symmetric dispersion curve). It is natural to call this case a thin waveguide. If $1<v<\sqrt{2}$ the waveguide can maintain one antisymmetric mode in

addition to the symmetric and asymmetric modes. If $v>\sqrt{2}$ a second asymmetric mode branches away from the antisymmetric mode. Since all fundamental modes appear it is natural to refer to the $v>\sqrt{2}$ case as thick waveguide. The domain $1<v<\sqrt{2}$ corresponds to intermediate thickness. Higher order modes appear whenever v exceeds a multiple of $\pi/2$.

SUMMARY

In this lecture we have outlined some striking new features of the optical waveguide system with nonlinear boundaries. The new physics comes from the fact that evanescent waves in a nonlinear medium differ drastically from evanescent waves in a linear medium (c.f. Eq. (8) and Eqs. (1)-(2)). Some of the new phenomena found are the following. Bistable behaviour of a given mode with hysteresis jumps between total reflection and self-focussing regimes; multistability as a consequence of multimode solutions and the lack of superposition principle in a nonlinear system; symmetry breaking, i.e. the appearance of asymmetric field patterns in an otherwise completely symmetric system. We were able to characterize the waveguide with a single parameter v describing the effective optical thickness in units of the wavelength. It is easy to see from our nonlinear eigenvalue equations that the predicted new phenomena occur in relatively thick waveguides. All the phenomena that we discussed in this paper are so-called low intensity phenomena, i.e. their experimental observation requires a relatively low input power (with present day nonlinear materials bistable switching can be expected to occur in the microwatt-milliwatt domain of the input power).

References

1. J. Bernstein, Spontaneous symmetry breaking, gauge theories, the Higgs mechanism and all that, Rev. Mod. Phys. 46:7 (1974).
2. P. W. Anderson, "Concepts in Solids", Benjamin, Reading, Ma. (1963).
3. D. Marcuse, "Theory of Dielectric Optical Waveguides", Academic Press, New York (1974).
4. J. Bergou, P. Dobiasch, M.O. Scully, and J.M. O'Hare, Theory of the Slab Waveguide with Nonlinear Boundaries: I. Guided TE Modes in a Symmetric Slab, submitted to Phys. Rev. A (1984).

NONEQUILIBRIUM STATISTICAL PHYSICS IN A DITHERED RING LASER GYROSCOPE
OR
QUANTUM NOISE IN PURE AND APPLIED PHYSICS

W. Schleich* and P. Dobiasch

Max-Planck-Institut für Quantenoptik
D-8046 Garching bei München
West Germany
and
*Center for Theoretical Physics
University of Texas at Austin
Austin, Texas 78712

V. E. Sanders
Rockwell Int. A.M.S.D.
3370 Miraloma Ave.
P. O. Box 49211
Anaheim, CA 92803

M. O. Scully
Institute for Modern Optics
University of New Mexico
Albuquerque, New Mexico 87131
and
Max-Planck-Institut für Quantenoptik
D-8046 Garching bei München
West Germany

I. INTRODUCTION

In the year 1851 Foucault demonstrated that the slow rotation of the plane of vibration of a pendulum could be used as evidence of the earth's own rotation. Nowadays high precision measurements of the earth's rotation are performed by using radio telescopes in Very Long Baseline interferometry [1]. However, a recent proposal [2] takes advantage of the ultra high sensitivity of a ring laser gyroscope [3] of 10m diameter to monitor changes in earth rate* or Universal time. The underlying principle of such a device is the optical analogue of the Foucault pendulum, the so-called Sagnac effect [5,6]. The frequencies of two counterpropagating waves in a ring interferometer are slightly different when the interferometer is rotating about an axis perpendicular to its plane. Since this frequency difference is proportional to the rotation rate it provides a direct measure of the rotation of the system.

* The first optical experiment to detect the earth's rotation was performed by Michelson and Gale [4] using an unusually large size for an interferometer: 4/10 miles x 2/10 miles.

Ring laser gyroscopes of this size would also allow tests [7] of metric gravitation theories [8]. In an ultrasensitive ring laser placed on the rotating earth, several "effective rotations" arise as indicated in Fig. 1. Apart from the "classic" Sagnac terms Ω_0 and $\Omega_{\oplus}$, due to the rotation of the interferometer about its own axis and about the earth's axis, the gyroscope is sensitive to general relativistic effects such as the existence of preferred frames in the universe, the curvature of space-time and Lense-Thirring effects. The size of these effects depends on the particular theory of metric gravity chosen to describe the experiment. One alternative to Einstein's theory of general relativity is the Ni cosmology [8]. This particular theory is based on the assumption that the universe is resting in a preferred frame thought of as the 3 K black body background. Since in general relativity no such preferred frame, no "aether" exists, the contribution Ω_{AETHER} vanishes when Einstein's theory is employed whereas in Ni's theory it does give a nonzero contribution. The proposed ring laser experiment would therefore test these two theories against each other.

Similarly, the terms $\Omega_{CURVATURE}$ and $\Omega_{LENSE-THIRRING}$ arising from the fact that space-time is curved by the presence of the mass of the earth and that rotating masses drag inertial frames - first discussed by Lense and Thirring [9] in 1918 - can be used for testing the competing relativistic theories against each other.

In order to detect these relativistic corrections a ring laser gyroscope device must be capable of measuring rotation rates as slow as 10^{-7} of earth's rotation rate. At this point it becomes important that the linear Sagnac relationship between frequency difference and rotation rate is not true in real life.

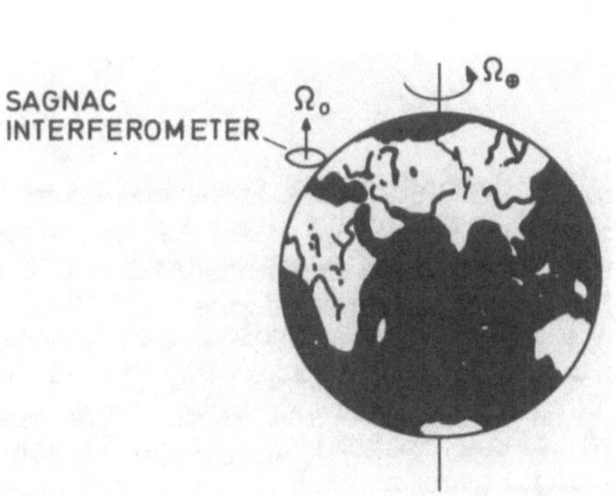

$$\Delta\omega = \frac{4A}{\lambdabar P}\,[\Omega_{\oplus} + \Omega_0 + \Omega_{AETHER} + \Omega_{CURVATURE} + \Omega_{LENSE-THIRRING}]$$

Fig. 1. A Sagnac ring laser interferometer fixed on the rotating earth is sensitive to its rotation rate Ω_0, the rotation rate of the earth $\Omega_{\oplus}$ and three effective rotation rates as indicated in the figure. In the equation, A denotes the area of the ring, $\lambdabar = \lambda/2\pi$ is the reduced wavelength and P the perimeter of the interferometer.

Due to backscattering from the laser mirrors the two counterpropagating beams couple in a nonlinear way [10]. As a result the frequency difference vanishes for small rotation rates. This so-called lock-in effect limits crucially the accuracy of ring laser gyroscopes. One method to overcome this problem consists of dithering [11] the ring laser, i.e. adding an external (controlled) rotation rate such that the gyroscope is operating most of the time outside of the dead band. With the use of this technique gyroscopes have reached a sensitivity such that the noise limit of these devices is only due to the quantum fluctuations [12], which arise from spontaneous emission of the laser atoms. Whereas all kinds of mechanical noise can be circumvented by some "tricky" techniques, there is no way around the quantum noise, which stems from the quantization of the electric field in the resonator. The final limitation on the performance of ring laser gyroscopes is thus given by the quantum noise. Therefore, it is extremely important to study this effect in detail.

In three papers [13-15], devoted to the discussion of quantum noise in an undithered gyroscope the spectrum and the mean beat frequency of the beat signal have been calculated. These considerations have been extended to a harmonically, i.e. sine-wave dithered ring laser gyroscope in a recent article [16]. Throughout these papers the authors have confined themselves to phase fluctuations, which determine the final accuracy of ring laser gyroscopes and have neglected intensity fluctuations which have been treated in detail in Ref. 17. In these lecture notes we give a brief review of quantum noise in ring laser gyroscopes and summarize our recent work [18,19] on the dithered gyro.

The notes are organized as follows: In Section II, we briefly review the basic elements of ring laser theory such as the Sagnac effect (Section II.A), the locking effect (Section II.B) and the influence of quantum noise on the mean beat frequency versus rotation rate (Section II.C). In Section III the Langevin equation for the phase difference between the counterpropagating waves in the presence of any periodic and time symmetric dither is cast into a form which allows a qualitative discussion of the resulting lock-in curve (Section IV) as well as an exact expression in terms of infinite matrix continued fractions [20]. The details of the transformation of the stochastic variable and the derivation of the exact expression for $\langle\langle\dot{\phi}\rangle_F\rangle_t$ may be found in the Appendices A and B, respecively. We conclude by presenting exact results for two special cases of the dithering function: the harmonic, i.e. sine-wave dither and the square-wave bias.

II. RING-LASER THEORY: A PICO-REVIEW

A ring laser [21] consists of a ring interferometer formed by three or more mirrors and a laser medium inside the cavity as shown in Fig. 2. In a linear, two mirror laser the amplitudes and phases of the counterpropagating waves are equal and the electric field is, therefore, a standing wave. In a ring laser configuration, however, the two oppositely directed running waves may have different amplitudes, phases and frequencies. In particular, a clockwise rotation of the laser about an axis perpendicular to the plane of the mirrors causes a frequency difference between the two counterpropagating waves - the so-called Sagnac effect [5,6]. Heterodyning the two waves and measuring the beat note as shown in Fig. 2, therefore, provides information about the rotation of the system. This leads to the use of ring lasers as rotation sensors, first demonstrated by Macec and Davis [22] in 1963.

In this section, we briefly review the basic elements of ring laser theory necessary for the understanding of the remainder of these notes.

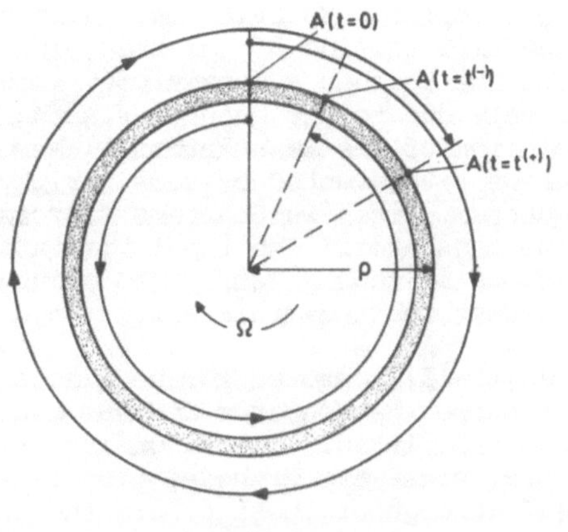

Fig. 2. The frequencies of two counterpropagating electromagnetic waves of a ring laser (laser medium inside the resonator) get slightly shifted when the interferometer is rotated with a rate Ω about an axis perpendicular to its surface. This frequency difference is measured by heterodyning the two beams.

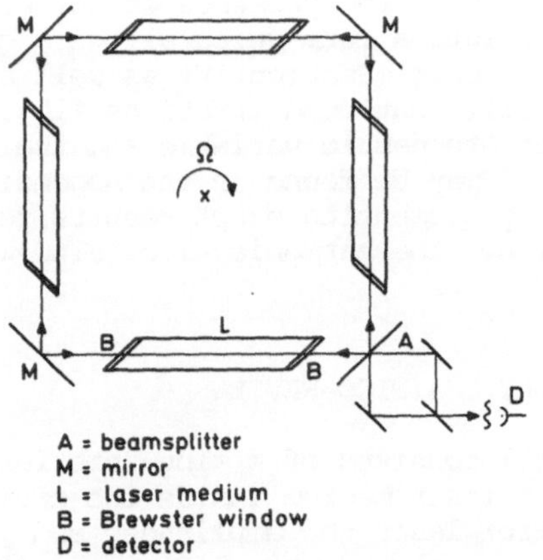

Fig. 3. Nonrelativistic explanation of Sagnac effect for a ring-interferometer of radius ρ rotating with rate Ω, depicted as a glass fiber. Consider two beams starting at time $t = 0$ from point A. Due to the rotation of the fiber the two beams arrive at different times $t^{(\pm)}$ at A and have traversed different pathlengths depending on their propagation direction.

Starting from a nonrelativistic explanation [6] of the Sagnac effect presented in Section II.A, the locking of the ring laser [10] caused by backscattering from the laser mirrors is explained in Section II.B. We conclude by discussing the influence of quantum noise due to spontaneous emission of the laser atoms on the locking curve [14]. In this review we do not derive the complete set of semiclassical, self-consistent phase and intensity equations but rather refer to the literature [21,23].

A. Sagnac Effect

In this section we give a simple explanation [6] of the optical analogue of the Foucault pendulum - the Sagnac effect.

For simplicity, we consider a circular ring interferometer of radius ρ rotating with a rate Ω around an axis perpendicular to its plane. Such an interferometer may be constructed using a glass fiber as indicated in Fig. 3. Consider two counterpropagating beams starting from point A at time $t = 0$. Since the interferometer rotates, the two beams have to traverse different path lengths $P^{(\pm)}$ in order to reach the point A again: the beam co-directional with the direction of rotation has to "catch up" with A and has thus to travel a distance slightly longer than $2\pi\rho$, i.e.

$$P^{(+)} = ct^{(+)} = 2\pi\rho + \rho\Omega t^{(+)} \quad . \tag{1}$$

(The round trip times for the two beams are denoted by $t^{(+)}$ and $t^{(-)}$, respectively). Similarly, we note from Fig. 3 that the beam propagating against the direction of rotation obviously has to traverse a distance $P^{(-)}$ slightly smaller than $2\pi\rho$, namely

$$P^{(-)} = ct^{(-)} = 2\pi\rho - \rho\Omega t^{(-)} \quad . \tag{2}$$

Since the laser medium is inside the resonator only oscillations with wavelength λ satisfying the resonance condition

$$n\lambda_n = P$$

can be sustained in the cavity. Here n is an integer and P denotes the effective length of the resonator. This relation rewritten in terms of resonant frequencies reads

$$\omega_n = 2\pi \frac{nc}{P} \quad . \tag{3}$$

Thus the different path lengths $P^{(\pm)}$ seen by the two counterpropagating beams cause a frequency difference

$$\dot{\phi} \equiv \Delta\omega \equiv \omega_n^{(-)} - \omega_n^{(+)}$$

between the two modes. From Eq. (3) we find using Eqs. (1) and (2)

$$\dot{\phi} = 2\pi nc \, \frac{P^{(+)} - P^{(-)}}{P^{(+)}P^{(-)}} \cong 2\pi \, \frac{nc}{P^2} \, \rho\Omega \, (t^{(+)} + t^{(-)}) \cong 4\pi \, \frac{nc}{P^2} \, \rho\Omega t$$

where in the last step we have approximated the sum of the travel times $t^{(\pm)}$ by twice the travel time $t = 2\pi\rho/c$ in the absence of rotation.

The final expression for the frequency difference between the two counterpropagating waves is thus given by

$$\dot{\phi} = \frac{4A\Omega}{\lambdabar P} \tag{4}$$

where $A = \pi\rho^2$ denotes the area enclosed by the light, P is the perimeter of the interferometer and $\lambdabar = \lambda/(2\pi)$ is the reduced wavelength.

From Eq. (4) we note that the frequency difference $\dot{\phi}$ induced by the rotation of the interferometer is proportional to the rotation rate Ω. Therefore measuring the frequency difference between the two counterpropagating beams in a ring interferometer by heterodyning the two beams as shown in Fig. 2 provides information about the rotation of the system. The obvious advantage of such an optical gyroscope lies in the fact that no moving parts, no spinning masses are involved.

We conclude this section by noting a historical reminiscence: When C. G. Sagnac experimentally* discovered this effect in Paris in the year 1913, he considered this phenomenon as a confirmation of the aether concept and gave the following pictorial explanation [5a]:

> "In clear conception, it ought to be regarded as a direct manifestation of the luminiferous aether. In a system moving as a whole with respect to the aether, the elapsed time of propagation between any two points of the system should be altered as though the system were immobile and subject to the action of aether wind which would blow away the light waves in the manner of atmospheric wind blowing away sound waves. The observation of the optical effect of such a relative wind of aether would constitute evidence for the aether just as the observation of the influence of the relative wind of the atmosphere on the speed of sound in a system in motion would (in the absence of a better explanation) constitute evidence of the existence of the atmosphere around the system in movement."

It is interesting to note that a year later H. Witte [5b] concluded that the effect observed by Sagnac does not prove the existence of the aether.

B. Locking Effect

In the last section we have derived the linear relationship between the Sagnac frequency difference $\dot{\phi}$ and the rotation rate Ω (see Eq. (4)). This derivation, however, assumes perfect mirrors which unfortunately do not exist. In particular, imperfections in the mirror surfaces scatter a small fraction of one of the beams back into the direction of propagation of the other [10]. Thus backscattering from the mirrors leads to a coupling of the counterpropagating beams. Despite the difference in the models describing this phenomenon [3,10,21] they all lead to the same equation of motion for the phase difference ϕ between the two counterpropagating waves,

$$\dot{\phi} = a + b \sin\phi \, . \tag{5}$$

Here $a \equiv 4A\Omega/(\lambdabar P)$ denotes the Sagnac frequency difference and b is the backscattering coefficient. (In the remainder of the paper we do not distinguish between the rotation rate Ω and a and call a the rotation rate of the interferometer).

* This effect was predicted already in 1904 by A. A. Michelson [24].

A detector measuring the beat signal between the two counterpropagating waves (see Fig. 2) averages over time and thus only the mean beat frequency $\langle\dot{\phi}\rangle_t$

$$\langle\dot{\phi}\rangle_t \equiv \lim_{t\to\infty} \{\frac{1}{t} \int_0^t dt' \, \dot{\phi}(t')\}$$

is of interest.

Equation (5) may be solved by separation of variables [21] to yield

$$\langle\dot{\phi}\rangle_t = \begin{cases} 0 & |a| \leqslant b \\ \sqrt{a^2 - b^2} & |a| \geqslant b \end{cases} \tag{6}$$

and

$$\langle\dot{\phi}\rangle_t (-a) = - \langle\dot{\phi}\rangle_t (a) \; .$$

We note that in the limit of large rotation rates, i.e. $|a| \gg b$, Eq. (6) reduces to

$$\langle\dot{\phi}\rangle_t \cong a - \frac{1}{2}\frac{b^2}{a}$$

and the gyroscope responds approximately in a linear way as is shown in Fig. 4a. However, for rotation rates a smaller than the backscattering coefficient b, the beat note vanishes even for nonzero rotation rates. This implies that the frequencies of the two counterpropagating waves are equal. Therefore, the phase difference between the two waves remains constant in time. This phenomenon, called locking of the laser, is the fundamental problem in ring laser gyroscope technology, since in this dead band (see Fig. 4a) the device is insensitive to rotation. Many techniques to overcome this problem have been developed and Section III is devoted entirely to the discussion of one particular method, namely dithering the gyroscope.

We conclude this section by noting that the phenomenon of locking was first observed in 1665 by Christiaan Huygens [25]. He noticed that two pendulum clocks hanging close together on the same wall would, after a short while, tick precisely together for an arbitrarily long period of time. He described the observation in a letter to his father [25]:

> "Being obliged to stay in my room for several days and also occupied in making observations on my two newly made clocks, I have noticed a remarkable effect, which no one could have ever thought of. It is that these two clocks hanging next to one another separated by one or two feet keep an agreement so exact that the pendulums invariably oscillate together without variation."

C. Quantum Noise

In the framework of semiclassical laser theory [21,26] usually employed to describe ring laser operation, the atoms are treated quantum

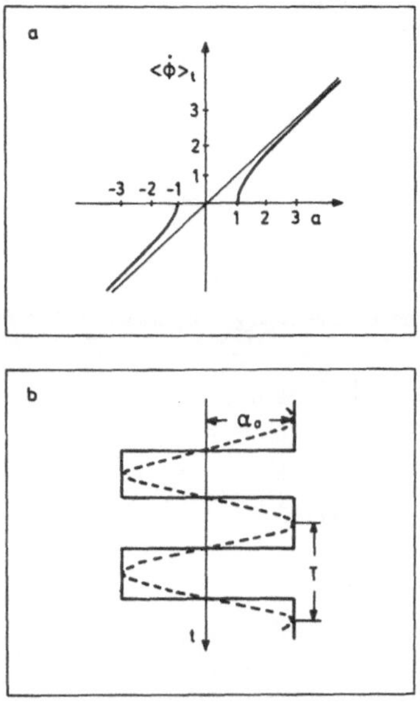

Fig. 4. (a) Lock-in curve of ring laser gyroscope. Due to backscattering from the mirrors the linearity between rotation rate a and mean beat frequency $<\dot{\phi}>_t$ as predicted by the Sagnac effect breaks down and a dead band around zero rotation rate emerges.

(b) To overcome the locking problem an external (controlled) rotation rate is added to the rate wanted to be measured. In order to minimize requirements on the stability of this artificial rotation rate, the bias is operating alternately in the positive and negative direction of rotation. Two typical examples for such a time dependent bias oscillating with period T are the sine-wave (broken line) and the square-wave (full line) dither.

mechanically whereas the electric field is considered to be a classical quantity. However, by the second quantization the field can become an operator giving rise to purely quantum mechanical effects such as vacuum fluctuations and spontaneous emission. The last effect is of special interest to us, because an excited atom can reach its ground state by spontaneously emitting a photon. The emitted electric field having a random phase superposes with the field present in the cavity whose phase is, therefore, no longer well determined: it becomes a stochastic quantity. A ring laser when employed as a rotation sensor measures the phase - or frequency difference between the counterpropagating waves. Quantum noise does, therefore, have an influence on the accuracy of determining the rotation rate. Moreover note that this limitation in accuracy is purely due to the quantum nature of light, i.e. due to second quantization.

In Ref. [13] it has been shown that the fluctuations in the phase difference between the two counterpropagating waves may be taken into account by adding a fluctuating force F to Eq. (5) which thus reads

$$\dot{\phi} = a + b \sin\phi + F(t) \; . \tag{7}$$

Here, F is assumed to be Gaussian with mean zero, i.e.

$$\langle F(t) \rangle = 0 \tag{8}$$

and the symbol < > denotes an average over the noise F. To fully describe this stochastic process the Markoff approximation is made, i.e. the noise is assumed to be delta-correlated, i.e.

$$\langle F(t)F(s) \rangle = 2D\delta(t-s) \ , \tag{9}$$

where $D = \nu/2Q\bar{n}$ denotes the diffusion constant of spontaneous emission [27]. (Here, ν is the frequency of light being in a resonator with quality factor Q and $\bar{n}$ is the average number of photons in the cavity.)

The phase diffusion in a ring laser gyroscope due to spontaneous emission of the laser atoms is thus determined* by the stochastic differential equation (7) together with the properties of F being Gaussian and satisfying Eqs. (8) and (9). Since similar stochastic differential equations also arise in a number of other physical situations, such as a laser with injected signal [29], Josephson junctions [30], self-locking in a laser [31], and in radio engineering [32], Eq. (7) has been studied in great detail [33].

Figure 5 shows the mean beat frequency averaged over the noise, i.e. $\langle\langle\dot{\phi}\rangle_F\rangle_t$ as a function of rotation rate in comparison with the corresponding deterministic locking curve, Eq. (6). We note that far away from the locking region ($|a| \gg b$) the two curves coincide. However, for rotation rates $a \cong b$ we note a new feature: due to noise, unlocking tends to occur at smaller rotation rates already.

Therefore noise reduces the width of the dead band. Moreover, in the laser gyro with noise, there is no sharp boundary as one goes from $|a| < b$ to $|a| > b$. Rather the system goes smoothly from the locked region to an essentially unlocked region. We conclude this section by emphasizing the useful role of quantum noise to unlock the gyro.

III. LANGEVIN EQUATION OF DITHERED GYROSCOPE

One technique to overcome the locking problem is the so-called stationary null-shift bias [10] where an external constant rotation rate f is introduced such that zero rotation rate corresponds to a point well outside the dead band. This method, however, requires a high degree of stability of the constant bias. By operating the bias alternately in the positive and the negative direction of rotation [11] (Fig. 4b) this requirement for the absolute stability of the magnitude of the bias is reduced. The Langevin equation for a ring laser gyroscope with oscillating bias (dithered ring laser gyroscope) reads

---

* To be more precise, semiclassical laser theory [21,26] provides a system of coupled differential equations for the phase of the electric field as well as for its intensity. In the present approach we have assumed the intensity to be constant and have therefore decoupled the system to treat the phase diffusion only. Intensity fluctuations have been discussed in detail in Ref. [17] whereas the complete set of deterministic phase and intensity equations has been investigated by R. Gauert [28].

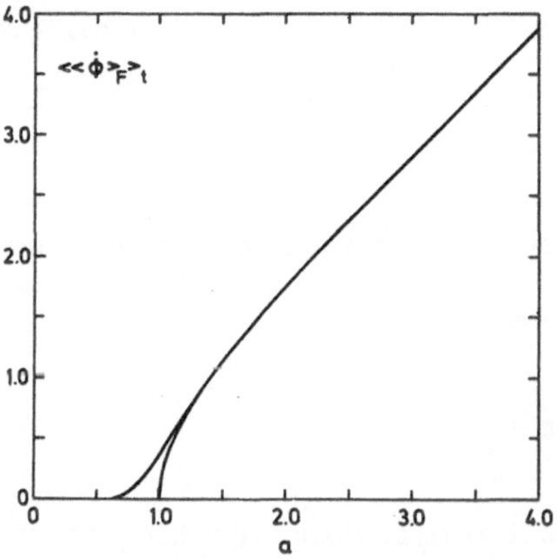

Fig. 5. Mean beat frequency $<<\dot{\phi}>_F>_t$ as a function of rotation rate a compared to the noiseless response curve Eq. (6). Noise clearly decreases the width of the dead band. ($b = 1$, $D = 10^{-1}$).

$$\dot{\phi} = a + b \sin\phi + f(t) + F(t) \tag{10}$$

where the dithering function f is assumed to be periodic

$$f(t + T) = f(t). \tag{11}$$

Here $\omega_D = 2\pi/T$ denotes the dither frequency. In commercial gyroscopes one typically finds $\omega_D \cong 400$ Hz [3]. In order to circumvent the stability requirements mentioned above the bias should not contain any constant components:

$$\int_0^T dt\, f(t) = 0 \; . \tag{12}$$

For simplicity, we consider only symmetric dithering functions

$$f(-t) = f(t) \; . \tag{13}$$

An extensively discussed [16,34-37] example of such a function f is the harmonic (i.e. sine-wave) dither

$$f_h(t) = \alpha_0 \cos\omega_D t \tag{14}$$

shown in Fig. 4b by the broken line. The parameter α_0 denotes the dither amplitude. However, this bias does not minimize the time the gyroscope spends in the locked region where it is insensitive to rotation. From

Fig. 4b (solid line) it is obvious that this requirement is best met by a periodic square wave [36]

$$f_s(t) = \begin{cases} \alpha_0 & -T/4 < t < T/4 \\ -\alpha_0 & T/4 < t < 3T/4 \end{cases} \tag{15}$$

satisfying Eqs. (12) and (13).

In order to treat the Langevin equation (10) in the presence of any (at least piece-wise continuous) periodic and symmetric dithering function $f = f(t)$ by one "universal" formalism [18] we transform the stochastic variable ϕ along the lines of Ref. 16 and 37.

Appendix A shows that by writing the rotation rate a in the form

$$a = r\omega_D + \tilde{a}, \qquad |\tilde{a}| < \omega_D$$

where $r = \ldots -2, -1, 0, 1, 2, \ldots$ and by introducing the new phase variable ψ_r via

$$\phi = r\omega_D t + \int_0^t dt'\, f(t') + \psi_r, \tag{16}$$

the Langevin equation (10) reduces to

$$\dot{\psi}_r = \tilde{a} + b \sum_{s=-\infty}^{\infty} M_{s-r}[f] \sin(s\omega_D t + \psi_r) + F(t) \tag{17}$$

where the functionals $M_k = M_k[f]$ are given by

$$M_k[f] = \frac{1}{2\pi} \int_{-\pi}^{\pi} d\zeta\, \exp[-ik\zeta + i\int_0^{\zeta/\omega_D} dt\, f(\tau)] \,. \tag{18}$$

Note that the structure of Eq. (17) is independent of the explicit form of the dither function f, which enters solely through the functionals M_k.

IV. MEAN BEAT FREQUENCY OF DITHERED RING LASER GYROSCOPE - QUALITATIVE DISCUSSION

In order to study the influence of quantum noise on the mean beat frequency, i.e. in order to perform the average of $\langle\dot{\phi}\rangle_t$ over the noise F, two (equivalent) approaches are possible: the Langevin and the Fokker-Planck approach [20,38]. In the Langevin approach the equation of motion for ϕ (respectively for ψ_r Eq. (17)) is solved in terms of integrals of the fluctuating force F and is substituted back into Eq. (17) to perform the average using Eqs. (8) and (9) together with the assumption of F being Gaussian. However, this approach fails in most of the cases, because one is unable to solve the equation of motion which, at least in problems of interest is nonlinear, see Eq. (17).

A more promising approach is the Fokker-Planck approach in which a linear partial differential equation for a probability distribution is derived. The average is then performed in the well known way. This technique works well because in many cases of interest, one is able to solve the Fokker-Planck equation in an exact way using infinite (matrix) continued fractions, a method developed by H. Risken and coworkers [20]. In particular, in the present problem of an arbitrarily dithered ring laser gyroscope a formally exact expression for $\langle\langle\dot{\phi}\rangle_F\rangle_t$ in terms of matrix continued fractions is derived in Appendix B. The results presented in Section V for the cases of a harmonically (i.e. sine-wave) or square-wave dithered gyroscope, Eqs. (14) and (15) respectively, are based on this technique.

Whereas the method of continued fractions is well suited for numerical analysis, it makes it difficult to get some insight into the functional dependence of the mean beat frequency on the rotation rate. This serves as a motivation for a qualitative discussion of the Langevin equation (17) and the resulting lock-in curve before the exact results (Eq. (B10)) are presented in Section V.

Limiting ourselves to rotation rates a close to multiples of the dither frequency, i.e. $|\tilde{a}| \ll \omega_D$ we can average Eq. (17) over the fast oscillations* to yield

$$\dot{\overline{\psi}}_r = \tilde{a} + b\, M_{-r}[f]\, \sin\overline{\psi}_r + F(t) \tag{19}$$

where $\overline{\psi}_r$ denotes the slowly varying component of ψ_r.

Comparing this equation with the corresponding Langevin equation of the undithered ring laser gyroscope Eq. (7) we observe two important modifications caused by the dither:

1. The integer r indicates that additional dead bands (gyroscope steps) appear at multiples of the dither frequency ω_D.

2. The backscattering coefficient b is changed to $b\, M_{-r}[f]$. Thus, in the absence of noise ($F = 0$, $D = 0$), the (half-)width of the r-th step is approximately given by $b|M_{-r}[f]|$ and the approximate mean beat frequency is then

$$\langle\dot{\phi}\rangle_t = \begin{cases} r\omega_D - \sqrt{\tilde{a}^2 - (bM_{-r})^2} & |bM_{-r}[f]| \leq |\tilde{a}| \quad \tilde{a} < 0 \\ r\omega_D & |\tilde{a}| < |bM_{-r}[f]| \\ r\omega_D + \sqrt{\tilde{a}^2 - (bM_{-r})^2} & |bM_{-r}[f]| \leq \tilde{a} \quad \tilde{a} > 0\ . \end{cases} \tag{20}$$

We conclude by noting that a comparison between Eqs. (7) and Eq. (19) shows that each gyroscope step is affected by the noise approximately in the same way as the undithered gyroscope and the clear distinction between locked and unlocked zones ceases to exist [14].

Note that in deriving the approximate Langevin equation (19), no use was made of the explicit form of the dither. Therefore, the appearance of the gyroscope steps is an unavoidable effect showing up for any bias f satisfying Eqs. (11), (12) and (13).

---

* This "method of averaging" well known in the theory of nonlinear oscillations may be found in Ref. [39].

V. HARMONIC DITHER VERSUS SQUARE-WAVE DITHER: EXACT RESULTS

In this section we apply the exact continued fraction method developed in Appendix B to the case of a harmonically dithered, i.e. sine-wave dithered ring laser gyroscope (Eq. (14)) as well as to the periodic square-wave dithering function f_s (Eq. (15)). Since in both cases the mean beat frequency versus rotation rate curve may be shown to be antisymmetric with respect to the origin [16], i.e.

$$\langle\langle\dot{\phi}\rangle_F\rangle_t(-a) = -\langle\langle\dot{\phi}\rangle_F\rangle_t(a)$$

we confine ourselves to positive rotation rates only.

For a sine-wave dithered ring laser gyroscope Eq. (14), the functionals $M_k[f]$ reduce to the Sommerfeld representation of the Bessel functions [40]

$$M_k[f_h] = J_k\left(\frac{\alpha_0}{\omega_D}\right) , \tag{21}$$

whereas in the case of the square-wave dithering the integrations arising in the definition of M_k Eq. (18) can be carried out analytically to yield

$$M_k[f_s] \equiv \mathcal{M}_k\left(\frac{\alpha_0}{\omega_D}\right) \equiv \frac{2}{\pi}\left(\frac{\alpha_0}{\omega_D}\right)\frac{\sin\left[\left(k-\frac{\alpha_0}{\omega_D}\right)\frac{\pi}{2}\right]}{k^2-\left(\frac{\alpha_0}{\omega_D}\right)^2} . \tag{22}$$

Thus the matrices $\underset{\approx}{B}_m$ and $\underset{\approx}{C}_m$ are fixed and the infinite matrix continued fraction $\underset{\approx}{R}_1$ determining $\langle\langle\dot{\phi}\rangle_F\rangle_t$ may be calculated. For details of the numerical evaluation of this kind of matrix continued fractions we refer to Refs. 16, 20, and 41 and limit ourselves to presenting the results only.

Figure 6 shows the mean beat frequency of a harmonically dithered ring laser gyroscope in the presence of noise $\langle\langle\dot{\phi}\rangle_F\rangle_t$ as a function of rotation rate a. Note that due to the nonlinearity of the equation of motion for the phase (Eq. (7)) the dithering causes additional dead bands - the so-called gyroscope steps - to appear at multiples of the dither frequency. (The behavior of $\langle\langle\dot{\phi}\rangle_F\rangle_t$ between the steps, which is more or less a straight line, is omitted for the sake of simplicity). The influence of quantum noise as a mechanism to "smear out" the clear distinction between the locked zones (gyroscope steps) and the unlocked regimes is obvious.

In Fig. 7 the locking curve around zero rotation rate is shown for increasing dither amplitude. Note that the dead band disappears for $\alpha_0/\omega_D = 2.4$. This is due to the fact that, in the absence of noise, its width is approximately given by $|bM_0[f_h]| = |bJ_0|$ and $J_0(2.4) \cong 0$. Moreover the width of the zeroth band decreases with increasing dither amplitude which can be understood by using the asymptotic expansion for the Bessel function [40]

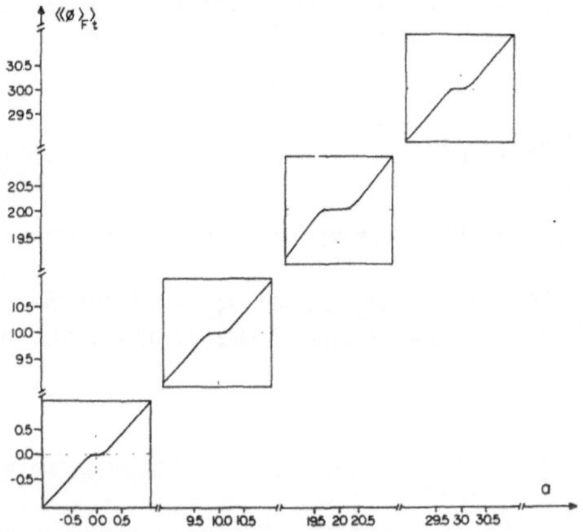

Fig. 6. Mean beat frequency $\langle\langle\dot{\phi}\rangle_F\rangle_t$ of harmonically, i.e. sine-wave dithered gyroscope as a function of rotation rate. Note the appearance of the gyroscope steps at multiples of the dither frequency. The almost linear behavior between the steps is omitted for simplicity. ($b = 1$, $D = 10^{-1}$, $\omega_D = 10$, $\alpha_0 = 30$).

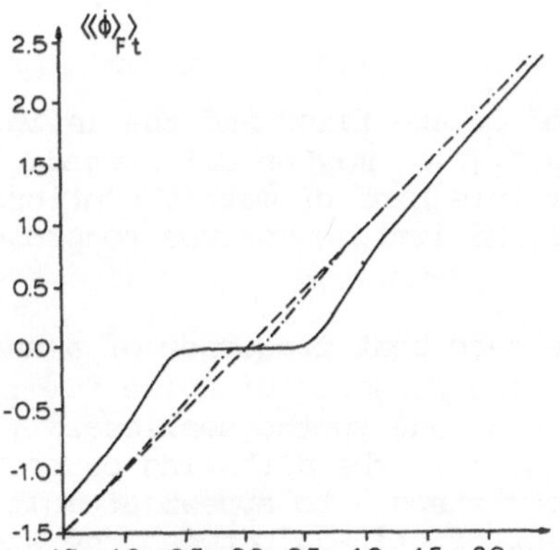

Fig. 7. Response curve of harmonically dithered gyroscope around zero rotation rate for increasing dither amplitude α_0. Note the decrease in the width of the zeroth dead band. ($b = 1$, $\omega_D = 10$, $D = 10^{-1}$) - $\alpha_0 = 10$; --- $\alpha_0 = 24.0$, -.- $\alpha_0 = 30$.

$$J_0\left(\frac{\alpha_0}{\omega_D}\right) = \sqrt{\frac{2}{\pi}\frac{1}{\alpha_0/\omega_D}} \cos\left(\frac{\alpha_0}{\omega_D} - \frac{\pi}{4}\right) . \tag{23}$$

Note that this expression decreases as the square root of the dither amplitude.

As already mentioned the appearance of the gyroscope steps is an unavoidable effect due to the nonlinearity of the equation of motion and shows up for any periodic dither. This step like behavior of the locking curve is also confirmed by the example of the square-wave dithering function f_s (see Fig. 8). However in this case the dead band around zero rotation rate vanishes already for $\alpha_0/\omega_D = 2$, since $\mathcal{M}_0\ (2) = 0$ as can be seen from Eq. (22). This is a hint that the width of the dead bands is indeed related to $|bM_{-r}[f]|$.

In order to check this conjecture as well as the approximate noise-free expression Eq. (20) the mean beat frequency of a square-wave dithered ring laser gyroscope was calculated in the limit of vanishing diffusion constant D.

It was found that when D is decreased the convergence of $\langle\langle\dot{\phi}\rangle_F\rangle_t$ based on the matrix continued fraction is somewhat slower for rotation rates well inside the gyroscope steps while outside the convergence is unchanged. Moreover, in the limit D = 0 the expression Eq. (B10) does not converge at all for rotation rates inside of the steps, but does give a converging result outside of the steps (Fig. 9). This enables us to compare the exact method with the approximate result Eq. (20) around zero rotation rate (Fig. 9a) and at the first gyroscope step (insets of Fig. 9b).

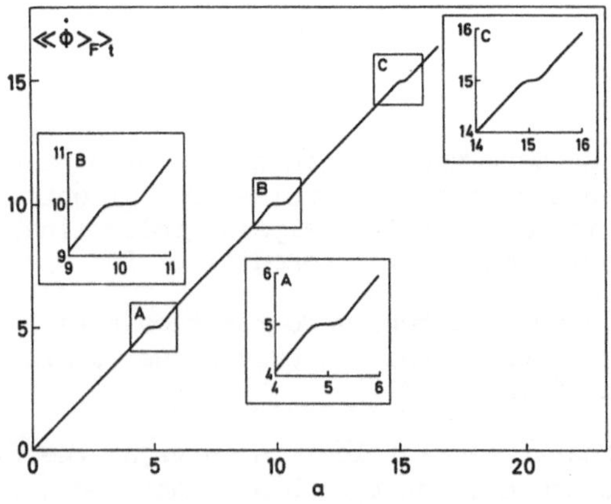

Fig. 8. Mean beat frequency $\langle\langle\dot{\phi}\rangle_F\rangle_t$ of a periodic square-wave dithered ring laser gyroscope as a function of rotation rate. The gyroscope steps arising also for this kind of dither are magnified in the insets. Note that the zeroth dead band is not present for this particular choice of parameters. (b = 1, $D = 10^{-1}$, $\alpha_0 = 10$, and $\omega_D = 5$).

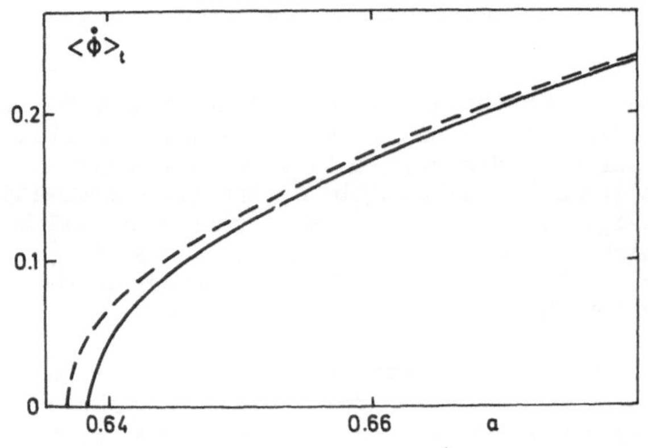

(a)

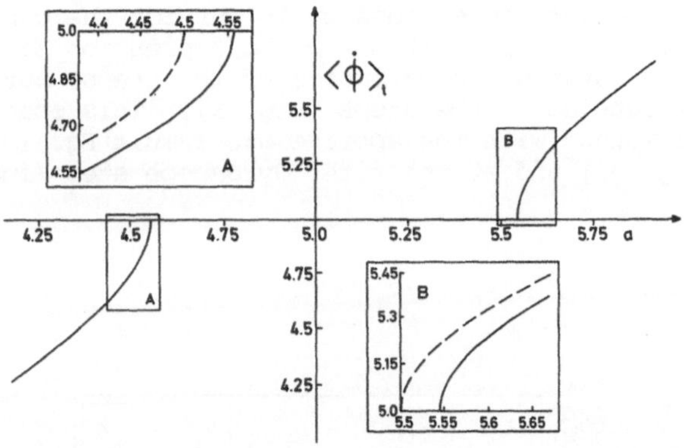

(b)

Fig. 9. Comparison between the lock-in curves of a square wave dithered gyroscope in the absence of noise (D = 0) obtained from the exact matrix continued fraction analysis (solid line) and the approximate expression Eq. (20) (broken line) for b = 1, $\alpha_0 = \omega_D = 5$.

(a) Lock-in curve around zero rotation rate. Note the slight disagreement between the two curves due to higher order corrections $(b/2\omega_D)^2$ in Eq. (19).

(b) Note that the exact lock-in curve at the first gyroscope step is not centered at $\omega_D = 5$ as suggested by the approximate expression Eq. (20) (broken line) but is shifted by an amount given by Eq. (24). Moreover corrections of higher order to the width of this dead band play an important role.

From 9a we note that the width of the dead bands is not completely determined by $b\mathcal{M}_{-r}$ but involves a small correction δ_r to it. Using the "method of averaging" outlined in Ref. 39 higher order corrections to Eq. (19) in powers of $b/2\omega_D$ are derived in Ref. 19 and δ_r can be shown to be of the order of $b(b/2\omega_D)^2$. Moreover the averaging technique clearly demonstrates that the r-th dead band ($r \neq 0$) is not centered around $r\omega_D$, but is slightly shifted as is confirmed by Fig. 9b. In Ref. 19 an explicit expression for this shift Δ_r in terms of the functionals $M_k[f]$ is found to be given by

$$\Delta_r = b\left(\frac{b}{2\omega_D}\right) \sum_{\substack{s=-\infty \\ s \neq 0}}^{\infty} \frac{M^2_{s-r}[f]}{s} \ . \tag{24}$$

Thus Eq. (20) has to be modified by the contributions δ_r and Δ_r discussed in more detail in Ref. 19.

We conclude this discussion by noting that in the case of a square-wave dithering function (Eq. (15)) the width of the zeroth gyroscope step is proportional to

$$\left|\mathcal{M}_0\left(\frac{\alpha_0}{\omega_D}\right)\right| = \left|\frac{\sin\left(\frac{\alpha_0}{\omega_D}\frac{\pi}{2}\right)}{\left(\frac{\alpha_0}{\omega_D}\frac{\pi}{2}\right)}\right| .$$

and decreases linearly with increasing dither amplitude whereas in the case of a harmonic (i.e. sine-wave) bias we note from Eq. (23) a decrease with the square root of the dither amplitude. This reflects the fact that using a square-wave dither the gyroscope practically spends no time in the locked regime.

VI. CONCLUSIONS

In this paper, the dithering technique designed to prevent the gyroscope from locking has been discussed assuming a rather general form of the dither. The central result of these notes is that by a transformation of variables the Langevin equation describing the diffusion of the phase difference between the two counterpropagating waves in the presence of quantum noise can be brought into a form such that an expression for the mean beat frequency $\langle\langle\dot{\phi}\rangle_F\rangle_t$ could be found using one "universal" formalism independent of the explicit shape of the dithering function. This derived result is an exact one and given by infinite matrix continued fractions. Moreover an approximate analytical expression in the absence of noise has been presented. The exact continued fraction treatment as well as the approximate method clearly show that additional dead bands must arise for any periodic dither. The appearance of gyroscope steps is thus an unavoidable effect due to the nonlinearity of the phase equation (10). Moreover, for the example of the square-wave dithering function we have demonstrated that these steps are not exactly centered around multiples of the dither frequency, but slightly shifted. An approximate

analytical expression for this shift in the presence of an arbitrary periodic and time-symmetric dither has been given. The transformation alluded to above also allowed us to express the approximate width of the gyro steps in the case of the square-wave dither in closed form in terms of simple trigonometric functions. Testing the approximate analytical results in the absence of noise against the exact method shows that additional terms of the order $(b/2\omega_D)^2$ are necessary to describe the width of the steps properly. Finally, we emphasize again the useful role of quantum noise in decreasing the tendency of the gyro to lock.

VII. ACKNOWLEDGMENTS

The authors consider it their pleasant duty to thank J. Bergou, C.-S. Cha, P. Meystre and S. Romani for useful and stimulating discussions. In particular, we wish to express our appreciation to H. Risken for introducing us to the powerful tool of matrix continued fractions. One of us (W. S.) would like to thank J. A. Wheeler for his encouragement and his support and for providing a fruitful atmosphere at the Center for Theoretical Physics at Austin, where part of this work was done. This article was partially supported for publication by NSF Grant #PHY 8205717.

APPENDIX A: TRANSFORMATION OF STOCHASTIC VARIABLE

In this appendix we derive Eq. (17) starting from the Langevin equation (10).

Substituting Eq. (16) into Eq. (10) yields

$$\dot{\psi}_r = \tilde{a} + \frac{b}{2i} \exp[ir\omega_D t + i\int_0^t dt' \, f(t') + i\psi_r] + \text{c.c.} + F(t) \, . \tag{A1}$$

From Eq. (11) and Eq. (12) it is easy to prove that the function

$$\mathcal{F}(t) \equiv \exp[i\int_0^t d\tau \; f(\tau)]$$

is periodic with period T and thus may be expanded into Fourier series

$$\exp[i\int_0^t d\tau \; f(\tau)] = \sum_{k=-\infty}^{\infty} M_k[f] e^{ik\omega_D t} \tag{A2}$$

where the coefficients $M_k = M_k[f]$ are given by

$$M_k[f] = \frac{1}{2\pi} \int_{-\pi}^{\pi} d\zeta \exp[-ik\zeta + i\int_0^{\zeta/\omega_D} d\tau \; f(\tau)]. \tag{A3}$$

Substituting Eq. (A2) into Eq. (A1) we find after minor algebra

$$\dot{\psi}_r = \tilde{a} + b \sum_{s=-\infty}^{\infty} \{Re(M_{s-r}[f]) \sin(s\omega_D t + \psi_r)$$

$$+ Im(M_{s-r}[f]) \cos(s\omega_D t + \psi_r)\} + F(t) \; . \tag{A4}$$

Since the functionals $M_k = M_k[f]$ are real for a time symmetric dithering function, i.e.

$$f(-t) = f(t) \; ,$$

as can be shown from Eq. (A3), Eq. (A4) reduces to

$$\dot{\psi}_r = \tilde{a} + b \sum_{s=-\infty}^{\infty} M_{s-r}[f] \sin(s\omega_D t + \psi_r) + F(t) \; .$$

Note, however, that for a dither containing no such symmetry in time all terms in Eq. (A4) have to be taken into account.

APPENDIX B: MEAN BEAT FREQUENCY - EXACT MATRIX CONTINUED FRACTION TREATMENT

In this appendix we derive an exact expression for the mean beat frequency of a dithered ring laser gyroscope in terms of infinite matrix continued fractions.

The Fokker-Planck equation corresponding to the Langevin equation (17) reads [20]

$$\frac{\partial P}{\partial t} = -\frac{\partial}{\partial \psi} \{ [\tilde{a} + b \sum_{s=-\infty}^{\infty} M_{s-r}[f] \sin(s\omega_D t + \psi)] P\} + D \frac{\partial^2 P}{\partial \psi^2} \tag{B1}$$

where we have suppressed the index r at ψ_r. Since ψ denotes the phase difference between the two counterpropagating waves periodic boundary conditions

$$P(t, \psi + 2\pi) = P(t, \psi)$$

are appropriate. Expanding P into a Fourier series

$$P(t,\psi) = \sum_{m=-\infty}^{\infty} c_m(t) e^{im\psi} \tag{B2}$$

and substituting into Eq. (B1) yields

$$\dot{c}_m = -[im\tilde{a} + m^2 D] c_m - \frac{mb}{2} \sum_{s=-\infty}^{\infty} M_{s-r} \, e^{is\omega_D t} \, c_{m-1}$$

$$+ \frac{mb}{2} \sum_{s=-\infty}^{\infty} M_{s-r} \, e^{-is\omega_D t} \, c_{m+1} \; .$$

Multiplying this equation by $e^{ik\omega_D t}$ we find for the time averaged coefficients

$$\mathcal{S}_m^k \equiv \langle c_m(t) e^{ik\omega_D t} \rangle_t \equiv \lim_{\tau\to\infty} \left\{ \frac{1}{\tau} \int_0^\tau dt \; c_m(t) e^{ik\omega_D t} \right\}, \tag{B3}$$

the algebraic recurrence relation

$$\begin{aligned} o = & -[i(m\tilde{a} - k\omega_D) + m^2 D]\mathcal{S}_m^k - m \frac{b}{2} \sum_{s=-\infty}^{\infty} M_{s-k-r} \mathcal{S}_{m-1}^s \\ & + m \frac{b}{2} \sum_{s=-\infty}^{\infty} M_{-s+k-r} \mathcal{S}_{m+1}^s \quad . \end{aligned} \tag{B4}$$

Here we have also used the fact that

$$\langle \dot{c}_m \, e^{ik\omega_D t} \rangle_t = - ik\omega_D \langle c_m \, e^{ik\omega_D t} \rangle_t$$

which is easy to prove on integration by parts.

Defining the vector [20,42]

$$(\underset{\sim}{S}_m)_k \equiv \mathcal{S}_m^k \tag{B5}$$

Equation (B4) may be cast into a three term vector recurrence relation

$$0 = \underset{\approx}{A}_m \underset{\sim}{S}_m + \underset{\approx}{B}_m \underset{\sim}{S}_{m+1} + \underset{\approx}{C}_m \underset{\sim}{S}_{m-1} \quad . \tag{B6}$$

The matrices $\underset{\approx}{A}_m$, $\underset{\approx}{B}_m$ and $\underset{\approx}{C}_m$ follow by inspection

$$(\underset{\approx}{A}_m)_{k,j} \equiv - [i(m\tilde{a} - k\omega_D) + m^2 D]\, \delta_{k,j} \tag{B7a}$$

$$(\underset{\approx}{B}_m)_{k,j} \equiv m \frac{b}{2} M_{k-j-r} \; [f] \tag{B7b}$$

$$(\underset{\approx}{C}_m)_{k,j} \equiv - m \frac{b}{2} M_{-k+j-r} \; [f] \quad . \tag{B7c}$$

Using the ansatz [20,42]

$$\underset{\sim}{S}_m = \underset{\approx}{R}_m \underset{\sim}{S}_{m-1} \tag{B8}$$

Equation (B6) yields the following continued fraction for the infinite matrix $\underset{\approx}{R}_m$

$$\underset{\approx}{R}_m = - (\underset{\approx}{A}_m + \underset{\approx}{B}_m \underset{\approx}{R}_{m+1})^{-1} \underset{\approx}{C}_m \quad . \tag{B9}$$

We have thus solved the time-averaged Fokker-Planck equation (B1) exactly, in terms of infinite matrix continued fractions. In a last step we now express the mean beat frequency $\langle\langle\dot{\phi}\rangle_F\rangle_t$ in terms of $\underset{\approx}{R}_m$. From Eqs. (8), (12), (16), and (17) follows

$$\langle\langle\dot{\phi}\rangle_F\rangle_t = r\omega_D + \tilde{a} + b \sum_{s=-\infty}^{\infty} M_{s-r}[f] \langle\langle \sin(s\omega_D t + \psi)\rangle_F\rangle_t.$$

The average over the stochastic force F may be performed using the Fourier expression (B2) of the probability distribution P together with the time averaged coefficients $\mathcal{S}_m^k$ Eq. (B3)

$$\begin{aligned}
\langle\langle\sin(s\omega_D t + \psi)\rangle_F\rangle_t &= - \frac{1}{2i} \langle e^{-is\omega_D t} \langle e^{-i\psi}\rangle_F\rangle_t + \text{c.c.} = \\
&= - \frac{1}{2i} \langle e^{-is\omega_D t} \int_0^{2\pi} d\psi\ P(\psi) e^{-i\psi}\rangle_t + \text{c.c.} = \\
&= - \frac{2\pi}{2i} \langle e^{-is\omega_D t} c_1(t)\rangle_t + \text{c.c.} = \\
&= - 2\pi \ \text{Im}\, \mathcal{S}_1^{-s} \quad .
\end{aligned}$$

Therefore the mean beat frequency may be expressed by

$$\langle\langle\dot{\phi}\rangle_F\rangle_t = r\,\omega_D + \tilde{a} - 2\pi b \sum_{s=-\infty}^{\infty} M_{-s-r}[f]\, \text{Im}\, \mathcal{S}_1^s \ .$$

From Eq. (B5) and (B8) we note the relation

$$\mathcal{S}_1^s = (\underset{\sim}{S}_1)_s = \sum_{j=-\infty}^{\infty} (\underset{\approx}{R}_1)_{sj} (\underset{\sim}{S}_0)_j.$$

The vector $\underset{\sim}{S}_0$ can easily be calculated from the definition (B3)

$$(\underset{\sim}{S}_0)_j = c_0 \delta_{j,0}$$

where the coefficient c_0 follows from the normalization of P

$$1 = \int_0^{2\pi} d\psi P(\psi) = 2\pi c_0 .$$

We therefore arrive at

$$\mathcal{S}_1^s = \frac{1}{2\pi} (\underset{\approx}{R}_1)_{s0}$$

and

$$\langle\langle\dot{\phi}\rangle_F\rangle_t = r\omega_D + \tilde{a} - b \sum_{s=-\infty}^{\infty} M_{-s-r} [f] \mathrm{Im}(\underset{\approx}{R}_1)_{s0} . \qquad \text{(B10)}$$

Therefore the mean beat frequency of an arbitrarily dithered ring laser gyroscope is expressed by a column of the infinite matrix $\underset{\approx}{R}_1$ defined via the matrix continued fraction (B9) and (B7). Note that the form of the dithering function does not influence the structure of the continued fraction (B9) but enters solely via the coefficients of the matrices $\underset{\approx}{B}_m$ and $\underset{\approx}{C}_m$.

REFERENCES

1. K. J. Johnson, J. H. Spencer, C. H. Mayer, W. J. Klepczynski, G. Kaplan, D. D. McCarthy, and G. Westerhout, in: "Time and the Earth's Rotation", D. D. McCarthy and J. D. H. Pilkington, eds., Reidel, Dortredit (1979).
2. J. G. Small, and W. W. Chow, Research proposal No. 104-113, University of New Mexico, November 1982.
3. W. W. Chow, J. Gea-Banacloche, L. M. Pedrotti, V. E. Sanders, W. Schleich, and M. O. Scully, The Ring Laser Gyro: A Tutorial Review, Rev. Mod. Phys., to be published.
4. A. Michelson, The Effect of the Earth's Rotation on the Velocity of Light, Part I, Astrophys. J. 61:137 (1925). A. Michelson, and H. Gale, The Effect of the Earth's Rotation on the Velocity of Light, Part II, Astrophys. J. 61:140 (1925).
5. a) G. Sagnac, The Luminiferous Aether Demonstrated by the Effect of the Relative Motion of the Aether in an Interferometer in Uniform Rotation, C. R. Acad. Sci. 157:708 (1913), English translation in: "The Einstein Myth and the Ives Papers", D. Turner and R. Hazelett, eds., Devin-Adair, Old Greenwich, (1979); b) See criticism by H. Witte, Deutsch. Phys. Gesell. Verh. 16:142 (1914).
6. E. J. Post, Sagnac Effect, Rev. Mod. Phys. 39:475 (1967).
7. M. O. Scully, M. S. Zubairy, and M. P. Haugan, Proposed Optical Test of Metric Gravitation Theories, Phys. Rev. A24:2009 (1981). W. Schleich and M. O. Scully, General Relativity and Modern Optics in: "Modern Trends in Atomic and Molecular Physics", Proceedings of the Les Houches Summer School, session XXXVIII, R. Stora and G. Grynberg, eds., North-Holland, Amsterdam (1984).
8. See for example: C. Misner, K. S. Thorne, and J. A. Wheeler, "Gravitation", Freeman, San Francisco (1974).
9. H. Thirring, "Über den Einfluss von fernen, rotierenden Massen in der Einsteinschen Gravitationstheorie, Phys. Zeits. 19:33 (1918). J. Lense, and H. Thirring, "Über den Einfluss der Eigenrotation von Himmelskörpern auf die Bewegung von Planeten und Monden nach der Einsteinschen Relativitätstheorie, Phys. Zeits. 19:156 (1918). An

English translation can be found in: B. Mashhoon, F. W. Hehl, and D. S. Theiss, On the Gravitational Effects of Rotating Masses: The Thirring-Lense Papers, Gen. Rel. Grav. 16;711 (1984).
10. See for example: F. Aronowitz, The Laser Gyro, in: "Laser Applications", Vol. 1, M. Ross, ed., Academic Press, New York (1977).
11. J. Killpatrick, The Laser Gyro, IEEE Spectrum 4;44 (1967).
12. T. A. Dorschner, H. A. Haus, M. Holz, I. W. Smith and H. Statz, The Laser Gyro at Quantum Limit, IEEE J. Quant. Elect. 16:1376 (1980). M. O. Scully, W. Chow, K. Drühl, V. Sanders, Quantum Noise Limitations and the Determination of Laser Frequency, to be published.
13. J. D. Cresser, W. H. Louisell, P. Meystre, W. Schleich, and M. O. Scully, Quantum Noise in Ring Laser Gyros. I. Theoretical Formulation of Problem, Phys. Rev. A25:2214 (1982).
14. J. D. Cresser, D. Hammons, W. H. Louisell, P. Meystre, and H. Risken, Quantum Noise in Ring Laser Gyros. II. Numerical Results, Phys. Rev. A25:2226 (1982).
15. J. D. Cresser, "Quantum Noise in Ring Laser Gyros. III. Approximate Analytical Results in unlocked Region, Phys. Rev. A26:398 (1982).
16. W. Schleich, C.-S. Cha, and J. D. Cresser, Quantum Noise in a Dithered Ring Laser Gyroscope, Phys. Rev. A 29:230 (1984).
17. S. Singh, Statistical Properties of Single Mode and Two-Mode Ring Lasers, Phys. Reports 108:217 (1984).
18. W. Schleich, and P. Dobiasch, "Noise Analysis of Ring Laser Gyroscope with Arbitrary Dither", Opt. Commun. to be published.
19. W. Schleich and P. Dobiasch, to be published.
20. H. Risken, "The Fokker-Planck Equation Methods of Solutions and Applications", in: Springer Series in Synergetics, H. Haken, ed., Springer, Berlin, (1984).
21. M. Sargent III, M. O. Scully, and W. E. Lamb, Jr., "Laser Physics", Addison-Wesley, Reading (1974).
22. W. M. Macec, and D. T. M. Davis, Rotation Rate Sensing With Traveling-Wave Ring Lasers, Appl. Phys. Lett. 2:67 (1963).
23. F. Aronowitz, Theory of a Traveling-Wave Optical Maser, Phys. Rev. 139:A635 (1965). L. Menogozzi, and W. E. Lamb, Jr., Theory of a Ring Laser, Phys. Rev. A 8:2103 (1973).
24. A. A. Michelson, Philosophical Magazine (London) 8:716 (1904).
25. See Ref. 21, p. 52.
26. W. E. Lamb, Jr., Theory of an Optical Maser, Phys. Rev. 134:A1429 (1964).
27. M. O. Scully, and W. E. Lamb, Jr., Quantum Theory of an Optical Maser. I. General Theory, Phys. Rev. 159:208 (1967).
28. R. Gauert, Die Lock-in Schwelle von Laserkreiseln als Funktion von Parametern des Ring Lasers, Optik 67:21 (1984).
29. W. W. Chow, M. O. Scully, and E. W. Van Stryland, Line Narrowing in a Symmetry Broken Laser, Opt. Commun. 15:6 (1975).
30. Yu. M. Ivanchenko, and L. A. Zil'berman, The Josephson Effect in Small Tunnel Contacts, Sov. Phys.-JETP 28:1272 (1969).
31. H. Haken, H. Sauermann, Ch. Schmidt, and H. D. Vollmer, Theory of Laser Noise in the Phase Locking Region, Z. Phys. 206:369 (1967).
32. R. L. Stratonovich, Synchronization of a Self-Excited Oscillator in the Presence of Noise, Radiotekh. i. Elektr. 3:497 (1958).
33. R. L. Stratonovich, "Topics in the Theory of Random Noise", Vol. 2, Gordon and Breach, New York (1967).
34. J. Killpatrick, see Ref. 10, p. 159.
35. T. J. Hutchings, and D. C. Stjern, Scale Factor Nonlinearity of a Body Dithered Laser Gyro, in: "Proceedings of the IEEE National Aerospace and Electronics Conference", X. Y. Unbekannt, ed., IEEE, New York (1978). J. J. Roland, and G. P. Agrawal, Opt. Laser Technol. 13:239 (1981). A. Bambini, and S. Stenholm, Theory of a

Dithered Ring Laser Gyro I: The Floquet Treatment, to be published.
36. A. Bambini, and S. Stenholm, Analysis of Nonlinear Response in a Body Dithered Ring Laser Gyro, Opt. Commun. 49:269 (1984).
37. W. Schleich, M. O. Scully, and V. E. Sanders, Quantum Noise in Ring Laser Gyroscopes, in: "Coherence and Quantum Optics V", L. Mandel and E. Wolf, eds., Plenum, New York (1984).
38. W. H. Louisell, "Quantum Statistical Properties of Radiation", John Wiley, New York (1973).
39. N. N. Bogoliubov, Y. A. Mitropolsky, "Asymptotic Methods in the Theory of Nonlinear Oscillations", Hindustan Publishing Corpn., Delhi (1961).
40. W. Magnus, F. Oberhettinger, and R. P. Soni, "Formulas and Theorems for the Special Functions of Mathematical Physics", Springer, Berlin (1966).
41. W. Schleich, P. Dobiasch, and H. Risken, Harmonic Frequency Mixing in a Cosine Potential, to be published.
42. H. Risken, and H. D. Vollmer, Solutions and Applications of Tridiagonal Vector Recurrence Relations, Z. Phys. B 39:339 (1980).

GENERAL

Here are a half-dozen papers on subjects ranging from Darwinian evolution to Josephson junctions.

DARWINIAN EVOLUTION IN PHYSICS AND BIOLOGY:

THE DILEMMA OF MISSING LINKS

F. A. Hopf

Optical Sciences Center
University of Arizona
Tucson, Arizona

and

F. W. Hopf

Department of Biology
Cornell University
Ithaca, New York

In the *Origin of Species*[1] Darwin presented a major dilemma of his own theory of evolution, which he called "the absence of transitional varieties" and which is currently referred to as the dilemma of missing links. The dilemma consists of two parts. The first part involves Darwin's conclusion that his theory predicts a continuous structure for living forms. In contrast to this continuum, the biota has a grainy structure due to the coalescence of organisms into species. The second part, which involves missing links in the geologic record, refutes Darwin's attempt to resolve the first part of the dilemma. In this paper, we make an analogy between laser operation and Darwin's theory of evolution which leads to Darwin's continuum. The analogy, however, is flawed insofar as biological reproduction usually involves two parents; lasers are more analogous to a biota with uniparental reproduction. We modify laser theory to make it consistent with biparental reproduction, which causes the laser output to coalescence into a few modes, which are analogous to discrete species. The temporal changes predicted by this modified model are consistent with the geologic record.

I. Natural Selection

Natural selection[1] is the fundamental principle of Darwinian evolution. It states, first of all, that organisms reproduce and grow in numbers according to the law of Malthus.[2] This reproduction produces a variety of offspring. The offspring are too numerous to be supported by the environment, and most will die due to

checks on the populations. Those offspring which are better adapted to their environment are more likely to survive and will thus tend to leave more descendants, out-competing and eventually replacing less capable forms. Natural selection works only on heritable characteristics, not on variations produced solely by differing environments.

Darwin built his theory in analogy with the known principles of artificial selection. An animal breeder selectively removes those animals which show the traits he desires to maintain and breeds them in isolation, in order to increase the population size of organisms with that trait. Experience has shown that substantial modifications in form can be generated this way, e.g., breeds of dogs.

Nonliving systems in which new components are generated from old ones by replication can obey Darwin's principle of natural selection.[3,4] We refer to such objects as replicators. We develop in some detail the case of the laser, in which the photon is the replicator and the replication process is stimulated emission. Lasers have close parallels in Bernard convection[5] and other processes in hydrodynamics. Of more relevance to life, RNA polymers[4] replicating through autocatalysis are also known to obey Darwinian principles.

Table 1 lists the components of natural selection in biology and analogous concepts in laser physics. Genetic variety in living organisms, i.e., the heritable

Table 1. Analogies between Darwin's evolutionary principles applied to living and nonliving systems. Lasers are used as an example.

	General	Life	Lasers
1)	Replication	Heritability ("like produces like")	Simulated emission $h\nu \rightarrow 2h\nu$)
2)	Variety	Genetic variety	Variety of frequencies ν
3)	Exponential growth	aw of Malthus	Gain and exponential growth
4)	Saturation	Checks on population	Saturation
5)	Competition	Competition	Competition

aspect of variation, is analogous to different photon frequencies. The law of Malthus is just a discrete form of exponential growth. Saturation and checks are synonymous and the concept of competition is common to both. Darwin[1] referred to items 3-5 in the table as the "struggle for existence," which replaced the deliberate action of breeders in artificial selection as the mechanism that brings about change in nature. What is omitted in this table is any discussion of the origin of novel forms in item 2, i.e., the fluctuations.[6]

The mathematics describing the dynamics of the simplest laser problem[7] can be written as

$$\frac{dS(\nu)}{dt} = \sigma(\nu)\ n\ S(\nu) - \kappa(\nu)\ S(\nu), \qquad (1)$$

where $S(\nu)$ is the spectral density, $\sigma(\nu)$ is the stimulated emission cross section, n is the population inversion and $\kappa(\nu)$ is the loss. For biology we label the replicator y and write

$$\frac{dN_y}{dt} = b_y\ R\ N_y - d_y\ N_y, \qquad (2)$$

where N_y is the density of individuals (number per unit area, not number per unit y), b_y is the birth rate, R the resources and d_y the death rate. The interrelation between the kinds of units used in these two equations is discussed in Appendix A. The saturation follows from conservation of energy or mass, and requires that n or R decrease as population increases. Taking the units of S, N, n, and R such that the unsaturated values of n and R are unity, we write the simplest saturation as

$$n = 1 - \int S(\nu)\ d\nu, \qquad (3)$$

$$R = 1 - \Sigma_y Ny. \qquad (4)$$

Restrict Eq. (4) to one population and substitute into Eq. (2). Then reparameterize using $r_y = b_y - d_y$, and $K_y = 1 - d_y/b_y$ which gives the Lokta-Volterra equation[8]

$$\frac{dNy}{dt} = N_y\ r_y\ (1 - \frac{N_y}{K_y}). \qquad (5)$$

We follow current terminology in using "ecology" in reference to competition between established species, as in Eqs. (2) and (4), and using "evolution" to refer to changes within species or the generation of new species from old ones.[8] Darwin made no such distinction. In this paper we deal with ecological difficulties that arise from natural selection.

In Fig. 1, we illustrate the outcome from integrating Eqs. (1) and (3) for long times. The spectrum narrows, and eventually becomes a delta function at the frequency for which $\sigma(\nu)/k(\nu)$ is a maximum. Similarly Eq. (5) results in a single surviving species which maximizes[9] K_y. This "winner takes all" result of natural selection occurs in a relatively short time for finite sized populations, and is thus capable of producing changes in the biota on geological timescales.

II. Effect of Inhomogeneous broadening

Equations (2) and (3) have oversimplified the problem insofar as n and R are treated as single variables. In reality, there are grades of resources on which biological replicators grow, and in lasers there is inhomogeneous broadening due, for example, to the Doppler effect. This means that each species or laser mode

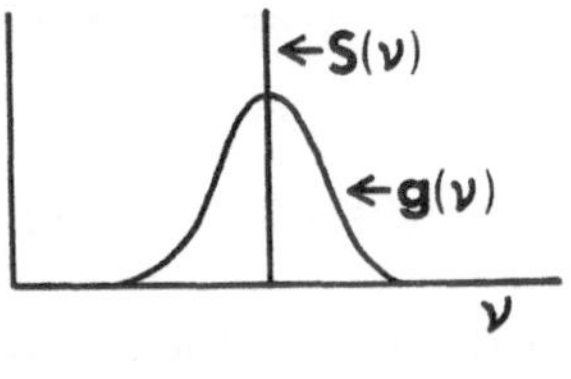

Fig. 1. Flux S(ν) as a function if ν obtained from a homogeneously broadened laser amplifier with gain g(ν) = σ(ν)n as sketched.

uses a slightly different set of resources than any other. The model used to describe inhomogeneous broadening is given in Appendix A. It is j st Lamb's[10] laser theory from which terms have been removed that describe processes such as population pulsations and four-wave mixing which have no analogy in biology.[11] The critical results of our development are known not to be qualitatively affected by the terms we have discarded.[12-15] Inhomogeneous broadening requires that we consider a distribution of atomic frequencies which we label ρ(ω). The frequency ω has the same units as ν but refers to the atomic resource distribution, not the replicators. For living organisms we assume that there is a variable s denoting a resource that has the same units as the species label. In Fig. 2, we illustrate an inhomogeneously broadened line, made up of a distribution of homogeneous lines. The homogeneous line determines the range of atomic frequencies that can generate gain for any one laser mode.

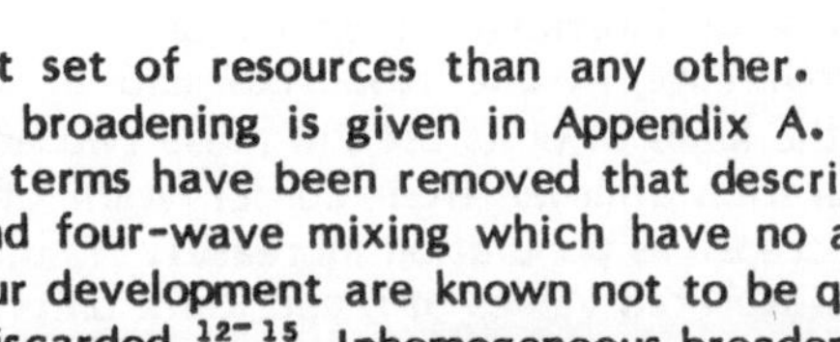

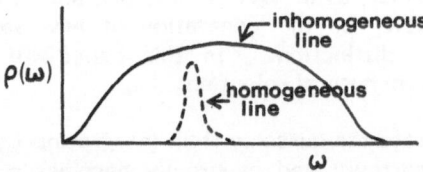

Fig. 2. Inhomogeneous gain line (solid curve) as a function of ω. Dotted insert indicates homogeneous line.

The laser cavity selects the modes that can oscillate. This selection has no counterpart in biology. To remove the influence of the cavity, imagine that the mirrors are taken to infinity. Alternatively, remove them altogether and examine high-gain laser amplifiers. The experimental[16] and theoretical[11-15] results of such a process are illustrated in Fig. 3. The outcome is no longer the single spike of

Fig. 2. Instead there is a continuum of frequencies, insofar as all laser modes continue to oscillate as they become increasingly dense. In the limit of high mode density, the average number of photons per laser mode decreases to zero.

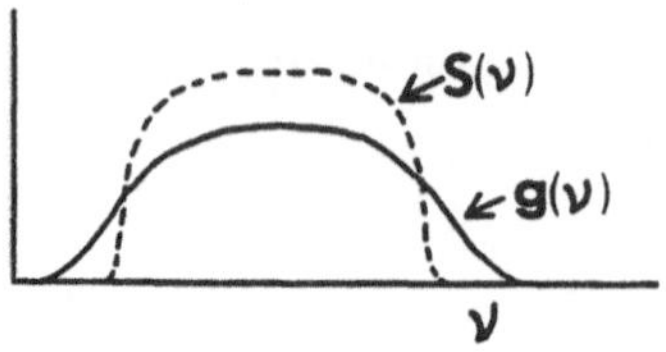

Fig. 3. Sketch of laser output vs. ν for an inhomogeneously broadened laser with widely separated mirrors.

III. Species Distributions

In order to evaluate Darwinian evolution in a biological context, one must consider observed species distributions. Figure 4a illustrates a census[17] of oak trees on a mountain near Tucson, and is a case with which we are familiar. The horizontal axis of the plot describes the resource grade which is parameterized by altitude. As altitude changes, there are grades in temperature and moisture which are important inhomogeneous elements in the resources of the trees. In Figure 4b, for viewing convenience, each species is plotted as a spike located at the resource point at which maximum species density occurs. Thus we label the oaks by the altitude at which they are the most abundant. In a continuum, the spikes are so dense that they are inconvenient to plot and we draw lines through the tops of the spikes as in Fig. 3.

Biological species distributions,[18,19] as displayed in Fig. 4, are discrete, meaning that when plotted as spikes, there are unoccupied resource points or gaps between existing species. In contrast, lasers show an infinitely dense series of spikes, i.e., a continuum. In addition, discrete distributions have many individuals per species, not just a few individuals per species as the laser analogy suggests.

It could be that the discreteness seen in biological communities is actually a transitional state on the way to a continuous final state. However studies of the geological record have shown that the number of forms tends to be constant for long times.[20,21] For example, the number of orders of land animals has varied by no more than 20% over the last 350 million years, even though many orders have gone extinct and have been replaced by others. The number of marine orders has remained roughly constant for a longer time. These observations are not decisive as to the constancy of the number of species; it is difficult to identify fossils so precisely. Nonetheless, the data gives no encouragement to the idea that there is an ongoing proliferation of species. A variation on this question is whether the data might be explained by the the slowing down that is expected as the

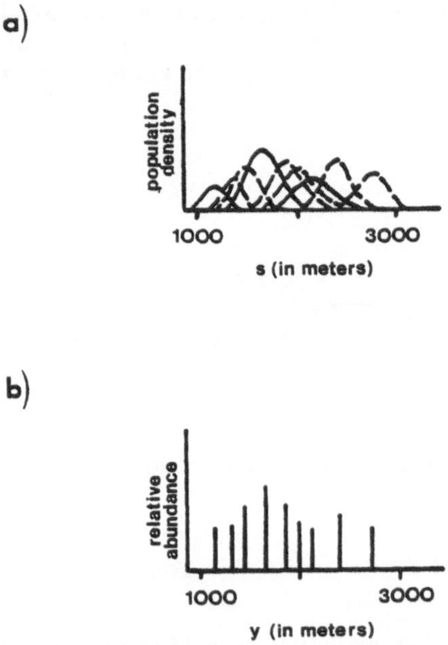

Fig. 4. a) Distribution of numbers of individuals per unit area of nine species of oak trees as a function of altitude up Mt. Lemmon near Tucson, Arizona. Data sketched from Ref. 17. b) Data replotted using a vertical line drawn at the point of maximum abundance of each species.

continuum is approached. However, the biota has changed substantially over the last several hundred million years, which is not indicative of slowing down. There are other difficulties with slowing down that are discussed in Appendix B. Hence it is unlikely that presently observed results are artifacts either of short times or of slowing down.

IV. The dilemma of missing links

The problem of accounting for the clustering of life into distinct and broadly homogeneous species is the heart of the Darwin's dilemma of the missing links. The part of the dilemma is the one given above; why are observed species distributions discrete as in Fig. 4, and not the continuum of Fig. 3? Darwin put it this way:

> "Why, if species have descended from other species by insensibly fine gradations, do we not everywhere see innumerable transitional forms? Why is not all nature in confusion instead of the species being, as we see them, well defined?" (p. 205)

Note that Darwin dismisses the possibility that the intermediate forms never existed. He notes that his theory claims that present species descended from earlier ones by small changes, and these ancestral species were surely different and probably intermediate to the ones found now. Hence, there were, at some time in the past, species that sat in the gaps between the spikes in Fig. 4. If the continuum solution, containing these intermediate forms, is the stable final state, why are they not found now? To resolve this part of the dilemma, Darwin argued that intermediate niches are small, and consequently intermediate species will have a small population size. He postulated that the rate of evolution is dominated by the rate at which useful varieties, e.g., novelties now attributed to mutations, arise in the population. If the probability of a useful novelty is the same for every individual, then populations with many individuals acquire useful novelties more rapidly than populations with few individuals. Darwin's version of natural selection implies that these novelties are rapidly fixed in the population. Therefore, rare populations improve more slowly than large populations, and are ultimately eliminated by competition.

Darwin realized that this solution is not totally satisfactory. First, in artificial selection, those individuals with favorable traits must be reproductively isolated[18] to increase the trait. Evolution in artificial selection occurs in small, not large, populations. Secondly, the resolution runs into difficulties with the geologic record. As Darwin put it:

"As by this theory innumerable transitional forms must have existed, why do we not find them embedded in countless numbers in the crust of the earth?" (p. 206).

The second part of the missing links dilemma is based on the expectation that abundant populations, which by the previous argument are the ones that are changing most rapidly, should leave the most fossils. Thus we should see "countless numbers" of intermediate forms in the geological record. Darwin knew that the geological record was inconsistent with this expectation, and appealed to its imperfection. The expectation was that, with improved data, evolution would be seen as the gradual process which Darwin expected based on his claim that the rate of evolution is dominated by the rate at which instabilities are nucleated.[6] We show, in Fig. 5a, the picture of temporal evolution that Darwin drew. However, the geological record remains largely inconsistent with continuous evolution, although a few fairly continuous cases, such as horses,[22] are now known. Figure 5b shows a more common finding: abrupt changes,[23] often simultaneously in different phylogenetic groups.[22,23] This type of change is called "punctuated" evolution.[23] From data like this, Simpson[22] concluded that evolution is most rapid in smaller, rarer populations, just as the analogy with artificial selection suggests.

V. Resolution of the dilemma

Rapid evolution in rare populations destroys Darwin's resolution of the continuum problem involving missing links in the space of habitats. Change the premises of his argument to make rare populations evolve more rapidly. Then any population that splits up into smaller units gains an advantage from an increased rate of evolution. Hence natural selection and more rapid evolution combine to

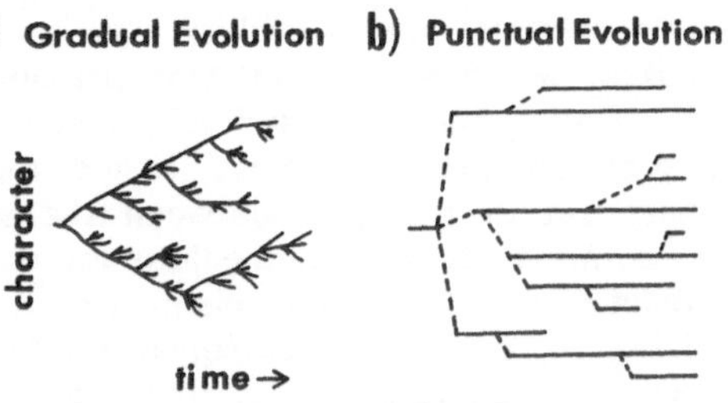

Fig. 5. Evolution of a character (vertical dimension) as a function of time (horizontal dimension). a) Continuous change as envisioned by Darwin. Traced from a portion of the original diagram.[1] b) Plot of elephant lineages over the last 6 million years. Solid lines indicate fossils. Dotted lines are educated guesses as to lineages. Highly simplified tracing from Fig. 4-11 of Ref. 23. Note that elephants fossilize quite well.

strongly destabilize discrete distributions. Slowing down no longer occurs and the continuum is reached with a vastly increased rate.

We propose, instead, that the resolution of the missing links dilemma lies in an obvious flaw in the third item of the analogy in Table 1. Stimulated emission is diagrammed as

$$h\nu + \text{resource} \rightarrow 2h\nu.$$

The analogous process in biology is

$$\text{parent} + \text{resource} \rightarrow \text{parent plus offspring.}$$

For most organisms this analogy is false. Reproduction involves two parents, i.e.,

$$2 \text{ parents} + \text{resource} \rightarrow \text{parents plus offspring.}$$

Laser gain arises from a two body collision. Biparental reproduction is a three body process, and the proper analogy is with two-photon gain. However, in lasers, two-photon processes are important at high photon densities. The three-body process in biology begins to saturate with the first encounter of male and female and is thus important at low densities where the encounters are infrequent. Hence we need to modify the laser equations so that the gain is exponential at higher densities but drops off at low densities.[24,25] We do this by taking b_y to be a function of N_y, and illustrate the function in Fig. 6.

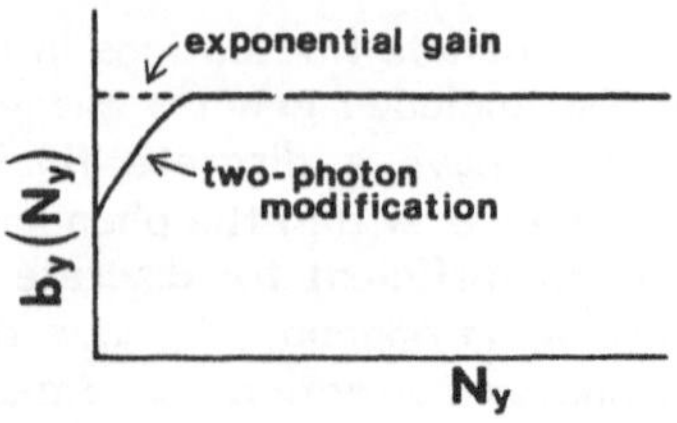

Fig. 6. Gain as a function of population density in the limit of small densities.

We illustrate the results of our numerical work with the example shown in Fig. 7. We take the homogeneous line to be a Gaussian proportional to $\exp(-s^2)$ which defines the units of s and y. The inhomogeneous line is taken to be constant

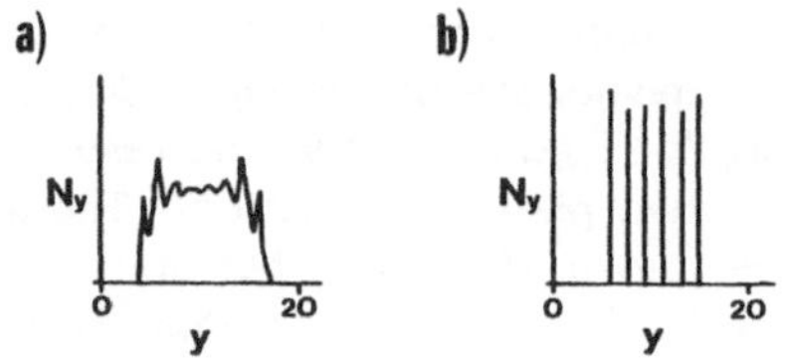

Fig. 7. a) Asymptotic solution for the model given in the appendix using exponential gain and conditions in the text. Output drawn as in Fig. 3. b) Asymptotic solution when gain modified as in Fig. 6. Output drawn as in Fig. 4b.

over a span of 9 such units. The edges of the inhomogeneous distribution drop linearly to zero over a span of 3 units, so the overall distribution is trapezoidal. All distributions in the numerical code are discrete, so the calculation is analogous to a laser with widely separated mirrors and hence high mode density. We show the case in which we compute 5 populations across every resource unit. Each species is seeded[6] with a population that is sufficiently large to give an unsaturated (set $R_y = 1$) two-photon gain that exceeds the loss. In Fig. 7a, there is no two-photon gain, i.e., the unsaturated gain is exponential. In that case, the result is a continuum like the one illustrated in Fig. 3. We then switch on the two-photon term and the distribution changes to the one shown in Fig. 7b, where all but six species have been driven to extinction.

We modified the calculation to include fluctuations in resources, starting times, and starting densities, and we also included genetic changes within the species.[11] All cases with a two-photon term have a discrete final distribution; all cases without it result in a continuum. Hence, within the phenomena we investigated, the two-photon process is necessary and sufficient for discreteness. An analysis of the case of a constant resource sketched in Appendix A shows that these results are not effected by how large the two-photon correction is. Small corrections slow down the rate at which discrete distributions are reached, but have minor effects on the number of species that survive. Our numerical results suggest that this result remains valid even if the environment is subject to large but statistically stationary fluctuations.

The recent ice ages are not statistically stationary phenomena and care is needed in applying our results to the present biota. We expect that organisms which always reproduce biparentally should show the greatest discreteness. Organisms for which uniparental reproduction is an option, often use it when they are at low density,[25] virtually eliminating the effect of the two-photon term. Hence, when biparental reproduction is optional, the distributions might be much less discrete.[19] Organisms which only reproduce uniparentally should approach continua. We discuss the latter case below.

Now let us consider the opposite limit of an extreme non-stationary environment in which the resources are varied systematically through time. Specifically, we slowly scan a homogeneous resource distribution back and forth along the resource axis of Fig. 7. Figure 8 shows the results, comparing the cases with (8b) and without (8a) the two-photon correction. The plot is parametric; the vertical axis labels the species that is the most abundant at time t, the horizontal axis labels the resource at time t. In the case with exponential gain (8a), there is a process of continuous competitive replacement as expected from Fig. 2. There is a slight temporal lag, insofar as the most abundant population is the one that was competitively superior in the recent past. With the two-photon correction, the replacement occurs in abrupt jumps.[23] Potential competitors are unable to overcome the combined difficulties of the decrease in gain from the two-photon process plus the diminishing of resources due to the abundant population. Replacement takes place only when the resource has changed character so much that the abundant population can no longer use it and would go extinct with or without competition.

In summary, the two-photon process results in a finite number of discrete species existing along a grade. It describes temporal change as involving discrete jumps, in which the competitive superiority of new forms plays a minor role. Old populations go extinct first; new forms replace them afterwards. Hence, by including the two-photon term, we resolve both parts of Darwin's dilemma at the same time. Our calculation does not, of course, address the issue of whether genetic changes[8] within a species are abrupt or gradual; that is beyond the scope of this discussion. The two-photon gain describes evolution as a series of first-order nonequilibrium phase transitions rather than the second-order processes associated with lasers having exponential gain. Our model describes the biota as being multistable. An abundant population is stable to invasion of a rare population. This is the case even if the latter would, at equal densities, be competitively superior. This stability is not simply a matter of fluctuations, since we have insured that the

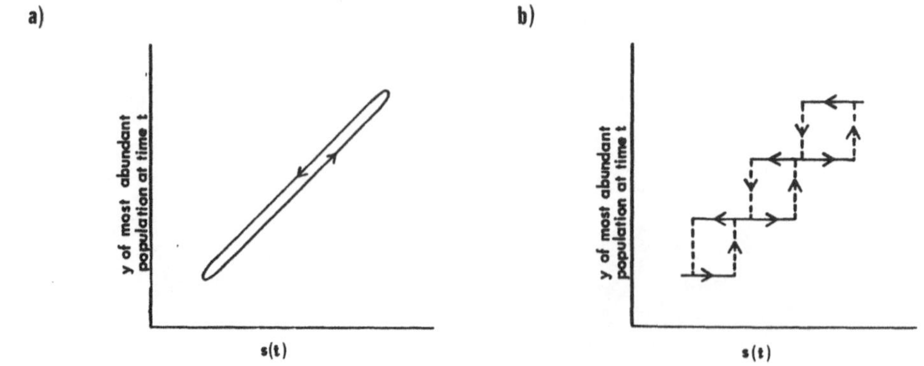

Fig. 8. Parametric plot of temporal change from the model in Appendix A in which a homogeneous resource $\rho(s) = \delta(s - (s_0 + vt))$ is scanned up at a velocity v and then down at a velocity -v across the resource axis of Fig. 7. Horizontal axis is $s = s_0 + vt$. Vertical axis is most abundant species. a) Exponential gain. b) Gain as in Fig. 6; dashed lines indicate jumps.

two-photon process alone is not, by itself, the reason why invasions fail. Every species is seeded such that it could invade an empty habitat. However, we find that the habitat must be empty for invasion to succeed. If already occupied, invasions normally fail. In contrast, if exponential gain is assumed, we find that invasions by competitively superior types always succeed, which we take to be a second-order nonequilibrium phase transition.

VI. Clonal populations

Clonal organisms have exponential gain and thus should have a continuous distribution in habitat space. The prediction that the clonal population encounters a slowing down as the continuum is reached means that the distribution of abundances in the population is largely a function of history. When we assume a Poisson statistic for favorable mutations in a clonal population, we find that the number of populations with N individuals is proportional to 1/N (see Appendix B). This clone distribution has a cutoff at some maximum N_{max} and is truncated at the minimum $N_{min}=1$. Unfortunately, clonal populations are not common and have not been studied in sufficient detail to evaluate the validity of this prediction.[26]

Parthenogenetic weevils in Scandinavia[27] (*Otiorrhynchus scaber*) are a well-studied case. In Figure 9 we show the Finnish distribution of two common clones that appear to be geographically discrete. In Figure 10 we examine the intermediate region where the two populations meet, and show all clonal populations that have been found so far. There one sees that there are many clonal populations living in the intermediate region, and many of these overlap the ranges of the more common clones in a haphazard fashion. This "clonal swarm" suggests the continuum found in the laser and the transitional forms in habitat that Darwin envisioned. There is no data to suggest that this clonal swarm is due to the effects we have described with our model. The results are merely consistent with our expectation that there should be no overall pattern of discreteness in occupancy of habitat.

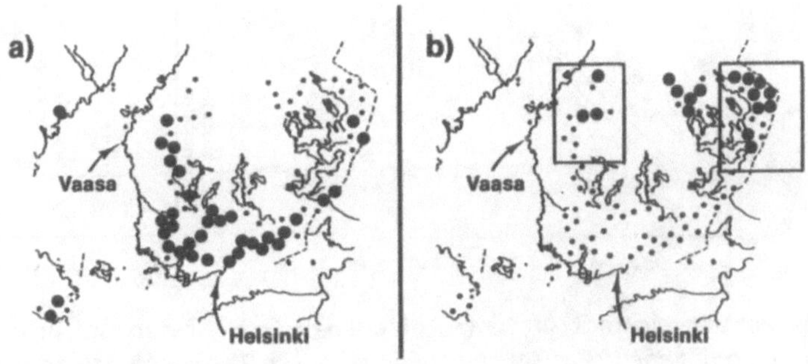

Fig. 9. Map of southern Finland summarizing part of census of clonal weevils.[27] Light dots, census locations. Heavy dots, locations where clone has been found. a) Clone 1 (labels 1 - 26 follow Ref. 27) also found in southern Sweden. b) Clone 16, which has also been found in Switzerland, northern Sweden, and Norway. Inserts in b) are areas shown in Fig. 10. Figures redrawn from originals in Ref. 27.

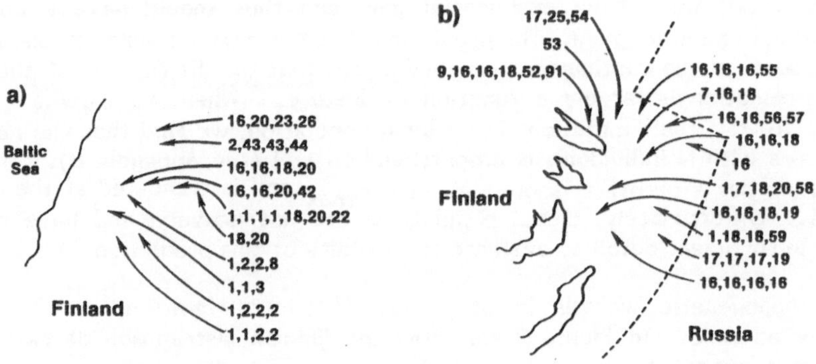

Fig. 10. Clones found in intermediate region between the two clones shown in Fig. 9. Locations shown in Fig. 9b. Labels higher than 27 are inventions of the authors. Data from Ref. 27.

VII. Clones vs. sexuals

Figure 7b illustrates a process of competitive exclusion insofar as some species are driven to extinction by competition. No such exclusion occurs in Fig. 7a. Our expectation based on this contrast is that competitive exclusion should play a much greater role in the ecology of sexual organisms than it does in clones. Comparisons of the ecology of clones and sexual organisms can test this prediction. While little data is available, most clone populations have existent sexual populations as likely ancestors, so such tests are possible and need not involve comparison of very different types of organisms. The only case we know of in which data addresses this issue involves the whiptail lizards (genus *Cnemidophorus*) in the southwestern United States and northern Mexico. In the neighborhood of Santa Fe, about a half dozen "species" of "virgin" whiptails live in the Rio Grande valley.

The predictions of our theory can take two forms. When more than one species of lizard coexist in an area, we predict that evidence for distinctness should be much more obvious within coexisting sexuals than within coexisting asexuals. Reference 28 examines such associations and finds this to be the case. Alternatively, if one examines associations without considering distinctness, one expects that there should be much less of a tendency within asexuals to show geographical exclusion than when sexuals are involved. We have examined data[29] on the Rio Grande associations, and find that the only species that shows evidence of systematic geographical exclusion is the sexual species that co-occurs with the clones. In contrast, the clonal associations are more in accordance with chance. Hence the data, meager as it is, is consistent with our expectations.

VIII. Difficulties with the theory

Our theory is not without its difficulties, and we wish to at least note what some of them are. Figure 7a shows a highly regular progression of species found along the resource axis. Figure 4 also shows a somewhat regular progression in the oaks. However, there is a serious question of whether these regular patterns are systematic. Real patterns might be random, from which examples showing regularity can be culled. Such issues are of intense current controversy.[30]

A second difficulty arising from the two photon correction in Fig. 5 is that it gives an advantage to asexuals. They have no decrease in gain. Why, then, do not asexuals always win? Why is sex maintained? The issue of the maintenance of sex[31] actually goes far beyond the two-photon correction which is, at most, a minor contribution to the overall disadvantage of sexual reproduction. Hence, the unsolved challenge of why sex persists is common to all evolutionary theories. Many attempted resolutions of the maintenance of sex involve arguments based on the relative rates of evolution in sexual and asexual organisms.[31] It is not clear that any of these arguments are sound. It would, however, be inconsistent for us to deny the role of the rate of evolution in the missing links dilemma and accept it as an explanation for the maintenance of sex.

A third objection questions whether the two photon term really occurs in nature. This is not so strong an objection since we predict that stable populations are sufficiently abundant to make the effect of the term very small in observed

populations. In addressing this issue, it is again useful to consider clonal populations and also populations in which biparental sex is optional. Here the data is generally consistent with the proposition that biparental reproduction is avoided when populations are rare,[25] and that asexuals are found in nature wherever biparental reproduction poses special difficulties. Hence the data is consistent with the proposition that the two photon term is real and can be competitively disadvantageous.

IX. Summary

We have made an analogy between the mechanisms of laser physics and of Darwinian evolution. In pursuing this analogy, we have rediscovered a dilemma of evolutionary theory which Darwin recognized without the help of mathematics or analogies. Darwin's evolutionary theory, which is based on the law of Malthus,[2] leads to the prediction that species should exist as an undifferentiated continuum, whereas they actually occur as discrete entities. By examining the analogy more closely, we find it needs to be modified to include explicitly the fact that most biological reproduction requires two parents. This demands that we abandon the law of Malthus, and modify the laser gain formula so that the gain decreases in the limit that the photon density is very low. This modification is roughly similar to putting a saturable absorber with a very narrow homogeneous linewidth into a laser and leads to a coalescence of the output into a few modes which are the analogy of existent discrete species. This model is in general accord with abrupt changes that are observed in the geological record, whereas Darwin's resolution of this problem is not. Just as the inclusion of a saturable absorber in a laser causes it to undergo first-order rather than second-order nonequilibrium phase transitions, we expect that most biological systems, those which are dominated by sexuals, should also undergo first-order transitions. Clones should obey second-order nonequilibrium phase transitions and should have much closer analogies with lasers.

Acknowledgement

We would like to thank S.B. Hopf for preparing the figures and for help with the manuscript.

References and footnotes

1. C. Darwin, *The Origin of Species by Means of Natural Selection* (Avenel Books, New York, 1979; facsimile of the first edition, 1859). To find our quotes in other editions look in the first two pages of Chapter VI.
2. Populations grow in time as a geometric series.
3. E. Schrodinger, *What is Life?* (Cambridge Press, Cambridge, 1944); P. Glansdorff and I. Prigogine, *Thermodynamic Theory of Structure, Stability, and Fluctuations* (J. Wiley and Sons, London, 1971); and H. Haken, *Synergetics* (Springer-Verlag, Berlin, 1977) have tried to develop such connections in the past. The work in Ref. (4) is much more relevant to the issues discussed here. H. Bernstein, H.C. Byerly, F.A. Hopf, R.E. Michod, and G.K. Vemulapalli, *Quart. Review of Biol.*, **58** (1983) have reviewed some of this material in a biological context.
4. M. Eigen, *Naturwissenschaften* **58** (1971); M. Eigen and P. Schuster, *The*

Hypercycle, a Principle of Natural Self-Organization (Springer-Verlag, Berlin, 1979). The latter contains a demonstration of the qualitative difference in the nature of asymptotic states in evolution that occurs when the two-photon correction is included (qualitative in the sense that it makes no difference how small the correction is).
5. M.G. Velarde, in *Evolution of Order and Chaos in Physics, Chemistry and Biology*, H. Haken editor (Springer-Verlag, Berlin, 1982).
6. The term fluctuations is potentially confusing in context. In ecology, the term usually refers to short-term environment change. In the conference the term was applied to mutations and in this usage, fluctuations in laser physics and biology are undoubtedly different. Dr. Landauer made the interesting suggestion that such fluctuations in biology be referred to as nucleations, implying a microscopic but unlikely event that is needed to trigger the instability. We have followed this suggestion, often using the synonym 'seeding'. Using this terminology, Darwin viewed evolutionary change as series of second-order nonequilibrium phase transitions whose character and rate of occurrence are dominated by nucleations. Our model predicts stablilty to such nucleations. Evolution is propelled by macroscopic events.
7. A. Yariv, *Quantum Electronics* (J. Wylie and Sons, New York, 1967) discusses simple forms of laser models.
8. J. Roughgarden, *Theory of Population Genetics and Evolutionary Ecology: an Introduction* (Macmillan, New York, 1979) gives a comprehensive overview of mathematical models in biology. A proof of the stability of the continuum solution that is more general than the one in Appendix A can be found here. The book contains a useful overview of genetics.
9. R.H. MacArthur, *Geographical Ecology: Patterns in the Distribution of Species* (Harper and Row, New York, 1972).
10. W.E. Lamb Jr., *Phys. Rev.* A1429, (1964).
11. F.A. Hopf and F.W. Hopf, *Theor. Pop. Biol.* (in press) contains more details of the model. This work and references therein refute a popular claim that short term environmental fluctuations alone can produce discrete distributions.
12. M.M. Litvak, *Phys. Rev.*, **A2**: 2107 (1970).
13. J.H. Parks, in *Fundamental and Applied Laser Physics*, M.S. Feld, A. Javan and N.A. Kurnit eds. (John Wylie and Sons, New York, 1972).
14. F.A. Hopf (unpublished, manuscript available on request).
15. L.N. Menegozzi and W.E. Lamb Jr., *Phys. Rev. A*, **17** (1978).
16. J.H. Parks, D.R. Rao and A. Javan, *Appl. Phys. Lett.*, **13** ,142 (1968). D. Schwamb and S. Smith, Phys. Rev. A, **21** 896 (1980).
17. R.H. Whittaker, in *Diversity and Stability in Ecological Systems*, Brookhaven Symposium in Biology **22**, Springfield, Va., National Bureau of standards, U.S. Department of Commerce, 178 (1969).
18. E. Mayr, *Animal Species and Evolution* (Harvard University Press, Cambridge, 1963) emphasizes the role of reproductive isolation in evolution. The entities labeled y in Eq. (2) refer to local populations that are reproductively isolated from other local populations. No population may contain distinguishable reproductively-isolated sub-populations. Most, but not all, biological communities can be divided this way. The oak example used in Fig. 4 is poor in this regard since they form hybrids with each other.
19. V. Grant, *Plant Speciation* (Columbia University Press, New York, 1971)

discusses botanical populations that are very difficult to sort into species. Most, if not all, involve substantial degrees of uniparental reproduction.
20. R.M. May, *Monographs in Population Biology*, **6**, (Princeton University Press, Princeton, N.J., 1973) introduces the idea that environmental fluctuations alone can cause discrete distributions (see, however, Ref. 11). This and the next reference contain data on statistically stationary features of the biota.
21. J.H. Brown and A.C. Gibson, *Biogeography* (The C.V. Mosby Company, Toronto, 1983) also has data on regular morphological progressions.
22. G.G. Simpson, *Tempo and Mode in Evolution.* (Columbia University Press, New York, 1944).
23. S.M. Stanley, *Macroevolution* (W.H. Freeman and Company, San Francisco, 1979).
24. M.J. Asmussen, *American Naturalist*, **114**, 796 (1979). This paper and the next discusses factors other than biparental reproduction that can modify the gain as in Fig. 6. Nearly all these factors are ultimately related to sexual reproduction.
25. H. Bernstein, H.C. Byerly, F.A. Hopf, and R.E. Michod, *Evolution*, (in press) is a non-mathematical companion paper to Ref. 11. It discusses data relevant to the biparental correction term.
26. G. Bell, *The Masterpiece of Nature: The Evolution and Genetics of Sexuality.* (University of California Press, Berkeley, 1982), and E.D. Parker Jr., *Amer. Zool.* **19**, 753 (1979) contain extensive reviews on clonal data. We are limited in our model[18] to describing asexual reproduction in which the daughters are, except for mutations, faithful replicas of the mother. Microorganisms sometimes reproduce sexually and are not suitable for testing predictions about clones.
27. A. Saura, J. Lokki, P. Lankinen, and E. Suomalainen, *Hereditas* **82**, 79 (1976) is the most extensive study of a group of local clones. About 400 individual weevils were studied out of a set of clonal populations that may include more than 10^{12} individuals.
28. T.J. Case, in *Lizard Ecology*, R.B. Huey, E.R. Pinaka and T.W.Schoener eds. (Harvard University Press, Cambridge, Mass, 1983) attributes the difference in degrees of morphological distinction to possible recent origin of the clones.
29. O. Cuellar, *Amer. Zool.* **19**, 773 (1979).
30. D. Simberloff and W. Boecklen, *Evolution*, **35**, 1206 (1981). R. Lewin, *Science*, **221** 636 (1983). F.A. Hopf and J.H. Brown, *Ecology*, in press. The fact that we have ignored parasites, predators etc. in our model is a serious defect. Including these factors can cause irregular spacing along the resource axis.
31. G.C. Williams, *Sex and Evolution* (Princeton University Press, Princeton, N.J., 1975), and J. Maynard Smith, *The Evolution of Sex* (Cambridge University Press, Cambridge, 1978) discuss the theoretical issues. Bell[26] has shown that some of the rate of evolution arguments are in conflict with data.
32. F.W. Preston, *Ecology* **43**, 185 (1972).

Appendix A

In this appendix we develop our model for dealing with inhomogeneous broadening. This model is developed in Refs. 8, 9, 11 and 20 in the context of ecology. Reference 12 develops the same model for lasers. The effects of terms left out of the laser model are found in Refs. 13-15. We use the biological version in the text since it deals with finite populations rather than densities and we prefer to introduce the correction for biparental reproduction in this form. The population density N(y) is related to the populations by

$$N(y) = \sum_{y'} N_{y'} \, \delta(y-y') \tag{A1}$$

The resource density is $\rho(s)$ and takes on the value $\rho_0(s)$ when it is not being used. The homogeneous line, which in biological terms refers to the capability of species y to utilize resource s, is denoted f(x). This is taken to be the same for all species and is normalized to unity.

The equation describing the change in N_y reads

$$\frac{dN_y}{dt} = b(y)\, R_y\, N_y - d_y\, N_y, \tag{A2}$$

where

$$R_y = \int_{-\infty}^{\infty} ds\, \rho(s) f(s-y). \tag{A3}$$

R_y denotes the total amount of resources available at any given time for the population of species y to grow. The effects of biparental reproduction are contained in $b_y(N_y)$. This function is assumed to be a continuous and monotonically increasing function of N_y and becomes constant in the limit that N_y becomes very large.

The density regulation takes place through the diminution of the resources by the populations. The saturation is given by

$$\rho(s) = \rho_0(s) - \sum_{y} f(s-y) N_y. \tag{A4}$$

If $b_y(N_y)$ is constant, f(x) is Lorentzian and $\rho_0(s)$ is Gaussian, Eqs. (A1)-(A4) describe a laser in which only the direct saturation terms influence the output.[12] The other terms[13-15] prohibit writing the equations in terms of fluxes and have no counterpart in biology.

Let us now analyze a special case of this model to look for stable final states. Take all d's and b's to be independent of y. Take $\rho_0(s) = 1$ to be independent of s and let the species be uniformly spaced along the resources with a spacing Y. With these assumptions, there is a tractable steady-state solution in which all species have the same density, denoted $\bar{N}$, which is given implicitly by

$$b(\bar{N})(1-\bar{N}P) - d = 0, \tag{A5}$$

where

$$P = \sum_{m=\infty,-\infty} f(mY). \tag{A6}$$

We now expand about the steady state to test for stability. We write

$$N_y = \bar{N} + n_y, \tag{A7}$$

where $n_y << \bar{N}$ is a perturbation. The expansion formally reads

$$\frac{dn_y}{dt} = D\, n_y\, (1-P\bar{N}) + b(\bar{N}) \sum_{y'} n_{y'} \frac{\partial R}{\partial n_{y'}}\Big|_{n=0}, \tag{A8}$$

where D is

$$D = \frac{db(x)}{dx}\Big|_{x=\bar{N}}. \tag{A9}$$

The second term in Eq. (A8) is the complicated one and is treated in detail in Ref. 20. The eigenvalues of this matrix are always negative definite. If b is constant, i.e., the first term in Eq. (A8) is zero, then the solution is stable no matter how small Y is. However, when Y becomes small compared to the width of f(x), the largest eigenvalue rapidly approaches zero, hence the slowing down.

The first term in Eq. (A8) describes the effect of biparental reproduction and is just a constant times n_y. Since this term is a constant times the unit matrix, it is not altered when the second term is diagonalized. The first term shifts the eigenvalues by a positive term $bD(1-P\bar{N})$. Reasonable b(x) have larger slopes for small x. Hence as Y decreases $\bar{N}$ decreases, D increases and the shift increases. The character of the two terms in Eq. (A7) guarantees that, for some finite Y, the largest eigenvalue is positive and the continuum solution becomes unstable.

Appendix B

In this appendix, we consider in greater detail the qualitative character expected of a community subject to the effects of slowing down. Let M denote the total number of populations (labeled y) in the community and let MP(N) be the number of populations with N individuals. Ecologists[32] use the series h_m where $h_1 = MP(1)$ and

$$h_{m+2} = M \sum_{k=2^m+1}^{2^{(m+1)}} P(k), \quad m = 0,1,2,\ldots, \tag{B1}$$

to plot census data of communities.

Slowing down occurs because all populations have the same relative growth rates independent of population size. Suppose there is some ancestral population from which a series of new improved populations arise, all with the same growth rate. Let T denote the interval of time over which these new populations double in size and let $t = 0$ denote the present time; then h_m counts the number of populations that originated in the interval $-(m-1)T > t > -mT$. A Poisson nucleation[6] process means that h_m is, on the average, independent of m and has no correlations. If one converts Eq. (B1) to an integral and takes P(N) to go as N^{-1},

then the resulting h's are independent of m, proving the result stated in Sec. VI. While this argument does not apply directly to our model, the result is in good agreement with our numerical calculations when we take the initiation process to be Poisson. In general, the expectation under slowing down is that individual distributions of h_m's should be haphazard, reflecting random occurrences in the past. Instead, observed communities tend to have log-normal distributions,[32] i.e., h_m goes as $\exp(-\alpha(m-m_0)^2)$ where α varies little[32] from one community to the next and m_0 is a constant measuring the census size. Observed communities have systematic structures, not the haphazard ones expected from slowing down. Hence, slowing down is an unlikely explanation for why biotic forms seem to be statistically stationary.

SELF-CONSISTENT MEAN FIELD THEORIES

A. Elçi

Institute for Modern Optics
Department of Physics and Astronomy
University of New Mexico
Albuquerque, NM 87131 USA

There is a variety of mean field theories which are also self-consistent, according to differing criteria. In the following discussion I will use the phrase "self-consistent mean field" in a narrowly defined sense. To define this sense, let us consider two systems A and B (Fig. 1) which are coupled. Each consists of many constituents that interact among themselves. The combined A+B system is isolated. In general, one can write the density matrix operator of the combined system in the form

$$\rho_{A+B}(t) = \rho_A(t) \otimes \rho_B(t) + \rho_{CORR}(t) \tag{1a}$$

where

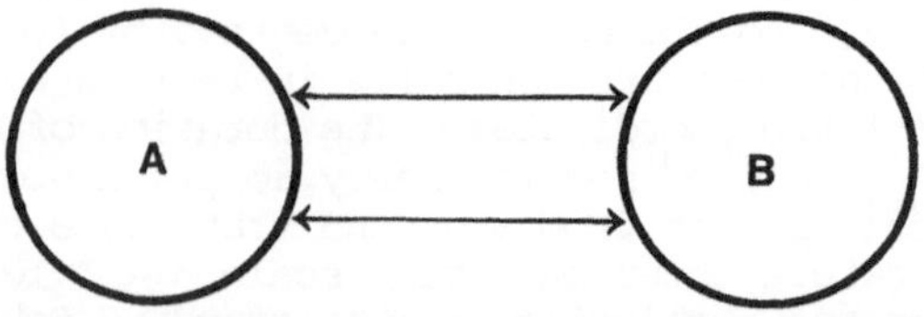

Fig. 1. The combined A + B system is isolated.

$$\rho_A(t) = \mathrm{Tr}_B\, \rho_{A+B}(t) \tag{1b}$$

$$\rho_B(t) = \mathrm{Tr}_A\, \rho_{A+B}(t) \tag{1c}$$

and ρ_{CORR} represents the off-diagonal correlation terms between the A and B system variables induced by the interaction of A and B. The approximation that neglects these off-diagonal terms, that is,

$$\rho_{A+B}(t) \cong \rho_A(t) \otimes \rho_B(t) \tag{2}$$

is called the self-consistent mean field approximation [1]. The following discussion is concerned with some of the basic features of the theories that use this approximation.

A supreme example of a self-consistent mean field theory in the sense defined above is the set of Maxwell-Bloch equations describing the coupling of N identical two-level atoms to a coherent radiation field [2]:

$$(c^2\nabla^2 - \partial_t^2)\vec{E} = 4\pi\partial_t^2\vec{P} \ , \tag{3a}$$

$$\hbar\partial_t\vec{P} = i\hbar(\omega - \omega_0)\vec{P} - 2i\vec{\mu}(\vec{\mu} \cdot \vec{E})W \ , \tag{3b}$$

$$|\vec{\mu}|^2\partial_t W = -2i\vec{\mu} \cdot \vec{E}[\vec{\mu} \cdot \vec{P} - (\vec{\mu} \cdot \vec{P})^*] \ . \tag{3c}$$

Here $\vec{E}$ and $\vec{P}$ are the complex electric and polarization fields, ω is the frequency of the field $\vec{E}$, ω_0 is the level separation of the atoms, $\vec{\mu}$ is their atomic dipole moment and W is the population inversion (i.e., $W = N(\rho_{11} - \rho_{00})$, where ρ_{11} is the upper state probability and ρ_{00} is the lower state probability). The Maxwell-Bloch equations, either in the form (3a-c) or in some other guise [3,4], have had significant impact on quantum optics.

A second example comes from nuclear physics and is related to the deep inelastic collisions of heavy ions [5] (Fig. 2). These experiments were undertaken with the hope of obtaining superheavy nuclei. That goal, however, has so far been elusive; instead, the experiments have produced some surprising results. The total degrees of freedom, that is the generalized coordinates, of the heavy ions involved in the scattering can be divided into two sets: collective (e.g., the c.o.m. positions and momenta of the outgoing fragments), and intrinsic (e.g., the coordinates of an individual nucleon). The experimental results, such as the nature and distribution of outgoing fragments, are consistent with having the intrinsic degrees of freedom in thermal equilibrium at a temperature $k_BT \sim 1$ MeV, in contrast to the collective degrees of freedom. In these experiments the incoming ions are generally in their ground states with respect to their intrinsic coordinates. The duration of a collision event is typically $\tau_{reaction} \sim 10^{-21}$ sec for heavy ions. Energy exchanges that bring the internal coordinates from a ground state to a thermal equilibrium at a high temperature occur on a time scale $\tau_{EX} \sim 10^{-22}$ sec. The rapidity of τ_{EX} makes the application of the standard Fokker-Planck approach to the problem inappropriate because the standard Fokker-Planck theory is based on small second order time derivatives in the phase space.

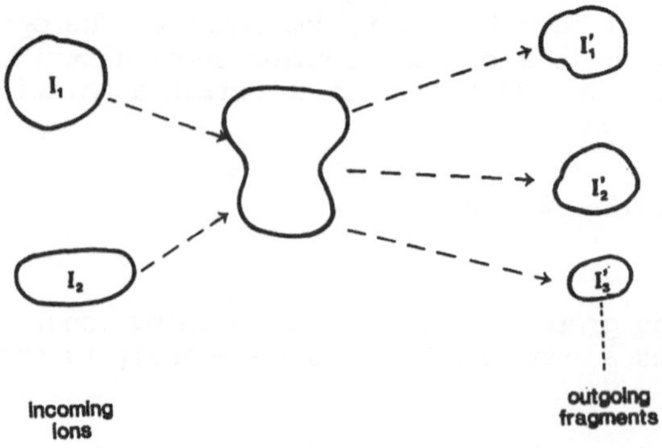

Fig. 2. Heavy ion collisions.

On the other hand, a self-consistent mean field approach does provide a unified treatment of the scattering process, both in the initial approach phase, when heavy ions form a scattering complex, and in the final break-up phase, when the intrinsic coordinates are in thermal equilibrium and favor particular channels of fragmentation [6]. Mukamel, et. al, transform the collective coordinates to a Wigner phase space representation which is classical, and treat the intrinsic coordinates quantum mechanically. The complete density matrix D(R,P,t) obeys an equation of motion of the form

$$\frac{d}{dt} D(R,P,t) = -\frac{i}{\hbar}[H(R,P), D(R,P,t)] - \frac{P}{M}\frac{\partial}{\partial R} D(R,P,t)$$

$$+ \frac{\partial}{\partial P}\left[\frac{\partial H}{\partial R}, D(R,P,t)\right]. \tag{4a}$$

R and P are symbolic and stand for the collective coordinates which are classical following a Wigner transform. D is factorized as in Eq. (2):

$$D(R,P,t) = \rho(t)\sigma(R,P,t) \tag{4b}$$

where

$$\rho(t) = \frac{1}{(2\pi\hbar)}\int dR\, dP\, D(R,P,t), \tag{4c}$$

$$\sigma(R,P,t) = \mathrm{Tr}\, D(R,P,t). \tag{4d}$$

ρ is a quantum mechanical density matrix dependent on the operators of the intrinsic coordinates. σ is a classical phase space distribution of the collective coordinates. The goal is then to obtain a solution of Eq. (4a) subject to the initial condition

$$D \underset{t\to-\infty}{\to} \delta\left(R - \frac{P}{M} t\right)\delta(P - P_0)|0\rangle\langle 0| , \tag{4e}$$

where $|0\rangle$ refers to the ground states of the incoming ions. To reach that goal, however, requires a considerably elaborate analysis and further approximations.

A third example comes from astrophysics and concerns the problem of stellar dynamics. This problem consists of the determination of the evolution of N stars (N is a large number) starting from an arbitrary initial configuration in the phase space [7]. The stars interact via long range, two-body gravitational forces. A Fokker-Planck approach to the problem is inadequate, because the long range of the gravitational force causes divergences of the Fokker-Planck coefficients. To avoid such unphysical results, Chandrashekhar and von Neumann [8,9] developed stochastic equations of motion for a given star.

A correct solution of the stellar dynamics problem should reflect the following physical considerations. Consider a given star (a probe star) in the collection (Fig. 3). One intuitively expects that the motion of the probe star is strongly influenced only by the stars

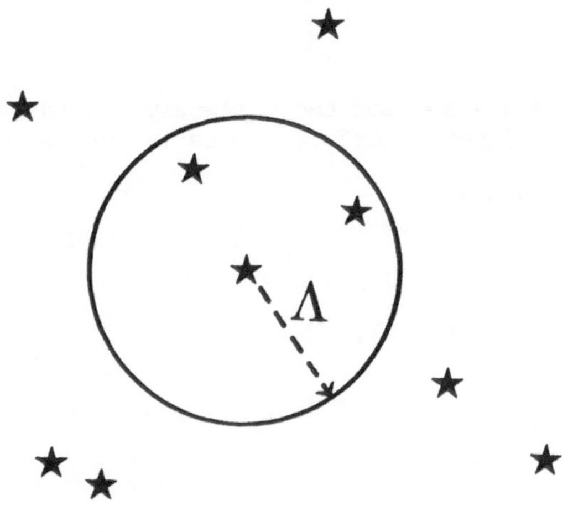

Fig. 3. Stellar Dynamics.

in its immediate neighborhood. One can visualize an imaginary sphere of radius Λ centered on the probe star, such that the stars that fall into this sphere (the near stars) can exert rapidly varying forces on the probe star and cause sometimes violent changes in its motion. On the other hand, those stars that are outside of this sphere (the far stars) need not affect the probe star at all; they may, however, exert sufficient influence on each other, as well as on the near stars, to maintain an overall configuration. In other words, they contribute to an average, or mean, gravitational field. The motion of the near stars causes fluctuations from this mean field experienced by the probe star.

A second point is that the phase space of this N-star system should be somewhat "lumpy". In other words, given an initial phase space configuration, the stars should not just diffuse away, eventually filling the ordinary space everywhere with equal probability. It is easy to visualize circumstances in which some subset of the stars might come together to form bound and stable configurations, that is, stellar clusters. Mathematically this means that the overall system should be unstable with respect to the formation of certain long-range correlations.

Such desired physical features as described above come out of a self-consistent mean field formulation of the stellar dynamics problem in a rather natural and direct manner [10]. One again factorizes the classical phase distribution function of the N-stars,

$$F_N = f_p \otimes G_{N-1} , \tag{5a}$$

where f_p is the phase space distribution of the probe star and G_{N-1} is the distribution of the remaining stars. f_p and G_{N-1} obey the Louiville equations,

$$\partial_t f_p + iL_p f_p + i\langle L'\rangle_{N-1} f_p = 0 \tag{5b}$$

$$\partial_t G_{N-1} + iL_{N-1} G_{N-1} - i\langle L'\rangle_p G_{N-1} = 0 \tag{5c}$$

where L's are the Louivillians defined in terms of the relevant Poisson brackets,

$$- iL_p = - \vec{v}_p \cdot \frac{\partial}{\partial \vec{r}_p} \tag{5d}$$

$$- iL_{N-1} = - \sum_{j=1}^{N-1} v_j \cdot \frac{\partial}{\partial \vec{r}_j} + \sum_{j\neq \ell} \frac{\partial W(|\vec{r}_j - \vec{r}_\ell|)}{\partial \vec{r}_j} \cdot \frac{\partial}{\partial \vec{v}_\ell} \tag{5e}$$

$$- iL' = \sum_{j=1}^{N-1} \frac{W(|r_p - r_j|)}{\partial \vec{r}_p} \cdot \frac{\partial}{\partial \vec{v}_p} + \frac{\partial W(|\vec{r}_j - \vec{r}_p|)}{\partial \vec{r}_j} \cdot \frac{\partial}{\partial \vec{v}_j} \tag{5f}$$

W is the two-body gravitational potential; the brackets <> indicate the appropriate averages in the phase space.

One result of the self-consistent mean field approach is the screening of the two-body gravitational force, that is, the replacement of the Newtonian force by a screened force (as far as the individual stars are concerned)

$$\frac{1}{R^2} \to \frac{1}{R^2(1 + R^2/\Lambda^2)} , \tag{6}$$

where Λ is on the order of the average interstellar distance. This is quite similar to the screening of charges in an ideal electron gas [11] and gets rid of the divergences arising from the long range of the Newtonian gravitational force. Another result is the introduction of "lumpiness" into the phase space. This is a general feature of self-consistent mean field theories. The self-consistency requirements introduce extra memory effects into the evolution of the subsystems in such a way as to restrict the class of trajectories in the phase space which are admissible. This feature will be clearer in the discussion of the next example, which is more detailed.

The final example concerns the spectrum and the optical interactions of atoms injected into an interacting medium [12,13]. This problem is of interest in condensed matter physics. It is also a useful vehicle in illustrating some of the basic properties of the self-consistent mean field theories mentioned earlier.

Let us consider the following idealized physical picture (Fig. 4). An atom enters, at time t_0, a medium which consists of neutral atoms that are strongly interacting among themselves. The entering atom is assumed

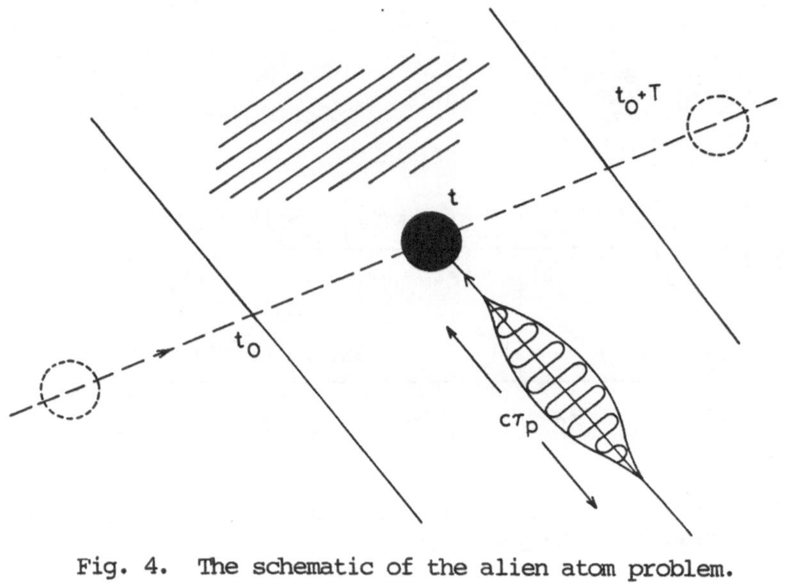

Fig. 4. The schematic of the alien atom problem.

to be distinguishable from the medium atoms. It will be referred to as the alien atom. In fact, it will be assumed that it is a member of an "ensemble" of alien atoms, although the other members are not shown in Fig. 4. The alien atom leaves the medium at time $t_0 + T$. While inside the medium, it interacts with an optical pulse. The duration of the pulse is much smaller than T. The optical pulse does not couple to the medium atoms, at least not directly. In the following we will discuss the implications of the self-consistent mean field approximation for the evolution of this three-part system in $t_0 \ll t \ll t_0 + T$, in particular, its implications for the "dressed" spectrum of the alien atom and how the "dressed" alien atom couples to the optical pulse.

The Hamiltonian for the problem [12] can be written as ($\hbar = 1$)

$$H = \sum_\eta E_\eta |\eta\rangle\langle\eta| + \sum_{\alpha\vec{K}} E^0_{\alpha\vec{K}} |\alpha\vec{K}\rangle\langle\alpha\vec{K}|$$

$$+ \sum_{\eta\eta'\alpha\alpha'\vec{K}\vec{q}} U_{\alpha\alpha'}(\vec{q})\sigma_{\eta\eta'}(\vec{q})\, |\eta\rangle\langle\eta'| \otimes |\alpha\vec{K}\rangle\langle\alpha'\vec{K}-\vec{q}|$$

$$- \sum_{\alpha\alpha'\vec{K}} A_0 \hat{e}_0 \cdot \vec{\mu}_{\alpha\alpha'} \cos\omega t\, |\alpha\vec{K}\rangle\langle\alpha'\vec{K}| \; . \tag{7a}$$

Here the sets $\{|\eta\rangle\}$ and $\{E_\eta\}$ represent the exact N-particle eigenstates and eigenenergies of the medium in the absence of the alien atom. The sets $\{|\alpha\vec{K}\rangle\}$ and $\{E^0_{\alpha\vec{K}}\}$ represent the eigenstates and eigenenergies of the isolated alien atom in vacuum. α stands for the internal quantum numbers of the alien atom; $\vec{K}$ is the momentum of its c.o.m. Here we are treating the alien atom as an elementary unit, i.e., viewing the coordinates of its electrons as internal degrees of freedom analogous to spin. $E^0_{\alpha\vec{K}}$ is given by

$$E^0_{\alpha\vec{K}} = E^0_\alpha + (2M_A)^{-1}\, \vec{K}^2 \tag{7b}$$

$U_{\alpha\alpha'}$ is the Fourier transform of a binary potential V_{AM} between the alien atom and a medium atom:

$$U_{\alpha\alpha'}(\vec{q}) = V_{o1}^{-1}\int d\vec{R}\, \exp(-i\vec{q}\cdot\vec{R})\langle\alpha|V_{AM}(\vec{R};\, S)|\alpha'\rangle \; . \tag{7c}$$

$\vec{R}$ represents the c.o.m. coordinate of the alien atom; S, the set of its internal coordinates. $\sigma_{\eta\eta'}$ is the matrix element of the density fluctuation operator of the medium:

$$\sigma_{\eta\eta'}(\vec{q}) = \langle\eta|\sum_{i=1}^{N} \exp(-i\vec{q}\cdot\vec{r}_i)|\eta'\rangle \; . \tag{7d}$$

We have assumed that the optical field is linearly polarized along $\hat{e}_0$, is coherent, and has frequency ω. A_0 represents its slowly varying envelope. $\vec{\mu}$ is the dipole moment operator of the alien atom. We neglect the recoil from absorption and emission of photons.

Note that the coupling between the alien atom and the medium as described by Eq. (7a) is such that the alien atom induces only longitudinal density fluctuations in the medium. This is the consequence

of the description of the interaction between the alien atom and the point-like medium atoms by a scalar potential. When a momentum exchange takes place between the medium and the alien atom, momentum is given to or taken from the c.o.m. of the alien atom. $\sigma(\vec{q})$ destroys a density fluctuation of momentum q in the medium; $\sigma(-\vec{q}) = \sigma(\vec{q})^{\dagger}$ creates it. Note also that the c.o.m. and internal motions of the alien atom are coupled because of the presence of the medium.

These interesting features of Eq. (7a) will not be explored in the following. Rather, in order to obtain analytic results, we will assume that the c.o.m. motion of the alien atom remains unchanged (or is negligibly changed). Only the $\vec{q} = 0$ components of $U_{\alpha\alpha'}$ and $\sigma_{\eta\eta'}$ survive under this assumption. Any energy exchange that can take place between the alien atom and the medium becomes similar to a heat exchange between two thermodynamic bodies.

Let us ignore the optical field for the moment and consider only the alien atom-medium coupling. As in Eq. (2), we make the self-consistent mean field approximation and factorize the complete density matrix: $\rho(t) = \rho_A(t) \otimes \rho_M(t)$. Next we calculate the linear response of the medium to an arbitrary $\rho_A(t)$. The substitution of this linear response into the equation of motion for $\rho_A(t)$ yields [12]

$$(i\partial_t - E^0_{\alpha\vec{K}} + E^0_{\alpha'\vec{K}'})\rho^{(A)}_{\alpha\vec{K},\alpha'\vec{K}'}(t) = \sum_{\alpha_1\alpha_2\alpha_3\vec{K}_1} \int_{-\infty}^{+\infty} dt' \, \chi_{MT}(t - t')$$

$$\times U_{\alpha_1\alpha_2} \rho^{(A)}_{\alpha_2\vec{K}_1,\alpha_1\vec{K}_1}(t') \left[U_{\alpha\alpha_3} \rho^{(A)}_{\alpha_3\vec{K},\alpha'\vec{K}'}(t) - U_{\alpha_3\alpha'} \rho^{(A)}_{\alpha\vec{K},\alpha_3\vec{K}'}(t) \right], \quad (8a)$$

where χ_{MT} is the linear response function of the medium in the time domain,

$$\chi_{MT}(t) = -i\theta(t)\mathcal{Z}^{-1}\sum_{\eta\eta'}|\sigma_{\eta'\eta}|^2[\exp(-i\Omega_{\eta\eta'}t) - \exp(i\Omega_{\eta\eta'}t)] \exp(-E_{\eta'}/k_BT_M) , \quad (8b)$$

$$\Omega_{\eta\eta'} = E_\eta - E_{\eta'} , \quad (8c)$$

$$\mathcal{Z} = \sum_\eta \exp(-E_\eta/k_BT_M) \quad (8d)$$

and $\theta(t)$ is the step function. The Fourier transform of χ_{MT} is the usual linear response function in the frequency domain,

$$\chi_M(\omega) = \mathcal{Z}^{-1}\sum_{\eta\eta'}|\sigma_{\eta'\eta}|^2[(\omega - \Omega_{\eta\eta'} + i\delta)^{-1} - (\omega + \Omega_{\eta\eta'} + i\delta)^{-1}] \exp(-E_{\eta'}/k_BT_M) , \quad (8e)$$

which will appear in the subsequent formulae. Equation (8a) is a generalization of the nonlinear Schrodinger equation which appears in the radiation reaction and spontaneous emission problems [14-17]. If the ensemble of the alien atoms is prepared in a pure state, Eq. (8a) reduces to a nonlinear Schrodinger equation. Clearly, the atomic system has become self-interacting through the reaction of the medium.

In order to determine the dressed spectrum of the alien atom, one can use the following fact. The elements of an arbitrary density matrix which is evolving freely are of the form

$$\rho_{jk} = (\text{constant})_{jk} \exp[-i(E_j - E_k)t] . \tag{9}$$

In other words, the off-diagonal elements oscillate exponentially with the corresponding transition frequencies, while the diagonal elements are constant. We therefore make the following ansatz:

$$\rho^{(A)}_{\alpha\vec{K},\alpha'\vec{K}'}(t) = \exp[-i(E_{\alpha\vec{K}} - E_{\alpha'\vec{K}'})t]$$

$$\times \{\exp[i(\theta_{\alpha\vec{K}} - \theta_{\alpha'\vec{K}'})] \cdot r_{\alpha\vec{K},\alpha'\vec{K}'}\} \tag{10}$$

Here the $E_{\alpha\vec{K}}$'s are the dressed eigenenergies of the alien atom in the medium. $\theta_{\alpha\vec{K}}$ and $r_{\alpha\vec{K},\alpha'\vec{K}'}$ are real. The first factor takes care of the exponential oscillations of $\rho^{(A)}_{\alpha\vec{K},\alpha'\vec{K}'}$. The terms in {} are restricted to only exponential decay and growth, or to variation in powers of t, i.e., they are postulated to be non-oscillatory. With this ansatz, Eq. (8a) is transformed into two sets of equations [12]. The first set describes the time evolution of $r_{\alpha\vec{K},\alpha'\vec{K}'}$:

$$\partial_t r_{\alpha\vec{K},\alpha'\vec{K}'} = -\Gamma_{\alpha\vec{K},\alpha'\vec{K}'} r_{\alpha\vec{K},\alpha'\vec{K}'} , \tag{11a}$$

where

$$\Gamma_{\alpha\vec{K},\alpha'\vec{K}'} = -\sum_{\alpha''}(1 - \delta_{\alpha\alpha''})|U_{\alpha\alpha''}|^2 r_{\alpha\vec{K},\alpha''\vec{K}} r_{\alpha''\vec{K},\alpha'\vec{K}}$$

$$\times (r_{\alpha\vec{K},\alpha'\vec{K}'})^{-1} \mathcal{Im}\, \chi_M(E_{\alpha\vec{K}} - E_{\alpha'',\vec{K}})$$

$$+ \sum_{\alpha''}(1 - \delta_{\alpha'\alpha''})|U_{\alpha''\alpha'}|^2 r_{\alpha\vec{K},\alpha''\vec{K}'} r_{\alpha''\vec{K}',\alpha'\vec{K}'}$$

$$\times (r_{\alpha\vec{K},\alpha'\vec{K}'})^{-1} \mathcal{Im}\, \chi_M(E_{\alpha''\vec{K}'} - E_{\alpha'\vec{K}'}) . \tag{11b}$$

The second set determines the transition frequencies of the alien atom inside the medium in terms of the asymptotic values of $r_{\alpha\vec{K},\alpha'\vec{K}'}$:

$$E_{\alpha\vec{K}} - E_{\alpha'\vec{K}'} = E_{\alpha\vec{K}} - E_{\alpha'\vec{K}'} + \chi_M(0)(U_{\alpha\alpha} - U_{\alpha'\alpha'})\sum_{\alpha''\vec{K}''} U_{\alpha''\alpha''} r_{\alpha''\vec{K}'',\alpha''\vec{K}''}(\infty)$$

$$+ \sum_{\alpha''}(1 - \delta_{\alpha\alpha''})|U_{\alpha\alpha''}|^2 r_{\alpha K,\alpha''K}(\infty) r_{\alpha''K,\alpha'K'}(\infty)$$

$$\times [r_{\alpha\vec{K},\alpha'\vec{K}'}(\infty)]^{-1} \mathrm{Re}\, \chi_M(E_{\alpha\vec{K}} - E_{\alpha\vec{K}''})$$

$$- \sum_{\alpha''}(1 - \delta_{\alpha'\alpha''})|U_{\alpha''\alpha'}|^2 r_{\alpha\vec{K},\alpha''\vec{K}'}(\infty) r_{\alpha''\vec{K}',\alpha'\vec{K}'}(\infty)$$

$$\times [r_{\alpha\vec{K},\alpha'\vec{K}'}(\infty)]^{-1} \mathrm{Re}\, \chi_M(E_{\alpha''\vec{K}'} - E_{\alpha'\vec{K}'}) \quad . \tag{11c}$$

The spectral equation (11c) is to be solved self-consistently with Eq. (11a). Note that the decay times $\{\Gamma_{\alpha\vec{K},\alpha'\vec{K}'}\}$ appear in the equations for the amplitudes of the density matrix elements. For the diagonal elements, Γ's are $(T_1)^{-1}$-type; for the off-diagonal elements, $(T_2)^{-1}$-type. Γ's depend on the imaginary part of the medium response function. $\mathrm{Im}\,\chi_M(\omega)$ is proportional to the density of states of the medium at energy ω. Hence it is a measure of the probability of the real transitions that can occur in the medium. The real transitions determine the asymptotic values of the amplitudes, which affect the spectral equation. The spectral equation also depends on $\mathrm{Re}\chi_M(\omega)$, the dispersive part of the medium response function, which plays a direct role in the formation of the dressed spectrum.

The spectral equation also depends on $\chi_M(0)$, which is related to the isothermal compressibility of the medium, κ_{iso}:

$$\chi_M(0) = -V_{01}^{-1} N^2 \kappa_{iso} \quad . \tag{12}$$

The appearance of $\chi_M(0)$ in Eq. (11c) has an intuitive appeal. The alien atom when it is inside the medium pulls or pushes the medium atoms, altering their average density in its vicinity. How much the alien atom can alter the medium density depends on the strength of its interaction with medium atoms and, further, on the strength of the interaction among the medium atoms themselves. The latter is directly reflected in the compressibility of the medium.

It is worthwhile to point out the similarities between the self-consistent mean field theory we are discussing here and the method of density functional formalism which is used in the impurity problem in charged liquids [18]. Stott and Zaremba used this formalism to develop a "quasi-atom" theory [19], with the assumption that the electron density in the vicinity of the impurity is uniform. This is analogous to our assumption that the energy exchanges between the alien atom and the medium are in the form of heat exchanges, independent of the position of the atom. $U_{\alpha\alpha'}(0)$ in Eq. (11a) is a potential which is averaged over all atomic positions:

$$U_{\alpha\alpha'}(0) = V_{o1}^{-1} \int d\vec{R} \langle\alpha|V_{AM}(\vec{R};S)|\alpha'\rangle \ . \tag{13}$$

Stott and his co-workers have also developed a nonlinear theory for the impurity problem [20,21], in which they calculate the response of the medium linearly, but modify the bare interaction potential to obtain an effective potential which accounts for the nonlinearity of the impurity+medium system. This is analogous to our approach. In Eq. (8a), although the response of the medium is calculated linearly, it is a response to an arbitrary, time-dependent atomic density matrix. This is equivalent to replacing the original bare interaction potential that the alien atom feels by a time-dependent effective potential.

The fact that the asymptotic values of $r_{\alpha\vec{K},\alpha'\vec{K}'}$ appear in the spectral equation is connected with some conservation law concerning the atomic

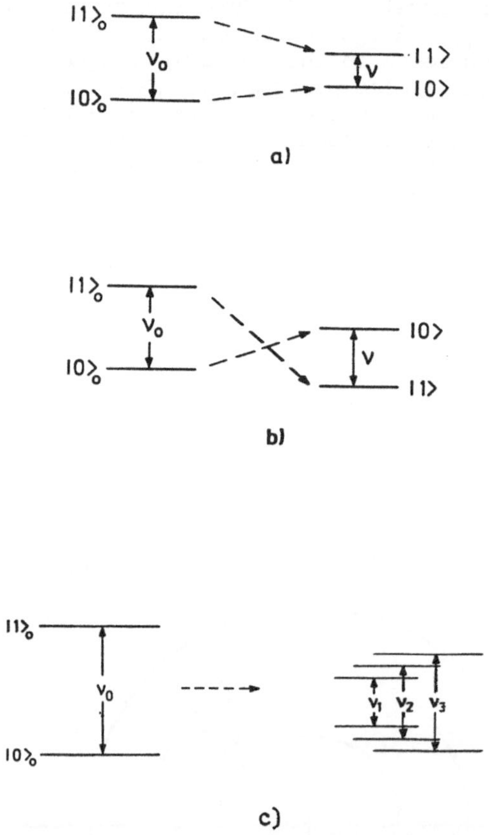

Fig. 5. Vacuum (left) and dressed (right) level configurations: a) normal, b) inverted, c) split.

system, which holds despite the nonlinear self-interaction induced by the medium. This becomes clear in the analysis of two-level systems [13]. For the two-level alien atoms, Eqs. (11a) and (11c) can be solved exactly, assuming that the atomic ensemble has a single momentum $\vec{K}$. We now consider these solutions.

We assume that the level separation is modified, while the relative orientation of the two states remains unchanged as shown in Fig. 5a. For the two-level atoms, one finds directly from Eq. (8a) that the quantity

$$\xi = \sqrt{(r_{11} - r_{00})^2 + 4r_{10}^2} \tag{14}$$

is conserved, i.e., is a constant of motion. ξ is a parameter that characterizes the statistical nature of the ensemble of the alien atoms. It is related to ρ_A by

$$\mathrm{Tr}\rho_A^2 = \frac{1 + \xi^2}{2} \leqslant 1 \ . \tag{15}$$

For a pure state-ensemble, $\xi = 1$. For a random mixture, which has no polarization (no coherence), $\xi = 0$. The conservation of ξ implies that in the phase space of r_{00} and r_{10}, which are related to the population distribution and the coherence in the ensemble, respectively, the atomic trajectories are half circles centered on $r_{00} = 1/2$ (see Fig. 6). This is an explicit example of one type of lumpiness in a phase space which arises from the self-consistent approach. The conservation of ξ arises entirely from the nonlinearity in the atomic motion. If the medium is treated as a reservoir, ξ is not conserved.

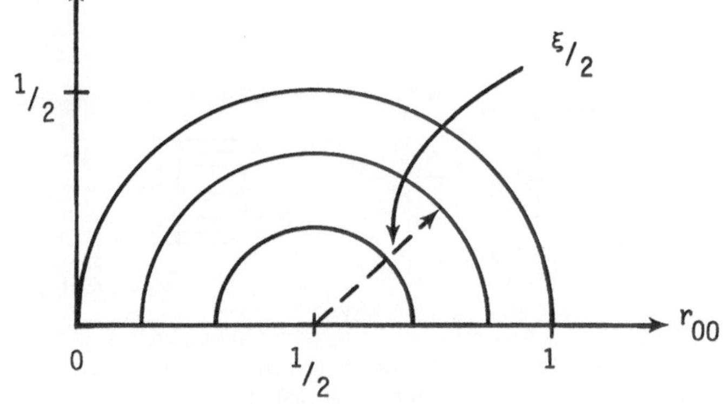

Fig. 6. Atomic trajectories in the phase space of the population (r_{00}) and coherence (r_{10}) coordinates.

If the initial conditions are specified at some time t_1 ($>t_0$), the solutions for r_{00} and r_{11} are given by

$$r_{00}(t) = (1 + \xi)/2 - R(t - t_1) , \tag{16a}$$

$$r_{11}(t) = (1 - \xi)/2 + R(t - t_1) , \tag{16b}$$

where

$$R(t - t_1) = \frac{\xi[\xi - 2r_{00}(t_1) + 1]e^{-\xi\Gamma(t-t_1)}}{\xi + 2r_{00}(t_1) - 1 + [\xi - 2r_{00}(t_1) + 1]e^{-\xi\Gamma(t-t_1)}} \tag{16c}$$

and

$$\Gamma = -2|U_{10}|^2 \mathcal{I}m\, \chi_M(\nu) . \tag{16d}$$

The off-diagonal amplitude is given by

$$r_{10}(t) = [R(t - t_1)]^{\frac{1}{2}}[\xi - R(t - t_1)]^{\frac{1}{2}} . \tag{16e}$$

Note that in the limit $t - t_1 >> (\xi\Gamma)^{-1}$, $R \rightarrow 0$ and

$$\rho_A(t) = \begin{pmatrix} \frac{1 - \xi}{2} & 0 \\ 0 & \frac{1 + \xi}{2} \end{pmatrix} . \tag{17}$$

Further insight into ξ and these solutions can be obtained by considering the initial atomic density matrix at time t_1:

$$\rho_A(t_1) = \begin{pmatrix} r_{11}(t_1) & r_{10}(t_1)e^{i(\theta-\nu t_1)} \\ \bar{r}_{10}(t_1)e^{-i(\theta-\nu t_1)} & r_{00}(t_1) \end{pmatrix} \tag{18}$$

Let $S(t_1)$ be a unitary matrix such that

$$S_{11} = (r_{00} - r_{11} - \xi)[(r_{00} + \xi - r_{11})^2 + 4r_{10}^2]^{-\frac{1}{2}} , \tag{19a}$$

$$S_{10} = 2r_{10}e^{i(\theta-\nu t_1)}[(r_{00} + \xi - r_{11})^2 + 4r_{10}^2]^{-\frac{1}{2}} , \tag{19b}$$

and

$$S_{00} = (r_{00} - r_{11} + \xi)[(r_{00} + \xi - r_{11})^2 + 4r_{10}^2]^{-1/2}. \tag{19c}$$

$S(t_1)$ diagonalizes $\rho_A(t_1)$:

$$S(t_1)^\dagger \rho_A(t_1) S(t_1) = \begin{pmatrix} \frac{1-\xi}{2} & 0 \\ 0 & \frac{1+\xi}{2} \end{pmatrix}, \tag{20}$$

which is the asymptotic solution given by Eq. (17). Thus, the asymptotic solutions are just the eigenvalues of the initial density matrix. It follows that the interaction between the alien atom and the medium can be viewed as an operational form of the diagonalization, in time, of the atomic density matrix. The diagonalization is achieved for t such that $(t - t_1) \gg (\xi\Gamma)^{-1}$.

The level separation of the dressed levels is determined by the spectral equation

$$\nu = \nu_0 + \frac{1}{2}\chi_M(0)(U_{11}^2 - U_{00}^2) + \frac{1}{2}\xi[-\chi_M(0)(U_{11} - U_{00})^2 + 2|U_{10}|^2 \mathrm{Re}\chi_M(\nu)] . \tag{21}$$

ν is a positive root of this implicit equation. The terms proportional to ξ in Eq. (21) arise entirely from the self-interaction of the alien atoms. They would be lacking in a reservoir approximation for the medium. Thus, in the self-consistent mean field approximation, the dressed spectrum depends on the statistical character of the atomic ensemble via ξ. This might have been anticipated, because the response of the medium to the presence of the alien atom can be expected to depend on the nature of the ensemble of the alien atoms. This response in turn determines the form of the atomic self-interaction.

If Eq. (21) has more than one positive root, then the original two levels are split in the medium. Figure 5c illustrates an example. Such level splitting is especially likely to occur if $\tilde{\nu}_0 \sim \Omega_R$, where Ω_R denotes a medium resonance frequency and

$$\tilde{\nu}_0 \equiv \nu_0 + \frac{1}{2}\chi_M(0)[(1 - \xi)U_{11}^2 - (1 + \xi)U_{00}^2 + 2\xi U_{11}U_{00}] . \tag{22}$$

The roots of Eq. (21) are given by the intersection of the line L: $\nu - \tilde{\nu}_0$ with the curve $\xi|U_{10}|^2\mathrm{Re}\chi_M(\nu)$ as shown in Fig. 7. As the line L moves from left to right for increasing $\tilde{\nu}_0$, one can have one, two, or three roots. The number of roots are further increased if there are other medium resonances nearby.

If there is not a positive root of Eq. (21), the level configuration may be inverted as shown in Fig. 5b. The level separation of the inverted configuration is determined by the positive root of the equation

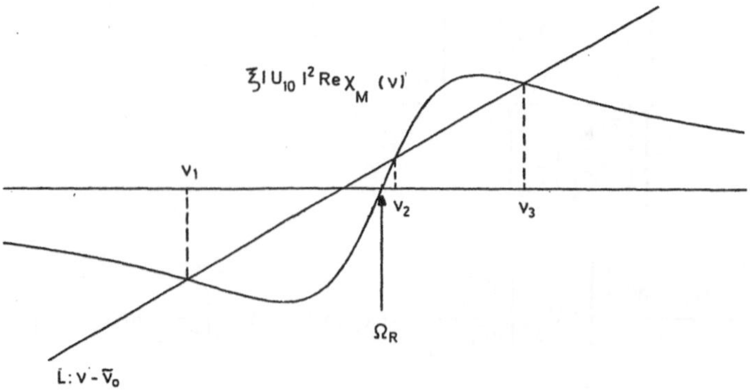

Fig. 7. Roots of (21) near a medium resonance.

$$\nu = \nu_0 - \frac{1}{2}\chi_M(0)(U_{11}^2 - U_{00}^2)$$

$$+ \frac{1}{2}\xi[-\chi_M(0)(U_{11} - U_{00})^2 + 2|U_{10}|^2 Re\chi_M(\nu)] \tag{23}$$

Equations (21) and (23) yield an interesting result concerning the dependence of ν on the pressure and the temperature of the medium. Let us assume that the medium has an equation of state relating P_M, T_M and $V_{o\ell}$, the pressure, the temperature and the volume of the medium respectively. Note that $U_{\alpha\alpha'}$ depends on $V_{o\ell}$, but the product $V_{o\ell}U_{\alpha\alpha'} = u_{\alpha\alpha'}$ is independent of the state variables of the medium. Using Eq. (12) and the definition of the isothermal compressibility,

$$\kappa_{iso} = -\left[V_{o\ell}\left(\frac{\partial P_M}{\partial V_{o\ell}}\right)_{T_M}\right]^{-1} \tag{24}$$

one finds from Eqs. (21) and (23) that

$$\left(\frac{\partial\nu}{\partial P_M}\right)_{T_M} = -\frac{A\left[\frac{\partial}{\partial P_M}\left(\frac{\partial P_M}{\partial V_{o\ell}}\right)_{T_M}\right]_{T_M}}{B\left[\left(\frac{\partial P_M}{\partial V_{o\ell}}\right)_{T_M}\right]^2}$$

$$+ \frac{\xi|u_{10}|^2}{B}\left[2V_{o\ell}\left(\frac{\partial V_{o\ell}}{\partial P_M}\right)_{T_M} \mathrm{Re}\chi_M(\nu) + V_{o\ell}^2\left(\frac{\partial}{\partial P_M}\mathrm{Re}\chi_M(\nu)\right)_{T_M}\right], \tag{25a}$$

$$\left(\frac{\partial\nu}{\partial T_M}\right)_{P_M} = -\frac{A\left[\frac{\partial}{\partial T_M}\left(\frac{\partial P_M}{\partial V_{o\ell}}\right)_{T_M}\right]_{P_M}}{B\left[\left(\frac{\partial P_M}{\partial V_{o\ell}}\right)_{T_M}\right]^2}$$

$$+ \frac{\xi|u_{20}|^2}{B}\left[2V_{o\ell}\left(\frac{\partial V_{o\ell}}{\partial T_M}\right)_{P_M} \mathrm{Re}\chi_M(\nu)\right.$$

$$\left. + V_{o\ell}^2\left(\frac{\partial}{\partial T_M}\mathrm{Re}\chi_M(\nu)\right)_{P_M}\right] \tag{25b}$$

where

$$A = \frac{1}{2} N\,[(u_{11}^2 - u_{00}^2)\ \mathrm{sgn} - \xi(u_{11} - u_{00})^2]\ , \tag{25c}$$

$$\mathrm{sgn} = \begin{cases} +1 & \text{normal configuration} \\ -1 & \text{inverted configuration}\ , \end{cases} \tag{25d}$$

and

$$B = 1 - \xi|u_{10}|^2\, V_{o\ell}^2 \frac{\partial}{\partial\nu}\mathrm{Re}\chi_M(\nu)\ . \tag{25e}$$

If the ξ-dependent terms are small, then the signs of $(\partial\nu/\partial P_M)$ and $(\partial\nu/\partial T_M)$ depend on the sign of A, which changes in going from the normal configuration to the inverted one. Thus, if the level separation increases with pressure and temperature in one configuration, it decreases in the other. Such contrasting behavior is encountered in narrow-gap semiconductor alloys [22].

Finally, let us introduce the optical field of intensity $I_0 = cA_0^2/4\pi$ into the two-level alien atom problem. One directly verifies from the

self-consistent mean field equations that ξ is still a constant of motion [13]. In other words, the optical field does not modify the statistical character of the atomic ensemble as defined by ξ (or $\mathrm{Tr}\rho_A^2$). Let us define

$$r_{00} = \frac{1+\xi}{2} - \Pi \tag{26a}$$

$$r_{11} = \frac{1-\xi}{2} + \Pi \tag{26b}$$

The off-diagonal matrix element of the atomic density matrix in the steady state is given by

$$\rho_{10}^{(A)}(t) = A_0\vec{e}_0 \cdot \vec{\mu}_{10}\left(\frac{\xi}{2} - \Pi\right)\left\{-\frac{e^{-i\omega t}}{\omega - \nu + \frac{2}{\xi}\Pi(\nu - \nu_0 - \Delta) + i\left(\frac{\xi}{2} - \Pi\right)\Gamma} + \frac{e^{i\omega t}}{\omega + \nu - \frac{2}{\xi}\Pi(\nu - \nu_0 - \Delta) - i\left(\frac{\xi}{2} - \Pi\right)\Gamma}\right\}, \tag{27a}$$

where

$$\Delta = \frac{1}{2}\chi_M(0)(U_{11}^2 - U_{00}^2) . \tag{27b}$$

Π itself satisfies the equation of motion

$$\frac{d}{dt}\Pi = -\Gamma\Pi(\xi - \Pi) + A_0^2|\hat{e}_0 \cdot \mu_{10}|^2\Gamma\left(\frac{\xi}{2} - \Pi\right)^2 \times \left\{\frac{1}{[\omega - \nu + \frac{2}{\xi}\Pi(\nu - \nu_0 - \Delta)]^2 + \left(\frac{\xi}{2} - \Pi\right)^2\Gamma^2} + \frac{1}{[\omega + \nu - \frac{2}{\xi}\Pi(\nu - \nu_0 - \Delta)]^2 + \left(\frac{\xi}{2} - \Pi\right)^2\Gamma^2}\right\} \tag{28}$$

The terms with denominators which have $\omega + \nu$ in Eqs. (27a) and (28) are the counter-rotating terms. If these terms are neglected, Π satisfies a biquadratic equation in the steady state:

$$\Pi^4 + a_1\Pi^3 + a_2\Pi^2 + a_3\Pi + a_4 = 0 , \tag{29}$$

where a's are complicated functions of ω, ν, ξ and I_0. The analysis of Eq. (29) shows [13] that there are always two real roots of Eq. (29). One of these is less than $\xi/2$ and stable. The other is greater than $\xi/2$ and unstable. The two other roots are usually complex. For the stable root, one finds

$$\lim_{I_0 \to 0} \Pi_s = 0 \tag{30a}$$

$$\lim_{I_0 \to \infty} \Pi_s = \frac{\xi}{2} \tag{30b}$$

The optical susceptibility of the alien atoms is readily obtained from Eq. (27a) with Π given by Π_s:

$$\chi_A(\omega) \simeq \frac{\left[-\frac{n_A}{2\pi}|\hat{e}_0 \cdot \vec{\mu}_{10}|^2 \left(\frac{\xi}{2} - \Pi_s\right)\right]}{\left[\omega - \nu + \frac{2}{\xi}\Pi_s(\nu - \nu_0 - \Delta) + i\left(\frac{\xi}{2} - \Pi_s\right)\Gamma\right]}, \tag{31}$$

where n_A is the density of the alien atoms. The absorption coefficient is given by

$$\alpha_A(\omega) \quad \frac{\left[\frac{2n_A}{c}\left|e_0 \cdot \vec{\mu}_{10}\right|^2 \omega\Gamma\left(\frac{\xi}{2} - \Pi_s\right)^2\right]}{[\omega - \nu + \frac{2}{\xi}\Pi_s(\nu - \nu_0 - \Delta)]^2 + \left(\frac{\xi}{2} - \Pi_s\right)^2\Gamma^2}. \tag{32}$$

For small optical intensities,

$$\lim_{I_0 \to 0} \alpha_A(\omega) \simeq \left(\frac{\omega\Gamma n_A\xi^2|\hat{e}_0 \cdot \vec{\mu}_{10}|^2}{2c}\right) \frac{1}{[(\omega - \nu)^2 + \frac{1}{4}\xi^2\Gamma^2]}. \tag{33}$$

That is, α_A has the usual Lorentzian form in this limit. For finite I_0, however, the resonance shifts from ν to ν_{eff}:

$$\nu_{eff} = \nu - \frac{2}{\xi}\Pi_s(\nu - \nu_0 - \Delta). \tag{34}$$

The absorption lineshape becomes asymmetric as shown in Fig. 8. Reχ_A, which contributes to the refraction of the field, is also asymmetric as seen in Fig. 9. As the optical field becomes very intense ($I_0 \to \infty$), ν_{eff} approaches $\nu_0 + \Delta$. In other words, the optical field penetrates the dressing of the alien atom provided by the medium and begins to see the bare atom, except for a constant shift Δ which depends on the homogeneous properties of the medium.

The asymmetry of the lineshapes at high optical intensities might seem surprising at first glance, in view of the Markovian, factorizability assumption for the density matrices of the alien atom and the medium in the self-consistent mean field approximation. However, an element of memory is introduced in the interactions of the atom by the fact that the response of the medium to a given atomic state is taken into account in the motion of the atom. The atom undergoes, through the response of

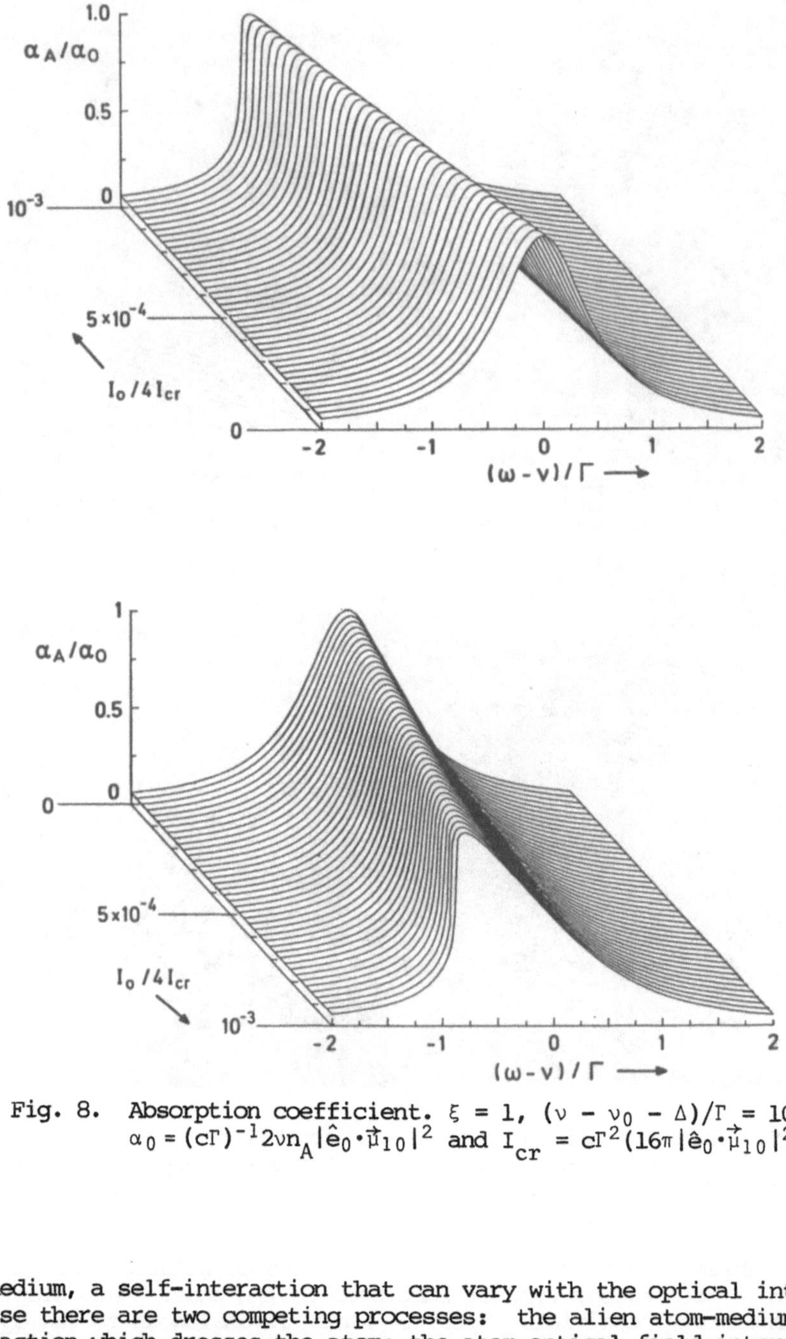

Fig. 8. Absorption coefficient. $\xi = 1$, $(\nu - \nu_0 - \Delta)/\Gamma = 100$, $\alpha_0 = (c\Gamma)^{-1}2\nu n_A|\hat{e}_0\cdot\vec{\mu}_{10}|^2$ and $I_{cr} = c\Gamma^2(16\pi|\hat{e}_0\cdot\vec{\mu}_{10}|^2)^{-1}$.

the medium, a self-interaction that can vary with the optical intensity, because there are two competing processes: the alien atom-medium interaction which dresses the atom; the atom-optical field interaction which probes the dressed spectrum. As a result, the atomic system in the self-consistent mean field approximation differs significantly from a Markovian system with a reservoir.

The element of memory referred to above is strikingly illustrated in the multistability exhibited by the alien atom in certain ranges of ω, ν, ξ and I_0. The two complex roots of Eq. (29) become real for these ranges

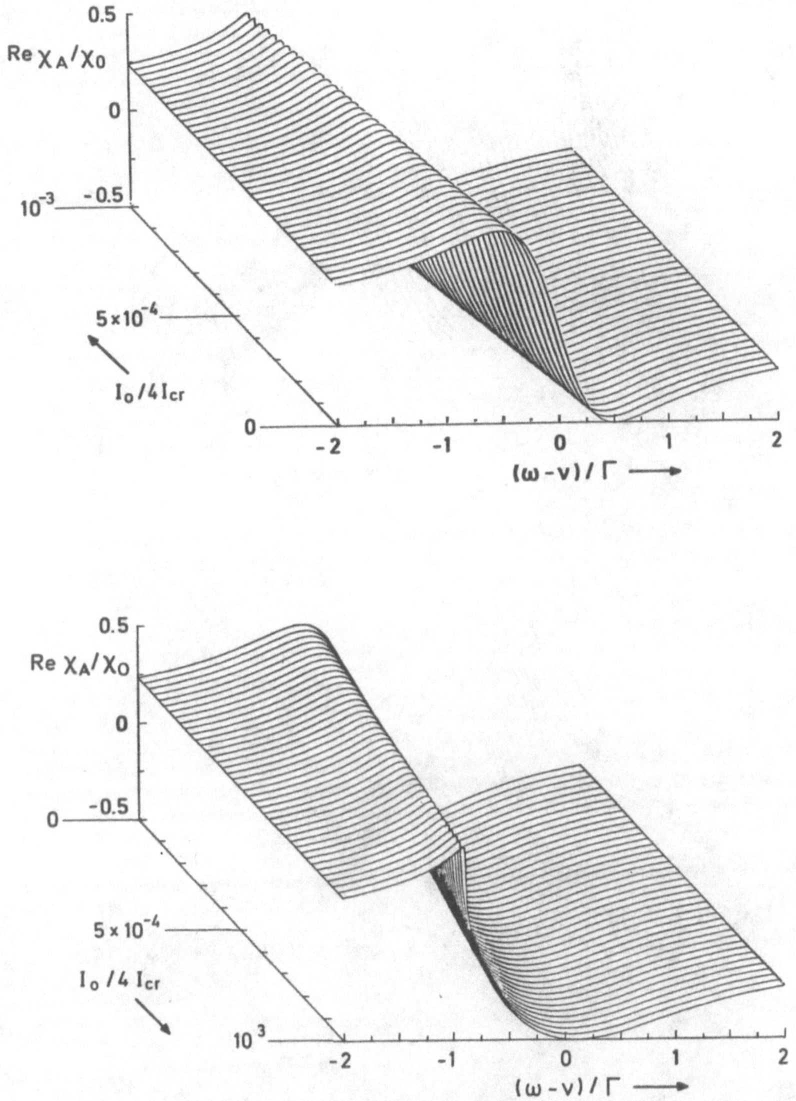

Fig. 9. Real part of the optical susceptibility.
$\chi_0 = (2\pi\Gamma)^{-1}n_A|\hat{e}_0\cdot\vec{\mu}_{10}|^2$, $\xi = 1$ and $(\nu - \nu_0 - \Delta)/\Gamma = 100$.

of the variables, and there is more than one stable solution for $\mathbb{I}$. This is an example of a second type of lumpiness in the phase space displayed by self-consistent mean field theories, which is analogous to the optical bistability observed in certain nonlinear cavities [23,24].

REFERENCES

1. R. H. Picard and C. R. Willis, Phys. Rev. A9 (1974) 343.
2. M. Sargent, M. O. Scully and W. Lamb, Laser Physics (Addison-Wesley, Reading 1974).
3. K. Ikeda, Opt. Comm. 30 (1979) 257.
4. P. Lee and M. O. Scully, Phys. Rev. B3 (1971) 769.

5. D. H. E. Gross and H. Kalinowski, Physics Reports 45 (1978) 176.
6. S. Mukamel, et. al, Nucl. Phys. A366 (1981) 1339.
7. H. E. Kandrup, Physics Reports 63, No. 1 (1980) 1.
8. S. Chandrashekhar, Principles of Steller Dynamics (Dover, New York 1942).
9. S. Chandrashekhar and J. von Neumann, Astrophys. J 95 (1942) 489 and 97 (1942) 1.
10. H. E. Kandrup, Astrophys. J. 244 (1981) 316.
11. M. Gell-Mann and K. Brueckner, Phys. Rev. 106 (1957) 364.
12. A. Elci, "Semiclassical electrodynamics of alien atoms in interacting media. I: Self-consistent mean field approximation", Ann. Phys. (N.Y.) (in press).
13. A. Elci, "Semiclassical electrodynamics of alien atoms in interacting media. II:: Two-level systems", Ann. Phys. (N.Y.) (in press).
14. E. Fermi, Rend. Lincei 5 (1927) 795.
15. J. E. Krizan, Phys. Rev. D3 (1971) 2333.
16. C. R. Stroud and E. T. Jaynes, Phys. Rev. A1 (1970) 106.
17. M. D. Crisp and E. T. Jaynes, Phys. Rev. 179 (1969) 1253.
18. N. D. Lang, Solid State Physics 28 (1973) 225.
19. M. J. Stott and E. Zaremba, Phys. Rev. B22 (1980) 1564.
20. S. Sjolander and M. J. Stott, Phys. Rev. B5 (1972) 2109.
21. D. Popovich, et. al, Phys. Rev. B13 (1976) 590.
22. R. Dornhaus and G. Nimitz, "The properties and applications of the $Hg_{1-x}Cd_xTe$ Alloy System", in Springer Tracts in Modern Physics, Vol. 78 (1976).
23. S. L. McCall, Phys. Rev. A9 (1974) 1515.
24. V. Benza and L. G. Lugiato, "Semiclassical and quantum statistical dressed mode description of optical bistability", in Optical Bistability (Plenum Press, New York 1980), eds. C. M. Bowden, M. Cifton and H. R. Robl.

ENTROPY, INFORMATION AND QUANTUM GEOMETRY

Eduardo R. Caianiello

Dipartimento Di Fisica Teorica E Sue Metodologie
Per Le Scienze Applicate
Universita' Di Salerno

1. - A White Horse is not a Horse

This Chinese apophthegm paraphrases here the equivalent one "Entropy is not Shannon Entropy". There are in fact at least two things wrong with Shannon entropy. The first is its name: had Claude Shannon not accepted John von Neumann's advice ("you should call it entropy and for two reasons: first, the function is already in use in thermodynamics under that name; second and more importantly, most people don't know what entropy really is, and if you use the word "entropy in an argument, you will win every time [1]") and just called it "uncertainty", endless confusion would have been spared. "Entropy" is psychologically tied with "thermodynamics" in a physicist's mind, so that the purely logical, far wider connotation of Shannon's concept escapes attention. Shannon entropy is simply and avowedly the "measure of the uncertainty inherent in a preassigned probability scheme"; as such it has nothing whatever to do with thermodynamical entropy, except in the case in which that probability distribution is known, or proven to be, "canonical".

Shannon entropy goes wrong in a second respect. Whenever one deals with a continuous probability distribution $\rho(x) \geq 0$

(1) $$\int_{-\infty}^{+\infty} \rho(x)\, dx = 1 \quad :$$

(2) $$H = -\int_{-\infty}^{+\infty} \rho(x) \lg \rho(x)\, dx \quad .$$

(When ρ = constant in a finite volume W (or for a finite number N of, say, letters of an alphabet), it reaches its maximum value, and one has the familiar formulae

$$H_{max} = k \lg N \quad (\text{"capacity" of the alphabet})$$

or

$$H_{max} = S = k \lg W ,$$

the introduction and identification of k being a matter of convenience). Shannon entropy diverges under honest attempts to deduce it as the limiting case of some finite distribution (it is also not invariant under change of variables $\rho(x) \to \rho(x(y)) \frac{dx}{dy}$). This divergence is a cause of horror in information theory. Subtle arguments which intend to show "how really a limit should be taken to suppress it" are tantamount to replacing Shannon entropy (2) with

$$(3) \quad H_c = -\int \rho(x) \lg \frac{\rho(x)}{\rho_o(x)} dx$$

(We change the sign here to conform to usage). This is the socalled "cross-entropy". It is exempt from all the troubles mentioned before, especially if we decide to call it (with one of the many uses of the term) "information". Contrary to Shannon entropy, which gives the amount of uncertainty connected with a single probability distribution(i.e. the average "self-information $-\lg\rho$", an essentially static concept), the cross entropy (3) measures the uncertainty connected to a posterior distribution $\rho(x)$, once a prior distribution $\rho_o(x)$ is known (ample room for typical statistical inference, and dynamics).

The classic works of Jaynes [(2)], and Tribus [(3)] and others maximize (2) subject to conditions chosen so as to obtain canonical (Boltzmann, Bose, Fermi, Gentile) statistics. When attention is paid to the inconvenience of doing formal work with the divergent expression (2), the remedy is to do exactly the same formal work on the expression (3), with identical results.

All known results have been reproduced in this way, with great formal economy. The erroneous notion that this manner of derivation belongs in the field of statistical thermodynamics, which has been engendered, as remarked, by the name "entropy", has much limited the interest in this approach.

2. - "Error" and "Noise" vs. "Uncertainty"

Support to the views that we propose to expound comes also from hitherto unsuspected quarters. In his recent works, R.E. Kalman (4) authoritatively voices and proves a claim which sounds revolutionary in the context of general systems theory. The notion that parameters or "indicators" in systemic models (e.g. social or econometric) should be measurable with arbitrary accuracy, provided one tries hard enough, had never been challenged before. Kalman maintains that all parameters are model-dependent (have therefore no "existence of their own": they "exist" only in the framework of a scheme of measurement based on some theory or model), and that "error" or "noise" may not be arbitrarily swept under the carpet in model building. He shows by actual computation that even the simplest two-dimensional linear models change drastically when this basic "uncertainty" is correctly taken into account. The consequences, in his words, will be profound "in all branches of science, including time-series analysis, economic forecasting, most of econometrics, psychometrics and elsewhere". The connections that we are going to explore here with a one-sided mind, may prove perhaps useful also in view of technological transfers between sciences that now belong in different universes.

3. - Kullback-Leibler Information, Jeffreys Divergence and Information Metric

Information, entropy, cross-entropy, etc. have been defined in a great many ways, most of them known only to specialists. Of particular interest to us here is the case of parametric distributions $\rho(x|z)$. Here $z \equiv \{z_{(1)}, z_{(2)}, \ldots z_{(m)}\}$ belongs to an R^m random space and $x \equiv \{x^{(1)}, x^{(2)}, \ldots, x^{(n)}\}$ to an R^m parameter space. For example, with the gaussian

$$p(x|z) = \frac{1}{\sqrt{2\pi}\,\sigma} \exp\left[-\frac{(z_{(1)}-\mu)^2}{2\sigma^2}\right] \qquad (4)$$

we read

$$z \equiv \{z_{(1)}\} \quad ; x \equiv \{x^{(1)} = f^{(1)}(\mu,\sigma) \quad , x^{(2)} = f^{(2)}(\mu,\sigma)\}$$

The cross-entropy (3) is called in this context "Kullback-Leibler information" [5] and written

$$H_c = I(1,2) = \int dz\, \rho(x_1|z) \lg \frac{\rho(x_1|z)}{\rho(x_2|z)} \tag{5}$$

Expression (5) measures a "distance" between distribution $\rho(x_2|z)$ and $\rho(x_1|z)$ (e.g. between different gaussians). Owing to its asymmetry it is called the directed distance 2 → 1. Its symmetrized form

$$J(1,2) = I(1,2) + I(2,1) \tag{6}$$

is called the J- divergence (after Jeffreys [6]) or "distance" between $\rho(x_2|z)$ and $\rho(x_1|z)$.
As numbers, (5) or (6) evaluate what their names indicate. The term "distance" however is not wholly appropriate. The triangle inequality applies only in the infinitesimal case, in which $(x_1 = x \;;\; x_2 = x + dx)$

$$2I(x, x+dx) = J(x, x+dx) = ds^2 = g_{hk}(x)\, dx^h dx^k \tag{7}$$

Here with the imitation $\partial_h = \frac{\partial}{\partial x^h}$ we have

$$g_{hk}(x) = g_{kh}(x) = \int dz\, \rho(x|z)\, \partial_h \lg \rho(x|z)\, \partial_k \lg \rho(x|z) \tag{8}$$

the well known information (or Fisher) metric [7] . Of course $\rho(x|z) \geq 0$ and

$$\int \rho(x|z)\, dz = 1 \tag{9}$$

The concepts of Fisher metric, information and infinitesimal distance have led to many classic developments in information and estimation theories, a field in active development. In this way

many concepts acquire then suggestive geometric connotations, with Riemannian geometry the "natural tool". "Cross-entropy", or "information distance", through (7) and (8) provides a natural metrization of otherwise affine space elements, prior to any introduction of physical concepts such as "energy", "temperature", etc.. Thermodynamics, as a particular instance, was treated successfully in this manner by F. Weinhold [8]. The work of Amari [9] and others shows, in fields other than physics, the fruitfulness of utilizing in statistical inference the logical principle of minimizing cross-entropy, i.e. choosing the least biased distribution permitted by the circustances. It has been also called the "Principle of Maximum Honesty" [10].

The ingredients in all work of this kind are, whatever the ideology behind it:

1) extremizing entropy, or cross-entropy, information;
2) doing so with appropriate constraints (Lagrange multipliers, etc.).

Step no.1 carries on, formally, regardless of whether the domain over which the integration is extended is the whole, or a small, "infinitesimal" region of it. Step no.2 requires instead that integrals be taken on the whole domain.

The end products are the well known distributions of mathematical statistics, thermodynamics, etc.. A typical estimation problem is, for instance, the detrmination of the numerical value of a parameter which specifies, within a given parametric family, the distribution that best fits some wanted requirements. (Canonical distributions have thermodynamical interpretations).

There is nothing, however, against adopting the same procedure with a different outlook, which we wish to emphasize here. One gives up, at the beginning, step no.2. Step no.1, alone, gives a metric G; a geometrical model can then be set up in the standard way by assigning a connection $\Gamma^{\alpha}_{\mu\beta}$ (not in general symmetric in μ and β), from which curvature is defined. The demands that previously were handled with step no.2 can now be met by a correct choice of G, Γ and what else geometry requires. This approach is standard in geometrical work on estimation [9].

At this stage, however, we can also look at this model as a standard source of wave equation, geodesics, and all that. We have shown [11] that there is a general geometric framework, best expressed in complex spaces, in which the natural objects to compute are "square roots of probabilities", i.e. wave func-

tions. Both the classic geometry of information theory mentioned before and the "quantum geometry" developed by us in phase space along these lines [12] appear as subcases.

4. - Information metric and probability amplitudes

The information metric (8) can be identically written

$$(10)\qquad g_{hk}(x) = 4\int dz\, \partial_h \sqrt{p(x|z)} \cdot \partial_k \sqrt{p(x|z)}\ ;$$

the most general connection compatible with information geometry is [13]

$$(11)\qquad \overset{\alpha}{\Gamma}_{ijk} = [i\,j\,k] - \frac{\alpha}{2}\int dz\, \partial_i \lg\rho \cdot \partial_j \lg\rho \cdot \partial_k \lg\rho$$

α arbitrary real. One has then, e.g. for the Gaussian distribution [9]

$$(12)\qquad G = \begin{pmatrix} \sigma^2 & -2\mu\sigma^2 \\ -2\mu\sigma^2 & 4\mu^2\sigma^4 + 2\sigma^4 \end{pmatrix}$$

and

$$(13)\qquad \overset{\alpha}{R}_{1212} = (1-\alpha^2)\,\sigma^6$$

If one wants a model in which parameter space is flat, one has to take either $\alpha = +1$ (Gaussian distributions become straight lines) or $\alpha = -1$ (the "dual" distributions to Gaussians, "mixture distributions", are then rendered as straight lines). Of interest to us here, however, is the choice $\alpha = 0$, the only one that guarantees the compatibility, or metricity of the connection with G, i.e. vanishing covariant derivatives of G.
Then one finds that

$$\overset{\circ}{\Gamma}_{ijk} = 4 \int dz\, \partial_i \partial_j \sqrt{p} \cdot \partial_k \sqrt{p} \tag{14}$$

$$\overset{\circ}{R}_{1212} = \sigma^6 \quad ; \tag{15}$$

the curvature tensor expresses our uncertainty, as it vanishes only when $\alpha = 0$. We see therefore that a Riemann tensor offers a natural way of expressing uncertainty.

We recally now Kalman's fundamental objection to extant systems theory. Quantum mechanics appears as a special case, privileged by the fact that Heisenberg's uncertainty relations are very neat and specific. We have used them indeed to define a curvature tensor in phase space (enlarged to t and E), from which all standard results of quantum mechanics are verified to derive, with several additions and generalizations. Such scheme, wide enough to accomodate both information and a quantum geometry, is suggested by the classic form (10). With some hindsight we generalize it to an hermitian metric, e.g. like

$$g_{hk}(x) = \overline{g_{kh}(x)} = \int dz\, \psi_h(x/z)\, \overline{\psi_k(x/z)} \tag{16}$$

if z is discrete, (14) becomes

$$g_{hk}(x) = \sum_p \psi_{(p)h}(x)\, \overline{\psi_{(p)k}(x)} \tag{17}$$

a viel-bein, which becomes real and holomonic if $\psi_{(\rho)h}(x) = \overline{\psi_{(\rho)h}(x)} = \partial_h \varphi_\rho(x)$ and clearly $\left(\int dz \equiv \sum_\rho \right)$ reduces then to the information metric.

The case of quantum mechanics is taken as a test case: no fooling allowed. Provided things do not go too wrong, one is then faced with new questions. We are essentially building a geometry of ignorance. In this light the "ignorance" that is at the very foundation of quantum mechanics would be of the same nature as that, say, which would accompany the "exact" determination of the price of the dollar at a given place and time (a measurement which would conceivably infinitely alter that same price). Models of nature work only up to a point.

5. - The Cramér-Rao inequalities

Mention is mode here of yet another connection of general systems with geometry. The information metric tensor just described statisfies the celebrated Cramér-Rao [4] inequalities. It suffices to recall here the one-dimensional case; going from the sophisticated language of information theory to that of physics, i.e. writing, if $\tilde{x}$ is an estimator, $\mathrm{var}_x(\tilde{x}) = \Delta x^2$, they reduce to (g_{11} is also called "Fisher information" [15])

$$(\Delta x)^2 \cdot g_{11} \geq 1 \tag{18}$$

If $\rho(x|z) = \rho(x-z)$, etc., the form (10) of g_{11} yields

$$(\Delta x)^2 \cdot 4 \int dz \left(\frac{\partial \sqrt{\rho}}{\partial x} \right)^2 \geq 1 \tag{19}$$

i.e. with $\sqrt{\rho} = \varphi \quad , \quad p_x = -i\hbar \frac{\partial}{\partial x}$

$$\Delta x \cdot \Delta p_x \geq \frac{\hbar}{2} \quad . \tag{20}$$

Uncertainty relations appear as special cases of the Cramér-Rao inequalities, not invented for physics. Text-book quantum mechanics could be obtained, as it is, from these inequalities. Our model, in which curvature in phase space expresses uncertainty, proves to be more general (is particular, it introduces naturally non-Abelian as well as Abelian gauge fields [16]).

6. - Cross-entropy and sign of ds^2

To conclude, consider ordinary Euclidean distance. A theory predicts average positions (or whatever): actual measurements give a scatter around them. Suppose that repeated measurements, according to book, yield gaussian distributions of observed z_x, z_y, z_z values around the corresponding average position ($\mu_x \equiv x$, $\mu_y \equiv y$, $\mu_z \equiv z$: our parameter space): dispersion (14) is supposed to be the same in all directions; then (4) now reads

$$(21)\quad p(x,y,z \mid z_x, z_y, z_z) = \frac{1}{(\sqrt{2\pi}\sigma)^3} \exp\left[-\frac{1}{2\sigma^2}\left\{(z_x-x)^2+(z_y-y)^2+(z_z-z)^2\right\}\right];$$

we find from (10)

$$(22)\quad ds^2 \equiv dH_c = |\mathrm{const.}|^2 \left(dx^2+dy^2+dz^2\right) \geq 0$$

In special relativity we would find

$$(23)\quad ds^2 \equiv dH_c = |\mathrm{const.}|^2 \left(dt^2 - \frac{1}{c^2}\left[dx^2+dy^2+dz^2\right]\right) \geq 0$$

and in our quantum geometry [11]

$$(24)\quad ds^2 = dH_c = |\mathrm{const.}|^2 \left\{ dt^2 - \frac{1}{c^2} d\vec{x}^2 + \frac{\hbar^2}{\mu^4 c^6}\left[\frac{dE^2}{c^2} - d\vec{p}^2\right]\right\} \geq 0$$

It appears that $dH_c \geq 0$, which is a consequence of the mathematical expression of the cross-entropy, not only imposes the known requirement for phenomena to be physical, but, read as ds^2, also for distances to describe physical particles. (22) and (20) need no comment; (24) is the condition that leads, in our quantum geometry, to the introduction of a maximal acceleration [17].

7. - Conclusion: on Wheeler's "measurements made"

The concepts presented here differ in some respects from those that are assumed, often implicity, in discussions on the foundations of quantum mechanics. Quantum laws are linked ultimately with actions, hamiltonians, lagragians, etc., which a physicist tends to regard as "fundamental" objects somehow connected with Nature itself. The present approach leads instead (cf. our works cited in the references) to a "quantum geometry" in which "ds^2" , and hence lagrangians etc., is an "information distance": all mathematical tools introduced derive from Jaynes's principle (as here re-interpreted), which belongs to "logic" , not to "dynamics" (as it was not "thermodynamical" before). We propose, that is, a universal logical paradigm that, through extremization of information or cross-entropy, should be applicable in principle to all sciences studying "phenomena" (rather than "noumena" as does classical mechanics). That the supplementary information needed to construct a model be given in the form of overall constraints, or by assignement of appropriate metrics and connections, is here a technical, not a fundamental issue.

We remark further that a quantum physicist, though well aware that if be knows "something" about momentum, he is limited in his knowledge of position by Heisenberg's relations, does not usually question the notion that both momentum and position have some reality of their own; his concern is how much can be known of one if something is known of the other. This view, entirely correct computationally, still leaves a substantive value to the spaces in which each of a pair of conjugate variables lives. The limitations imposed by Heisenberg's uncertainty relations seem to regard only our ability to gain access to a more or less exact knowledge of conjugate variables, as observers.

The conceptual basis is here, instead, that if I know "something" about a quantity, that is all I know. The measurement of another quantity creates both that quantity and the value that uncertainty permits for it.

In other words, notions such as "space", or "value of the dollar here and now" are, alike, model-dependent: they have no existence or meaning of their own before measurement, and different measurements may generate quite different notions and values. Measurement creates the thing it measures.

If our connection with information and inference theory is correct, then it necessarily follows from it that the only way to speak about reality and measurements is that so clearly enunciated by J.A. Wheeler: (18) "the only measurements one can talk or think about are measurements made".

Acknowledgments

In conclusion, it is my privilege to thank most particularly, among the many people who have taken interest in this approach, J.A. Wheeler and E.T. Jaynes.

References

1. M. Tribus : "Rational Descriptions decisions and designs", Pergamon Press,Oxford,1969.
2. E.T. Jaynes : Phys.Revs 106,620 (1952);in "Brandeis Theor. Phys.Lectures on Statistical Physics"Vol.3 (New York)
3. M.Tribus : Journ.Appl.Mech. 28,1 (1961).
4. R.E.Kalman : "Proc.Int.Symp on Dynamical Systems",ed.A. Bednarek ; "Current Developments in the Interface : Economics,Econometrics,Mathematics",ed.,S.M.Hazenwinkel and A.H. G. Rinnoy Kan (Dordrecht,1982)
5. S.C. Kulback :"Information Theory and Statistics", J.Wiley,New York,1959.
6. H. Jeffreys :"Theory of Probability" 2nd ed.,Clarenton Press,Oxford,1948.
7. R.A. Fisher :" Phil.Trans.Roy.Soc. 222A,309 (1921)"Proc. Cambridge Phil.Soc. 122,700 (1925).
8. F.Weinhold : Journ.Chem.Phys.63,2479/2496 (1975).

9. S.Amari : Raag.Reps. 106,Febr.1980; Techn.Reps.Fac.Eng.Univ. of Tokyo,METR 81-1,April 1981;ib,METR 84-1,Jan.1984.
B.Efron: Ann.Statist. 3,1189 (1975); 6,362 (1978).
A.P.Dawid Ann.Statistic 3,1231 (1975); 5,1249 (1977).
10. J.N.Kapur:Journ.Math.Phys.Sci. 17,103 (1983).
11. E.R.Caianiello:Lett.Nuovo Cimento 25,225 (1979); 27,89 (1980); 38,539 (1983); 35 ,381 (1982);Nuovo Cimento 59B 350 (1980);Proc.IV Conf. on Quantum Theory and the Structure of Space and Time,Tutzing (1980);Proc. VI JINR Int. Conf.on Problems of Quantum Theory,Alushta (1980); (with G.Vilasi):Lett.Nuovo Cimento 30,469 (1981);(with G.Vilasi and S.De Filippo) ib, 33,555 (1982); (with G. Marmo and G.Scarpetta), ib, 36,487 (1983);several other papers inprint.
12. E.R.Caianiello: Lett.Nuovo Cimento, 38,539 (1983).
13. N.N. Chentzov: "Statistical Decision rules and Optimal Conclusions" (in Russian) Moscow,1972.
14. C.R.Rao: "Linear statistical Inference and its applications" J. Wiley,New York,1973,and papers quoted therein.
15. R.A. Fisher : cf.Ref.7,b.
16. E.R.Caianiello,G.Marmo,G.Scarpetta: "(Pre) quantum Geometry", Nuovo Cimento,in print.
17. E.R.Caianiello : Lett.Nuovo Cimento 32,65 (1981);(with S.De Filippo,G.Marmo;G.Vilasi), ib, 34,112 (1982).
18. J.A. Wheeler: "Frontiers of Time " LXXII Course Varenna,1977.

PHASE TRANSITIONS IN NONEQUILIBRIUM SYSTEMS: DYE LASERS AND LASERS WITH SATURABLE ABSORBERS

J. F. Scott

Condensed Matter Laboratory
Department of Physics
University of Colorado
Boulder, Colorado 80309

ABSTRACT

Since the pioneering work of Graham and Haken, and of Lamb, Scully and DeGiorgio developing the analogy between lasers near threshold and second-order transitions in systems at thermal equilibrium, nearly a hundred papers have been published which use or expand upon this idea. In 1975, Sargent, Cantrell, and the present author extended this approach to show that lasers with saturable absorbers behaved as mathematical analogs to equilibrium systems with first-order phase transitions, complete with latent heats (or their analogs) and bistability. The latter aspect -- bistability -- is of course now an important specialty in optics in its own right and has provided renewed attention to these rather simple theories. In the present paper we review the work along these lines since 1975 and suggest some possible new extensions, in particular, to analogs of the Pippard relationship for phase transitions in equilibrium systems.

I. INTRODUCTION

The early work of Graham and Haken[1] and independently, of Scully and Lamb[2] which showed the mathematical analogy between lasers near threshold and equilibrium systems near conventional phase transition points has proved to be pedagogically useful in part because it relates rather complicated nonequilibrium dynamics to free energies and simple thermodynamic relationships which most physicists find more familiar and vastly simpler than Fokker-Planck equations. In 1975, Scott, Sargent and Cantrell[3] extended this approach by showing that lasers with saturable absorbers were mathematically analogous to first-order phase transitions in equilibrium systems, complete with an analog of latent heat and bistability. The possible presence of a tricritical point in the phase space defined by laser parameters E (optical field strength), σ (population inversion), and S (injected signal strength) was also pointed out.[4] Subsequently it was

observed that this first-order theory explained phenomena not only in systems such as He-Ne lasers with Ne absorbers, for which the experimental work of Chebotayev et al.[5] required explanation, but also in dye laser systems, in which the triplet states serve as "built-in" saturable absorbers.[6-9] We review these developments in the section which follows. More recently there has been a great interest in the general phenomenon of optical bistability and this has in turn caused continued use of the general approach of Reference 3. In addition, we find free energies of the same curious logarithmic form as occur in the laser problem reappearing in the symmetry-breaking early-universe cosmology[10] This approach is therefore seen to span a wide range of physics problems of current interest, and its heuristic simplicity and pedagogical advantages over more rigorous (e.g., Fokker-Planck) descriptions should not be underestimated.

In the section which follows we review the literature in this field between 1975 and 1983. In the final section we make some suggestions about future extensions; for example, lasers near threshold have an analog to specific heat, and for equilibrium systems Pippard, Garland and Janovec[11-13] developed a number of relations between seemingly independent thermodynamic quantities (such as specific heat and sound velocities). It will be interesting to look for optical analogs of the Pippard relationship.

II. THEORY

In References 1 and 2 the theory of a laser near threshold was analogized to an equilibrium system near a transition point in phase space. The parameter analogous to the Gibbs free energy in Landau theory[14] of the latter was shown to be

$$G\left(E\right) = -\frac{1}{2}A\left(\sigma - \sigma_T\right)E^2 + \frac{1}{4}B\sigma E^4 + \ldots \quad (1)$$

where E is the expectation value of the (optical) electric field in the laser; σ is the population inversion; σ_T is the threshold population inversion; and A and B are positive coefficients independent of σ and E.

In Reference 3 the effect of a saturable absorber was first demonstrated by simply inserting a saturable loss of form

$$\sigma_T\left(E^2\right) = \sigma_{TO} + \sigma_S\left(1 + E^2/I_S\right)^{-1} \quad (2)$$

into equation (1). Here σ_{TO} represents field-independent losses due to diffraction, mirror transmission, etc., and σ_S describes the saturable absorber.

If one substitutes Equation (2) into Equation (1) one finds

$$G\left(E\right) = -\frac{1}{2}A\left(\sigma - \sigma_{TO}\right)E^2 + \left(\frac{1}{4}B\sigma - \frac{1}{2}A\sigma_S/I_S\right)E^4 + \ldots (3)$$

from which it is apparent that the coefficient of E^4 can be negative; as is well known'[14] this yields a first-order

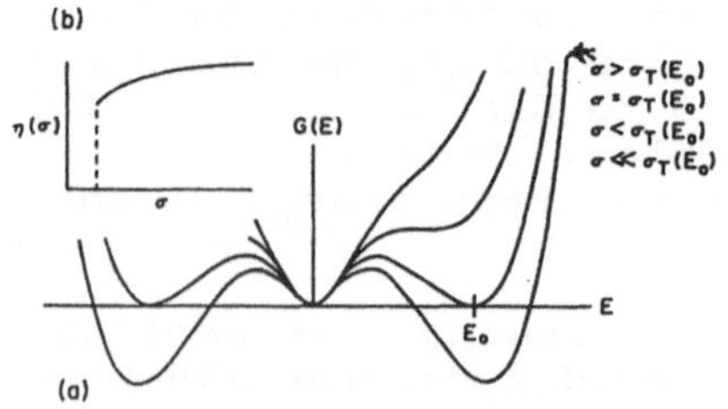

Fig. 1. Laser "free energy" G(E) versus E for population inversion σ well above threshold ($\sigma = 3\sigma_T$); at threshold ($\sigma = 1.82\sigma_T$); at threshold for formation of a metastable state ($\sigma = 1.82\sigma_T$); and well below threshold ($\sigma = 1.30\sigma_T$).

INSET: Order parameter (expectation value of electric field) versus population inversion σ (Ref. 3).

transition -- or analog -- with discontinuities and bistability. (This is shown graphically in Figure 1, from Reference 3.) However, Equation (3) is not quite right, because Equation (1) was derived with saturation of the loss ignored. If we go back to the "equation of motion" for the laser[2]

$$\dot{E} = \frac{\partial G}{\partial E} \tag{4}$$

and extend the relationship for the case of no absorber

$$\dot{E} = A\left(\sigma - \sigma_T\right)E - B\sigma E^2 \tag{5}$$

as

$$E = A\left(\sigma - \sigma_T - \frac{\sigma_S}{1 + E^2/I_S}\right) E - B\sigma E^3 \tag{6}$$

(an expression for the running-wave case, e.g., a unidirectional ring laser) we obtain:

$$G\left(E\right) = -\frac{1}{2} A\left[(\sigma - \sigma_{TO}) E^2 - \sigma_S I_S \log\left(1 + E^2/I_S\right)\right] + \frac{1}{4} B\sigma E^4 \tag{7}$$

where the strong-signal formula for the loss has been used since we are interested in the case $E^2 \geq I_S$.

The loga-rithmic term in (7) can be expanded to yield

$$G\left(E\right) \simeq -\frac{1}{2} A\left[\sigma - \sigma_{TO} - \sigma_S\right]E^2 + \left(\frac{1}{4} B\sigma - \frac{1}{4} A\sigma_S/I_S\right)E^4 + \ldots \tag{8}$$

This result is nearly the same as in Equation (3). Note that as I_S decreases below $A\sigma_S/B\sigma$ the threshold "transition" goes from continuous to discontinuous. This is a tricritical point[4] -- or its analog. Note also the presence of the logarithmic term in the free energy, a characteristic of many systems.[10]

We emphasize in reviewing this section that similar conclusions were reached independently by others,[15,16] in part motivated by the Ne absorber studies in He-Ne lasers.[5] Notable were the works of Saloma and Stenholm[15] and of Kazantsev, Rautian and Surdutovich.[16] Antoranz et al.[9] have pointed out that the building of a Landau potential, as above, and the demonstration that the system has a state corresponding to a minimum, is not enough to prove that it is stable; in addition, one must show that the Landau potential is a Lyapunov functional of the system, and in general this is not easy.

In addition to the application to dye lasers, considered in the following section, a number of other developments relate to the first-order phase transition analogy developed above. Especially interesting are the papers by Haken and Ohno,[17] which show that the transition from cw laser emission to the ultrashort pulses can be first- or second-order depending upon laser length. It is also of interest to note that a detailed theoretical treatment by Roy[18] yields a final result for the chemical potential of a laser with saturable absorber (Equation 2.31 of Reference 18) in agreement with Equation (7) above. In addition, Mandel[19] has shown that under some circumstances the bistability predicted in Reference 3 may actually be tri-stable.

Application of the simple "thermodynamic" theory has been rather widespread, including such direct extensions as lasers with output feedback,[20] but also to general non-laser systems which have non-equilibrium transitions. The latter include the Rayleigh-Benard instability,[21] stimulated Cerenkov radiation,[22] freezing and nonlinear diffusion,[23] and coherent boson descriptions of ferromagnetism.[24] Thus, while most applications of this simple theory are to optical **bistability,**[25-31] some authors have found it useful in characterizing other non-equilibrium phase transitions, such as Josephson junctions[32,33] chemical reactions,[34] and other systems,[35] such as supercooled liquids.[36]

III. LASERS WITH SATURABLE ABSORBERS

Studies continue on bistable behavior of lasers with saturable absorbers. In addition to the early work on Ne absorbers in He-Ne laser systems, some particularly good work has been done[5,27] on CO_2 with SF_6.

Figure 2 shows the laser flux in W/cm^2 for a CO_2 laser having an SF_6 saturable absorber. The small signal gain was found to be proportional to the SF_6 pressure. The data extrapolate to a critical point at zero flux and SF_6 pressure of 0.01mTorr. For this value, $I_S = A\sigma_S/B\sigma$ in Equation (8). A

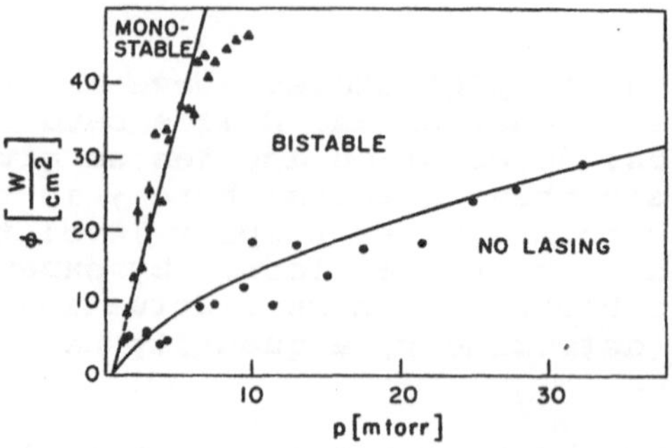

Fig. 2. Laser flux versus SF_6 pressure in a CO_2 laser system (Ref. 5).

greater SF_6 pressure lowers the saturation current I_S and produces bistability. The line between bistable and monostable operation is a function of laser flux; higher flux allows the monostable operation to be reached for a given value of I_S. This figure is compatible with the earlier prediction of Scott,[4] shown in Figure 3. In Figure 3 transitions through the dashed curve are from non-lasing to bistable, whereas transitions through the solid curves are to mono-stable regimes. Figure 3 shows why the point in phase space where transitions go from second-order to first-order is called a tricritical point: phase boundaries meet at that point. In Figure 4 we show[7] plots of phonon number versus pump parameter for both second-order and first-order cases.

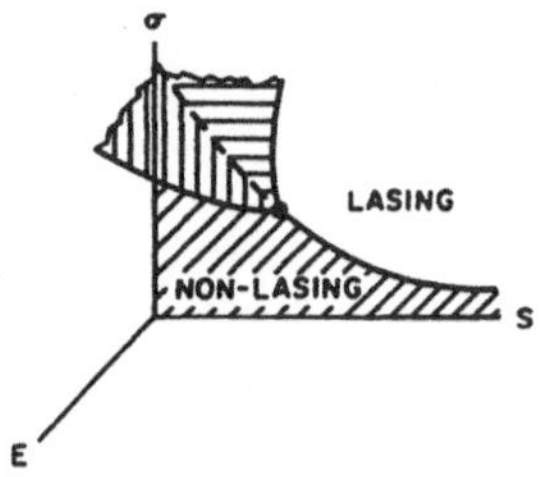

Fig. 3. Tricritical point in the phase space of a laser with saturable absorber (Ref. 4). Here E is the expectation value of the laser field; σ is the population inversion; and S is the signal strength.

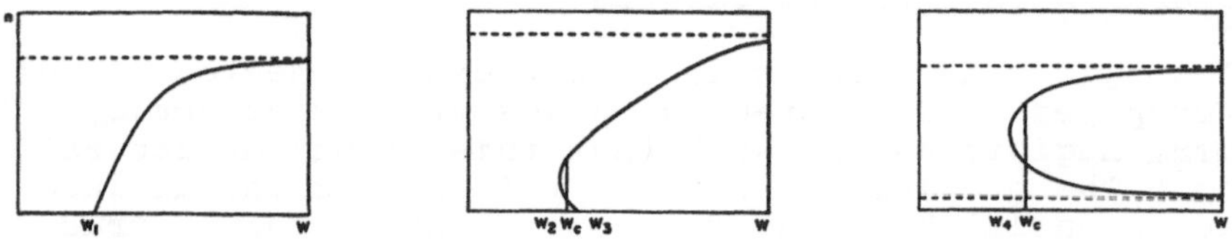

Fig. 4. Phase diagram for second-order threshold (left) and first-order thresholds (middle and right). Photon number n is plotted versus pumping parameter (Ref. 7).

IV. DYE LASERS

In dye lasers the triplet states serve as "built-in" saturable absorbers. Schaefer and Willis have shown that the same phenomenological description applies as given above. In Figure 5 we see their phase diagram; here γ is the fractional return of molecules in the upper triplet level to the lower, and α is proportional to triplet loss. By experimentally measuring the phase boundary (where fluctuations diverge) and knowing α, one can determine γ, a quantity not easily measured via other techniques.

Let us note that the behavior of laser field E upon population inversion σ at a critical point may be expressed in general as

$$E \sim \left(\sigma - \sigma_T\right)^{\beta} \qquad (9)$$

Lasers may be described as mean-field, and at most critical points in their phase diagram, $\beta = 1/2$ for mean field systems. However, exactly at a tricritical point,[14] $\beta = 1/4$. This point has been emphasized by Dembinski and Kossakowski,[7] but does not seem to have been tested experimentally.

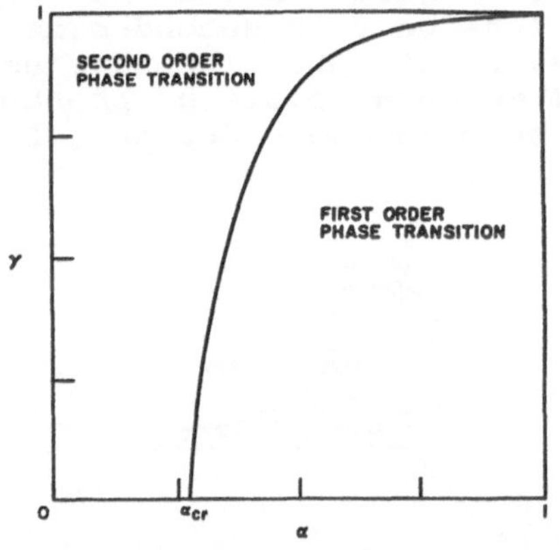

Fig. 5. Phase diagram for dye lasers. Triplet recombination parameter γ is plotted versus triplet loss α (Ref. 6).

V. LASERS WITH INJECTED SIGNALS

A single mode laser with an optically injected signal was also proposed as a first-order phase transition analog[4,37]; however, Lugiato has shown[38] that this is not an accurate analog. The problem is that for this system the defined Landau potential has a saddle point, rather than a true minimum. This does not invalidate the use of injected signal strength as a variable of interest in the case of lasers with saturable absorbers, but it shows that there are no discontinuities at threshold in lasers without saturable absorbers, regardless of injected signal strength.

VI. NEW EXTENSIONS OF THE THEORY: PIPPARD RELATIONS

Those who study "real" phase transitions often utilize the Pippard relations.[11-13] Originally derived by Pippard for helium, these thermodynamic relations assume that certain functions in phase space near critical points can be approximated as cylindrical. Buckingham and Fairbank[11] put this theory on more rigorous footing; Garland[12] extended it to solids and elasticity; and Janovec[13] provided the most complete tensorial form. Basically, the theory shows that the divergent (or "critical" part) of certain functions are proportional near the critical point. For example, the anomalous increase in specific heat versus temperature is proportional to the change in longitudinal sound velocities (the latter are negative); the increase in pyroelectric tensor is proportional to the change in piezoelectric tensor. In general, a variety of parameters have their increases or decreases near T_c in proportion.

What are the analogs of the Pippard relations in a laser near threshold? If the laser is mean field, its electronic specific heat shows a discontinuity at threshold. If it is not mean field, its electronic specific heat shows a divergence. What measurable parameters are proportional to this specific heat?

REFERENCES

1. R. Graham and H. Haken, Z. Phys. 213:420 (1968); Z. Phys. 237:31 (1970); H. Haken, Rev. Mod. Phys. 47:67 (1975).
2. M. O. Scully and W. E. Lamb, Jr., Phys. Rev. 159:208 (1967); V. DeGiorgio and M. O. Scully, Phys. Rev. A 2:1170 (1970); J. C. Goldstein, M. O. Scully and P. A. Lee, Phys. Lett. 35A:317 (1971); M. O. Scully, Proc. 3rd Roch. Conf. on Coherence and Quantum Optics, Rochester, New York, May 1972, edited by L. Mandel and E. Wolf (New York: Plenum Publishing Co., 1973).
3. J. F. Scott, M. Sargent III and C. D. Cantrell, Optics Commun. 15:13 (1975).
4. J. F. Scott, Optics Commun. 15:343 (1975).
5. V. P. Chebotayev, I. M. Beretov, and V. N. Lisitsyn, IEEE J. Quant. Elec. QE4:788 (1968). Perhaps the best experimental example is the CO_2/SF_6 data of S. Ruschin and S. H. Bauer, Chem. Phys. Lett. 66:100 (1979).
6. R. B. Schaefer and C. R. Willis, Phys. Lett. 58A:53 (1976); Phys. Rev. A 13:1874 (1976).
7. S. T. Dembinski and A. Kossakowski, Z. Phys. B24:141 (1976); A. Baczynski, A. Kossakowski, and T. Marszalek, Z. Phys. B23:205 (1976); S. T. Dembinski and A. Kossakowski, Z. Phys. B25:207 (1976); S. T. Dembinski, A. Kossakowski and L. Woliewicz, Z. Phys. 27:281 (1977).
8. G. Marowsky and W. Heudorfer, Optics Commun. 26:381 (1978); Applied Physics 17:181 (1978).
9. J. C. Antoranz, J. Gea, and M. G. Verlarde, Phys. Rev. Lett. 47:1895 (1981); F. Casagrande and L. A. Lugiato, Nuovo Cim. 2D:287 (1978); M. L. Asquini and F. Casagrande, Nuovo Cim. 2D:917 (1983).
10. K. T. Mahanthappa and M. A. Sher, Phys. Rev. D22:1711 (1980).

11. A. B. Pippard, Phil. Mag. 1:473 (1956); M. J. Buckingham and W. M. Fairbank, Progr. Low Temp. Phys. 3:80 (1961).
12. C. W. Garland, J. Chem. Phys. 41:1005 (1964).
13. V. Janovec, J. Chem. Phys. 45:1874 (1966).
14. C. D. Landau, Phys. Z. Sowjetunion 11:26 (1937).
15. R. Salomaa and S. Stenholm, Phys. Rev. A8:2695 (1973); R. Salomaa, J. Phys. A7:1094 (1974).
16. A. P. Kazantsev, S. G. Rautian, and G. I. Surdutovich, Sov. Phys. JETP 27:756 (1968); Sov. Phys. JETP 29:1075 (1969); Sov. Phys. JETP 31:133 (1970).
17. H. Haken and H. Ohno, Optics Commun. 26:117 (1978); Optics Commun. 16:205 (1976).
18. R. Roy, Phys. Rev. A20:2093 (1979).
19. P. Mandel, Z. Phys. B33:205 (1979); Phys. Lett. A83:207 (1981).
20. H. Inaba, Phys. Lett. 86A:452 (1981).
21. M. G. Velarde and J. C. Antoranz, J. Stat. Phys. 24:235 (1981); Phys. Lett. 72A:123 (1979).
22. H. Dekker, Physica 90C:283 (1977).
23. T. Munakata, J. Phys. Soc. Japan 45:749 (1978).
24. B. S. Shastry, G. S. Agarwal, and I. R. Rao, Pramana 11:85 (1978).
25. T. S. Dlodlo, Physica 111C:353 (1981).
26. A. Schenzle and H. Brand, Optics Commun. 31:401 (1979).
27. A. Jacques and P. Glorieux, Optics Commun. 40:455 (1982); K. Tomita, T. Todani, and H. Kidachi, Physica 84A:350 (1976); Phys. Lett. 51A:483 (1975).
28. A. R. Bulsara and W. C. Schieve, Optics Commun. 26:384 (1978); G. Haag, M. Munz and G. Marowsky, IEEE J. Quant. Elec. JQE17:349 (1981).
29. L. A. Lugiato, P. Mandel, S. T. Dembinski, and A. Kossakowski, Phys. Rev. A18:238 (1978); C. Harder, IEEE J. Quant. Elec. JQE18:1351 (1982).
30. R. Bonifacio, M. Gronchi, and L. A. Lugiato, Phys. Rev. A18, 2266 (1978); Phys. Rev. A18:1129 (1978).
31. R. Graham and A. Schenzle, Phys. Rev. A23:1302 (1981); S. S. Hassan, P. D. Drummond and D. F. Walls, Optics Commun. 27:480 (1978); S. A. Collins, Opt. Eng. 19:478 (1980).
32. M. Milani, R. Bonifacio, and M. Scully, Nuovo Cim. Lett. 26:353 (1979).
33. S. R. Shenoy and G. S. Agarwal, Phys. Rev. Lett. 44:1524 (1980).
34. A. Nitzan and J. Ross, J. Chem. Phys. 59:241 (1973); G. Nicolis and I. Prigogine, "Self Organization in Non-Equilibrium Systems" (New York: Wiley, 1977).
35. M. Agu and Y. Teramachi, J. Appl. Phys. 49:3645 (1978).
36. R. E. Heist and H. Reiss, J. Chem. Phys. 59:665 (1973).
37. W. W. Chow, M. O. Scully and E. W. van Stryland, Optics Commun. 15:6 (1975).
38. L. A. Lugiato, Lett. Nuovo Cim. 23:609 (1978). Also see C. R. Willis, Optics Commun. 23:151 (1977) and R. Bonifacio and L. A. Lugiato, Lett. Nuovo Cim. 21:510 (1978).

NONEQUILIBRIUM PHENOMENA AT INCOMMENSURATE PHASE TRANSITIONS

J. F. Scott

Condensed Matter Laboratory
Department of Physics
University of Colorado
Boulder, Colorado 80309

ABSTRACT

In addition to the large number of nonequilibrium analogs to phase transitions, such as lasers near threshold, there exists a class of true phase transitions in crystals which display qualitatively nonequilibrium phenomena: incommensurate structural phase transitions. In this paper we review the diverse data for incommensurate crystal structures, with special emphasis upon $BaMnF_4$, and show that most experimental phenomena can be explained by the assumption -- independently proved -- that the dynamics of the material near the phase transition temperature are not those of a system in thermal equilibrium. This explains several qualitative features of the transition in $BaMnF_4$: 1) $V_{ij} \neq V_{ji}$ for transverse sound velocities, where ij subscript denotes propagation along i with polarization along j; 2) linear birefringence is proportional to the square of the order parameter, in agreement with theory, whereas both optical activity and monoclinic distortion angle (4' of arc) are temperature independent, in qualitative disagreement with intrinsic theories; 3) large hysteresis effects are observed even if cycling is confined to the incommensurate phase; 4) the incommensurate a-axis lattice constant is temperature independent in the incommensurate phase, contrary to expectations and contrary to the temperature dependence of the b- and c-axes; 5) the temperature dependence of the incommensurate wave vector satisfies the predictions of the extended Ising model calculation of Yamada and Hamaya, $2a^*/5 > q_o > 5a^*/13$ for trajectories in phase space, although that model does not permit as T = 0 ground state either an incommensurate structure or a $2a^*/5$ phase, in contrast with observations.

Introduction:

Incommensurate structural phase transitions are systems of considerable interest at present; they exist in metals,[1] semiconductors, and insulators,[2] including intercalcated compounds.[3] By "incommensurate", one designates a structure whose primitive unit cell below a transition temperature T_I is not an integer multiple of (or even a rational fraction of) the unit cell above T_I. Incommensurate structures were first analyzed in the case of spiral magnets by Dzyaloshinskii,[4] and a theoretical formalism was developed which explained their stability in terms of gradient terms in a Landau free energy. A later and independent development[5] in metal physics theory by Overhauser led to the understanding of incommensurate structures in semiconductors and metals in terms of charge density waves and spin density waves. Incommensurate structures are mathematically pathological: As explained below, they are not members of any of the 230 space groups,[6] although they may be perfectly translationally ordered; in addition, they have infinitely degenerate ground states (within the usual approximation that a crystal contains infinitely many atoms). As a result of these considerations, the Goldstone boson which describes excitation from one of these ground states to another is a phase mode (quantized as "phasons") which corresponds to a "sliding" of the modulation along a specified axis. Very recently a number of seemingly unrelated phenomena have been reported for incommensurate insulators. We show in the sections which follow that effects are indeed related and may be explained by the hypothesis that the systems are far from thermal equilibrium. Indeed, in the cases of $BaMnF_4$ and $RbH_3(SeO_3)_2$, emphasized below, they appear to arise from the combined effects of nonequilibrium chiral strains and the presence in each crystal of a screw axis.

The basic cell multiplication possibilities which can result for the primitive unit cell of a crystal are sketched schematically in Figure 1. In Figure 1a we see a crystal having rocksalt structure above the transition temperature T_0 and a distortion which is exactly in-phase in adjacent unit cells. If such a distortion becomes static below T_0, the low-temperature structure develops a spontaneous polarization and is ferroelectric (or at least pyroelectric); however, its unit cell remains unchanged. In Figure 1b we see an example in which the distortion is precisely out-of-phase in adjacent cells. Below T_0 this yields an antiferroelectric phase transition with no net spontaneous polarization and a doubled primitive cell. We note parenthetically that SnTe/GeTe/PbTe compounds are examples[7] of real crystals exhibiting the transition diagrammed in Figure 1a. There are no simple diatomic lattices which are well-studied and which exhibit the transition diagrammed in Figure 1b, although SnSe under high hydrostatic pressure may be quite similar[8].

In Figure 1c we see a third possibility: The distortion of equivalent nearest neighbor ions has an arbitrary phase. If this distortion is static, i.e., "frozen in", the resulting lattice below T_I is not in a commensurate relationship with that above T_I. That is, the translational symmetry along the z-axis is not characterized by a finite lattice constant **c**; it requires a infinite translation to find an ion equivalent to the one chosen at the origin. This infinite translational

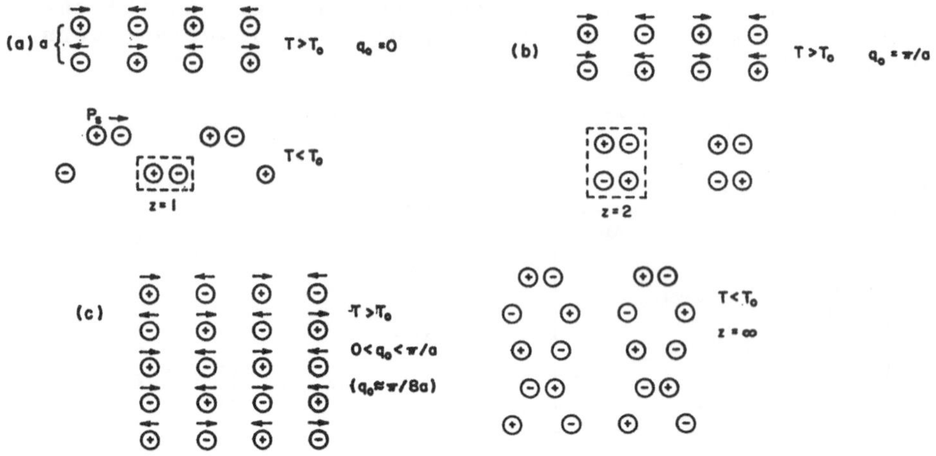

Fig. 1. Distortions: **a)** Ferroelectric distortion in a diatomic lattice; **b)** antiferroelectric distortion; **c)** incommensurate distortion, Z is the number of diatomic formula groups per primitive cell.

repeat distance is the reason the resulting low temperature phase cannot be characterized as one of the 230 space group symmetries for periodic structures.

It is important to note that if we go "far enough" along the z-axis in Figure 1c we will find an ion at every possible phase position in the modulation wave characterizing the incommensurate distortion. This has a unique (and pathological) result in crystal physics: It means that it is possible to "slide" the phase of the distortion modulation an infinitesimal amount along c without changing the energy of the lattice, integrated over all c. This implies that such a sliding distortion is a zero-energy Goldstone boson for this system -- a phase-wave quantized into "phasons".

Unexplained Dynamical Phenomena:

The first evidence of nonequilibrium dynamical phenomena in incommensurate insulators was provided by Hamano et al.[9] at Tokyo Institute of Technology. They made careful measurements of the dielectric constants of such materials and found strong hysteresis effects as functions of heating and cooling cycles, **even when** the cycles were constrained to lie within the incommensurate phase. This, then, could not be due to the usual hysteresis encountered for first-order phase boundaries, due to latent heat, but only to non-intrinsic, non-equilibrium origins. Figures 2a and 2b show the kind of dielectric and wave-vector hysteresis found experimentally; and Figure 2c shows the result of sitting at one temperature over a long period (50-100 hours) of time. Both effects seem to be due to the pinning effects of extrinsic defects on the phase modulation of the lattice.

Some other unexplained phenomena were measured by several investigators on other incommensurate insulators. In both

$BaMnF_4$ and $RbH_3(SeO_3)_2$ it was found (Figures 3a and 3b) that the transverse sound velocities V_{ij} were not invariant upon interchange of subscripts. That is, $V_{ij} \neq V_{ji}$ in the incommensurate phase.[10,11] This was a most puzzling observation, for one naively assumes that

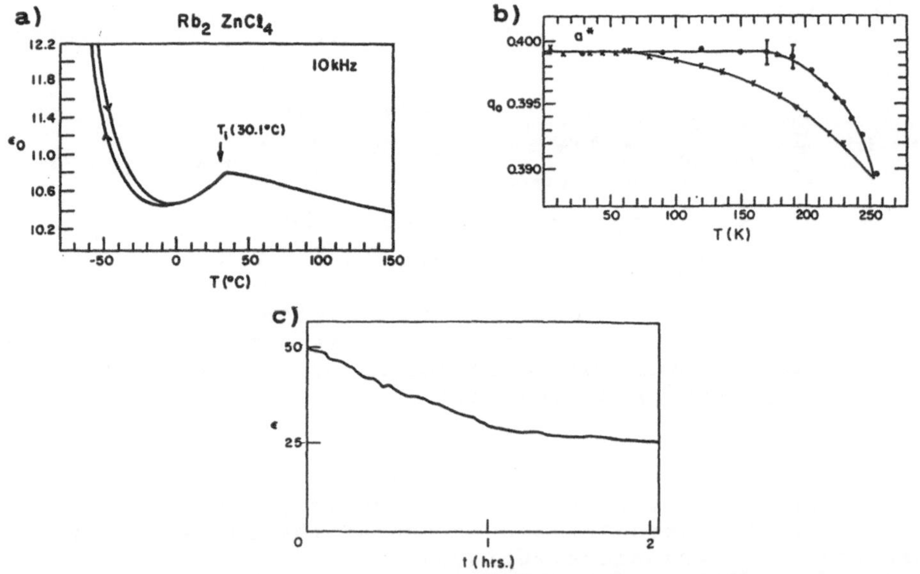

Fig. 2. **a) Dielectric constant temperature hysteresis (Ref. 9). b) Translation wave vector q_0 hysteresis (Ref. 24) in incommensurate lattice. c) Time dependence of dielectric constant in incommensurate Rb_2ZnCl_4 (after Prof. Hamano, see Ref. 9).**

$$V_{ij} \sim (C_{ij}/d)^{1/2} \tag{1}$$

where C_{ij} is the elastic coefficient, and d is the density. Since it is well-known that the elastic coefficients C_{ij} are invariant under interchange of subscripts

$$C_{ij} = C_{ji} \tag{2}$$

an apparent paradox exists. It is important to note that Equation (2) follows from purely thermodynamic arguments and does not rely upon any microscopic arguments.

Both Scott[12] and Dvorak and Esayan[13] suggested that this paradox might be reconciled by dispersion. Equation (2) is true at zero frequency, whereas the sound velocities V_{ij} are measured at finite frequency and wave vector. Thus, $V_{ij} \neq V_{ji}$ might arise from the influence of finite wave vector and energy; however, variation of probe frequency by a factor of a thousand, via ultrasonic and Brillouin experiments, does not confirm this hypothesis. Therefore we are led to suspect the validity of Equation (2). In fact, $C_{ij} = C_{ji}$ only under conditions of thermal equilibrium; in particular, it is false in the presence of external stresses. The dielectric data of Hamano et al.[9] described above suggest that external stresses

are present in the incommensurate phase of crystals and anneal out only on time scales which are long compared with the duration of normal experiments. Therefore, it is reasonable to question the validity of Equation (2).

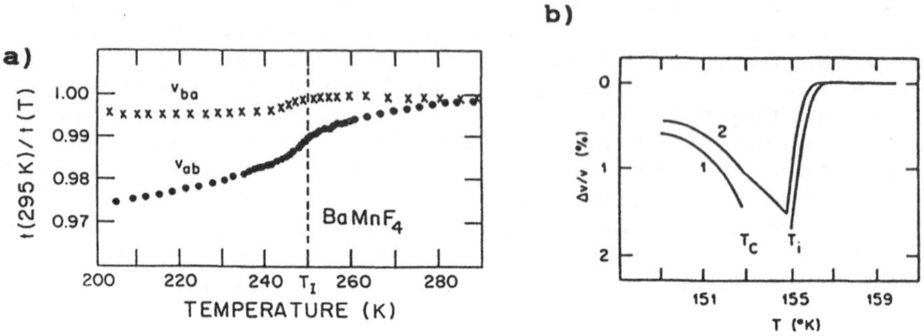

Fig. 3. $V_{ij} \neq V_{ji}$ sound velocities in: a) $BaMnF_4$ (Ref. 10); b) $RbH_3(SeO_3)_2$, (Ref. 11).

Chiral Characteristics:

It is most important to note that in both $BaMnF_4$ and $RbH_3(SeO_3)_2$ there is a screw axis.[14,15] We also find in the optical studies of critical phenomena in $BaMnF_4$ by Pisarev et al.[16] the surprising and unexplained observation that whereas the linear birefringence is proportional to the square of the order parameter over a wide range of temperature below T_I, in accord with theoretical predictions,[17] the optical activity tensor component and the monoclinic distortion angle of 4' (more precisely, the rotation angle for the optical ellipsoid) are temperature independent. This experimental observation shows that phenomena which are explicitly chiral (e.g., optical activity) are not in accord with intrinsic equilibrium theories; the role of the screw axis is immediately suggested. We not parenthetically that $V_{ij} \neq V_{ji}$ asymetry has never been observed in an incommensurate crystal lacking a screw axis; such experiments are in progress in the author's laboratory.[18]

The remaining puzzle which is purely experimental is the observation that the lattice constant along the incommensurate axis is temperature independent.[19] Or, since it is difficult to define a lattice constant along the incommensurate axis, a more precise statement might be that the thermal expansion coefficient along this axis in $BaMnF_4$ is nearly zero. This is in contrast to the behavior along the two orthogonal axes, and it is also in contrast with normal theories, which predict linear expansion with temperature in this temperature range.

Other aspects of $BaMnF_4$ statics and dynamics which lead to the conclusion that the system is not in thermal equilibrium for the measurements performed involve the disagreement with data and an extended Ising model calculation. To explain these discrepancies one must first explain the extended Ising model, describe its successes in describing incommensurate phase transitions, and show in what ways it both explains phenomena in $BaMnF_4$ and at the same time pinpoints nonequilibrium characteristics.

Extended Ising Model:

As first developed by Bak and von Boehm[20] and others,[21] the extended Ising model is an attempt to explain the trajectory of transitions in phase space which is encountered in many real crystals having incommensurate intermediate phases. Yamada and Hamaya[22] have been most successful in applying such a model to real incommensurate crystals, namely the A_2BX_4 family. Allowing third nearest neighbor interactions and assuming J_1 and J_2 antiferroelectric and J_3 small and ferroelectric ($J_3 = -0.1J_1$), they obtain the phase diagram shown in Figure 4. Here the ordinate is reduced temperature and the abscissa is J_2/J_1. Trajectories in phase space corresponding to real sequences of phase transitions encountered upon lowered temperatures are shown as slightly non-vertical lines; the slope is due to the allowance for temperature dependent exchange -- $J_1(T)$, $J_2(T)$.

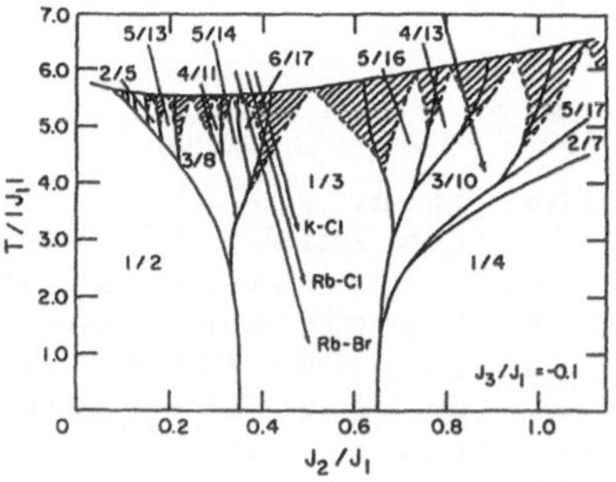

Fig. 4. **Theoretical phase diagram for A_2BX_4 incommensurates (Ref. 22). Ordinate is proportional to temperature; abscissa is J_2/J_1 from an extended Ising model.**

The shaded regions are incommensurate structures. The ground states correspond to structures in which the translational symmetry has stayed the same, tripled, or doubled. With such a model Yamada and Hamaya were able to explain 22 out of 22 observed phase transition sequences in A_2BX_4 structures. Those for A_2BCl_4 are shown in three cases in Figure 4.

A primary success of this model is the prediction, in accord with experiment, of strange fractional changes in the commensurate structures, e.g., 6a*/17 or 5a*/13. In the conventional Landau theory such phases would require[23] unrealistic (large) interaction terms in the free energy of form $Q^{13}P$ or $Q^{17}P$, where Q is the order parameter and P the spontaneous electric polarization.

The diagonal trajectory on the upper right part of the diagram represents $BaMnF_4$. The incommensurate phase is approached from high temperatures near the 5a*/13 phase and exits at 2a*/5. Thus, this trajectory predicts an incommensurate wave vector which varies from slightly greater than 5a*/13 at T_I to 2a*/5 at the lower "lock-in" transition. This is perfectly in accord with the latest inelastic neutron scattering data from Grenoble,[24] which show

$$2a^*/5 > 0.399 \geq q_o \geq 0.389 > 5a^*/13 \qquad (3)$$

with a lock-in transition at 60K upon cooling or 170K upon heating and a high-temperature transition near 250K.

Although the extended Ising model of Yamada and Hamaya explains the observed end point values of incommensurate wave vector q_o in $BaMnF_4$, the model raises and obvious question: Since the incommensurate phase and the $2a^*/5$ phase are both unstable at absolute zero, why doesn't the crystal phase have another transition to the predicted $a^*/2$ ground state? This may in fact be viewed as additional evidence that the material is not in true equilibrium. Both $Ba_2NaNb_5O_{15}$ and $BaMnF_4$ have[25] apparently stable phases within $0.01a^*$ of commensurate over very wide temperature ranges (about 200K). It seems likely that these are extrinsically stabilized.

Summary:

A variety of unexplained and seemingly unrelated phenomena in incommensurate insulators can be reconciled by two assumptions: That they are not in thermal equilibrium; and 2), that they have screw axes. This combination of characteristics permits chiral strains. Such strains are independently confirmed through dielectric measurements by Hamano et al. or the hysteresis effects in inelastic neutron studies of Barthes-Regis et al., although neither of these shows the chirality directly. The chiral strains would permit $C_{ij} \neq C_{ji}$ and hence $V_{ij} \neq V_{ji}$. They would also permit the temperature independence of optical activity below the transition temperature while allowing linear birefringence (which is not chiral) to scale as the square of the order parameter. Finally, they would explain the generally good agreement between the experimental data and the temperature dependence of the incommensurate wave vector q_o found in the extended Ising model to predict the observed gound state, while at the same time allowing insight into the failure of this model to predict the observed ground state. Even the temperature independence of the a-axis lattice constant in the incommensurate phase of $BaMnF_4$ can be qualitatively understood from such a model: as temperature increases the helical pitch of the screw axis would change from $5a/2 = 2.5a$ at low

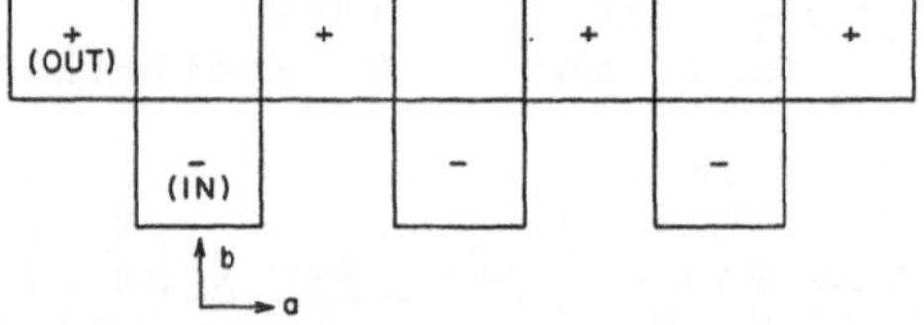

Fig. 5. $BaMnF_4$ structure (Ref. 14) showing displacement in incommensurate phase -- a rotation of MnF_6 octahedra out of the ab-plane.

temperatures to 13a/5 = 2.6a at T_I = 250K if there were no lattice expansion; however, in the real crystal this thermal strain can be accommodated by increasing the rotation of MnF_6 octahedra out of the ab-plane, as shown in Figure 5. In fact, in this way the lattice contraction encountered with decreasing temperature can serve as a stabilization of the incommensurate wave vector.

To Be Done:

There are no microscopic models from which one can calculate J_1, J_2, J_3 in the extended Ising model, for any crystal. Moreover, there is no quantitative theory which explains in tems of specific defect models the magnitude of $C_{ij} - C_{ji}$ or the temperature independence of optical activity or lattice constant in $BaMnF_4$.

REFERENCES

1. F. J. DiSalvo, Jr., and T. M. Rice, Physics Today 32:32 (1979); Surf. Sci. 58:297 (1976).
2. R. Blinc, Physics Reports 79:331 (1981).
3. M. S. Dresselhaus, Physics Today 37:60 (1984).
4. I. E. Dzyaloshinskii, Zh. Eksp. Teor. Fiz. 46:1420 (1964) [Translation: Sov. Phys. 19:960 (1964)].
5. A. W. Overhauser, Phys. Rev. 167:691 (1968); Phys. Rev. B 3:3173 (1971).
6. A. Janner and T. Jannssen, Phys. Rev. B 15:643 (1977).
7. W. Cochran, Phys. Rev. Lett. 3:521 (1959); G. S. Pawley, W. Cochran, R. A. Cowley, and G. Dolling, Phys. Rev. Lett. 17:753 (1966).
8. J. A. Kafalas and A. N. Mariano, Science 143:952 (1964); G. S. Pawley, J. Physique Suppl. 29:C4 (1969).
9. K. Hamano, Y. Ikeda, T. Fujimoto, K. Ema and S. Hirotsu, J. Phys. Soc. Japan 49:2278 (1980).
10. I. J. Fritz, Phys. Lett. 51A:219 (1975).
11. S. Kh. Esayan, V. V. Lemanov, N. Mamatkulov, and L. A. Shuvalov, Kristallografia 26:1086 (1981) [Translation: Sov. Phys. Crystallogr. 26:619 (1983)]..
12. J. F. Scott, Ferroelectrics 47:33 (1983).
13. V. Dvorak and S. Kh. Esayan, Sol. St. Commun. 44:901 (1982).
14. E. T. Keve, S. C. Abrahams, and J. L. Bernstein, J. Chem. Phys. 54:3185 (1971).
15. R. Tellgren, D. Ahmad, and R. Liminga, J. Sol. St. Chem. 6:250 (1973).
16. R. V. Pisarev, B. B. Krichevtzov, P. A. Markovin, O. Yu. Korshunov, and J. F. Scott, Phys. Rev. B 28:2677 (1983).
17. J. Fousek and J. Petzelt, Phys. Stat. Sol. A55:11 (1979); G. Gehring, J. Phys. C10:531 (1977).
18. W. L. Oliver, T. Yagi, and J. F. Scott (1984, unpublished).
19. D. E. Cox, S. M. Shapiro, R. A. Cowley, M. Eibschutz, and H. G. Guggenheim, Phys. Rev B 19:575 (1979).
20. P. Bak and J. von Boehm, Phys. Rev B 21:5297 (1980).
21. J. VIllain and M. B. Gordon, J. Phys. C13:3117 (1981); F. Axel and S. Aubry, J. Phys. C14:5433 (1981).
22. Y. Yamada and N. Hamaya, J. Phys. Soc. Japan 52:3466 (1983).

23. C. J. Pater, J. D. Axe, and R. Currat, Phys. Rev. B 19:4684 (1979).
24. M. Barthes-Regis, R. Almairac, P. St.-Gregoire, C. Filippini, U. Steigenberger, J. Nouet, and Y. Gesland, J. Physique Lett. 19:L829 (1983).
25. J. Schneck and F. Denoyer, Phys. Rev. B 23 (1982); J. Schneck, J. C. Toledano, C. Joffrin, J. Aubree, B. Joukoff, and A. Gabelotaud, Phys. Rev. B 25:1766 (1982).

ON THE QUASIPARTICLE- AND SUPER-CURRENT AT THE

FINITE TEMPERATURE JOSEPHSON JUNCTION

Alfred Rieckers

Institute for Theoretical Sciences
Department of Physics and Astronomy
University of New Mexico
Albuquerque, NM 87131, USA
and
Institute for Theoretical Physics
University of Tübingen
Tübingen, West Germany

INTRODUCTION

In the temperature dependent reconstructed quantum mechanics over the limiting Gibbs state for two weakly coupled superconductors the effective dynamics is worked out and the current operator is expressed in terms of gauge covariant quasi-particle and condensed Cooper pair fields, where only the latter exhibit macroscopic quantum features.

THE STANDARD CALCULATION PROCEDURE

The current between two weakly coupled superconductors has first been calculated by Josephson [1] and then by various authors using different methods (cf. the references in [2]). The most representative approach seems to be that employed and carefully analyzed in [2]. There one starts from the conventional tunneling Hamiltonian

$$h = \sum_{kl\sigma} (T_{kl} c^*_{k\sigma a} c_{l\sigma b} + T^*_{lk} c^*_{l\sigma b} c_{k\sigma a}) \tag{1}$$

where the T_{kl} are complex numbers (proportional to the one electron overlap integral) and $c_{l\sigma b}$ is the electron annihilation operator for the momentum l, the spin σ, and the superconductor b. The star denotes either the complex conjugate or the Hermitian adjoint. The time derivative of the number operator $N_a = \sum_{k\sigma} c^*_{k\sigma a} c_{k\sigma a}$ is evaluated up to first order perturbation theory in the interacting reference state. The anomalous expectation values $\langle c_{-k\downarrow x} c_{k\uparrow x}\rangle$ for the superconductor $x \in \{a,b\}$ are assumed to be non-vanishing complex numbers with the phases $\theta_x(t)$. In this way the expectation for the current from a to b is obtained in the form

$$\langle I\rangle_t = \langle I_n\rangle + \langle I_{J1}\rangle \cos\phi(t) + \langle I_{J2}\rangle \sin\phi(t) \tag{2}$$

where

$$\phi(t) = \theta_a(t) - \theta_b(t) = 2eVt + \phi(0), \tag{3}$$

with V the voltage difference. For V = 0 only the third term survives and is interpreted as the pair tunneling current. The first term is the normal current, whereas the second is described as a "quasi-particle tunneling process which involves concomitant destruction and creation of pairs on different sides, therefore involving phase coherence effects" ([2], p. 39). The experimental significance of this cosϕ-term led to a controversial discussion ([2], p. 47 ff).

In our opinion the "cosϕ-problem" is an indication for the principal insufficiency of the theoretical treatment outlined above. First of all it is not evident that the physical nature of the current terms may be clarified by means of one expectation value alone. A discussion in terms of current operators would cover a much wider class of experimental situations. It is, in fact, confusing that the slightest disturbance of the macroscopic state (as well as the inclusion of higher order terms), resulting in a totally different form for the current expression, would necessitate a new interpretational discussion. Also the fundamental difference between a weak and a strong coupling model is not set forth by the standard calculation procedure.

In any case, a clear theoretical picture of the fascinating quantum mechanical collective phenomena at the Josephson junction may only be expected from a careful extrapolation of the traditional quantum mechanics to the thermodynamical limit. For finitely many degrees of freedom the equilibrium ensembles have always the same symmetries as the Hamiltonians and no anomalous expectation values may arise. It is the purpose of this note to outline the basic steps and qualitatively new features, which are involved in constructing an extrapolated quantum mechanics with a nontrivial superimposed macroscopic structure.

THE FINITE SUPERCONDUCTOR

For a finite volume the superconductor x, $x \in \{a, b\}$, has only finitely many electron momenta k in a shell around the Fermi surface, and we call such a set Λ_x. For a given Λ_x we consider here the most simple BCS-Hamiltonian

$$H_{\Lambda_x} = \sum_{\sigma, k \in \Lambda x} \varepsilon_x c^*_{k\sigma x} c_{k\sigma x} - (g/|\Lambda_x|) \sum_{kl \in \Lambda x} b^*_{kx} b_{lx} \tag{4}$$

where $|\Lambda_x|$ is the cardinality of Λ_x, $b_{kx} = c_{-k\downarrow x} c_{k\uparrow x}$, and where the kinetic and interaction energies are momentum independent. (In [3] arguments are provided for the qualitative persistence of the resulting picture if these restrictions are dropped.) With N_{Λ_x} the particle number operator confined to Λ_x and μ_x the chemical potential (fixed to its zero point value) the reduced Hamiltonian is $H_{\Lambda x} = H_{\Lambda x} - N_{\Lambda x}$. The electron algebra

$$\mathcal{A}_{\Lambda x}: = * - \text{alg.span}\{c_{k\sigma x};\ k \in \Lambda_x,\ \sigma \in \{\uparrow, \downarrow\}\} \tag{5}$$

is a finite dimensional full matrix algebra. The gauge transformations of the first kind in the Heisenberg picture are given by

$$\alpha_{\Theta_x}(A) := \exp(i\Theta_x N_{\Lambda x}) A \exp(-i\Theta_x N_{\Lambda x}) \tag{6}$$

where $A \in \mathcal{A}_{\Lambda x}$ and $\Theta_x \in [0,2\pi)$. In spite of the momentum space representation we call $\mathcal{A}_{\Lambda x}$ a "local algebra".

THE QUASILOCAL ELECTRON ALGEBRA

If we let increase the volume of the superconductor in discrete steps the electron momenta populate the shell around the Fermi surface in a dense but countable manner. For given Λ_x and Λ'_x there exists a Λ''_x which contains the former sets. We introduce an injection of $\mathcal{A}_{\Lambda x}$ into $\mathcal{A}_{\Lambda'' x}$ by mapping $c_{k\sigma x} \in \mathcal{A}_{\Lambda x}$ onto $c_{k\sigma x} \in \mathcal{A}_{\Lambda'' x}$ for all $k \in \Lambda_x$ (which is accompanied by a change of the position space wave function). This so-called inductive system of local algebras leads in a natural manner to a smallest mathematically nice algebra, which contains all $\mathcal{A}_{\Lambda_x}$ and may be written as

$$\mathcal{A}_x := \overline{\bigcup_{\Lambda x} \mathcal{A}_{\Lambda x}}^{\,\text{norm}} \; . \tag{7}$$

The construction of the so-called quasilocal algebra $\mathcal{A}_x$ does not depend on any Hilbert space and constitutes an abstract, normclosed, distributive algebra with $||A^*A|| = ||A||^2$ for all $A \in \mathcal{A}_x$, that is a C^*-algebra (cf., e.g., [4]). Beside the c-numbers there are no elements in $\mathcal{A}_x$ which commute with all other elements of $\mathcal{A}_x$, i.e. the center of $\mathcal{A}_x$ is trivial. From this we learn that infinitely many degrees of freedom alone don't lead to non-trivial classical observables, which by definition are to be compatible with all locally approximable quantum observables.

In spite of $N_{\Lambda x}$ escaping $\mathcal{A}_x$ for $\Lambda_x \to \infty$ (6) remains well defined in this limit and may be extended to an invertible mapping $\alpha_{\Theta x}: \mathcal{A}_x \to \mathcal{A}_x$, which preserves all algebraic compositions: the gauge transformations are realized in $\mathcal{A}_x$ in terms of $*$-automorphisms.

THE LIMITING GIBBS-STATE

The most difficult step of a model discussion in the thermodynamic limit is the calculation of the limiting equilibrium state which can be only carried through for the most simple Hamiltonians. In the case of (4) this calculation has been published only for the quasi-spin formulation [5], which refers to a subalgebra of $\mathcal{A}_x$, whereas we use here the the full electron algebra. Also in this more realistic model one can show [6] that for every Λ_x and every $A \in \mathcal{A}_{\Lambda_x}$ it holds

$$\lim_{\Lambda'_x \to \infty} \operatorname{tr}_{\Lambda'_x} \{\exp(-\zeta - \beta H^r_{\Lambda'_x}) A\} = : \langle \omega^\beta_x ; A \rangle$$
$$= \int_0^{2\pi} \langle \omega^{\beta\theta_x} ; A \rangle d\theta_x / 2\pi \; , \tag{8}$$

where $\operatorname{tr}_{\Lambda'_x}$ is the unnormalized trace on the matrix algebra $\mathcal{A}_{\Lambda_x}$. This means firstly that the limiting expectation value functional ω^β_x, which we still call a "state", exists, no matter in which way Λ_x tends to infinity, and secondly, that it has an integral decomposition over an angle θ_x.

Obviously the states $\omega^{\beta\theta_x}$, which we call "pure phases", are less mixed than ω^{β}. They satisfy

$$<\omega^{\beta\theta_x};\alpha_{\Theta_x}(A)> = <\omega^{\beta\theta_x+2\Theta_x};A> , \tag{9}$$

and we have a spontaneous break down of gauge invariance whereby the pair phase θ_x comes into play. One can show that (8) is the so-called central decomposition of the state ω^{β}, which is the unique demixture of this ensemble into disjoint (= macroscopically different) and factorial (= macroscopically pure) sub-ensembles. The anomalous expectation value is written in our terminology now $<\omega^{\beta\theta_x}; b_{kx}>$. It is different from zero if the pure phases $\omega^{\beta\theta_x}$ are different from each other and this holds for β larger than a certain critical β_{cx}.

THE COMPOSITE UNCOUPLED SYSTEM

The quasilocal algebra of the composite system is the antisymmetric tensorproduct $\mathcal{A} := \mathcal{A}_a \hat{\otimes} \mathcal{A}_b$, which may be generated by the linear combinations of commuting tensorproducts. Since it is sufficient to know the action of a state functional on the commuting tensorproducts we obtain for the composite uncoupled limiting Gibbs state

$$\omega^{\beta} = \omega_a^{\beta} \otimes \omega_b^{\beta} = \int \omega^{\beta\theta} d\theta, \tag{10}$$

where the "$\otimes$" inbetween state symbols indicates the product property for expectations of commuting tensorproducts, $\theta := (\theta_a, \theta_b)$, $d\theta := d\theta_a d\theta_b/(2\pi)^2$, and $\omega^{\beta\theta} := \omega^{\beta\theta_a} \otimes \omega^{\beta\theta_b}$.

According to the so-called GNS-construction [4] there are a unique (up to unitary equivalence) representation π_β, Hilbert space $\mathcal{H}_\beta$ and cylic vector $\Omega_\beta \in \mathcal{H}_\beta$ such that $\mathcal{H}_\beta = \overline{\pi_\beta(A)\Omega_\beta}$ (closure in the vector norm) and $<\omega_\beta;A> = (\Omega_\beta, \pi_\beta(A)\Omega_\beta)$ for all $A \in \mathcal{A}$. One knows (because $\mathcal{A}$ is a simple algebra) that $\pi_\beta(\mathcal{A})$ is isomorphic to $\mathcal{A}$. There are, however, certain sequences of physical interest in $\pi_\beta(\mathcal{A})$ as, e.g.,

$$\text{s-}\lim_{\Lambda_x} \sum_{k\in\Lambda_x} \pi_\beta(b_{kx})/|\Lambda_x| =: s_x , \tag{11}$$

which converge in the strong operator topology of $\mathcal{H}_\beta$ but not in the norm topology. We have, therefore, need for the representation dependent extension

$$\mathcal{N}_\beta := \overline{\pi_\beta(\mathcal{A})}^{s} , \tag{12}$$

which is even closed in the strong operator topology (which is weaker than the norm topology) and constitutes a so-called von Neumann algebra. For $\beta > \beta_{cx}$, $x = a$ or $x = b$, the center $\mathcal{Z}^\beta$ of $\mathcal{N}^\beta$ is non-trivial and contains s_x. Thus, it is crucial for many body physics to discriminate between representation dependent and independent limits in the observable algebra. The unique polar decomposition of s_x has the form

$$s_x = \exp(-2\Theta_x^\beta)\, w_x(\beta) , \tag{13}$$

where $w_x(\beta) > 0$ for $\beta > \beta_{cx}$ and zero otherwise, and where $\theta_x^\beta \in \mathcal{Z}^\beta$ is self-adjoint. The non-triviality of s_x is connected with the fact, that $\mathcal{H}_\beta$ has a direct integral decomposition over θ_x for $\beta > \beta_{cx}$ as a consequence of (10). $2\theta_x^\beta$ is then the multiplication operator by θ_x.

Since ω^β is gauge invariant we have a unique unitary representation $U_{\theta x}^\beta$ of the separate gauge transformations for the superconductor x which leaves Ω_β stable and satisfies for all $A \in \mathcal{A}$ and all $\theta_x \in [0,2\pi)$

$$\pi_\beta(\alpha_{\Theta_x}(A)) = U_{\theta_x}^\beta \pi_\beta(A) U_{-\Theta_x}^\beta \quad . \tag{14}$$

Its selfadjoint generator N_x^β is to be considered as the appropriate, renormalized particle number operator for the side x in the temperature representation (and differs essentially from the Fock particle number operator). It contains a differentiation to the macroscopic phase Θ_x^β (cf. [7], [15]) which leads to the commutation relations

$$[\Theta_x^\beta, N_x^\beta]_- = i\pi_\beta(1) \tag{15}$$

and

$$[N_x^\beta, s_x]_- = -2s_x \quad . \tag{16}$$

From the last relation we infer that s_x is the pair-annihilation operator of a special kind of particles, which we interpret as the condensed Cooper pairs. In fact, the defining averaging relation (11) imitates mathematically the loss of individual properties a particle-like structure experiences if incorporated into a condensate.

The relation (15) expresses a most characteristic feature of macroscopic quantum phenomena: Quantum condensation leads to peculiar observables as Θ_x^β, which are classical by being compatible with all locally approximable observables, but which exhibit also quantum properties if confronted with "exterior observables" like N_x^β. The genuine physical observable N_x^β is outside the scope of usual treatment since its spectral projections are no more in $\mathfrak{N}^\beta$ (it is not "affiliated" with $\mathfrak{N}^\beta$). The relation (15) is the origin for macroscopic incompatibility and macroscopic quantum coherence.

The consideration of the s_x, automatically given by the reconstructed quantum mechanics, raises the model discussion to a new stage. They enter the definition of the quasi-particle operators

$$\gamma_{k0x} := u_{kx}\pi_\beta(c_{b\uparrow x}) - r_{kx}s_x\pi_\beta(c^*_{-k\downarrow x})$$

$$\gamma_{k1x} := u_{kx}\pi_\beta(c_{-k\downarrow x}) + r_{kx}s_x\pi_\beta(c^*_{k\uparrow x}) \tag{17}$$

which diagonalize the renormalized Hamiltonian of the uncoupled limiting dynamics in $\mathcal{H}_\beta$ (cf. [7], [3]), if u_{kx} and v_{kx} are the usual coefficients of the Bogoliubov-Valatin transformation and $r_{kx} := |v_{kx}|/w_x(\beta)$. Observe that the $\gamma_{k\sigma x}$ are in $\mathfrak{N}^\beta$, but not in $\pi_\beta(\mathcal{A})$. Their gauge covariance and particle interpretation are secured by

$$[N_x^\beta, \gamma_{k\sigma x}]_- = -\gamma_{k\sigma x} \,. \tag{18}$$

In contrast to the electron-operators they don't generate $\mathcal{H}_\beta$ from Ω_β but have to be supplemented by the s_x (and by their Hermitian adjoints). The mentioned limiting dynamics in the Heisenberg picture exists in $\mathfrak{N}^\beta$ (but not in $\pi_\beta(\mathcal{A})$) and is characterized by

$$\Theta_x^\beta(t) = \Theta_x^\beta + \mu_x t \pi_\beta(1) \tag{19}$$

giving

$$s_x(t) = \exp(-i\mu_x t) s_x \tag{20}$$

and by

$$\gamma_{k\sigma x}(t) = \exp(-it(E_x + \mu_x))\gamma_{k\sigma x} \tag{21}$$

with

$$E_x := (\varepsilon_x^2 + g_x^2\, w_x^2(\beta))^{1/2} \,.$$

Equations (19) and (20) demonstrate that the BCS-Hamiltonian is able to move even classical observables in the thermodynamic limit.

THE WEAK COUPLING

Let us assume that the two superconductors are so close that even electron pairs may tunnel through the barrier. Then also fourth order terms in the $c_{k\sigma x}^{(*)}$ are relevant for the coupling Hamiltonian. Reversing (17) we may express any interaction in terms of the $\gamma_{k\sigma x}^{(*)}$ and $s_x^{(*)}$, if a certain temperature is given. One can show [8], that the spectrum (20) (27) is stable against all interactions which are bounded functions of the $\gamma_{k\sigma x}^{(*)}$ and $s_x^{(*)}$ (but which are not necessarily relatively compact with respect to the uncoupled Hamiltonian). This strong stability property is a many body effect (since the Connes theory for type-III$_\lambda$-factors comes essentially into play [8]) and classifies a rather large set of interactions as "weak". It gives physical significance to the exchange of these stable excitations in spite of their temperature dependence. Since the averaging procedure (11) applied to quasi-particles gives zero, we conclude that these particle structures cannot condense. Thus, the interaction cannot include terms of the type $s_a^*\gamma_b\gamma_b$ or $s_a^*\gamma_a\gamma_b$. The break up of ground pairs, which is important for some physical effects (cf., e.g., [9], Sect. 5.1) goes via the normal electron states and has been taken into account by the introduction of the $\gamma_{k\sigma x}$-operators (cf. the second term on the r.h.s. of (17)). We derive, therefore, from the physical interpretation of the developed formalism the following form for the effective interaction

$$h_\beta = h_\beta^s + h_\beta^q$$

$$= g_s(s_a^* s_b + s_b^* s_a) + \sum_{kl\sigma} (g_{kl}\gamma_{k\sigma a}^* \gamma_{l\sigma b} + g_{lk}^* \gamma_{l\sigma b}^* \gamma_{k\sigma a}) \tag{22}$$

where the gkl are assumed to give rise to a strongly convergent series in $\mathcal{H}_\beta$ implying that $h_\beta^q \in \mathcal{N}^\beta$. By (11) we have a natural local approximation for the s_x, by (17) also for the $\gamma_{k\sigma x}$ and by (22) (making a finite sum-approximation) also for h_β, which will be denoted by $h_{\beta,\Lambda}$, $\Lambda := (\Lambda_a, \Lambda_b)$. From our reasoning it is clear that $h_{\beta,\Lambda}$ is not an entirely microscopically motivated coupling term, but is to approximate the main contribution of the interaction processes at the temperature β. (Observe that also (1) contains severe physical assumptions!)

THE COMPOSITE INTERACTING SYSTEM

Given the family of local interactions for all pairs (Λ_a, Λ_b) we may apply the methods of [10] and [7] to prove also in our model the convergence of the interacting local grand-canonical equilibrium states to the limiting state ω'^β with the central decomposition

$$\omega'^\beta = \int \omega'^{\beta\theta} \, d\mu^\beta(\theta) \tag{23}$$

where

$$d\mu^\beta(\theta) = \exp(-\xi_\beta - d_\beta \cos(\theta_a - \theta_b)) d\theta \tag{24}$$

with ξ_β, $d_\beta \in \mathbb{R}$ and the abbreviations of (10). The pure phase states are coupled by h^q. Our weak coupling assumptions have the decisive consequence that the interacting GNS-representation is unitarily equivalent to that of ω^β, and that ω'^β may be realized by a vector $\Omega'_\beta \in \mathcal{H}_\beta$. Because of this we have also for the interacting case the particle number operators N_x^β for the subsystems. This prerequisite for the Josephson current (see below) is lost for $\beta \to +\infty$, where

$$d\mu^\beta(\theta) \to \delta(\theta_a - \theta_b) d\theta_a \, d\theta_b \; . \tag{25}$$

At the zero point the h_β^s-coupling is strong enough to make the macroscopic phase uniform.

The existence of the interacting limiting dynamics is proved as in [7] by means of a norm-convergent Dyson series (again a consequence of the weak coupling assumption). Since Θ_x^β commutes with h_β, (19) is still valid. The arise of coupling terms in the phase dynamics as in [11] and [12] is an indication for the coupling not being an element of $\mathcal{N}^\beta$ (which is typical for a strong coupling). For a quite arbitrary weak coupling we have as part of the Heisenberg dynamics in $\mathcal{N}^\beta$

$$\Theta_a^\beta(t) - \Theta_b^\beta(t) = \Theta_a^\beta - \Theta_b^\beta + (\mu_a - \mu_b)t\pi_\beta(1) \; . \tag{26}$$

From (13) follows that the condensed Cooper pairs are time dependent, but not dressed by the interaction. The quasi-particles $\gamma'_{k\sigma x}(t)$ are dressed by means of h^{q}_{β}. This leads in the Schrodinger representation to the quantum dynamical description how small perturbations of ω'^{β} or ω^{β} evolve in time, that is an example of the most detailed form of nonequilibrium quantum statistical physics.

THE CURRENT OPERATORS

Because the condensed Cooper pairs are not dressed by any weak interaction one would not expect a supercurrent flow between the subsystems. So far we have discussed, however, the dynamics in $\mathcal{N}^{\beta}$, to which s_x belongs, but not N^{β}_{x}. In full anology to [7], we could introduce a natural time dependence of N^{β}_{x} by means of the Tomita-Takesaki theory and then calculate the time-derivative, whereby the standard calculation procedure leading to (2) would have been imitated in an improved form. Here we evaluate directly (and equivalently)

$$\begin{aligned} J^{\beta}(t) :&= \lim_{\Lambda\to\infty} \pi^{\beta}(i[h_{\beta,\Lambda}, N_{\Lambda a}]) \\ &= J^{\beta}_{s}(t) + J^{\beta}_{q}(t) , \end{aligned} \tag{27}$$

where the supercurrent is an element of the center $\mathcal{Z}^{\beta}$, which satisfies the Josephson relation in operator form

$$J^{\beta}_{s}(t) = 4g_s w_a w_b \sin(2\Theta^{\beta}_{a}(t) - 2\Theta^{\beta}_{b}(t)) , \tag{28}$$

and the quasi-particle current has the all order perturbation theoretical expression

$$\begin{aligned} J^{\beta}_{q}(t) = \sum_{kl\sigma} & (g_{kl}\gamma'^{*}_{k\sigma a}(t)\, \gamma'_{l\sigma b}(t) \\ & - g^{*}_{lk}\gamma'^{*}_{l\sigma b}(t)\gamma'_{k\sigma a}(t)). \end{aligned} \tag{29}$$

Under our assumptions for the interaction Hamiltonian we are thus able to derive two clearly separated and interpretable parts of the current operator. Even if we had formally included into h^{β} an $s - \gamma^{*}\gamma^{*}$ break up, no additional dependence on the phase angle difference $\Theta^{\beta}_{a} - \Theta^{\beta}_{b}$ would have emerged.

There are other model discussions (cf. [13]), where an alternating current is deduced from an increasing coupling constant via a Hopf-bifurcation for reservoir driven superconductors. Experimentally the coupling energy at a Josephson junction is, however, small. Let us mention in this connection that the use of grand canonical ensembles in our treatment is essential to obtain a macroscopic phase rotation. Thus, we deal also here with a total system, which has a weak particle exchange with the surrounding.

THE ACTUAL CURRENTS

The actual current flow between the two superconductors depends on the prepared state. By means of the spectral projections $F(I)\in\mathcal{Z}^{\beta}$, which filter the values ϕ of $\Theta_a^{\beta} - \Theta_b^{\beta}$ into the interval $I\subset\mathbb{R}$, we consider quite general situations in terms of the state vectors

$$\Psi_i := F(I_i)P_sP_q\Omega_{\beta}^{(')}\in\mathcal{H}_{\beta},\quad i = 1,2\ , \tag{30}$$

where $P_{s,q}$ stands for an arbitrary polynomial in $s_x^{(*)}$ (resp. $\gamma_{k\sigma x}^{(*)}$). That is, we describe here a fixing of the phase difference to the two intervals I_1 and I_2, which should be small and disjoint. (The finite length of the I_i is to avoid mathematical difficulties.) Since we did not include the electromagnetic field into our model, this determination of the phase difference may not be achieved by preparing a certain magnetic flux through the barrier but only by imposing an appropriate supercurrent for a while. After this, it is realistic not to stick to the exact coupled or incoupled equilibrium states $\Omega_{\beta}^{(')}$ but allow for a rather arbitrary change in the particle numbers as in (30). The SQUID provides a device, where both possibilities of (30) may be realized simultaneously. Let us formalize this situation by a state vector

$$\Psi_3 = c_1\Psi_1 + c_2\Psi_2,\quad |c_1|^2 + |c_2|^2 = 1\ . \tag{31}$$

As is explained in [14] a product structure of $\mathcal{H}_{\beta}$ leads to the following current expectation value in the SQUID state

$$(\Psi_3,\ J^{\beta}(t)\Psi_3) = \sum_{i=1}^{2} |c_i|^2(F(I_i)\Omega_{\beta}^{(')},\ J_s^{\beta}(t)F(I_i)\Omega_{\beta}^{(')})d_1$$

$$+ (P_q\Omega_{\beta}^{(')},\ J_q^{\beta}(t)P_q\Omega_{\beta}^{(')})d_2 \tag{32}$$

where the real constants $d_{1/2}$ depend on P_s, P_q. Thus we see that the supercurrent is a superposition of the two Josepheson currents with (almost) specified phase differences, whereas the value of the quasiparticle current does not depend on any phase manipulation (in contradistinction to $J_q^{\beta}(t)$ iteself). If one takes $\mathcal{N}^{\beta}$ as the algebra of observables then Ψ_1 and Ψ_2 are separated by phase superselection rules. If one takes also N_x^{β} into account, then these superselection rules are removed and Ψ_3 is not more mixed than the $\Psi_{1/2}$: We have quantum coherence on a macroscopic level [14]. Altogether our discussion may give an impression, in which way the temperature dependent, reconstructed quantum mechanics makes possible a detailed analysis of the interplay between microscopic and macroscopic dynamical effects.

ACKNOWLEDGMENTS

This work has been supported in part by the U. S. Air Force Office of Scientific Research.

REFERENCES

1. B. D. Josephson, "Possible new effects in superconductive tunneling", Phys. Lett. 1, 251 (1962).
2. A. Barone and G. Paterno, "Physics and applications of the Josephson effect", J. Wiley and Sons, New York (1982).
3. A. Rieckers, "On the two-fluid picture for two weakly coupled superconductors", in preparation.
4. O. Bratteli and D. W. Robinson, "Operator algebras and quantum statistical mechanics I, II" Springer, New York (1979), (1981).
5. W. Thirring, Commun. Math. Phys. 7, 1981 (1968); W. Fleig, Acta. Phys. Austriaca 55, 135 (1983).
6. A. Rieckers, unpublished, W. Fleig, Ph. D. Thesis, in preparation.
7. A. Rieckers and M. Ullrich, "On the microscopic derivation of the finite temperature Josephson relation in operator form", preprint, Tubingen (1984).
8. A. Rieckers, "Spectral stability at the Josephson Junction", in preparation.
9. L. Solymar, "Superconductive tunneling and applications", Chapman and Hall LTD, London (1972).
10. U. Ullrich, "Calculation of the limiting Gibbs states for weakly coupled macroscopic quantum systems with application to the Josephson oscillator", preprint, Tubingen (1984).
11. R. P. Feynman, R. B. Leighton, and M. Sands, "The Feynman lectures on physics", Vol. III, Addison-Wesley (1965).
12. D. Rogovin and M. O. Scully, "Does the two-level atom picture of a Josephson junction have a theoretical foundation in B.C.S.?", Ann. Phys. 88, 377 (1974).
13. K. Hepp, "Two models for Josephson oscillators", Ann. Phys. 90, 285 (1975).
14. A. Rieckers, "Macroscopic quantum coherence at the Josephson junction", in preparation.
15. A. Rieckers, "On the classical part of the mean field dynamics for quantum lattice systems in grand canonical representations" to appear in J. Math. Phys. (1984).

CONTRIBUTORS

J. M. Aguirregabiria
Laboratoire de Physique Theorique, Institut H. Poincare, Paris, France (pp. 259-268).

V. Ambegaokar
Institute for Theoretical Physics, University of California, Santa Barbara, California; and Laboratory of Atomic and Solid State Physics, Cornell University, Ithaca, New York (pp. 231-239).

A. Aspect
Institut d'Optique Theorique et Appliquee, Centre Universitaire d'Orsay, Orsay Cedex, France (pp. 163-183, 185-189).

L. Banyai
Institut fur Theoretische Physik der Universitat Heidelberg, West Germany (pp. 241-257).

A. O. Barut
Max-Planck Institut fur Quantenoptik, Garching, West Germany; and Department of Physics, University of Colorado, Boulder, Colorado (pp. 119-128, 139-143).

L. Bel
Laboratoire de Physique Theorique, Institut H. Poincare, Paris, France (pp. 259-268).

G. P. Beretta
Massachusetts Institute of Technology, Cambridge, Massachusetts; and Politecnico di Milano, Milano, Italy (pp. 193-204, 205-212).

J. Bergou
Max-Planck Institut fur Quantenoptik, Garching, West Germany (pp. 375-383).

E. R. Caianiello
Dipartimento di Fisica Teorica e sue Metodologie per le Scienze Applicate, Universita' di Salerno, Italy (pp. 453-464).

L. Cohen
Hunter College of the City University, New York, New York (pp. 97-117).

P. Dobiasch
Max-Planck Institut fur Quantenoptik, Garching, West Germany (pp. 385-408).

A. Dorsel
Sektion Physik, Universitat Munchen, Garching, West Germany (pp. 361-374).

A. Elci
Institute for Modern Optics, Department of Physics and Astronomy, University of New Mexico, Albuquerque, New Mexico (pp. 431-451).

W. J. Firth
Department of Physics, Heriot-Watt University, Edinburgh, United Kingdom (pp. 361-374).

A. Guzman de Garcia
Max-Planck-Institut fur Quantenoptik, Garching, West Germany (pp. 361-374).

W. Guttinger
Institute for Information Sciences, University of Tubingen, West Germany (pp. 57-82).

F. A. Hopf
Optical Sciences Center, University of Arizona, Tucson, Arizona (pp. 411-429).

F. W. Hopf
Department of Biology, Cornell University, Ithaca, New York (pp. (pp. 411-429).

E. T. Jaynes
St. John's College and Cavendish Laboratory, Cambridge, United Kingdom; and Department of Physics, Washington University, St. Louis, Missouri (pp. 33-55).

M. Kleber
Institute for Theoretical Sciences, University of New Mexico, Albuquerque, New Mexico; and Physik Department, T. U. Munchen, West Germany (pp. 213-220).

W. E. Lamb, Jr.
Department of Physics and Optical Sciences Center, University of Arizona, Tucson, Arizona, 85721 (pp. 3-23).

G. Leuchs
Joint Institute for Laboratory Astrophysics, University of Colorado and National Bureau of Standards, Boulder, Colorado (pp. 329-360).

R. L. Liboff
Schools of Applied Physics and Electrical Engineering, Cornell University, Ithaca, New York (pp. 221-229).

F. Marquis
Max-Planck Institut fur Quantenoptik, Garching, West Germany (pp. 361-374).

J. D. McCullen
Department of Physics, University of Arizona, Tucson, Arizona, (pp. 361-374).

P. Meystre
Max-Planck Institut fur Quantenoptik, Garching, West Germany (pp. 139-143, 361-374).

M. M. Nieto
Theoretical Division, Los Alamos National Laboratory, Los Alamos, New Mexico (pp. 287-307).

D. P. J. O'Brien
Max-Planck-Institut fur Quantenoptik, Garching, West Germany (pp. 361-374).

R. F. O'Connell
Department of Physics and Astronomy, Louisiana State University, Baton Rouge, Louisiana (pp. 83-95).

A. Rieckers
Institute for Theoretical Sciences, Department of Physics and Astronomy, University of New Mexico, Albuquerque, New Mexico; and Institute for Theoretical Physics, University of Tubingen, West Germany (pp. 483-492).

G. Reiner
Max-Planck-Institut fur Quantenoptik, Garching, West Germany (pp. 361-374).

V. E. Sanders
Rockwell Int. A.M.S.D. Anaheim, California (pp. 385-408).

M. Sargent III
Max-Planck Institut fur Quantenoptik, Garching, West Germany; and Optical Sciences Center, University of Arizona, Tucson, Arizona (pp. 361-374).

W. Schleich
Max-Planck Institut fur Quantenoptik, Garching, West Germany; and Center for theoretical Physics, University of Texas at Austin, Austin, Texas (pp. 385-408).

J. F. Scott
Condensed Matter Laboratory, Department of Physics, University of Colorado, Boulder, Colorado (pp. 465-472, 473-481).

M. O. Scully
Max-Planck Institut fur Quantenoptik, Garching, West Germany; Center for Advanced Studies and Institute for Modern Optics, University of New Mexico, Albuquerque, New Mexico (pp. 131-138, 375-383, 385-408).

K. Ujihara
Max-Planck-Institut fur Quantenoptik, Garching, West Germany; and University of Electro-Communications, Chofu, Tokyo, Japan (pp. 361-374).

E. Vignes
Department of Physics, UCSD, La Jolla, California (pp. 361-374).

D. F. Walls
Institute for Theoretical Physics, University of California, Santa Barbara, California (pp. 309-328).

H. Walther
Sektion Physik, Universitat Munchen and Max-Planck-Institut fur Quantenoptik, Garching, West Germany (pp. 271-285, 361-374).

J. A. Wheeler
Center for Theoretical Physics, Univ. of Texas at Austin (pp. 25-32).

E. M. Wright
Max-Planck-Institut fur Quantenoptik, Garching, West Germany (pp. 361-374).

W. H. Zurek
Theoretical Astrophysics, Los Alamos National Laboratory, Los Alamos, New Mexico; and Institute for Theoretical Physics, University of California, Santa Barbara, California, (pp. 145-149, 151-161).

INDEX